Iodine Catalysis in Organic Synthesis

Iodine Catalysis in Organic Synthesis

Edited by Kazuaki Ishihara and Kilian Muñiz

WILEY-VCH

Editors

Prof. Kazuaki Ishihara
Nagoya University
Graduate School of Engineering
B2-3(611), Furo-cho
Chikusa
Nagoya 464-8603
Japan

Prof. Kilian Muñiz
Inst. of Chemical Research of Catalonia
Av. Països Catalans 16
Tarragona 43007
Spain

Cover Design: Wiley
Cover Image: © LuYago/Shutterstock

Library of Congress Card No.: applied for

British Library Cataloguing-in-Publication Data
A catalogue record for this book is available from the British Library.

Bibliographic information published by the Deutsche Nationalbibliothek
The Deutsche Nationalbibliothek lists this publication in the Deutsche Nationalbibliografie; detailed bibliographic data are available on the Internet at <http://dnb.d-nb.de>.

Print ISBN: 978-3-527-34829-9
ePDF ISBN: 978-3-527-82955-2
ePub ISBN: 978-3-527-82957-6
oBook ISBN: 978-3-527-82956-9

Typesetting Straive, Chennai, India
Printing and Binding CPI Group (UK) Ltd, Croydon, CR0 4YY

Printed on acid-free paper

C108778_211221

Contents

Preface

Iodine (symbol **I** and atomic number 53) is the heaviest of the stable halogens and exists as a lustrous, purplish-black nonmetallic solid under ambient conditions. It was discovered by the French chemist Bernard Courtois in 1811, and was named two years later by Joseph Louis Gay-Lussac, after the Greek *ἰώδης* "violet-colored."

The human body needs iodine because iodine plays an essential role in the biosynthesis of thyroid hormones. The iodine needed by the body must come from the diet. Most food contains very little iodine. However, processed food typically contains more iodine due to the addition of iodized salt. Most of the world's iodine is found in the ocean, where it is concentrated by sea life, especially seaweed, and in certain rocks and sediments. The dominant producers of iodine today are Chile and Japan.

Iodine occurs in many oxidation states, including iodide (I^-), iodate (IO_3^-), and various periodate anions. Iodine-containing compounds are useful as oxidative reagents and catalysts, due to their oxidative ability and ease of attachment to organic compounds. Furthermore, iodine-containing compounds are useful as Lewis acid catalysts, due to their ability as halogen-bonding(XB) donors. The XB interactions follow the general trend $F < Cl < Br < I$, with iodine normally forming the strongest interactions.

Over the past decade, the expansion of hypoiodite(I)- and organoiodine(III or V)-catalyzed oxidative transformations has been unremitting. Iodine-containing compounds are now some of the most important oxidative catalysts in organic synthesis. Both intramolecular and intermolecular enantioselective oxidative coupling reactions have recently been developed using chiral iodine catalysts. Moreover, iodine-catalyzed oxidative reactions and Lewis acidic reactions are highly attractive from an environmental point of view. Thus, we decided to publish a new book titled "Iodine Catalysis in Organic Synthesis."

This book focuses on different areas of iodine chemistry and catalysis, which have been selected because they have developed significantly, and, in some cases, completely, over the past few years. Each author is very knowledgeable in his/her particular field of chemistry and can provide a valuable perspective.

It is our strong hope that this book will be an invaluable guide for synthetic chemists in both academic and industrial laboratories.

Kazuaki Ishihara
Nagoya

Kilian Muñiz
Tarragona

PS:
We launched this book project in January 2020. Unfortunately, Prof. Muñiz passed away on 16 March 2020. However, we inherited his spirit and were able to keep pushing until the book was complete. I pray for Kilian's soul.

Kazuaki Ishihara

1

Historical Introduction

Viktor V. Zhdankin

Department of Chemistry and Biochemistry, University of Minnesota Duluth, Duluth, MN, USA

1.1 Discovery of Iodine and Early Studies

Iodine was originally isolated and recognized as a new element early in the nineteenth century. The French industrial chemist Bernard Courtois had noticed that the addition of concentrated sulfuric acid to seaweed ashes resulted in the emission of a violet vapor that could be condensed to the deep purple solid with the metal-like appearance of the crystals. These observations were first published in November 1813 in the *Annales de Chimie* [1]. In this original paper, the name "iode," derived from the Greek word *ἰώδης* for violet, was first used for the new substance "due to the beautiful violet color of its vapor." Soon after, J. L. Gay-Lussac published his initial results on the chemical properties of iodine demonstrating that this was a novel element capable of forming compounds with other elements [2, 3].

Numerous inorganic compounds of polyvalent iodine in the oxidation states +3, +5, and +7 were prepared early in the nineteenth century. In particular, iodine trichloride was first prepared by J. L. Gay-Lussac by reacting iodine or iodine monochloride with an excess of chlorine under gentle heating [4]. In the same 1814 paper [4], the preparation of potassium iodate by the action of iodine on hot potash lye was described. The history of the inorganic chemistry of iodine was summarized in numerous common textbooks and reference sources [5–9]. Iodine history, chemistry, and applications were discussed in detail in the review commemorating two centuries of iodine research [10].

The preparation of first polyvalent organoiodine compound, (dichloroiodo) benzene **1** (Figure 1.1) was reported by the German chemist C. Willgerodt in 1886 in the *Journal fuer Praktische Chemie* [11]. Many other organic iodine(III) and iodine(V) compounds were prepared during the 1890s and at the beginning of the twentieth century. In particular, (diacetoxyiodo)benzene **2**, iodosylbenzene **3**, and iodylbenzene **4** were reported in 1892 [12]; 2-iodoxybenzoic acid (IBX, **5**) in 1893 [13]; and the first example of diaryliodonium salts **6** were reported by C. Hartmann and V. Meyer in 1894 [14].

Iodine Catalysis in Organic Synthesis, First Edition. Edited by Kazuaki Ishihara and Kilian Muñiz.

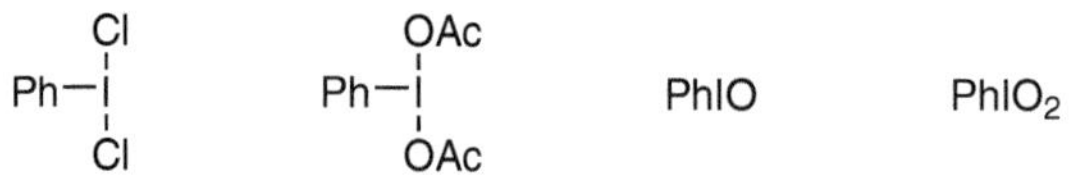

1 2 3 4

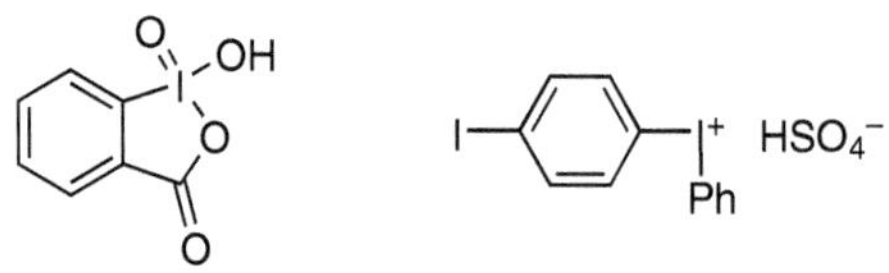

5 6

Figure 1.1 First examples of polyvalent organoiodine compounds reported in 1886–1894.

The early research on organoiodine chemistry was summarized by C. Willgerodt in 1914 in a comprehensive book *Die Organischen Verbindungen mit Mehrwertigen Jod* describing nearly 500 polyvalent organoiodine compounds [15].

1.2 Iodine Research in the Twentieth Century

During the period between 1914 and 1970s, research activity in the area of organic chemistry of iodine compounds was relatively low and represented mainly by valuable contributions from the laboratories of I. Masson, R. B. Sandin, F. M. Beringer, K. H. Pausacker, A. N. Nesmeyanov, and O. Neilands. This research, dealing mostly with various iodonium compounds, was summarized in the reviews of R. B. Sandin [16] and D. F. Banks [17] published in 1943 and 1966, respectively. A comprehensive list of known at that time iodine(III) and iodine(V) compounds with their physical properties was published by F. M. Beringer and E. M. Gindler in *Iodine Abstracts and Reviews* in 1956 [18].

Starting from the 1970s, the interest in the chemistry of iodine compounds had significantly increased. At that time, numerous new classes of polyvalent organoiodine compounds were discovered and many useful synthetic applications of these compounds were developed. The concept of hypervalent molecules was originally proposed by J. I. Musher in 1969 [19], and shortly after, the terminologies "hypervalent iodine," "hypervalent iodine reagents," "hypervalent iodine compounds," and "organohypervalent iodine reagents" were broadly introduced in the works of J. C. Martin, R. M. Moriarty, and several other researchers. During the 1980s, polyvalent iodine compounds have achieved the status of valuable synthetic reagents known under the common name of hypervalent iodine reagents.

The foundation of modern hypervalent iodine chemistry was established in the 1980s in the groundbreaking works of G. F. Koser, J. C. Martin, R. M. Moriarty, P. J. Stang, A. Varvoglis, Y. Kita, M. Ochiai, and N. S. Zefirov. The twentieth-century iodine research was summarized in two books published by A. Varvoglis in 1992 and 1997: a comprehensive monograph *The Organic Chemistry of Polycoordinated Iodine* [20] and a book on the applications of hypervalent iodine compounds in organic synthesis [21]. Numerous general reviews [22–27], book chapters [28–33], and specialized reviews on phenyliodine(III) carboxylates [34, 35], [hydroxy(tosyloxy)iodo]benzene [36], the chemistry of alkynyliodonium salts [37], electrophilic perfluoroalkylations [38], application of hypervalent iodine in the carbohydrate chemistry [39], carbon–carbon bond formation via hypervalent iodine [40], hypervalent iodine oxidations [41, 42], hypervalent iodine compounds as free radical precursors [43], synthesis of heterocyclic compounds using organohypervalent iodine reagents [44], and the chemistry of benziodoxoles [45] were also published during the 1980s and 1990s.

1.3 Iodine Research in the Twenty-first Century

During the first two decades of the twenty-first century, iodine chemistry has experienced explosive development. Six books [46–51] and hundreds of reviews summarizing various aspects of iodine chemistry and applications have been published between 2001 and 2020. Thousands of research works utilizing iodine reagents in organic and inorganic synthesis are currently published every year. Starting from the beginning of the twenty-first century, the International Conference on Hypervalent Iodine Chemistry (ICHIC) is regularly convened in different countries, the Society of Iodine Science (SIS) holds annual meetings in Japan, and the American Chemical Society presents the National Award for Creative Research and Applications of Iodine Chemistry biennially. The World Iodine Association (www.worldiodineassociation.com) was officially registered in 2017 as an international nonprofit organization established with the main goal of providing information about the purposes, uses, and applications of iodine and its derivatives.

Current surging interest in iodine chemistry is mainly explained by the very useful oxidizing properties of hypervalent iodine reagents, combined with their benign environmental character and commercial availability. Iodine(III) and iodine(V) derivatives are now routinely used in organic synthesis as reagents for various selective oxidative transformations of complex organic molecules. The discovery and utilization of similarities between the transition metal chemistry and the hypervalent iodine chemistry, and in particular, the development of the highly efficient and enantioselective catalytic systems based on the iodine redox chemistry, has added a new dimension to the field of hypervalent iodine chemistry and initiated a major surge of research activity, which is expected to continue in the future.

1.4 Brief History of Iodine Catalysis

A quick SciFinder search on the concept "iodine catalysts" reveals over 10 000 papers published between 1890 and 2020, including about 300 older papers published during the first half of the twentieth century. The first reliable observations of inorganic and organic reactions catalyzed by iodine date back to the early 1900s. For example, J. Brode [52] and J. H. Walton [53] reported in 1904 the catalytic decomposition of hydrogen peroxide in the presence of iodine involving hypoiodite as a key intermediate. L. Bruner in 1902 published a mechanistic study of the catalytic action of iodine on the bromination of benzene, in which the catalytic effect of iodine is explained by the formation of iodine bromide, IBr, as active species [54].

The vast majority of papers on iodine catalysis published between 1900 and 2005 did not involve the oxidation of iodine to the hypoiodite or hypervalent iodine-active species. These papers almost exclusively deal with the catalytic application of iodine as a Lewis acid or as a radical initiator. In some catalytic reactions, iodide anion is utilized as a strong nucleophile, generating organic iodides RI as intermediate species incorporating the excellent iodide-leaving group. Such catalysis involving the generation of organic iodides has been employed in the important Monsanto–Cativa industrial process for the production of acetic acid. In this process, the catalytic hydroiodic acid converts the methanol feedstock into methyl iodide, which undergoes Rh- or Ir-catalyzed carbonylation. Hydrolysis of the resulting acetyl iodide regenerates hydroiodic acid and gives acetic acid [55]. The carbonylation of methanol, catalyzed by rhodium in the presence of hydroiodic acid, was originally invented by Monsanto company in the 1960s. In the 1990s, Eastman Chemical Company developed a modification of this industrial process based on the use of lithium iodide as the cocatalyst [56].

A new era in iodine catalysis was opened in 2005 by the discovery of reactions catalyzed by hypervalent iodine species. The similarities between hypervalent iodine species and transition metal-organic complexes had been widely recognized in the works of many researchers since the end of the twentieth century. At that time, the terminologies "oxidative addition," "reductive elimination," "ligand exchange," and "ligand coupling" became common in mechanistic discussions of the reactions of hypervalent molecules [57]. However, catalytic reactions, typical of transition metals, remained unknown for hypervalent iodine compounds until the beginning of the twenty-first century.

In 2005, Kita and Ochiai independently reported the catalytic use of aryl iodides, in the presence of stoichiometric *m*-chloroperoxybenzoic acid, to perform oxidative dearomatization of phenolic substrates **7** (Scheme 1.1) [58], or α-acetoxylation of carbonyl compounds **8** (Scheme 1.2) [59], respectively. These reactions involved selective generation of the highly reactive hypervalent iodine(III) species (e.g. **9** in Scheme 1.2) *in situ* from aryl iodide and terminal oxidant.

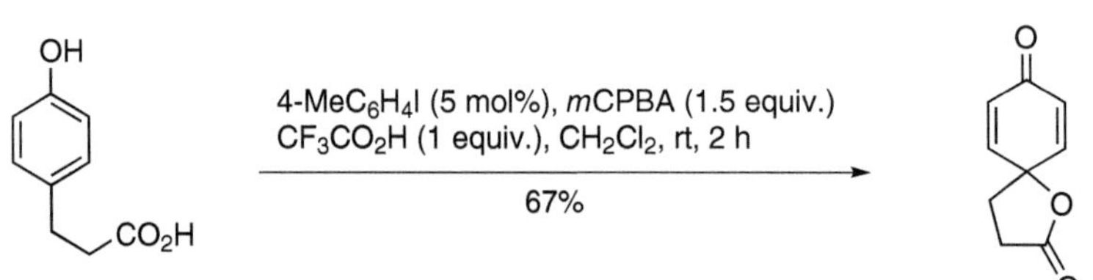

7

Scheme 1.1 Catalytic oxidative spirocyclization reaction of a phenolic substrate in the presence of 4-iodotoluene.

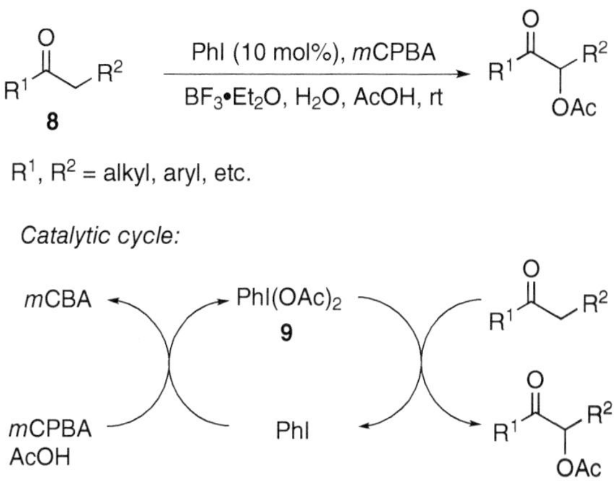

Scheme 1.2 Catalytic α-acetoxylation of carbonyl compounds in the presence of iodobenzene.

First examples of the catalytic application of the iodine(V) species in the oxidation of alcohols using Oxone® ($2KHSO_5{\bullet}KHSO_4{\bullet}K_2SO_4$) as a stoichiometric oxidant at 70 °C were independently reported by the groups of Vinod [60] in 2005 and Giannis [61] in 2006. A few years later, Ishihara and coworkers discovered that 2-iodoxybenzenesulfonic acid (IBS) can act as an extremely active catalyst for the selective oxidation of alcohols to aldehydes, ketones, carboxylic acids, and enones with Oxone [62]. While chemical reactions catalyzed by hypervalent iodine species were discovered only in 2005, the electrochemical generation of iodine(III) species *in situ* from catalytic amounts of iodoarenes (0.05–0.2 equiv), and the use of these species as the in-cell mediators in electrochemical fluorination reactions, had been known since 1994 [63].

First examples of enantioselective reactions catalyzed by chiral aryl iodides [64–69] or chiral ammonium iodides [70] were reported by several research groups in 2007–2010. These groundbreaking initial reports were followed by a huge wave of publications describing various enantioselective oxidative transformations catalyzed by hypervalent iodine species. A brief historical overview of enantioselective iodine catalysis can be found in the 2019 review of Muñiz and coauthors [71].

References

1 Courtois, B. (1813). Découverte d'une substance nouvelle dans le Vareck. *Ann. Chim.* 88: 304–310.

2 Gay-Lussac, J.L. (1813). Sur la combinaison de l'iode avec l'oxygène. *Ann. Chim.* 88: 319–321.

3 Gay-Lussac, J.L. (1813). Sur un nouvel acide formé avec la substance découverte par M. Courtois. *Ann. Chim.* 88: 311–318.

4 Gay-Lussac, J.L. (1814). Sur l'iode. *Ann. Chim.* 91: 5–160.

5 Greenwood, N.N. and Earnshaw, A. (1997). *Chemistry of the Elements*. Oxford: Butterworth-Heinemann.
6 Mellor, J.W. (1922). *A Comprehensive Treatise on Inorganic and Theoretical Chemistry*. London: Longmans, Green and Co.
7 (1933). *Gmelins Handbuch der Anorganischen Chemie, 8 Auflage*. Berlin: Verlag Chemie.
8 Brasted, R.C. (1954). Iodine. In: *Comprehensive Inorganic Chemistry*, vol. 3 (eds. M.C. Sneed, J.L. Maynard and R.C. Brasted). Princeton, NJ: D. Van Norstrand Co., Inc.
9 Gutmann, V. (1967). *Halogen Chemistry*. New York: Academic Press.
10 Kuepper, F.C., Feiters, M.C., Olofsson, B. et al. (2011). Commemorating two centuries of iodine research: an interdisciplinary overview of current research. *Angew. Chem. Int. Ed.* 50: 11598–11620.
11 Willgerodt, C. (1886). Découverte d'une substance nouvelle dans le Vareck. *J. Prakt. Chem. (Weinheim, Germany)* 33: 154.
12 Willgerodt, C. (1892). Zur Kenntniss aromatischer Jodidchloride, des Jodoso- und Jodobenzols. *Ber. Dtsch. Chem. Ges.* 25: 3494–3502.
13 Hartmann, C. and Meyer, V. (1893). Ueber Jodobenzoësäure. *Ber. Dtsch. Chem. Ges.* 26: 1727–1732.
14 Hartmann, C. and Meyer, V. (1894). Ueber eine neue Klasse jodhaltiger, stickstofffreier organischer Basen. *Ber. Dtsch. Chem. Ges.* 27: 426–432.
15 Willgerodt, C. (1914). *Die Organischen Verbindungen mit Mehrwertigen Jod*. Stuttgart: Ferdinand Enke Verlag.
16 Sandin, R.B. (1943). Organic compounds of polyvalent iodine. *Chem. Rev.* 32: 249–276.
17 Banks, D.F. (1966). Organic polyvalent iodine compounds. *Chem. Rev.* 66: 243–266.
18 Beringer, F.M. and Gindler, E.M. (1956). Organic compounds of polyvalent iodine. *Iodine Abstr. Rev.* 3: 70. pp.
19 Musher, J.I. (1969). Chemistry of hypervalent molecules. *Angew. Chem. Int. Ed.* 8: 54–68.
20 Varvoglis, A. (1992). *The Organic Chemistry of Polycoordinated Iodine*. New York: VCH Publishers, Inc.
21 Varvoglis, A. (1997). *Hypervalent Iodine in Organic Synthesis*. London: Academic Press.
22 Varvoglis, A. (1997). Chemical transformations induced by hypervalent iodine reagents. *Tetrahedron* 53: 1179–1255.
23 Merkushev, E.B. (1987). Organic compounds of polyvalent iodine. Derivatives of iodosobenzene. *Russ. Chem. Rev.* 56: 826.
24 Moriarty, R.M. and Prakash, O. (1986). Hypervalent iodine in organic synthesis. *Acc. Chem. Res.* 19: 244–250.
25 Stang, P.J. and Zhdankin, V.V. (1996). Organic polyvalent iodine compounds. *Chem. Rev.* 96: 1123–1178.

26 Kitamura, T. and Fujiwara, Y. (1997). Recent progress in the use of hypervalent iodine reagents in organic synthesis. A review. *Org. Prep. Proced. Int.* 29: 409–458.

27 Wirth, T. and Hirt, U.H. (1999). Hypervalent iodine compounds. Recent advances in synthetic applications. *Synthesis*: 1271–1287.

28 Nguyen, T.T. and Martin, J.C. (1984). Heterocyclic rings containing halogen. In: *Comprehensive Heterocyclic Chemistry*, vol. 1 (eds. A.R. Katritzky and C.W. Rees), 563. Oxford: Pergamon Press.

29 Koser, G.F. (1983). Hypervalent halogen compounds. In: *The Chemistry of Functional Groups, Suppl. D: Chemistry of Halides, Pseudo-Halides, Azides* (eds. S. Patai and Z. Rappoport), 721–811. Chichester: Wiley-Interscience.

30 Koser, G.F. (1983). Halonium ions. In: *Chemistry of Halides. Halides, Pseudo-Halides Azides*, vol. 2 (eds. S. Patai and Z. Rappoport), 1265–1351. Chichester: Wiley-Interscience.

31 Koser, G.F. (1995). Halonium ions. In: *Chemistry of Halides, Pseudo-Halides and Azides, Suppl. D2* (eds. S. Patai and Z. Rappoport), 1173–1274. Chichester: Wiley-Interscience.

32 Zhdankin, V.V. and Stang, P.J. (1999). Polycoordinate iodine compounds. In: *Chemistry of Hypervalent Compounds* (ed. K.-Y. Akiba), 327–358. New York: VCH Publishers.

33 Ochiai, M. (1999). Organic synthesis using hypervalent organoiodanes. In: *Chemistry of Hypervalent Compounds* (ed. K.Y. Akiba), 359–387. New York: VCH Publishers.

34 Varvoglis, A. (1981). Aryliodine(III) dicarboxylates. *Chem. Soc. Rev.* 10: 377–407.

35 Kirschning, A. (1998). (Diacetoxyiodo)benzene DIB. A multi-talented oxidant in organic synthesis. *J. Prakt. Chem.* 340: 184–186.

36 Moriarty, R.M., Vaid, R.K., and Koser, G.F. (1990). [Hydroxy(organosulfonyloxy)iodo]arenes in organic synthesis. *Synlett*: 365–383.

37 Zhdankin, V.V. and Stang, P.J. (1998). Alkynyliodonium salts in organic synthesis. *Tetrahedron* 54: 10927–10966.

38 Umemoto, T. (1996). Electrophilic perfluoroalkylating agents. *Chem. Rev.* 96: 1757–1778.

39 Kirschning, A. (1998). Hypervalent iodine and carbohydrates. A new liaison. *Eur. J. Org. Chem.*: 2267–2274.

40 Moriarty, R.M. and Vaid, R.K. (1990). Carbon-carbon bond formation via hypervalent iodine oxidations. *Synthesis*: 431–447.

41 Moriarty, R.M. and Prakash, O. (1999). Oxidation of carbonyl compounds with organohypervalent iodine reagents. *Org. React.* 54: 273–418.

42 Kita, Y., Takada, T., and Tohma, H. (1996). Hypervalent iodine reagents in organic synthesis: nucleophilic substitution of *p*-substituted phenol ethers. *Pure Appl. Chem.* 68: 627–630.

43 Muraki, T., Togo, H., and Yokoyama, M. (1997). Hypervalent iodine compounds as free radical precursors. *Rev. Heteroatom Chem.* 17: 213–243.

44 Moriarty, R.M. and Prakash, O. (1998). Synthesis of heterocyclic compounds using organohypervalent iodine reagents. *Adv. Heterocycl. Chem.* 69: 1–87.

45 Zhdankin, V.V. (1997). Chemistry of benziodoxoles. *Rev. Heteroatom Chem.* 17: 133–151.
46 Wirth, T. (ed.) (2003). *Hypervalent Iodine Chemistry: Modern Developments in Organic Synthesis, Topics in Current Chemistry*, vol. 224.
47 Preedy, V.R., Burrow, G.N., and Watson, R.R. (eds.) (2009). *Comprehensive Handbook of Iodine: Nutritional, Biochemical, Pathological and Therapeutic Aspects*. Elsevier Inc.
48 Zhdankin, V.V. (2013). *Hypervalent Iodine Chemistry: Preparation, Structure, and Synthetic Applications of Polyvalent Iodine Compounds*. Chichester UK: Wiley.
49 Kaiho, T. (ed.) (2015). *Iodine Chemistry and Applications*. Chichester, UK: John Wiley & Sons, Inc.
50 Wirth, T. (ed.) (2016). *Hypervalent Iodine Chemistry: Modern Developments in Organic Synthesis, Topics in Current Chemistry*, vol. 373. Springer.
51 Olofsson, B., Marek, I., and Rappoport, Z. (eds.) (2019). *Patai's Chemistry of Functional Groups: The Chemistry of Hypervalent Halogen Compounds*. Wiley.
52 Brode, J. (1904). Oxidation of the iodine ion to hypoiodite as an intermediate stage in several reactions. *Z. Phys. Chem.* 49: 208–216.
53 Walton, J.H. (1904). Catalysis of hydrogen peroxide by iodine ions. *Z. Phys. Chem.* 47: 185–222.
54 Bruner, L. (1902). Mechanism of the catalytic action of iodine on the bromination of benzene. *Bull. Int. Acad. Sci. Cracovie*: 181–200.
55 Jones, J.H. (2000). The Cativa process for the manufacture of acetic acid. *Platin. Met. Rev.* 44: 94–105.
56 Zoeller, J.R., Agreda, V.H., Cook, S.L. et al. (1992). Eastman Chemical company acetic anhydride process. *Catal. Today* 13: 73–91.
57 Akiba, K.-Y. (ed.) (1999). *Chemistry of Hypervalent Compounds*. New York: VCH Publishers.
58 Dohi, T., Maruyama, A., Yoshimura, M. et al. (2005). Versatile hypervalent-iodine(III)-catalyzed oxidations with m-chloroperbenzoic acid as a cooxidant. *Angew. Chem. Int. Ed.* 44: 6193–6196.
59 Ochiai, M., Takeuchi, Y., Katayama, T. et al. (2005). Iodobenzene-catalyzed alpha-acetoxylation of ketones. *In situ* generation of hypervalent (diacyloxyiodo)benzenes using m-chloroperbenzoic acid. *J. Am. Chem. Soc.* 127: 12244–12245.
60 Thottumkara, A.P., Bowsher, M.S., and Vinod, T.K. (2005). *In situ* generation of o-iodoxybenzoic acid (IBX) and the catalytic use of it in oxidation reactions in the presence of oxone as a co-oxidant. *Org. Lett.* 7: 2933–2936.
61 Schulze, A. and Giannis, A. (2006). Oxidation of alcohols with catalytic amounts of IBX. *Synthesis*: 257–260.
62 Uyanik, M., Akakura, M., and Ishihara, K. (2009). 2-Iodoxybenzenesulfonic acid as an extremely active catalyst for the selective oxidation of alcohols to aldehydes, ketones, carboxylic acids, and enones with oxone. *J. Am. Chem. Soc.* 131: 251–262.
63 Fuchigami, T. and Fujita, T. (1994). Electrolytic partial fluorination of organic compounds. 14. The first electrosynthesis of hypervalent iodobenzene difluoride

derivatives and its application to indirect anodic gem-difluorination. *J. Org. Chem.* 59: 7190–7192.

64 Richardson, R.D., Page, T.K., Altermann, S. et al. (2007). Enantioselective α-oxytosylation of ketones catalyzed by iodo arenes. *Synlett*: 538–542.

65 Dohi, T., Maruyama, A., Takenage, N. et al. (2008). A chiral hypervalent iodine(III) reagent for enantioselective dearomatization of phenols. *Angew. Chem. Int. Ed.* 47: 3787–3790.

66 Altermann, S.M., Richardson, R.D., Page, T.K. et al. (2008). Catalytic enantioselective α-oxysulfonylation of ketones mediated by iodoarenes. *Eur. J. Org. Chem.*: 5315–5328.

67 Quideau, S., Lyvinec, G., Marguerit, M. et al. (2009). Asymmetric hydroxylative phenol dearomatization through *in situ* generation of iodanes from chiral iodoarenes and *m*-CPBA. *Angew. Chem. Int. Ed.* 48: 4605–4609.

68 Uyanik, M., Yasui, T., and Ishihara, K. (2010). Enantioselective Kita oxidative spirolactonization catalyzed by *in situ* generated chiral hypervalent iodine(iii) species. *Angew. Chem. Int. Ed.* 49: 2175–2177.

69 Uyanik, M., Yasui, T., and Ishihara, K. (2010). Chiral hypervalent iodine-catalyzed enantioselective oxidative Kita spirolactonization of 1-naphthol derivatives and one-pot diastereo-selective oxidation to epoxyspirolactones. *Tetrahedron* 66: 5841–5851.

70 Uyanik, M., Okamoto, H., Yasui, T., and Ishihara, K. (2010). Quaternary ammonium (hypo)iodite catalysis for enantioselective oxidative cycloetherification. *Science* 328: 1376–1379.

71 Flores, A., Cots, E., Berges, J., and Muñiz, K. (2019). Enantioselective iodine(I/III) catalysis in organic synthesis. *Adv. Synth. Catal.* 361: 2–25.

2

Concepts in Iodine Catalysis

Kazuaki Ishihara

Graduate School of Engineering, Nagoya University, Nagoya, Japan

2.1 Introduction

Iodine can form inorganic and organic derivatives in various oxidation states (−1, 0, +1, +3, +5, +7), and the structural features and reactivity patterns of these iodine compounds are, in many respects, similar to those of the derivatives of heavy transition metals. However, in contrast to heavy metals, iodine is both environmentally friendly and relatively inexpensive. About 30 000 tons of iodine are produced annually, and the world's total reserves are estimated to be 15 million metric tons, located mainly in Chile and Japan [1].

Iodine readily disproportionates into iodide and hypoiodites as well as other iodine–oxygen species in basic aqueous solutions according to the Latimer diagrams in Scheme 2.1 [2, 3]. In contrast, iodine is stable against disproportionation in acidic solutions, near pH 0. For neutral solutions around pH 7, which are probably closest to most experimental conditions, a very unfavorable equilibrium constant of 2.0×10^{-13} has been determined for the reaction between water and iodine [3]. Therefore, water and alcohols can react with molecular iodine to give hydroiodic acid (HI).

Iodine compounds are used in industry. About 16% of the global production of iodine is utilized in industrial catalysis [4]. Hydroiodic acid is used as a cocatalyst for the production of acetic acid by the Monsanto process using rhodium catalyst and the Cativa process using iridium catalyst, which are the main industrial processes for the production of acetic acid. In the Cativa process (Scheme 2.2), hydroiodic acid converts the methanol feedstock into methyl iodide, which then undergoes iridium-catalyzed carbonylation [5]. Hydrolysis of the resulting acetyl iodide regenerates hydroiodic acid and gives acetic acid. The role of iodine as a cocatalyst for the Monsanto and Cativa processes is based on the strong nucleophilicity of the iodide anion and the high reactivity as a leaving group of intermediates. Please refer to Chapter 13 for the industrial application of iodine catalysis.

In recent years, hypervalent organoiodine derivatives and hypoiodite salts have attracted significant attention as versatile and environmentally benign catalysts for

Iodine Catalysis in Organic Synthesis, First Edition. Edited by Kazuaki Ishihara and Kilian Muñiz.

I(0)	I_2: iodine	I^-: iodide
I(I)	I–OH: hypoiodous acid	$I–O^-$: hypoiodite
I(III)	O=I–OH: iodous acid	$O=I–O^-$: iodite
I(V)	$(O=)_2I–OH$: iodic acid	$(O=)_2I–O^-$: iodate
I(VII)	$(O=)_3I–OH$: periodic acid	$(O=)_3I–O^-$: periodate

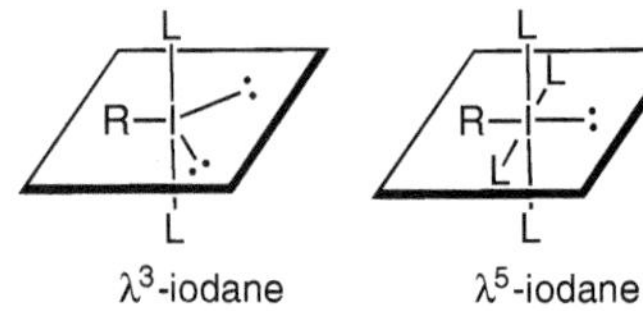

Latimer diagrams summarizing the standard potentials for iodine

Basic solution (pH 14)

$$(O=)I(OH)_3(O^-)_2 \xrightarrow{0.65\ V} (O=)_2I{-}O^- \xrightarrow{0.15\ V} I{-}O^- \xrightarrow{0.42\ V} I_2 \xrightarrow{0.54\ V} I^-$$

Acidic solution (pH 0)

$$(O=)I(OH)_5 \xrightarrow{1.60\ V} (O=)_2I{-}O^- \xrightarrow{1.13\ V} I{-}OH \xrightarrow{1.44\ V} I_2 \xrightarrow{0.54\ V} I^-$$

Scheme 2.1 Iodine compounds.

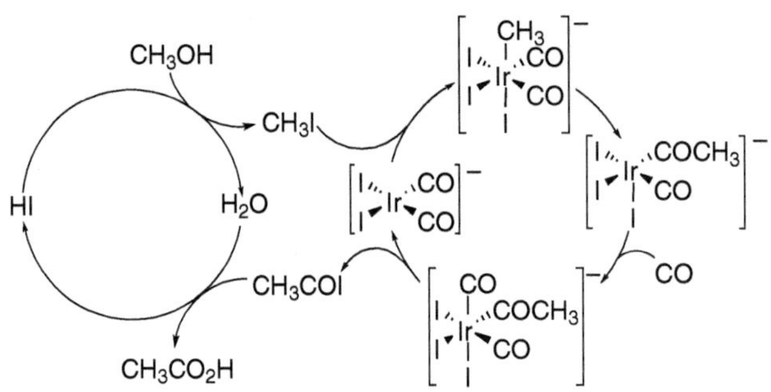

Scheme 2.2 Production of acetic acid by the Cativa process.

organic synthesis [6–8]. One of the most impressive recent achievements in this area has been the discovery of the catalytic activity of iodine derivatives in numerous oxidative transformations leading to the formation of new C–O, C–N, and C–C bonds in organic compounds, as shown in Scheme 2.3a. These transformations involving (i) oxidation to hypoiodites, (ii) nucleophilic addition of Nu^1–H to hypoiodites, and (iii) reductive elimination are similar to the transition-metal-catalyzed coupling reactions involving (i) oxidative addition, (ii) transmetalation, and (iii) reductive elimination steps, as shown in Scheme 2.3b. However, the reductive elimination step of iodine intermediates is mechanistically different from that of transition metal intermediates. In the former case, umpolung of Nu^1 due to the electron-deficient iodine atom of the I–Nu^1 bond is a driving force for the Nu^1–Nu^2 coupling step (reductive elimination).

The choice of stoichiometric cooxidants [O] is very important to achieve the catalytic use of hypervalent organoiodine derivatives or hypoiodite salts for oxidative transformations (Scheme 2.3a). Oxygen gas, hydrogen peroxide, alkyl hydroperoxides such as *tert*-butyl hydroperoxide (TBHP) and cumene hydroperoxide (CHP), peracids such as *m*-chloroperbenzoic acid (*m*-CPBA) and Oxone®, and electrophilic

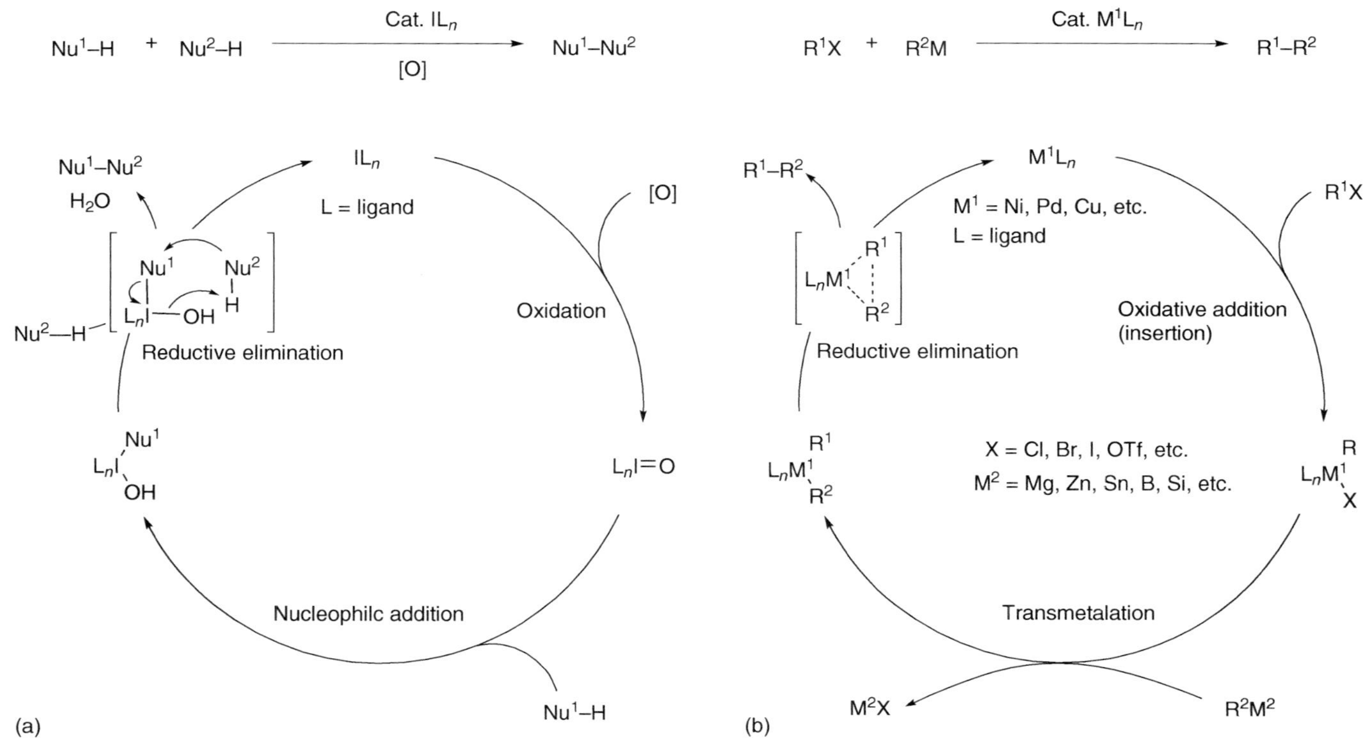

Scheme 2.3 (a) Iodine-catalyzed coupling reaction. (b) Transition-metal-catalyzed coupling reaction.

halides such as Selectfluor® and *N*-bromosuccinimide (NBS) have been used as cooxidants for the *in situ* generation of active iodine species. More recently, electrochemical methods have also been developed. Please refer to Chapter 12 for cooxidants in iodine catalysis.

Halogen-bonding (XB) interactions of electron-deficient iodine derivatives as soft Lewis acid catalysts have also attracted significant research activity. Dihalogens (X_2) and organohalides (X–L) are capable of acting as XB donors and follow the general trend: F < Cl < Br < I, with iodine normally forming the strongest interactions.

This chapter focuses on iodine catalysis for organic reactions. Please refer to Chapter 1 for historical aspects of iodine catalysis.

2.2 Halogen-bonding Catalysis

Dihalogens (X_2) tend to form strong halogen bonds (XBs) [9]. If the halogen (X) is bonded to an electron-withdrawing group (L), it is more likely to form stronger XBs. For example, iodoperfluoroalkanes are well-designed for XB in crystal engineering. XBs are strong, specific, and directional interactions that give rise to well-defined structures. XB strengths range from 5 to 180 kJ mol^{-1}. The strength of XB allows it to compete with hydrogen bonds (HBs), which are slightly weaker. XBs tend to form at 180° angles. Another factor that contributes to the strength of an XB comes from the short distance between the halogen (Lewis acid, XB donor) and Lewis base (XB acceptor). The attractive nature of XBs makes the distance between the donor and acceptor shorter than the sum of their van der Waals radii. The halogen-bonding interaction becomes stronger as the distance between the halogen and Lewis base decreases. For example, the crystal structure of molecular iodine and triphenylphosphine sulfide is shown in Scheme 2.4 [10]. Molecular iodine is activated by double

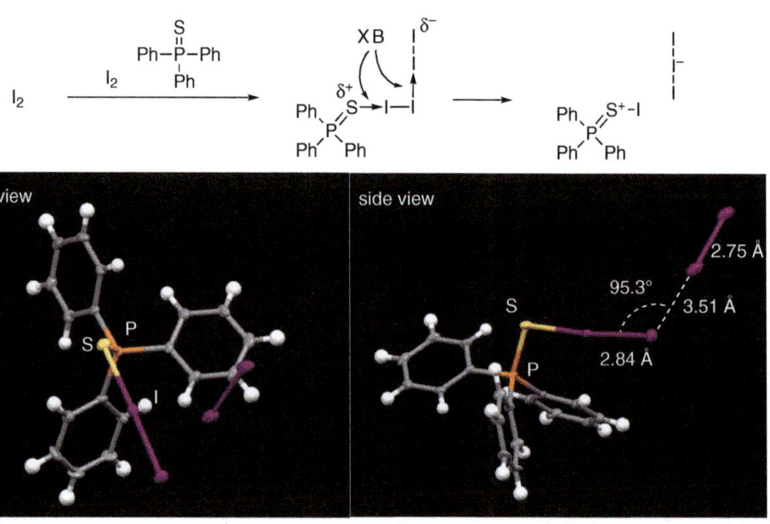

Scheme 2.4 Activation of molecular iodine through double halogen-bonding interaction.

halogen-bonding interaction of Lewis acidic another molecular iodine and Lewis basic triphenylphosphine sulfide. This method for the activation of iodine is highly effective for enantioselective iodocyclization using chiral Lewis base catalyst and halosuccinimide [10, 11]. Please refer to Chapter 3 for details on catalysis by XB based on iodine.

2.3 Iodine(I) Catalysis

Iodide salts are often used as precatalysts in the presence of appropriate cooxidants such as hydrogen peroxide, alkyl hydroperoxides, sodium percarbonate, Oxone, potassium persulfate, and *m*-CPBA. Depending on the coupling partners, oxidants, and reaction conditions, these reactions may proceed *via* an ionic or radical mechanism. In 2010, Ishihara et al. developed a quaternary ammonium hypoiodite catalysis for the oxidative coupling reactions [12, 13]. Hypoiodite salts, the catalytic active species, are generated *in situ* from the corresponding iodides in the presence of relatively inexpensive and mild oxidants such as hydrogen peroxide or alkyl hydroperoxides (Scheme 2.5) [14]. This catalytic oxidation system proceeds under milder conditions, and water or alcohol is the only side product derived from the oxidant used.

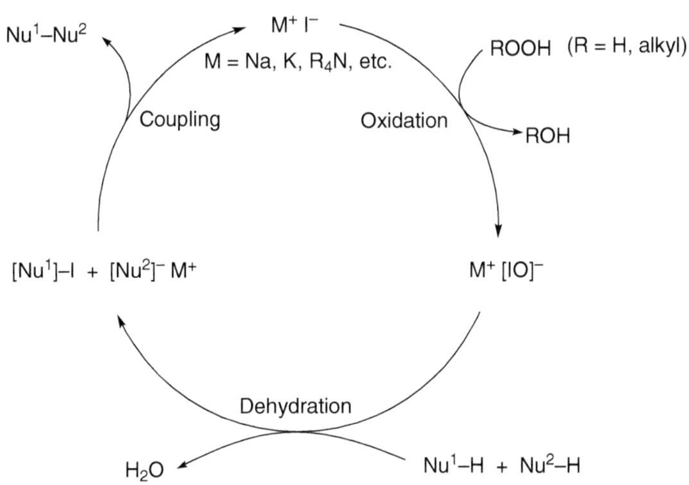

Scheme 2.5 General catalytic cycle of hypoiodite catalysis for oxidative coupling.

As an example, in 2020, Ishihara et al. reported the hypoiodite-catalyzed dearomative peroxycyclization of homotryptamine derivatives to peroxytetrahydropyridoindolenines under mild conditions (Scheme 2.6) [15]. This is the tandem reaction of the intramolecular oxidative aminocyclization of homotryptamines at the C-2 position and the subsequent oxidative coupling of tetrahydropyridoindoles with TBHP at the C-3 position.

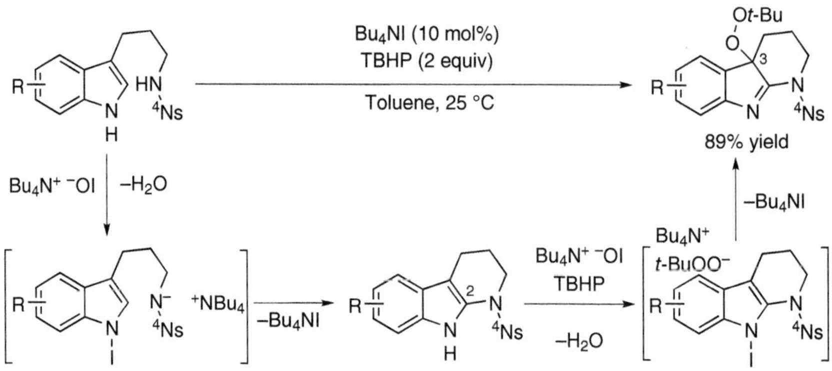

Scheme 2.6 Hypoiodite-catalyzed tandem oxidative cyclization/coupling reactions of homotryptamine derivatives.

Interestingly, a tandem oxidative cyclization/epoxidation proceeds in the presence of 2,2,6,6-tetramethylpiperidine 1-oxyl (TEMPO) as an additive to give epoxyindolenines (Scheme 2.7). In the latter case, a preferential electrophilic addition of $TEMPO^+$ to tetrahydropyridoindole, a common intermediate, is generated *in situ* by the hypoiodite-catalyzed oxidation of TEMPO, at the C-3 position followed by elimination and epoxidation.

In contrast, the tandem oxidative cyclization of indole derivatives tethered to aniline sulfonamides using a catalytic amount of tetrabutylammonium iodide (TBAI) and TBHP as an oxidant gives the corresponding indolo[2,3-*b*]quinolines as

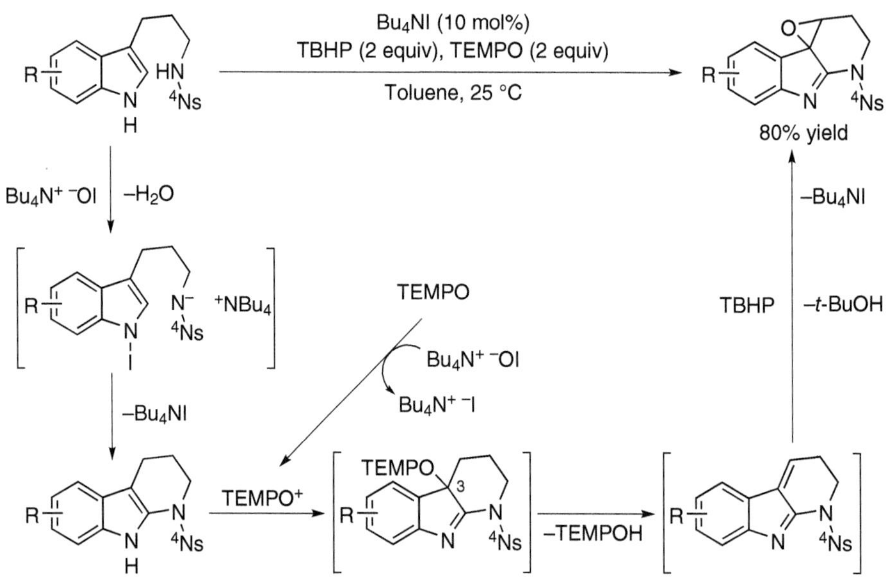

Scheme 2.7 Hypoiodite-catalyzed tandem oxidative cyclization/epoxidation reactions of homotryptamine derivatives (ionic pathway).

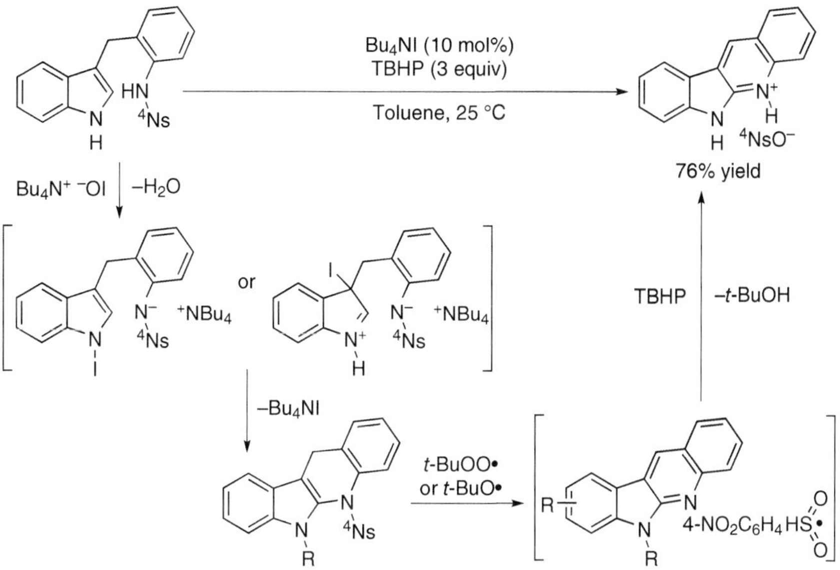

Scheme 2.8 Hypoiodite-catalyzed tandem oxidative cyclization/epoxidation reactions (radical pathway).

sulfonate salts (Scheme 2.8) [16]. These results mean that oxidative cyclization of indole derivatives proceeds with hypoiodite as a catalytic species *via* an ionic process, and radical species might play a role in oxidative aromatization/desulfonylation of the intermediate to give indoquinoline sulfonate salts. Not only hypoiodite species but also alkoxy or alkylperoxy radical species are generated from TBAI and TBHP (Scheme 2.9). Importantly, both ionic and radical active species would be generated *in situ* under these mild conditions for the corresponding oxidative transformation to proceed in a chemoselective manner. Please refer to Chapters 4, 5, and 10 for details on iodine(I)-catalyzed oxidative transformations *via* an ionic or radical pathway.

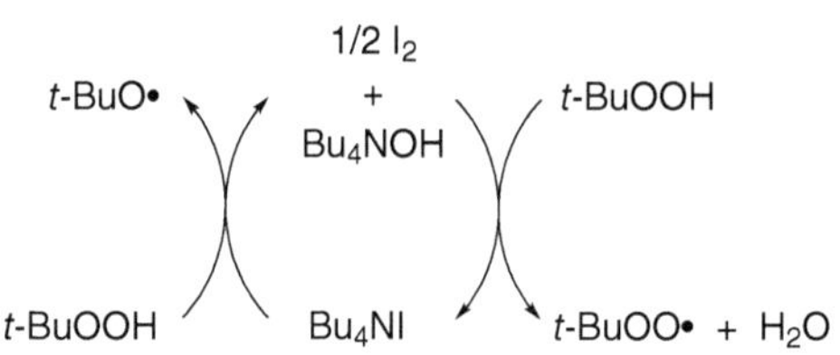

Scheme 2.9 Iodine-catalyzed generation of alkoxy and alkylperoxy radicals.

2.4 Iodine(III) Catalysis

Over the past three decades, hypervalent organoiodines(III and V) have attracted great attention due to their mild and chemoselective oxidizing properties and environmentally benign characteristics in contrast to heavy metal-based oxidants [17]. In particular, the reactivities of organoiodines(III) resemble those of mercury(II), thallium(III), and lead(IV). However, the use of hypervalent iodines as stoichiometric oxidants has been limited because some are shock sensitive or explosive and/or show poor solubility in organic solvents. Therefore, a catalytic application of hypervalent iodine is strongly needed for both economic and environmental reasons. The first catalytic use of organoiodine(III) compounds with chemical oxidants was reported by Ochiai et al. [18] and Kita et al. [19] independently in 2005. In these reactions, the active iodine(III) species can be generated *in situ* from a catalytic amount of iodoarene and *m*-CPBA as a stoichiometric oxidant in the presence of Lewis or Brønsted acids as a coactivator (Scheme 2.10). Since these breakthroughs, rapid progress has been made in the development of hypervalent iodine catalysis [6, 7].

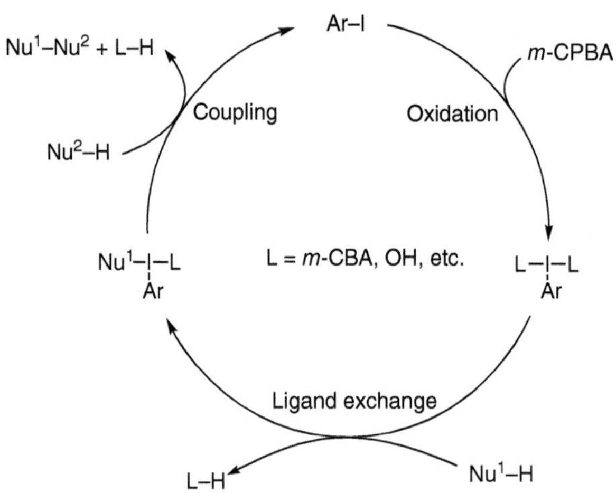

Scheme 2.10 General catalytic cycle of organoiodine(III) catalysis for oxidative coupling.

In 2010, Ishihara et al. reported a highly catalytic enantioselective Kita oxidative spirolactonization catalyzed by *in situ*-generated chiral hypervalent iodine(III) species in the presence of *m*-CPBA as a cooxidant (Scheme 2.11) [20–23]. The rational design of conformationally flexible C_2-symmetric iodosylarene catalyst is highly effective for inducing high enantioselectivity [24]. Please refer to Chapters 6–8 and 10 for details on organoiodine(III)-catalyzed oxidative transformations.

Organoiodine(III) compounds can activate the C_{sp2}–C_{sp2} π-bond for alkene difunctionalization reactions. Generally, coordination of organoiodine(III) compounds to alkenes proceeds *via* dissociative substitution to generate an electrophilic π-complex (Scheme 2.12). Ligand (X) exchange with Brønsted acids such as hexafluoroisopropanol (HFIP) or the coordination of Lewis acids to the ligand X accelerates this dissociative substitution step. On the other hand, the ligand exchange with alkanols such as ethanol suppresses this dissociative substitution step.

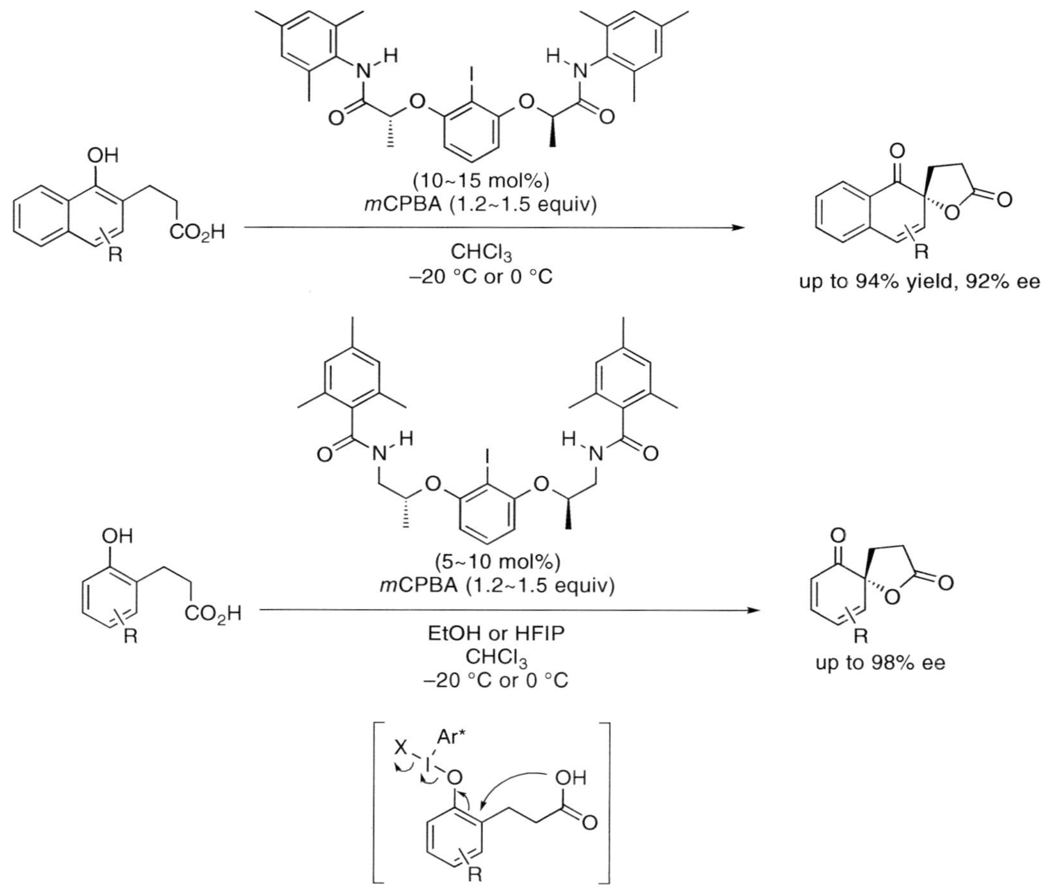

Scheme 2.11 Enantioselective Kita oxidative spirolactonization reactions.

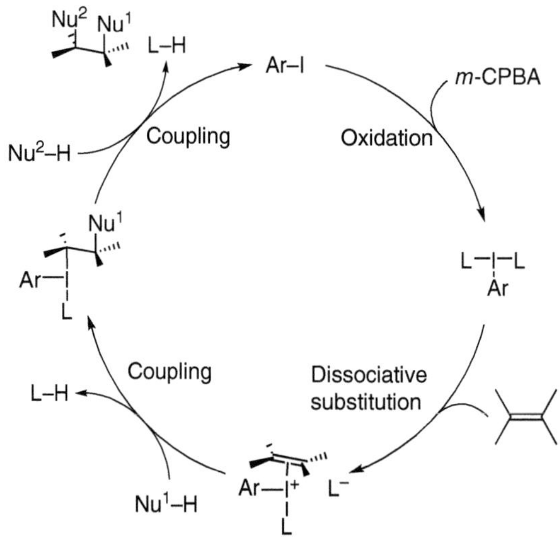

Scheme 2.12 General catalytic cycle of organoiodine(III) catalysis for alkene difunctionalization reactions.

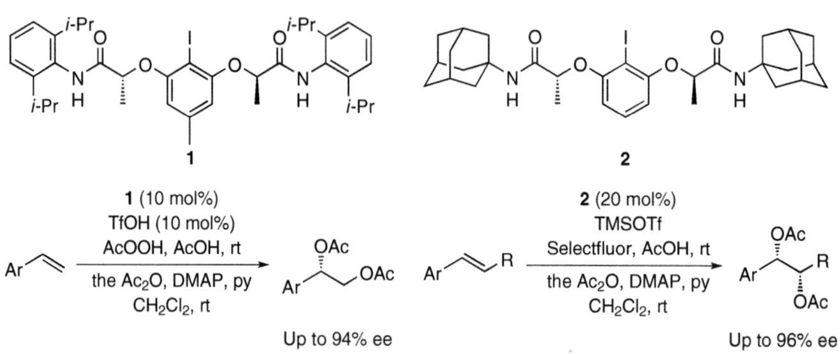

Scheme 2.13 Catalytic enantioselective diacetyoxylation of styrenes reported by Muñiz et al.

For example, Muñiz et al. developed highly enantioselective vicinal diacetoxylation of styrenes using chiral lactic acid-derived Ishihara-type precatalysts such as **1** and **2** and *m*-CPBA or Selectfluor® as a stoichiometric oxidant (Scheme 2.13) [25, 26]. Please refer to Chapter 9 for details on organoiodine(III)-catalyzed alkene difunctionalization reactions.

As mentioned above, since pioneering works by Ochiai and Kita [18, 19], a stoichiometric cooxidant such *m*-CPBA has been used for organoiodine(III) catalyses, thus generating an equimolar quantity of *meta*-chlorobenzoate (*m*CBA). A more desirable alternative is the use of molecular oxygen that only generates aqueous waste. Since the single electron oxidation potential of iodobenzene is much higher (half peak potential: $E_{p/2} = 2.17\,V$ vs. Standard calomel electrode [SCE]) [27] than the reduction potential (−0.33 V vs. Standard hydrogen electrode [SHE], for single-electron reduction at pH 7) [28], resorting to an electron-transfer

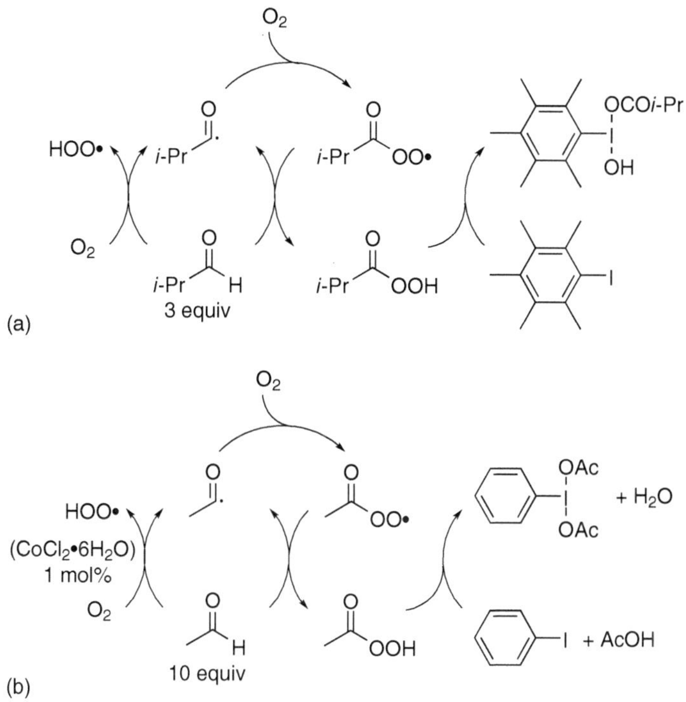

Scheme 2.14 Aerobic oxidation protocols from iodoarenes to organoiodine(III) compounds. (a) Miyamoto and Uchiyama. (b) Powers.

mediator (ETM) is mandatory. The strategy was eventually implemented in 2017 by Miyamoto and Uchiyama et al. [29] and in 2018 by Powers et al. [30] who independently reported the first examples of aerobic oxidation of iodoarenes (Scheme 2.14). Both approaches are based on the formation of a suitable oxidant during the O_2-mediated autooxidation of aldehydes, although the protocols slightly differ. The Miyamoto–Uchiyama protocol requires a sterically hindered aldehyde and was applied to the glycol scission of 1,2-diols and the Hofmann rearrangement of primary amides with 3–5 mol% of isobutyraldehyde and 5–20 mol% of pentamethyliodobenzene (Scheme 2.14a) [29]. Powers' method uses 10 equivalents of acetaldehyde and 1 mol% of $CoCl_2$ as the initiator to promote the synthesis of (diacetoxy)iodoarenes from the corresponding aryl iodides in acetic acid (Scheme 2.14b) [30]. Despite the spectacular improvement that these aerobic strategies bring to iodoarene catalysis, they still require a rather large excess of aldehyde and, in some cases, even a stoichiometric amount of the iodobenzene catalyst.

In 2021, Cariou et al. reported that using the spirocyclization of amides as a benchmark reaction, aerobic iodoarene catalysis can be enabled by relying on a pyrylium photocatalyst under blue light irradiation (Scheme 2.15) [31]. This dual organocatalytic system allows the use of low catalytic loading of both catalysts under mild operating conditions. Please refer to Chapter 12 for cooxidants in iodine catalysis.

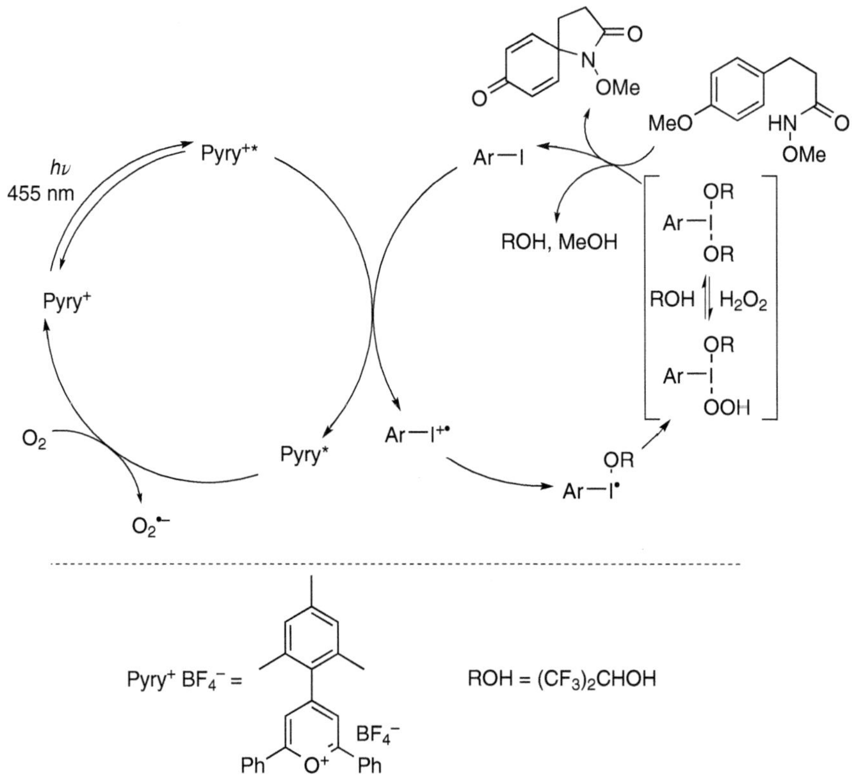

Scheme 2.15 Photoinduced aerobic iodoarene catalysis for spirocyclization.

2.5 Iodine(V) Catalysis

Two representative iodine(V) compounds, 2-iodoxybenzoic acid (IBX) and Dess–Martin periodinane (DMP), have been heavily studied as oxidative agents as shown in Scheme 2.16. These reagents are frequently used in the total synthesis of complex natural products due to their high chemoselectivity. However, hypervalent iodine reagents generally have hazardous characteristics, which hamper their use in large-scale applications.

Vinod et al. reported a method for the oxidation of alcohols with the catalytic use of 2-iodobenzoic acid in the presence of stoichiometric Oxone [32]. While many iodine(V)-based catalysts are variants of the IBX core, in 2009, Ishihara et al. reported

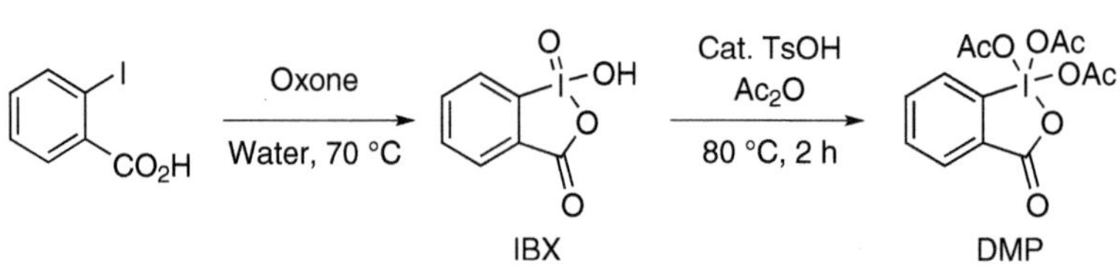

Scheme 2.16 Synthetic approach toward 4-(dimethylamino)pyriidine (DMAP) using an IBX intermediate.

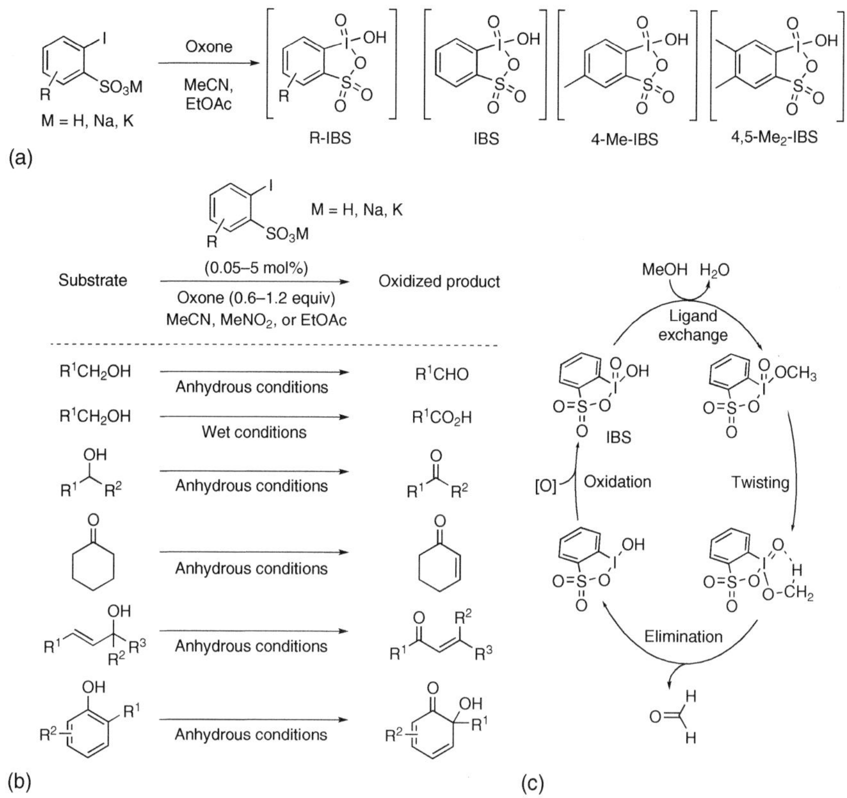

Scheme 2.17 IBS-catalyzed oxidation of alcohols and arenols. (a) *In situ* preparation of R-IBS. (b) R-IBS-catalyzed oxidation. (c) Proposed mechanism of IBS catalysis.

a highly efficient catalyst that relied on 2-iodoxybenzenesulfonic acid (IBS), rather than IBX [33, 34] (Scheme 2.17a). The efficacy of *in situ*-generated IBS was outstanding for a range of oxidative transformations when it was used in combination with tetrabutylammonium hydrogen sulfate and Oxone as the external oxidant of choice. For example, several alcohols were selectively oxidized in the presence of 0.05–5 mol% of *in situ*-generated IBS with Oxone (Scheme 2.17b). Under the reaction conditions, tertiary alcohols were found to undergo an appealing oxidative rearrangement that led to enone formation [35]. Recently, IBS was isolated and its X-ray crystal structure was determined by Zhdankin, Yusubov, et al. [36]. The catalytic cycle of IBS and the oxidation mechanism of methanol with IBS were proposed by Ishihara et al. (Scheme 2.17c) [33, 37]. Please refer to Chapter 11 for details on organoiodine(V)-catalyzed oxidative transformations [38].

Powers' method shown in Scheme 2.14b was further refined for the formation of iodine(V) reagents [39]. Aerobic oxidation of iodobenzenes is coupled with disproportionation of the initially generated I(III) compounds to generate I(V) reagents (Scheme 2.18). The aerobically generated I(V) reagents exhibit substrate oxidation chemistry analogous to that of DMP. Please refer to Chapter 12 for cooxidants in iodine catalysis.

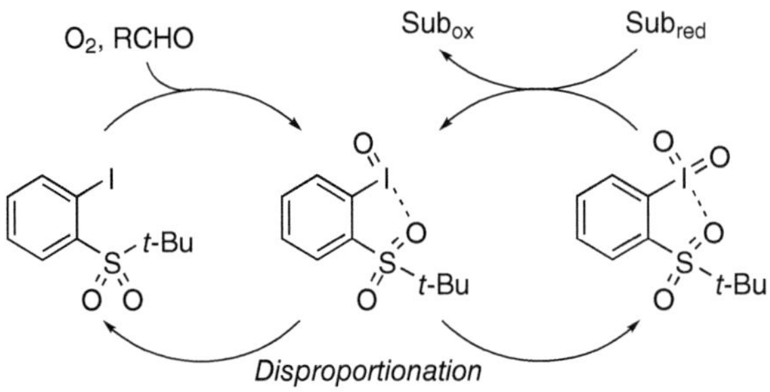

Scheme 2.18 Proposed aerobic catalytic cycle of I(V) species.

References

1 Tremblay, J.-F. (2011). Earthquake rattles iodine market. *Chem. Eng. News* 89 (49): 22–24.

2 Sharpe, A.G. (1967). Halides containing multicentered metal-metal bonds. In: *Halogen Chemistry*, vol. 1 (ed. V. Gutmann), 1–39. London: Academic Press.

3 Holleman, A.F. and Wiberg, E. (1995). *Lehrbuch der anarganischen Chemie, 101. Aufl. ed.* Berlin: de Gruyter.

4 Kaiho, T. (2014). *Iodine Chemistry and Applications.* Chichester, UK: Wiley.

5 Sunley, G.J. and Watson, D.J. (2000). High productivity methanol carbonylation catalysis using iridium: the Cativa™ process for the manufacture of acetic acid. *Catal. Today* 58: 293–307.

6 Finkbeiner, P. and Nachtsheim, B.J. (2013). Iodine in modern oxidation catalysis. *Synthesis* 45: 979–999.

7 Yusubov, M.S. and Zhdankin, V.V. (2015). Iodine catalysis: a green alternative to transition metals in organic chemistry and technology. *Resource-Efficient Technologies* 1: 49–67.

8 Claraz, A. and Masson, G. (2018). Asymmetric iodine catalysis-mediated enantioselective oxidative transformations. *Org. Biol. Chem.* 16: 5386–5402.

9 Breugst, M. and von der Heiden, D. (2018). Mechanisms in iodine catalysis. *Chem. A Eur. J.* 24: 9187–9199.

10 Lu, Y., Nakatsuji, H., Okumura, Y. et al. (2018). Enantioselective halo-oxy- and halo-azacyclizations induced by chiral amidophosphate catalysts and halo-Lewis acids. *J. Am. Chem. Soc.* 140: 6039–6043.

11 Nakatsuji, H., Sawamura, Y., Sakakura, A., and Ishihara, K. (2014). Cooperative activation with chiral nucleophilic catalysts and *N*-haloimides: enantioselective iodolactonization of 4-arylmethyl-4-pentenoic acids. *Angew. Chem. Int. Ed.* 53: 6974–6977.

12 Uyanik, M., Okamoto, H., Yasui, T., and Ishihara, K. (2010). Quaternary ammonium (hypo)iodite catalysis for enantioselective oxidative cycloetherification. *Science* 328: 1376–1379.

13 Uyanik, M., Hayashi, H., and Ishihara, K. (2014). High-turnover hypoiodite catalysis for asymmetric synthesis of tocopherols. *Science* 345: 291–294.

14 Uyanik, M. and Ishihara, K. (2011). Catalysis with *in situ*-generated (hypo)iodite ions for oxidative coupling reactions. *ChemCatChem* 4: 177–185.
15 Uyanik, M., Tanaka, H., and Ishihara, K. (2020). Hypoiodite-catalyzed chemoselective tandem oxidation of homotryptamines to peroxy- and epoxytetrahydropyridoindolenines. *Org. Lett.* 22: 8049–8054.
16 Uyanik, M., Tanaka, H., and Ishihara, K. (2021). I^+/TBHP catalysis for tandem oxidative cyclization to indolo[2,3-*b*]quinolines. *Asian J. Org. Chem.* 10: 164–169.
17 Yoshimura, A. and Zhdankin, V.V. (2016). Advances insynthetic application of hypervalentiodine compounds. *Chem. Rev.* 116: 3328–3435.
18 Ochiai, M., Takeuchi, Y., Katayama, T. et al. (2005). Iodobenze-catalyzed α-acetoxylation of ketones. *In situ* generation of hypervalent (diacyloxyiodo)benzenes using *m*-chloroperbenzoic acid. *J. Am. Chem. Soc.* 127: 12244–12245.
19 Dohi, T., Maruyama, A., Yoshimura, M. et al. (2005). Versatile hypervalent-iodine(III)-catalyzed oxidations with *m*-chloroperbenzoic acid as a cooxidant. *Angew. Chem. Int. Ed.* 44: 6193–6196.
20 Uyanik, M., Yasui, T., and Ishihara, K. (2010). Enantioselective Kita oxidative spirolactonization catalyzed by *in situ* generated chiral hypervalent iodine(III) species. *Angew. Chem. Int. Ed.* 49: 2175–2177.
21 Uyanik, M., Yasui, T., and Ishihara, K. (2013). Hydrogen bonding and alcohol effects in asymmetric hypervalent iodine catalysis: enantioselective oxidative dearomatization of phenols. *Angew. Chem. Int. Ed.* 52: 9215–9218.
22 Uyanik, M., Ishizaki, S., and Ishihara, K. (2021). Synthesis of chiral organoiodine catalyst for enantioselective oxidative dearomatization reactions: *N*,*N*'-(2*S*,2*S*')-(2-iodo-1,3-phenylene)bis(oxy)bis(propane-2,1-diyl)bis(2,4,6-trimethylbenzamide). *Organic Synth.* 98: 1–27.
23 Uyanik, M., Ishizaki, S., and Ishihara, K. (2021). Chiral organoiodine-catalyzed enantioselective oxidative dearomatization of phenols. *Organic Synth.* 98: 28–50.
24 Uyanik, M. and Ishihara, K. (2019). Designer C_2-symmetric chiral diamide-type organoiodine catalysts. *TCIMAIL* 182: 2–14.
25 Haubenreisser, S., Wöste, T.H., Martinez, C. et al. (2016). Structurally defined molecular hypervalent iodine catalysis for intermolecular enantioselective reactions. *Angew. Chem. Int. Ed.* 55: 413–417.
26 Muñiz, K. and Wöste, T. (2016). Enantioselective vicinal diacetoxylation of alkenes under chiral iodine(III) catalysis. *Synthesis* 48: 816–827.
27 Roth, H.G., Romero, N.A., and Nicewicz, D.A. (2016). Experimental and calculated electrochemical potentials of common organic molecules for applictaions to single-electron redox chemistry. *Synlett* 27: 714–723.
28 Wood, P.M. (1988). The potential diagram for oxygen at pH 7. *Biochem. J.* 253: 287–289.
29 Miyamoto, K., Yamashita, J., Narita, S. et al. (2017). Iodoarene-catalyzed oxidative transformations using molecular oxygen. *Chem. Commun.* 53: 9781–9784.
30 Maity, A., Hyun, S.-M., and Powers, D.C. (2018). Oxidase catalysis *via* aerobically generated hypervalent iodine intermediates. *Nat. Chem.* 10: 200–204.

31 Habert, L. and Cariou, K. (2021). Photoinduced aerobic iodoarene-catalyzed spirocyclization of *N*-oxy-amides to *N*-flused spirolactams. *Angew. Chem. Int. Ed.* 60: 171–175.

32 Thottumkara, A.P., Bowsher, M.S., and Vinod, T.K. (2005). *In situ* generation of *o*-iodoxybenzoic acid (IBX) and the catalytic use of it in oxidation reactions in the presence of oxone as a co-oxidant. *Org. Lett.* 7: 2933–2936.

33 Uyanik, M., Akakura, M., and Ishihara, K. (2009). Iodoxybenzenesulfonic acid as an extremely active catalyst for the selective oxidation of alcohols to aldehydes, ketones, carboxylic acids, and enones with oxone. *J. Am. Chem. Soc.* 131: 251–262.

34 Uyanik, M., Mutsuga, T., and Ishihara, K. (2017). 4,5-Dimethyl-2-iodoxybenzenesulfonic acid catalyzed site-selective oxidation of 2-substituted phenols to 1,2-quinols. *Angew. Chem. Int. Ed.* 56: 3956–3960.

35 Uyanik, M., Fukatsu, R., and Ishihara, K. (2009). IBS-catalyzed oxidative rearrangement of tertiary allylic alcohols to enones with oxone. *Org. Lett.* 11: 3470–3473.

36 Mironova, I.A., Postnikov, P.S., Yusubova, R.Y. et al. (2018). Preparation and X-ray structure of 2-iodoxybenzenesulfonic acid (IBS) – a powerful hypervalent iodine(V) oxidant. *Beilstein J. Org. Chem.* 14: 1854–1858.

37 Jiang, H., Sun, T.-Y., Wang, X. et al. (2017). A twist of the TWIST MEchanism, 2-iodoxybenzoic acid (IBX)-mediated oxidation of alcohol revisited: theory and experiment. *Org. Lett.* 19: 6502–6505.

38 Uyanik, M. and Ishihara, K. (2010). 2-Iodoxybenzenesulfonic acid (IBS) catalyzed oxidation of alcohols. *Aldrichimica Acta* 43: 83–91.

39 Maity, A., Hyun, S.-M., Wortman, A.K., and Powers, D.C. (2018). Oxidation catalysis by an aerobically generated Dess–Martin periodinane analogue. *Angew. Chem. Int. Ed.* 130: 7205–7209.

3

Catalysis by Halogen Bonding Based on Iodine

*Revannath L. Sutar and Stefan M. Huber**

Fakultät für Chemie und Biochemie, Ruhr-Universität Bochum, Bochum, Germany

3.1 Introduction

Iodine and its hypervalent compounds have shown excellent performance as versatile and environmentally benign reagents in organic chemistry. Starting from its use as a catalyst (0.01 mol%) in the efficient dehydration of diacetone alcohol [1], iodine has attracted attention in the synthetic community as a sustainable and relatively inexpensive alternative to transition metal catalysis [2]. Although the origin of its catalytic activity was rather unclear for a long time, this discovery has eventually opened a new era of halogen-based Lewis acid catalysis, which induces numerous transformations for the creation of C–O, C–N, C–H, and C–C bonds. Three different rationales are proposed to explain the observed catalytic activity of iodine: (i) Brønsted acid catalysis by HI formed after the decomposition of iodine, (ii) Lewis acid catalysis through halogen-bonding (XB) interaction, and (iii) catalysis by I^+, resulting from the splitting of iodine. Here we will only focus on the Lewis acid catalysis, and for other types of activations, the readers are directed to further review articles [2].

The formation of halogen bonding in solution is evident in supramolecular chemistry since Colin's discovery of a molecular complex between ammonia and iodine in 1814 [3]. For a more clear understanding of this interaction in the solution phase, particularly with regard to the modes [4] and strengths of association [5, 6], a halogen-bond basicity scale of common Lewis bases with I_2 was later established [5]. Although these interactions can be very strong in many cases (for example, with quinuclidine: $pK_{BI2} = 5.22$, $\Delta G = 29.8\,kJ\,mol^{-1}$), the complexation energies for weakly Lewis basic compounds are relatively low. For example, typical organic electrophiles like ketones, aldehydes, or esters usually show interaction energies of only 0–5 kJ mol^{-1} ($0 < pK_{BI2} < 1$). Nevertheless, these are sufficient for the observation of catalytic effects in many cases as the catalysis usually relies on the coordination to a more Lewis basic transition state.

Iodine Catalysis in Organic Synthesis, First Edition. Edited by Kazuaki Ishihara and Kilian Muñiz.

3.2 Catalysis by Molecular Iodine

The solvolysis of tertiary alkyl (*t*-butyl and *t*-amyl) iodides gets promoted by iodine in aqueous ethanol and aqueous hexafluoropropan-2-ol [6a]. This catalytic activity of iodine was even observed with the less reactive *t*-butyl bromide as the substrate. However, the proposed mechanism was rather complex and involved a combination of halogen-bonding interaction between iodine and alkyl halide, ionization of the alkyl halide, ion-pair separation, and the reaction of reversibly formed intimate ion pairs with the solvent. In the iodine-induced activation of primary and secondary alcohols (Scheme 3.1a), the involvement of halogen bonding was indirectly supported based on the disproportionation of bis(benzhydryl)ethers [6b, c]. In alternative settings, a large variety of transesterification reactions can be promoted by substoichiometric amounts of iodine [2i], where again Brønsted acid catalysis by HI was ruled out by comparison experiments [2d, 7], but still, the exact role of iodine is not clearly elucidated yet in both cases.

The problem of substrate-specific decomposition of catalytic iodine can be reduced when the substrates are less Lewis basic, which minimizes the possibility of catalysis by these decomposition products. For example, the Diels–Alder-type cycloaddition of unnatural fatty acid analogues and different dienophiles (Scheme 3.1b) proceeded efficiently in the presence of a radical scavenger, 6-di-*tert*-butyl-4-methylphenol (BHT), rendering both Brønsted acid catalysis and radical pathways unlikely [8]. In another example, iodine catalyzes cycloadditions between imines and enol ethers, where the Lewis acid activation of the substrate was postulated based on a small shift of the C = N stretching vibration (from 1628 to 1636 cm^{-1}) and the shift of the imine proton in ^{1}H nuclear magnetic resonance (NMR) (from 8.48 to 8.50 ppm) [8c]. A similar mechanism has also been proposed for the iodine-catalyzed synthesis of *cis*-fused furanobenzopyrans [8e] and for related thermal cyclizations of butadienyl isocyanates [9]. It was further supported by computational investigations performed on the iodine-catalyzed Diels–Alder reaction between isoprene and acrolein [10].

This activation of α,β-unsaturated carbonyls can be utilized to promote several Michael-type additions [2a–f]. The mechanism of these reactions involving Lewis acid activation was studied using a typical example, the addition of indols to chalcones, as the model reaction [11]. The initially used 10 mol% of iodine in ethanol solvent [11a] was later replaced by 1 mol% under solvent-free conditions (Scheme 3.1c) [11b], was found to be better than typical Lewis and Brønsted acids, and prevented polymerization reactions. The use of HI as the catalyst in the latter case did not result in product formation, which again suggests halogen-bonding-based catalysis. This was later supported by Breugst and coworkers based on a significant reduction in the activation-free energies, with the strongest one being observed for the reaction between indol and *trans*-crotonophenone ($\Delta\Delta G^{\ddagger} = -32\ kJ\ mol^{-1}$) [11e]. An almost identical catalytic activity of iodine over time and also for several runs further strengthens the proposal of halogen-bonding catalysis [2h]. Finally, based on *in situ* infrared (IR) spectroscopy and reaction calorimetry in different solvents, the reaction order in the catalyst was found to be close to one.

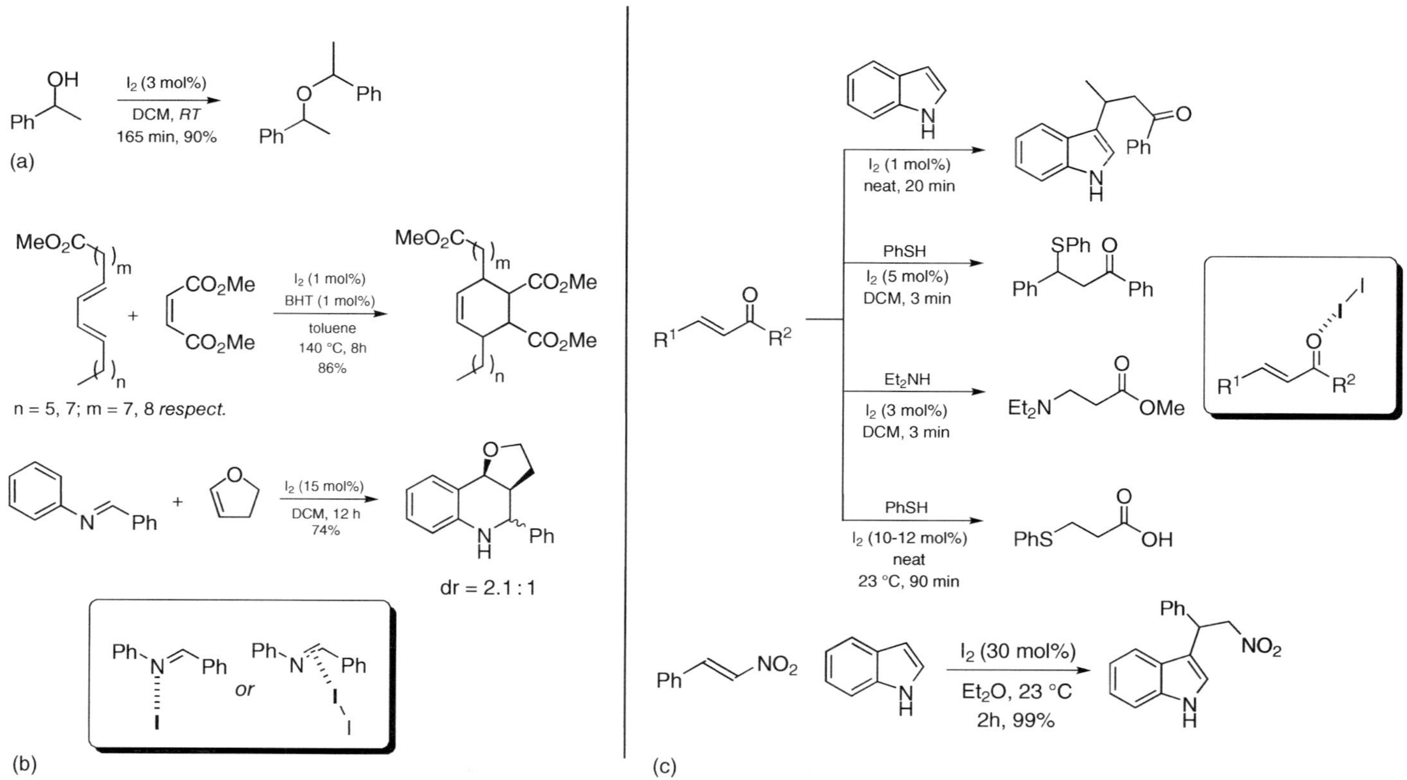

Scheme 3.1 Iodine-catalyzed reactions. (a) Dehydration of alcohols, (b) Diels-Alder reaction, (c) Michael addition reaction.

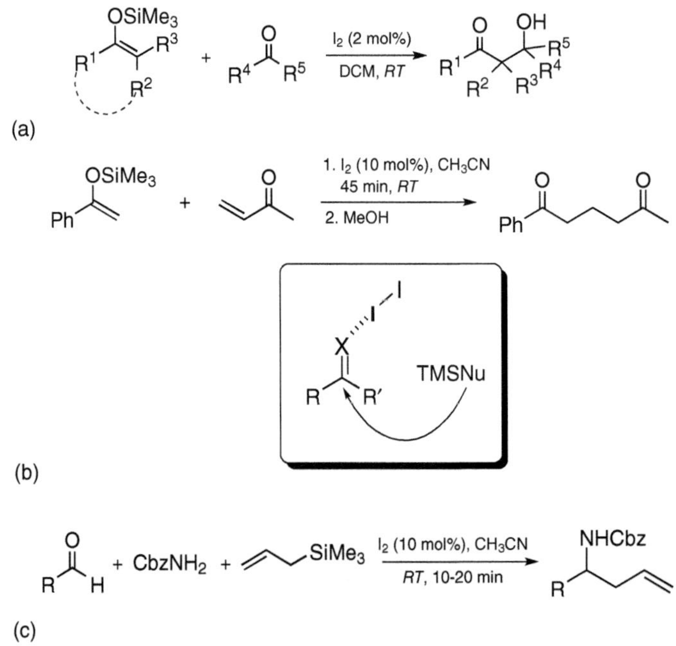

Scheme 3.2 Iodine-catalyzed additions of silyl nucleophiles. (a) Mukaiyama-aldol reaction, (b) Mukaiyama-Michael reaction, (c) Three-component condensation.

Brønsted acid catalysis was also ruled out in another study *via* the unsuccessful reaction between thiophenols and α,β-unsaturated carboxylic acids when carried out in the absence of catalytic iodine (I_2) [12]. However, the exact mechanism of this reaction remained unclear, and the formation of HI addition products in small amounts was also observed, indicating at least partial decomposition of the catalyst. Molecular iodine also catalyzes the Michael additions to nitroolefines [11c, e]. In comparison to the earlier examples, the reactions are quite slow and need higher loading of iodine (30 mol%) which is also supported by theoretical calculations demonstrating a comparably small lowering of the activation barrier ($\Delta\Delta G^{\ddagger} = 8\,\text{kJ mol}^{-1}$) [11e]. Also, this finding is corroborated by the weaker Lewis basicity of the nitro compounds compared to the carbonyl derivatives [5]. In all the Michael addition reactions discussed above, molecular iodine was found to be a similar or superior catalyst compared to typical Lewis acids.

Other nucleophiles such as silyl enol ethers and allyl silanes could also be added to carbonyl derivatives activated by iodine. Three variants of these nucleophilic addition reactions include Mukaiyama aldol reactions [13a, c], Mukaiyama–Michael reactions [13d], and Mannich-type reactions (Scheme 3.2) [13b]. These reactions, which proceed smoothly in acetonitrile, benefit from the addition of methanol, which typically promotes desilylation of the product. The Lewis acid mode of activation of the carbonyl compounds by iodine was supported by theoretical calculations performed on both the Mukaiyama aldol [13c] and the Mukaiyama–Michael reaction [13d]. Two mechanisms (concerted and stepwise) are possible, which are strongly influenced by the nature of the substituents on the reactants. Thus, a stepwise mechanism is more feasible when TMS-enolates are involved in the reaction.

The activation of carbonyls by iodine can also be used to affect acetal formation [14], for example, in the protection of various carbohydrates as their di-O-isopropylidene derivatives [14a]. Similarly, iodine can be used for the introduction and removal of the THP group [14c]. In the same way, thioketals can be prepared from the corresponding thiols [15], and geminal dihydroperoxides can be synthesized by the addition of hydroperoxides [16]. Although it seems that halogen bonding could be involved in these reactions, no mechanistic investigations have been reported yet. However, the involvement of HI as the active species is always speculated about, based on the frequent use of an excess of alcohol in these transformations.

Other types of reactions that are promoted through at least transient halogen bonding with iodine are cyclization reactions, especially Prins [17] and Nazarov cyclizations (Scheme 3.3) [18]. In the former reaction, a very different type of mechanism is proposed, where, in addition to the Lewis acid activation of the aldehyde through halogen bonding, iodine undergoes splitting to form Hypoiodous acid [17b], generating the reactive oxocarbenium ion (Scheme 3.3a). High *syn*-selectivity is achieved in this reaction through the preferential formation of the *E*-oxocarbenium ion, and this has been supported by theoretical calculations [17c]. Also, the Nazarov cyclization of a variety of divinyl ketones occurs in the presence of a substoichiometric amount of molecular iodine with moderate to very good yields (Scheme 3.3b) [18b]. Mechanistic studies based on experimental and theoretical means indicated the involvement of halogen bonding between iodine and the carbonyl group of the substrate, and a reaction order of one in iodine was noted. Several functional groups are tolerated under the reaction conditions and it can be performed in a variety of polar and apolar solvents. Similarly, a very related cycloisomerization of conjugated dienals to yield substituted 2-cyclopentenones (iso-Nazarov cyclization) can be carried out using iodine as the catalyst (Scheme 3.3c) [18a]. Its performance was far better than that of several other Lewis and Brønsted acids or even light irradiation. The mechanism has been supported by theoretical calculations indicating that the involvement of two iodine molecules is necessary to sufficiently lower the transition state energy barrier. In addition, for this cyclization to happen, the substrate should have a substituent at the α-position – otherwise, the cyclization energy barrier is too high to result in product formation even after the coordination of two iodine molecules. This has been further supported by experimental results. This method also allowed a new synthesis of dihydrojasmone – an aroma compound.

In many of the above transformations, iodine performed similarly or sometimes even better than transition metal catalysts and with regard to environmental concerns, it is more eco-friendly and inexpensive. However, elemental iodine does not provide any opportunity for further structural modifications of the catalyst. Also, there are several functional groups and molecules which are reactive toward molecular iodine, which further limits its synthetic applicability. On the other hand, by now a series of experimental and theoretical studies have shown that the interaction

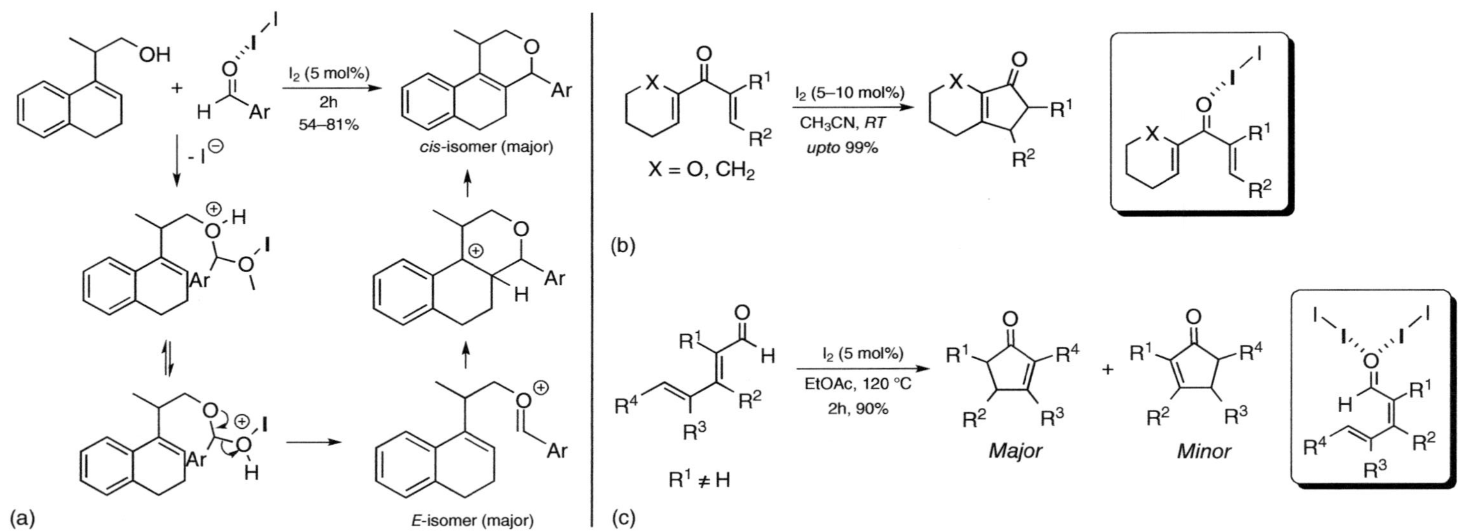

Scheme 3.3 Iodine-catalyzed cyclizations through halogen bonding. (a) Prins cyclization, (b) Nazarov cyclization, (c) Iso-Nazarov cyclization.

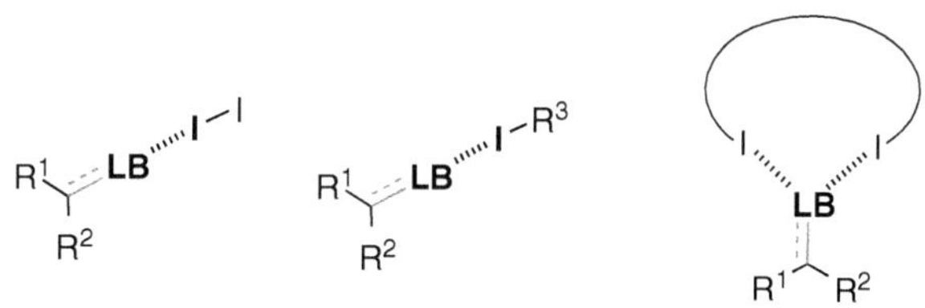

Figure 3.1 General representation of halogen-bonding catalysis.

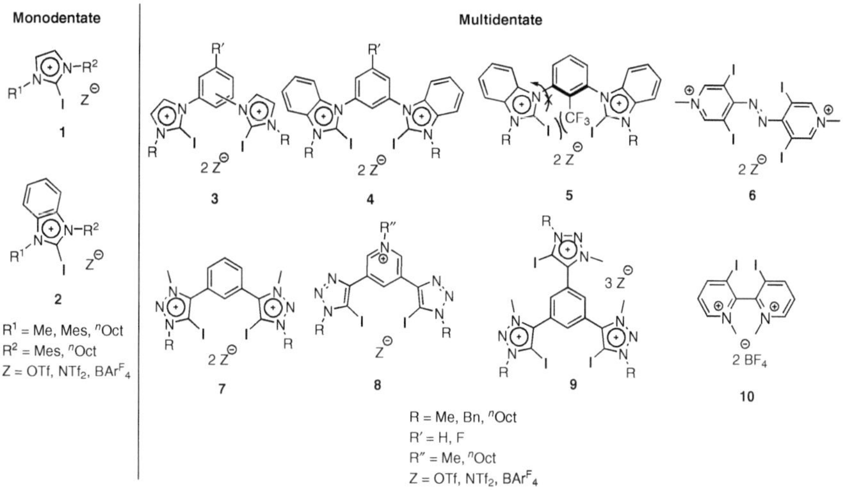

Figure 3.2 Cationic XB donors.

between polarized halogen substituents of an organic molecule and a Lewis basic unit (i.e. halogen bonding, XB) may induce the activation of substrates [19]. Thus, these organohalogen molecules promise the abovementioned catalytic function of molecular iodine. The utilization of this phenomenon in the field of organocatalysis is continuously increasing [2i, 20].

In order to polarize the otherwise Lewis basic halogen substituent of organic molecules, it needs to be attached to an electronegative or otherwise polarizing group (either cationic or neutral) (Figure 3.1). Out of these groups, the cationic ones (Figures 3.2 and 3.4) are mainly the derivatives of triazoles or (benz)imidazoles and the neutral ones often contain polyfluorinated alkyl/aryl moieties, alkyne or amido substituents (Figure 3.3). Considering the ease of synthesis and stability, very appealing among these are the derivatives of 1,2,3-triazoles [21]. It is customary to study the binding properties of newly synthesized XB donors and to understand the modes and strengths of their interactions with common Lewis bases using techniques such as UV-vis, IR, NMR, ITC, Raman, and ESR [5, 22]. In a recent study, for instance, the group of Franz has quantified the strengths of a series of cationic and neutral halogen bond donors using ^{31}P NMR titrations with triethyl phosphine oxide as Lewis base [23]. In general, besides the solvent [5], the binding strength of the XB donor obviously depends largely on its electronic structure [24]. Two recently reported uncommon XB donors include "*anti*-electrostatic" variants, having overall negative charge on the molecule [25a, b] and iodolium species with

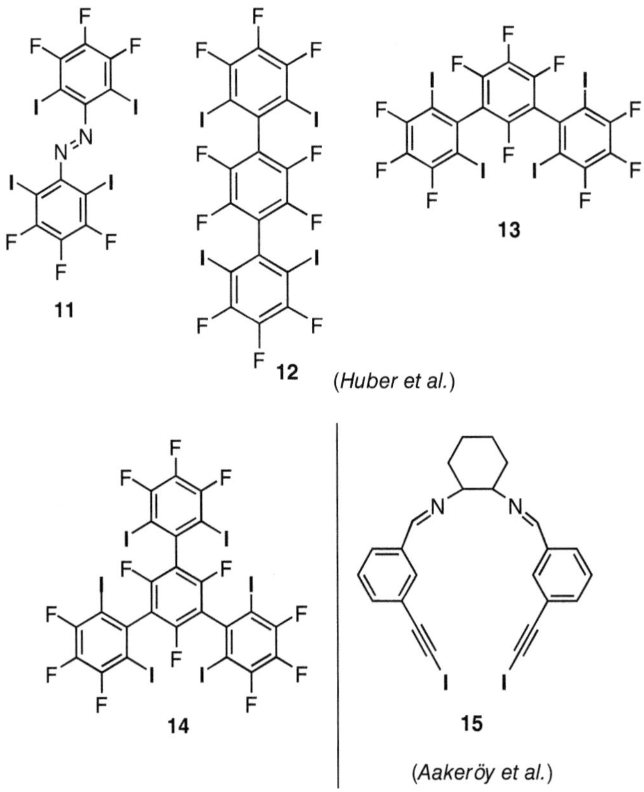

Figure 3.3 Neutral multidentate XB donors.

two perpendicular electrophilic axes [25c]. Overall, cationic XB donors featuring more electron-withdrawing groups are more Lewis acidic than neutral (poly- or perfluorinated) ones [26]. However, the stabilities and solubilities are often reverse. For enhancing the binding strengths, multiple binding sites and particularly those that are *syn*-preorganized are advantageous [27]. The latter also helps achieve asymmetric induction due to the formation of a rigidly confined space [27e].

3.3 Catalysis by Organic Halogen Bond Donors

Bolm and coworkers first suggested the involvement of halogen bonding for the activation of otherwise unreactive molecules in an organic reaction [28]. This initiated studies toward the understanding of the exact mode of activation and of the feasibility to use this interaction for promoting benchmark organic reactions.

The coordination of a Lewis basic unit of the substrate or the catalyst to the halogen bond donor results in the activation of the substrate (as well as the halogen bond donor). Based on the strength of the XB interaction and the stability of the formed adduct, XB can be transient or non-transient. In this chapter, we will mostly focus on the latter and will provide only selected examples for the former.

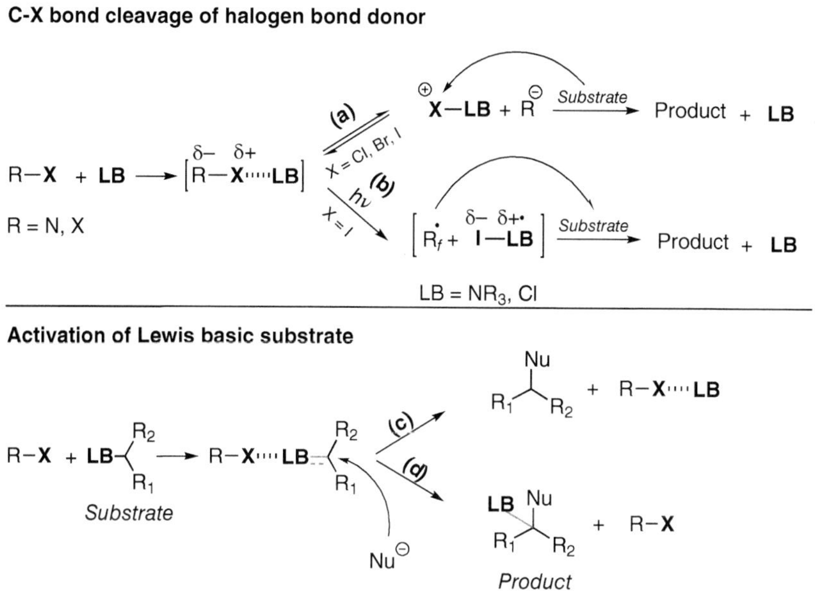

Scheme 3.4 Transient (top a, b) and non-transient (bottom c, d) XB in organic synthesis.

Transient halogen bonding leads to the cleavage of the halogen atom from the XB donors and generates highly reactive intermediates (Scheme 3.4a) like the ones observed in classical halogenation [1c] and halocyclization reactions [29]. In these reactions, the addition of Lewis bases has been shown to enhance the reactivity and selectivity of the halogenating reagent. In halocyclization reactions in particular, this aspect has been extensively investigated and has been demonstrated to have a decisive role. For example, based on Raman spectroscopy, Ishihara and coworkers proposed the involvement of XB between Lewis acidic *N*-haloimides (*N*-iodosuccinimide [NIS], *N*-chlorophthalimide [NCP], and *N*-chlorosuccinimide [NCS]) and iodine in generating the active iodinating species during the chiral phosphate catalyzed enantioselective iodolactonization reaction of 4-arylmethyl-4-pentenoic acids [30a] (Scheme 3.5a). The method was further extended to an efficient, enantioselective, and site-selective (in the presence of multiple olefins) iodo- and bromocycloetherification using chiral Lewis basic amidophosphate catalysts and halo-Lewis acids (X–L). A highly reactive species was proposed to form *via* the interaction of the catalyst, elemental iodine, and *N*-bromosuccinimide (NBS) [30b]. Moreover, the broad applicability of this catalytic system not only for halo-oxycyclization but also for halo-azacyclization reactions was demonstrated. The XB mode of activation was further corroborated by Ishihara *et al.* through extensive NMR studies and density functional theory (DFT) calculations on the formation of 1:1:1 adducts between thiourea derivatives (as Lewis bases), elemental iodine, and *N*-iodosuccinimide (NIS) (Scheme 3.5a) [30c]. A similar thiourea-I_2-based strategy was employed by the authors in the iodochlorination of alkenes with *in situ*-generated iodine chloride [30d]. Some

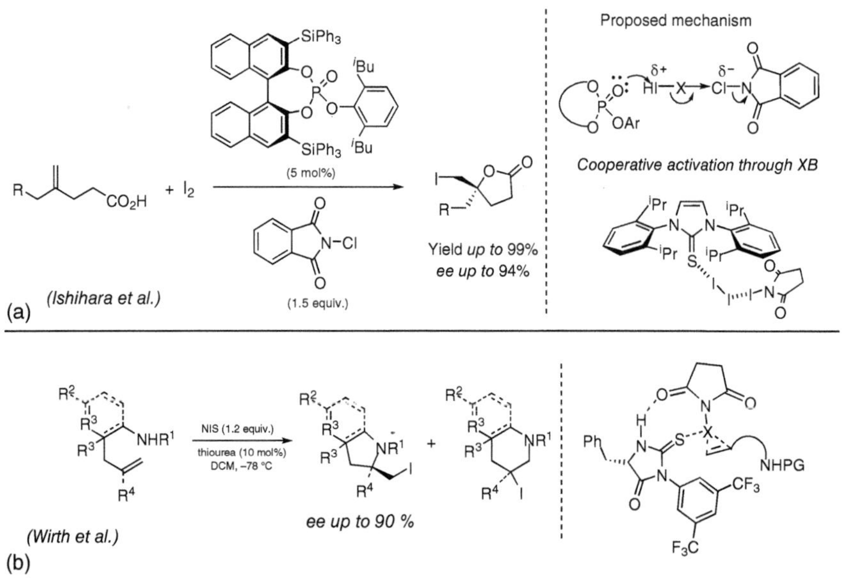

Scheme 3.5 Enantioselective halocyclization reactions involving halogen bonding. (a) Chiral amidophosphate catalyzed halocyclization, (b) Chiral thiourea catalyzed iodoamination of alkenes.

form of XB is also involved in the thiourea-directed iodoamination of alkenes by Wirth *et al.*, using said chiral thiourea catalyst and NIS (Scheme 3.5b) [30e]. The outcome of this reaction is decisively influenced by additives, with KBr promoting endocyclization and KI favoring exocyclization. Further examples featuring XB in iodocyclization reactions include iodolactonizations by Jacobsen *et al.* [30f] (using a tertiary aminourea and an *N*-iodophthalimide derivative) and Arai *et al.*, who report the interplay of hydrogen and halogen bonding in a zinc acetate-catalyzed asymmetric reaction [30g].

In another variant of transient halogen bonding (Scheme 3.4b), otherwise stable (non-transient) XB adducts of perfluoroalkyl iodides and Lewis bases undergo C–I bond cleavage on photo-irradiation and generate (highly) reactive perfluoroalkyl radicals [27e] which initiate a variety of radical reactions. In both of these cases (Scheme 3.4 a and b), the halogen substituent of the XB donor gets detached and thus it acts as a reagent. These reactions do not work in the absence of Lewis bases, indicating the need for XB bond formation for the activation of the XB donor. A variety of transformations, including enantioselective and cascade ones, are reported in the literature based on these approaches [19f, 29]. These reactions are not catalytic in iodine-based halogen bonding and thus will not be covered herein in more detail.

However, under suitable conditions, the XB donor still remains intact after coordination with the Lewis basic substrate while still being able to activate the latter. Depending on the extent of the resulting polarization of the substrate, this can result in the complete detachment of a Lewis basic group (for example, a halide) from the substrate (Scheme 3.4c) or it can promote a nucleophilic addition to this polarized functional group (Scheme 3.4d). The former case results in the formation of a stable

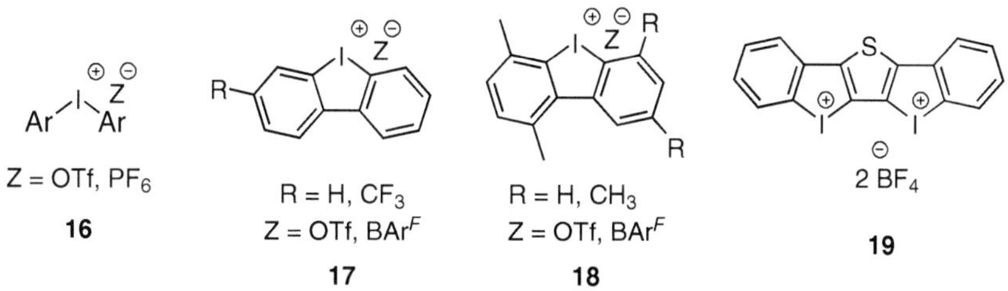

Figure 3.4 Iodine (III)-derived XB donors.

XB adduct as a side product and thus the XB donor is typically needed in stoichiometric amounts to induce the desired transformation. The first example of this was reported by Huber *et al.* in the solvolysis of secondary alkyl halides (benzhydryl bromide) [31, 32a, b, e], using specially designed XB donors (Scheme 3.6a). The mechanism of this Lewis acid activation was thoroughly supported by control experiments. Recently, a computational study claimed that this activation is attributed to the combined effect of halogen bonding and lone pair···π interactions [31c]. Furthermore, the nucleophilic addition of isopropanol to a glycosyl chloride [32c] was achieved using an XB donor as the stoichiometric promoter (Scheme 3.6b). Several multidentate cationic (triflate salts of **3**, **6**, **7**, **10**; Figure 3.2) and neutral XB donors (**11**, **15**; Figure 3.3) promoted these halide abstraction reactions more efficiently than their monodentate analogues [26, 31, 32]. For structurally similar XB donors, the cationic ones are more active than neutral ones [26] and often for XB donors featuring a benzene core; *meta*-substituted dicationic donors are more active than their *para*-substituted counterparts (for example **3**) due to a more suitable bite angle. However, in the challenging solvolysis of benzhydryl chlorides, iodine(III) derivatives (like **17**) showed more activity than **3** (for the chemical structure, see Figure 3.4) [31b]. One of the most promising electrophilic hydrogen-bonding organocatalysts (Schreiner's thiourea) was severely outperformed by the XB donor catalysts in this reaction.

The higher activity of cationic XBs is primarily due to charge assistance. However, an increased cationic charge sometimes reduces their Lewis acidity due to strong ion pairing with the counterion. For example, bidentate dicationic (**7**) and monocationic XB donors (**8**) showed almost comparable activity in a Friedel–Crafts alkylation reaction (Scheme 3.6c) [32d]. The binding of these two Lewis acids to halides was studied and the comparable halogen-bonding strength was found to be due to the compensation of a more favorable enthalpic contribution in **7** with a more favorable entropic contribution in **8**. In addition, because of the probable coordination of anions to the Lewis acidic halogen substituents, a strong counter-anion dependency of the activity is always noted for cationic XB donors (with $BAr^F{}_4{}^- > NTf_2{}^- > OTf^-$). Moreover, to enhance the solubility of the multicationic XBs, they often need to be equipped with nonpolar alkyl substituents. Interestingly, variations in the length of these substituents typically affect the catalytic activity only negligibly.

In another study, the Lewis basic halogen substituent of the substrate forms a transient XB interaction with more polarized halogen bond donors, e.g. NIS or

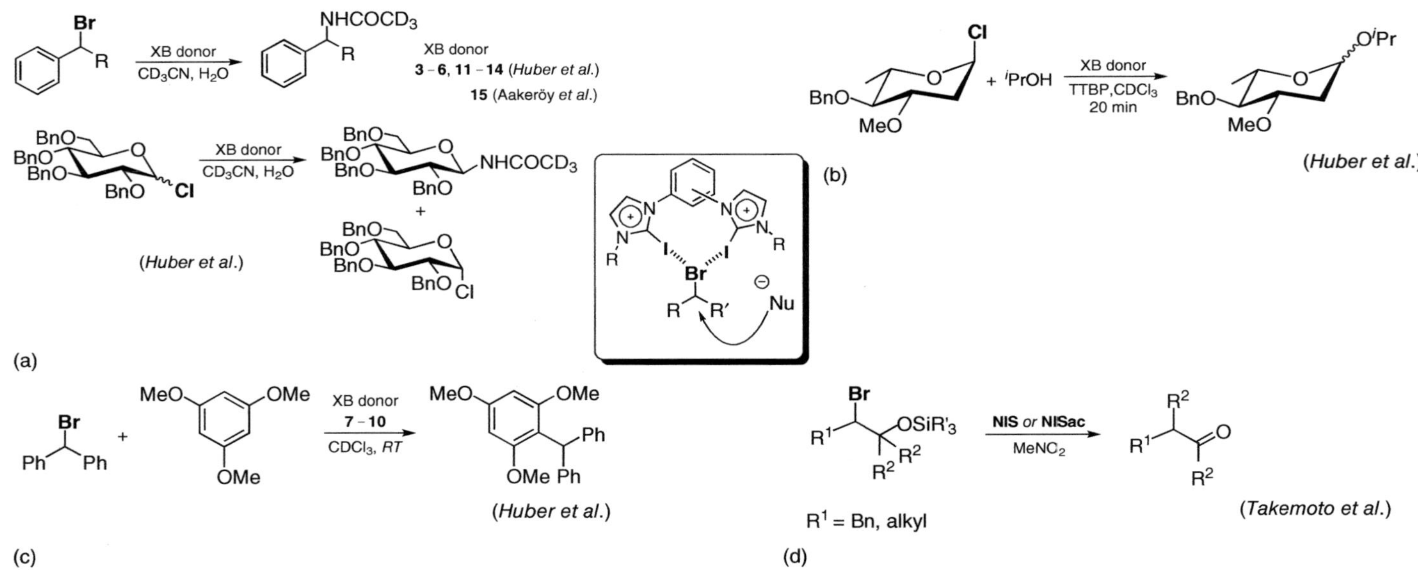

Scheme 3.6 XB-induced halide abstraction reactions. (a) Ritter-type reaction, (b) Koenigs-Knorr-type glycosylation, (c) Friedel–Crafts alkylation, (d) Semipinacol rearrangement.

N-iodosaccharin (NISac). This interaction leads to the removal of the halogen substituent, which then induces a desilylative semipinacol rearrangement of a variety of cyclic and acyclic α-silyloxyhalides (Scheme 3.6d) [33]. This operationally simple reaction occurs with a broad substrate scope (R^1 = Bn, alkyl) in nitromethane and needs stoichiometric amounts of NIS due to the irreversible formation of Br–I. Unfortunately, this reaction doesn't occur in the presence of cationic XB donors making the catalytic protocol difficult.

The inhibition of the XB donor in the above reactions could be overcome by *in situ* dissociation of the XB-halide adduct during the reaction. As silyl nucleophiles generate stoichiometric amounts of the silyl cations during the reaction, they act as a potential trap for the halide anions from these XB-halide adducts (Scheme 3.7a) [27b, 34a]. The first example of this concept was the reaction between silyl enol ether and 1-chloroisochromane, which proceeds efficiently in the presence of catalytic amounts of several monocationic (salts of **1a**, **1b**, **2**), bidentate neutral (**12**, **13**, **14**), or dicationic (**3**, **6**, **7**, and **5**) XB donors, out of which the *syn*-preorganized one (a derivative of **5**) was the most active even at 0.5 mol% loading [27b]. The high activity of this bidentate catalyst – attributed particularly to *syn*-preorganization – was further supported by ITC titrations and the catalytic performance of structurally related *syn*-preorganized and non-preorganized XB donors [27e, f].

In a related study, Takemoto's group reported the catalytic activation of the hydroxyl groups of alcohols via XB-bonded hypervalent trimethylsilyl intermediates. Efficient dehydrogenative coupling of activated alcohols (benzylic alcohols and hemiaminals) with allyltrimethylsilane and trimethylsilyl cyanide was achieved through the combined use of the catalytic amount of XB donor **1a** and a halide source (I_2) (Scheme 3.7b) [34b]. Later, Chan and Yeung used TMSCl to induce the *in situ* formation of *N*-acyliminium ions from *N*,*O*-aminals which then react with a variety of nucleophiles (Scheme 3.7c) [34c]. The catalyst is regenerated after the capture of the chloride anion of the XB complex by the counterparts of the nucleophile. Interestingly, the imidazolium-based monodentate XB donors (**1**) performed better than the bidentate (**3**) and the benzimidazolium-derived ones (**2**), thus demonstrating the dependency of XB-based catalysis on several factors. It should be noted that this reaction is not catalyzed by a thiourea catalyst, which further stresses the potential of halogen-bonding organocatalysis. A counter-anion-directed asymmetric catalysis using **1b** was attempted and high diastereoselectivity and moderate enantioselectivity (44% enantiomeric excess [ee]) was achieved when chiral phosphate anions were used.

Polymerization reactions can be affected by catalytic amounts of halogen bond donor if the monomeric substrate contains a Lewis basic group. The first example of this approach used an iodine-based interhalogen compound (ICl_3) to catalyze the highly selective ring-opening polymerization of *L*-lactide with 11-bromo-1-undecanol at room temperature, producing polymers of predictable molecular weights and narrow molecular weight distribution (Scheme 3.8a) [35a]. The proposed mechanism of this polymerization involves activation of both alcohol and *L*-lactide through hydrogen- and halogen-bonding interactions, respectively. Later, a purely halogen-bonding-induced controlled cationic living polymerization

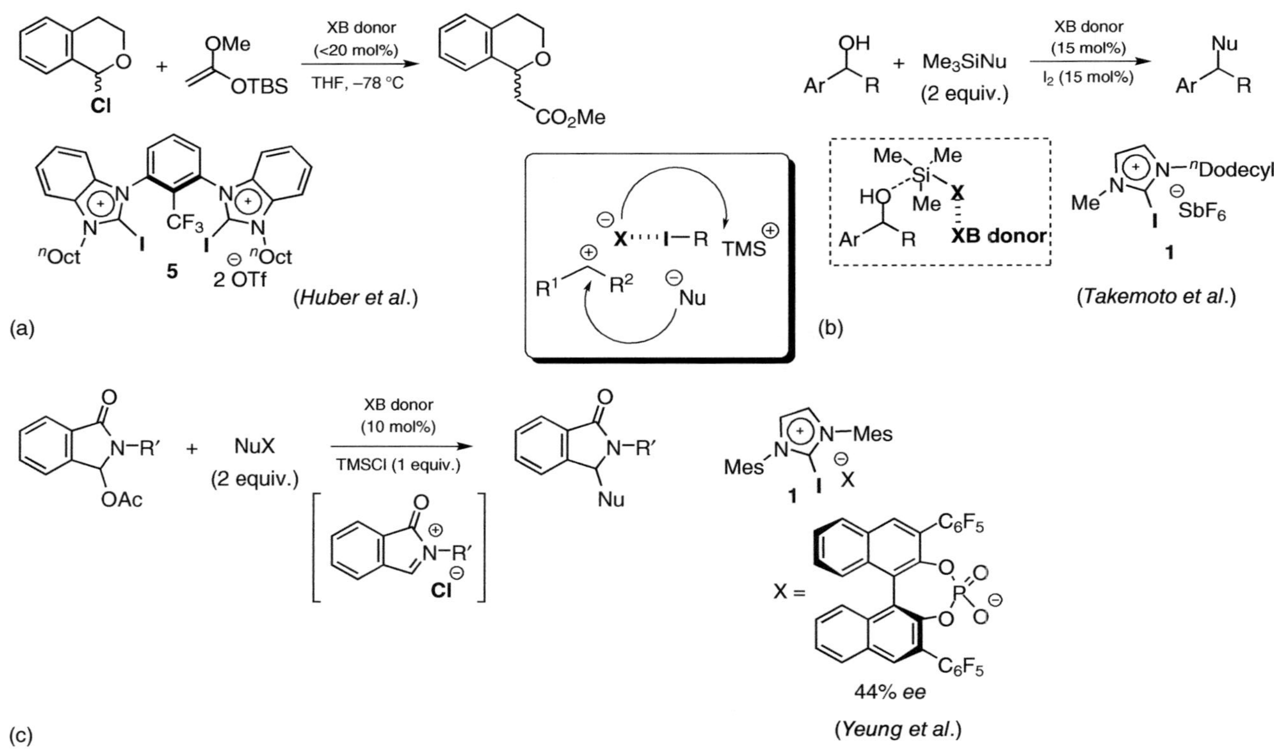

Scheme 3.7 Reactions involving XB-catalyzed halide abstraction. (a) Halide abstraction/C–C bond formation, (b) Dehydroxylative coupling reaction, (c) Functionalization of *N*, *O*-aminals.

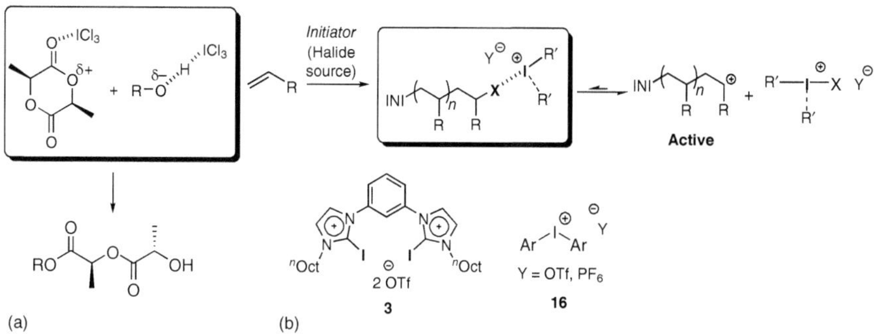

Scheme 3.8 XB-mediated polymerizations. (a) Neutral group activation, (b) Halide abstraction.

of isobutyl vinyl ether (IBVE) was reported using cationic bidentate XB donor **3** (Scheme 3.8b) [35b]. XB-promoted halide abstraction from the initiator (IBVE-HCl adduct) initiates the polymerization at −10 °C. The addition of $^{n}Bu_4NCl$ as an additive has a beneficial effect as it competes with the halide from the propagating end and decreases the concentration of active species. This positively improves the polymerization process and the molecular weight of the polymer. Recently, this polymerization reaction has been achieved again with predictable molecular weights and very narrow molecular weight distributions by using catalytic amounts of diaryliodonium salts (**16**) in the presence of $^{n}Bu_4NI$ as the cationogen which typically generates a carbon–halogen-propagating end. By a series of experiments, the mechanism was thoroughly supported as Lewis acid catalysis rather than the usual behavior of these compounds as a photoinitiator [35c]. Using this method, several vinyl ethers (including those with polar groups) and styrene monomers were efficiently polymerized in a controlled manner.

As discussed above, with Lewis basic functional groups of organic compounds being relatively weak XB acceptors compared to anions, they bind weakly to XB donors. However, coordination of the latter to the usually more Lewis basic transitions state may be sufficient to induce the desired reaction, without XB inhibition. Thus, catalysis is possible unless the nucleophilic species/reaction parameters deactivate the XB donor. On the other hand, due to the weak binding, the activation of organic molecules is sometimes challenging and strong XB donors are often needed. Figure 3.5 shows a typical example of the XB-induced activation of a carbonyl-derived substrate, but in addition, there are several other Lewis basic functional groups and even π-systems which were activated through this interaction [19].

The first example of catalysis using organic halogen bond donors – as briefly mentioned above – was the iodoperfluoroalkane (e.g. $C_8F_{17}I$)-catalyzed reduction of quinoline derivatives (Scheme 3.9a) [28]. The involvement of XB interaction was postulated considering the shifts of NMR signals. Later, Tan's group studied this reaction using chiral cationic XB donors and the involvement of XB as the driving force was again supported by several NMR experiments [36]. A chiral bidentate iodoimidazolinium-derived XB donor (**21**) showed strong performance as a catalyst although it did not yield any enantioselectivity. Contrary to the above results, recent

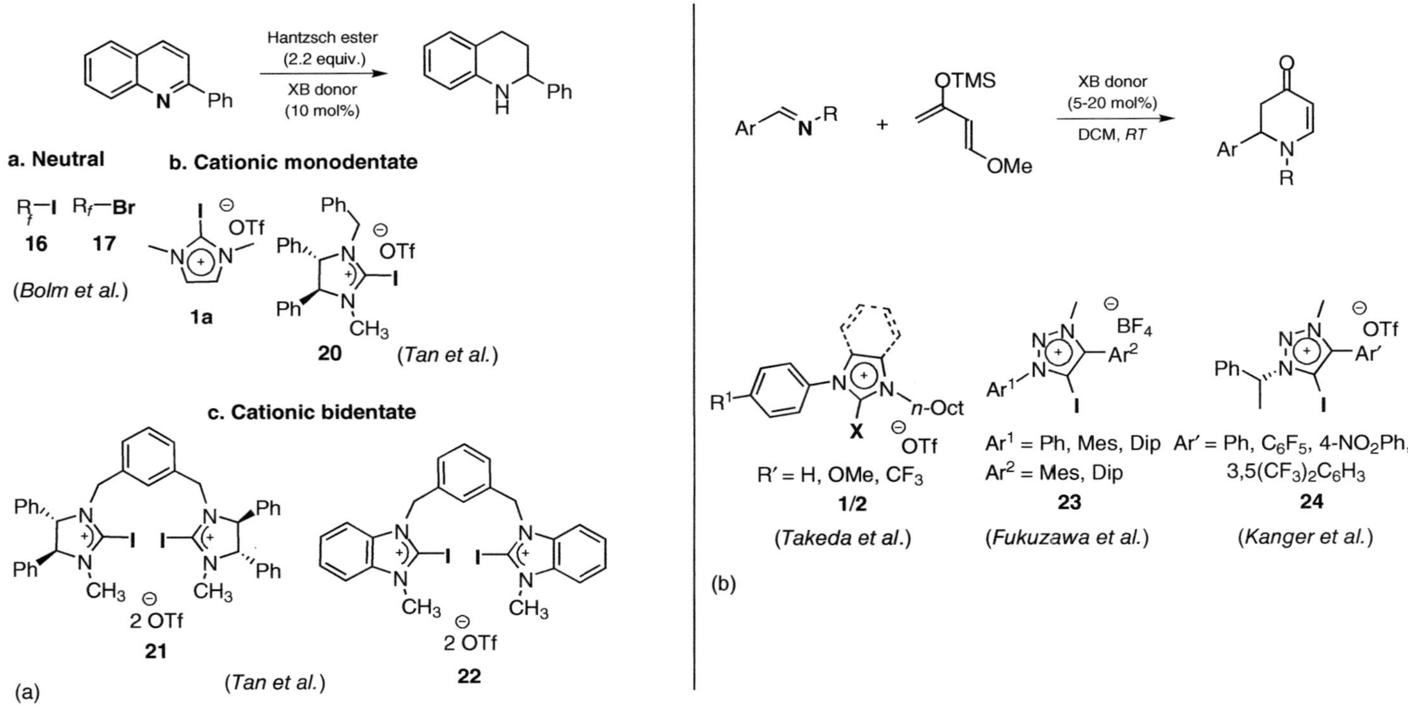

Scheme 3.9 XB activation of *N*-based Lewis basic functional groups in organic reactions. (a) Reduction of quinoline, (b) Aza-Diels-Alder reaction.

Figure 3.5 Activation of organic functional groups through halogen bonding.

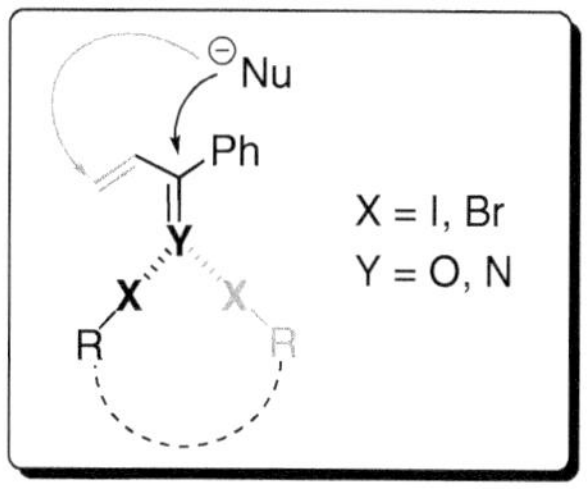

DFT studies of Wong and coworkers suggest that the role of iodoimidazolinium XB donors is not to activate quinoline but to stabilize the conjugate anion of the employed Hantzsch ester through halogen bonding [20b].

Another type of imine activation involved the aza-Diels–Alder reaction between Danishefsky's diene and aldimines with monodentate iodo(benz)imidazolium triflates as XB donors (Scheme 3.9b) [37a]. The proposed mechanism of Lewis acid activation was supported by experimental evidence. However, the observed trend regarding the catalytic activity of differently halogenated XB donors was more complex (Br > Cl > I) than the routinely observed and theoretically expected one (I > Br > Cl). Later, Scheiner's group performed quantum chemical calculations using $H_2C = NH$ and *cis*-1,3-butadiene as the model substrates [37b] and showed that the lowering of the activation barrier occurs both with imidazolium and haloimidazolium catalysts through hydrogen and halogen bonding, respectively. Indeed, the predicted trend for the catalytic activity of (halo)imidazolium salts (I > Br ≈ H > Cl) was found roughly in accordance with the one experimentally observed in other reactions. Fukuzawa's group used an O-TBS variant of the diene and halo-triazolium salts as XB donors (**23**) to study this reaction [37c] (Scheme 3.8b). They noted significant alterations in the catalytic efficiency with an increase in the steric hindrance of the *N*-mesityl substituent around the iodine centers of the halo-triazolium unit. This causes orientation of the aromatic rings perpendicular to the triazolium ring (for reduction of the steric hindrance and π-conjugation) and enhances the catalytic activity. Also, the authors used triazolium-based chiral monodentate XB donors (**24**) of the earlier reported bidentate variant [32a]. However, enantioselectivity was not achieved [37d]. Their detailed mechanistic studies – involving the identification of intermediates by Heteronuclear multiple bond correlation (HMBC) and HRMS – suggested the preferential involvement of a stepwise pathway through a Mukaiyama–Michael-type addition followed by elimination rather than the expected pericyclic one.

Considering the importance of the activation of carbonyl groups for nucleophilic additions, Huber *et al.* reported the Diels–Alder reaction between methyl vinyl ketone (MVK) and cyclopentadiene using dicationic halogen bond donors with non-coordinating counterions **3** (Scheme 3.10a) [38a]. The activation of the carbonyl group of MVK through halogen bonding was rigorously supported by multiple control experiments. More recently, Wang and coworkers, based on DFT calculations using a tripodal neutral XB donor as a model, found that in this and related reactions, hydrogen- and halogen-bonding catalysis are competitive to each other [37b]. Later, the catalytic activity of variously substituted iodine(III) compounds was exploited in this Diels–Alder reaction as well [31b]. These compounds (e.g. **17**, **18**) showed equal activity than that of the bidentate dicationic organoiodine(I) XB donor **3**. Recently, a bidentate iodine(III)-based XB donor (**19**) was even markedly more active in a more challenging variant of this reaction (between 1,3-cyclohexadiene and MVK) [39].

In the next study, the Michael addition reaction of indole to *trans*-crotonophenone was tested and, consistent with earlier results on halide abstraction organocatalysis [27b], the preorganized *syn*-isomer of XB donor **5b** performed the best (Scheme 3.10b) [40a]. Meanwhile, the screening of salts of various monodentate iodobenzimidazolium derivatives (OTf, NTf_2, BF_4, BPh_4, and $BAr^F{}_4$ salts of **2**) by Breugst's group [40b] somewhat unexpectedly suggested the possibility of using coordinating anions (triflate) in the activation of α,β-unsaturated carbonyl compounds. Another important finding was reported by the group of Scheidt. They used chiral XB donors (**24**, **25**) and an achiral XB donor (**23**) in combination with chiral anions as catalysts for this reaction. Although these catalysts promoted the expected reaction, enantioselectivity was not obtained [20a]. Based on additional experiments, they doubtfully suggested the preference of a Brønsted-acid-catalyzed pathway over the XB activation for their catalysts.

Similarly, electron-rich aromatic heterocycles (thiophene derivatives) react in Michael addition fashion with α,β-unsaturated carbonyl compounds in the presence of XB donors, out of which the previously established bis(iodoimidazolinium) compound **21** (Scheme 3.9) produced the best results in DCE as solvent (Scheme 3.10c) [41]. Based on ^1H NMR studies and DFT calculations, the authors proposed a mechanism where in addition to XB catalysis, Brønsted acid catalysis also shows some limited involvement. Once again the iodobenzimidazolium-based bidentate XB donor **20** was found not to be very active in this reaction. In a further example, Toy's group used an axially chiral bidentate 2,2′-bipyridine-based XB donor to promote the Michael addition of furan derivates to various electron-deficient α,β-unsaturated carbonyl compounds. The addition products were obtained in high yields in acetonitrile [41b]. The temperature of the reaction depends on the reactivity of the Michael acceptors – for less reactive substrates, higher temperatures (50 °C) were required.

Recently, the activation of a doubly unsaturated ketone was performed through XB catalysis (Scheme 3.10d) [42]. Screening of several mono- and bidentate neutral and cationic XB donors indicated that the best catalyst for this reaction is *syn*-preorganized dicationic XB donor with non-coordinating $BAr^F{}_4$ anions (*syn*-**5b**) [26b], which is consistent with earlier results on carbonyl activation. In this reaction, the main role of the XB donor seems to be to induce the electrocyclization and not the keto-enol tautomerization.

Similar to the Michael additions just mentioned, 1,2-additions of nucleophiles to aldehydes and ketones were also achieved. For example, bidentate XB donor **4a** catalyzes the addition of indol to aldehydes and ketones in acetonitrile at 70 °C with typically a low catalyst loading of 1 mol% (Scheme 3.11a) [43]. Involvement of halogen bonding was supported by control experiments, and catalysis by molecular iodine – a

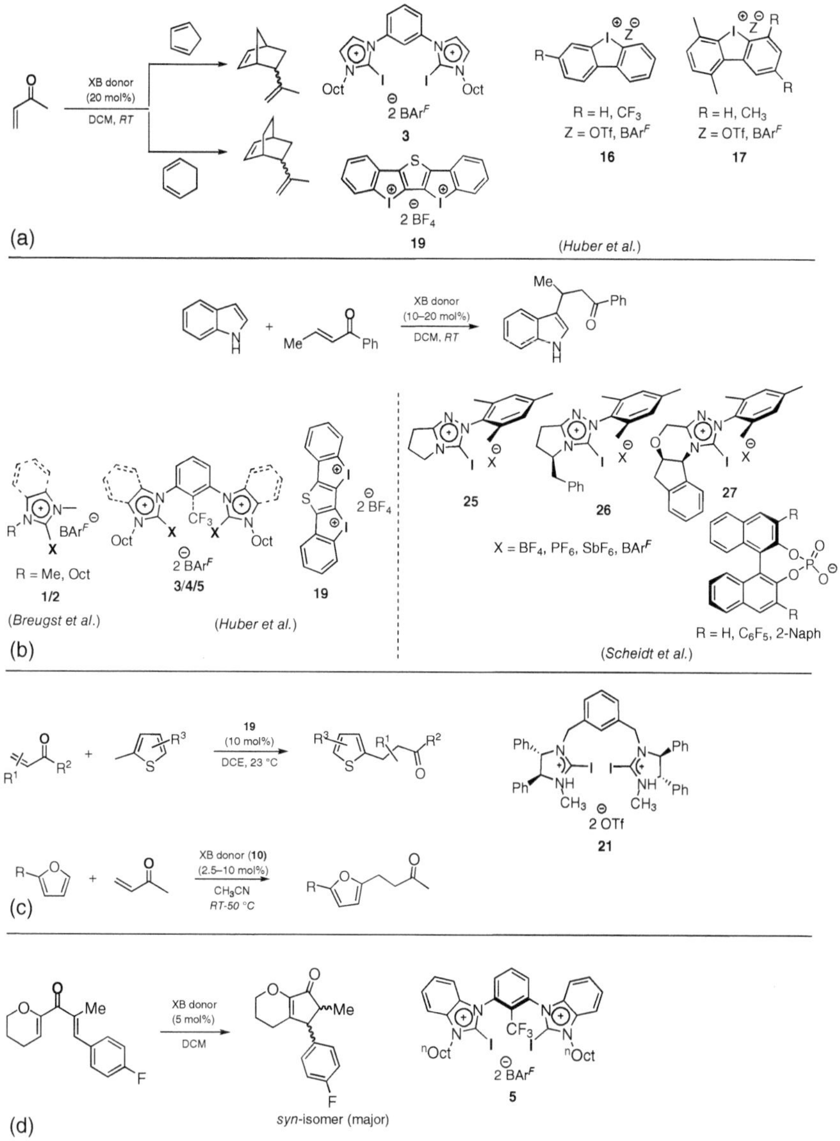

Scheme 3.10 XB-catalyzed reactions of conjugated carbonyl compounds. (a) Diels-Alder reaction, (b) Michael addition of indols, (c) Michael addition of thiophenes and furans, (d) Nazarov cyclization.

possible decomposition product of XB donor **4** at this higher temperature – was ruled out primarily based on ^{1}H NMR analysis. Also, the addition of organic bases such as pyridine or Hünig's base did not affect the yield dramatically, while the addition of $^{n}Bu_4NI$ significantly inhibited the reaction. Interestingly, in this reaction, once again a prominent thiourea-based hydrogen-bonding catalyst and even a trityl cation were not active and did not induce product formation. Recently, this reaction has also been

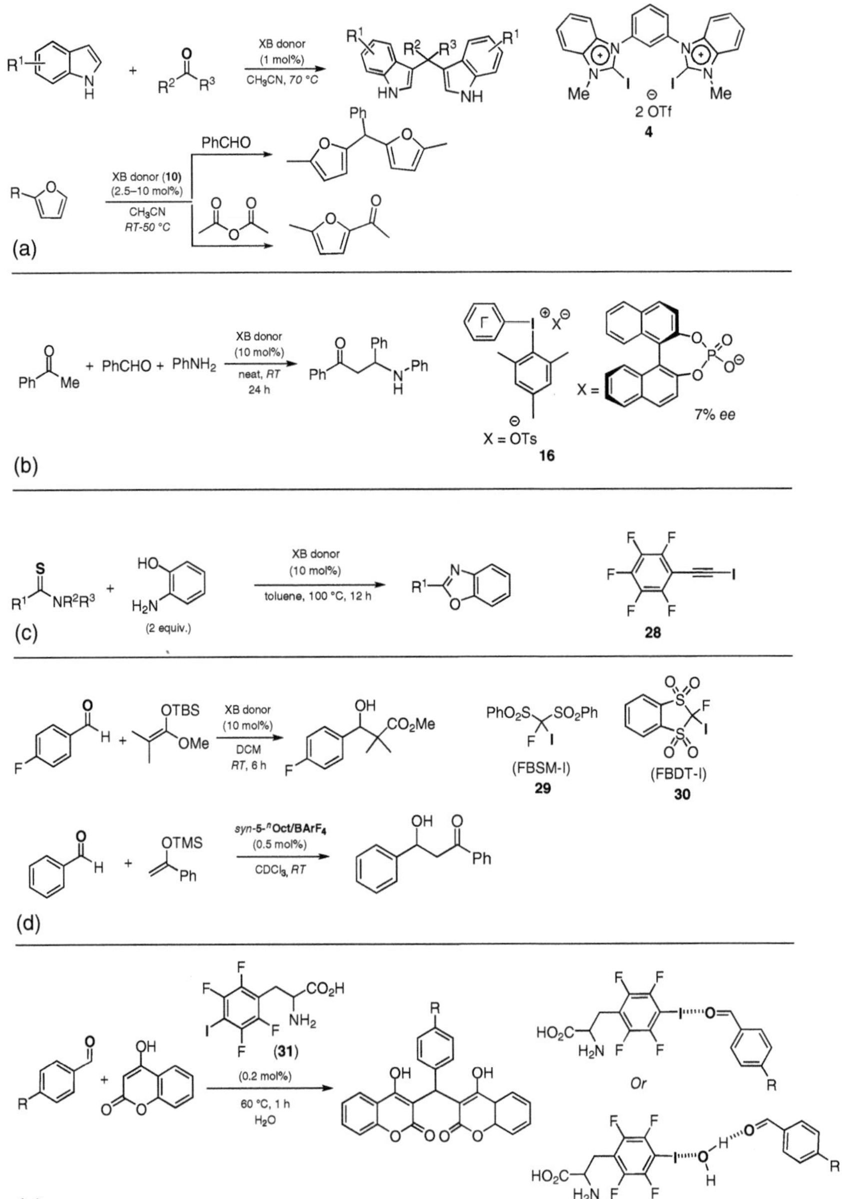

Scheme 3.11 XB-induced 1,2-additions to carbonyl derivatives. (a) Friedel Crafts alkylation, (b) Three component Mannich reaction, (c) Activation of thioamides, (d) Mukaiyama-aldol reaction, (e) Dehydrative coupling reaction.

carried out at a relatively low temperature (50 °C) with the 2,2′-bipyridine-based bidentate XB donor by Toy *et al.* mentioned above [41b].

In another example, Han and Liu successfully used hypervalent iodine compound **16** as a halogen bond donor catalyst for the solvent-free reaction between a ketone, an aldehyde, and an amine [44]. Two routes, a Mannich- or a Michael-type pathway, are possible for this reaction (Scheme 3.11b). However, the exact mode of activation by these iodine(III) compounds was not clearly mentioned by the authors. As mentioned above for the Michael addition reactions with indols, once again very low enantioselectivity was achieved when using a chiral phosphate anion.

Matsuzawa and coworkers showed that a neutral monodentate iodoalkyne featuring a C(sp)–I bond and pentafluorophenyl group (**28**) can catalytically activate thioamides in their reaction with 2-aminophenol to generate benzoxazoles (Scheme 3.11c) [45]. This catalyst (**28**) was found superior to other monodentate XB donors such as pentafluoroiodobenzene and cationic iodobenzimidazolium derivatives (like **2**). The involvement of XB was supported by mechanistic studies and control experiments and the catalyst was found to be stable under reaction conditions (100 °C). This provides further possibilities for XB-based catalysis at elevated temperatures.

As discussed in Section 3.2, one of the promising reactions catalyzed by iodine are nucleophilic additions involving silyl nucleophiles. The first example of an organic XB donor-catalyzed Mukaiyama-aldol reaction (Scheme 3.11d) was reported by using powerful neutral XB donors bearing a Csp^3–I bond attached to a strongly electron-withdrawing fluorobis(sulfonyl)-methane scaffold (**29** and **30**) [46a]. The halogen-bonding ability of these XB donors was confirmed by NMR titrations, single-crystal X-ray diffraction (XRDs), and DFT calculations. In addition, these XB donors catalyzed hydrogen-transfer reductions under neutral and mild conditions (see Scheme 3.9a).

Recently, this reaction was reported using unprecedentedly low loading of a bidentate XB donor (*syn*-**5**) as the catalyst at room temperature [46b]. The involvement of halogen-bonding interaction by the catalyst (and not traces of elemental iodine) was confirmed by various comparison experiments and theoretical calculations. Interestingly, this catalyst can be reused several times without any loss in catalytic activity.

In another study, Metrangolo *et al.* developed the amino acid-based XB donor **31** having both XB donor properties and good water solubility (Scheme 3.11e) [47a]. It promotes the synthesis of bis(heterocyclic)methanes through a dehydrative coupling reaction of the substrates in water at a higher temperature (> 60 °C). Two modes of XB activation were proposed based on mechanistic investigations by NMR studies and on XRD of co-crystals. One approach involves the direct coordination of the XB donor to the carbonyl group while the other comprises acidification of a water molecule which then interacts with the substrate. Recently, a 2-iodo-imidazolium-based tetrapodal halogen bond donor has shown very strong binding properties in aqueous media which shows strong potential for further developments for the XB catalysis in aqueous media [47b].

Strain-release glycosylation for the preparation of *O*,*N*-glycosides with excellent anomeric selectivity was achieved by Loh's group using *syn*-**5** as the XB donor.

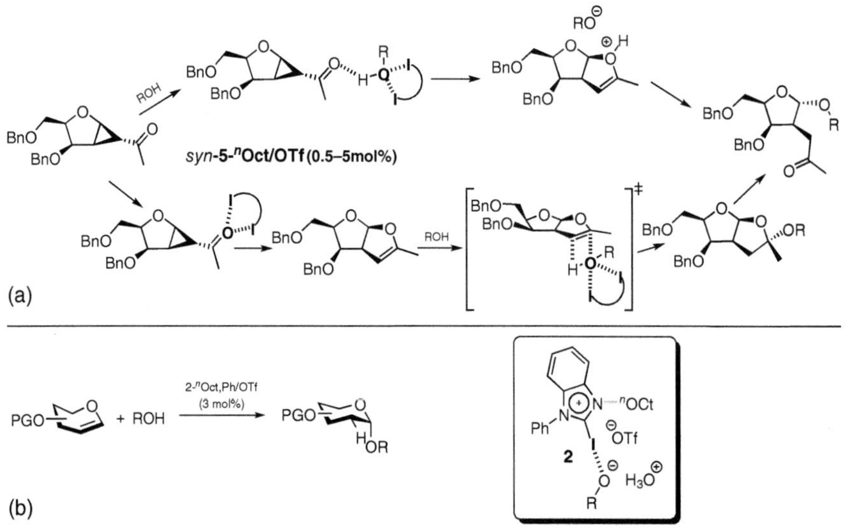

Scheme 3.12 Glycosylation through XB interaction. (a) Multistage strain-release glycosylation, (b) Deoxyglycosylation with quantum tunneling.

(Scheme 3.12a) [48a].The selectivity was much better than with thiourea-based HB catalysis and the method allowed the efficient synthesis of a new class of hedgehog-signaling glycoside inhibitors. The activation of the substrate by the XB donor can occur by two different modes and based on this, a divergent mechanism involving multistage XB interaction was postulated. Recently, the same group has achieved the XB **2a**-catalyzed 2-deoxyglycosylation of several primary and secondary glycosyl acceptors (Scheme 3.12b) [48b]. Broad substrate scope, moderate-to-good yields, and excellent anomeric selectivity are the highlights of this method. Similar to several earlier reports, XB catalysis shows better performance and improved substrate tolerance than thiourea-based hydrogen-bonding catalysis. Based on mechanistic studies, the authors suggest an off/on switching of the catalysts through halogen swapping. In addition, quantum tunneling of the rate-determining step is noted.

A soft electrophilic halonium species involved in routine Lewis base-catalyzed halocyclization reactions can be generated in the presence of an XB catalyst. For instance, in the presence of *N*-methyl 4-iodopyridinium triflate (**32**) as XB donor 1,3-dibromo-5,5-dimethylhydantoin (DBDMH) generates such species and can be used for the synthetically useful and highly chemoselective bromo-carbocyclization of *N*-cinnamyl sulfonamides and *O*-cinnamyl phenyl ethers (Scheme 3.13) [49]. The XB donor coordinates to the carbonyl group of DBDMH and enhances the electrophilicity of the bromine substituent toward the formation of the bromonium ion intermediate. Other neutral XB donors, molecular iodine, Trimethylsilyl trifluoromethanesulfonate (TMSOTf), Lewis bases, hydrogen-bonding catalysts (thiourea and squaramide), and Brønsted bases (iPr$_2$NEt, 4-dimethylaminopyridine [DMAP]) and acids do not catalyze this reaction. Mechanistic studies suggest an autocatalytic process in which hydantoin side products such as 5,5-dimethylhydantoin (DMH) or mono-*N*-bromo-5,5-dimethylhydantoin (MBDMH) accelerate the reaction. Excellent diastereoselectivities (dr > 99:1) and regioselectivities are attained by this method, allowing the synthesis of tetrahydroquinolines and chromanes which are valuable drug cores and natural product scaffolds.

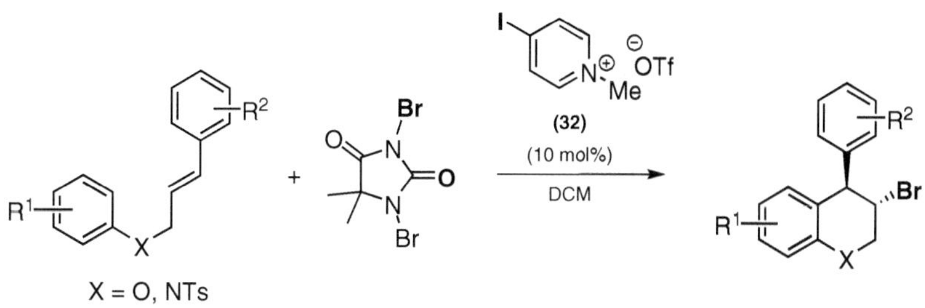

Scheme 3.13 Bromocarbocyclization via halogen bonding.

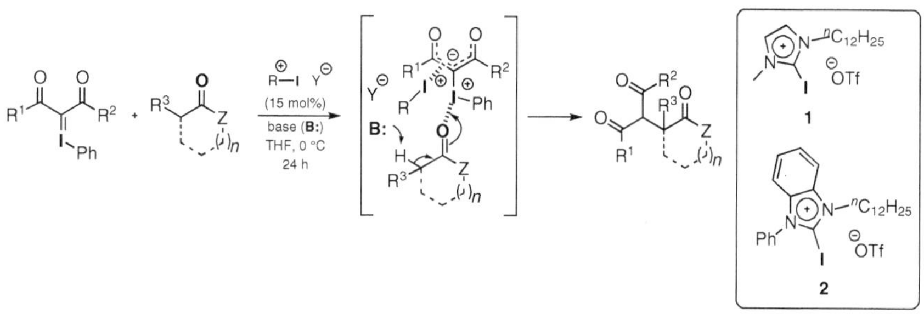

Scheme 3.14 Umpolung C–C bond formation involving an iodolium ylide-halogen bonding.

In an interesting example involving the umpolung alkylation of silyl enol ethers with iodonium(III) ylides, both the iodine(I)-based XB donor and an iodine(III)-containing substrate participate in the transition state through halogen bonding (Scheme 3.14) [50]. The involvement of XB between the monodentate halo(benz)imidazolium catalyst and iodonium ylide was supported by theoretical calculations. Other prospective catalysts such as Schreiner's thiourea, phosphoric acid, and metal-based Lewis acids are ineffective in this XB-induced alkylation method, which results in the synthesis of 1,4-dicarbonyl compounds. Thus, once again the XB donor catalyst was found to be superior to a thiourea catalyst.

Lately, a further functional group, the nitro group, was activated by a rather strong bidentate iodine(III)-based XB donor (**19**) (Scheme 3.15) [39]. Thus, the Michael addition between 5-methoxyindole and nitrostyrene could be efficiently catalyzed at 5 mol% catalyst loading at room temperature. In contrast, the *syn*-preorganized iodine(I)-based XB donor **5** (Scheme 3.7) is completely inactive in this case.

Similar to the nonbonding electron pairs of heteroatoms discussed so far, π-electrons are also efficient halogen bond acceptors. In the solid state and in

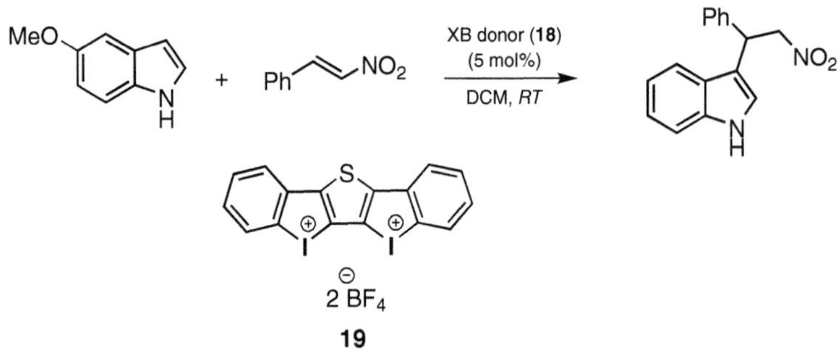

Scheme 3.15 Halogen-bonding-catalyzed nitro-Michael addition.

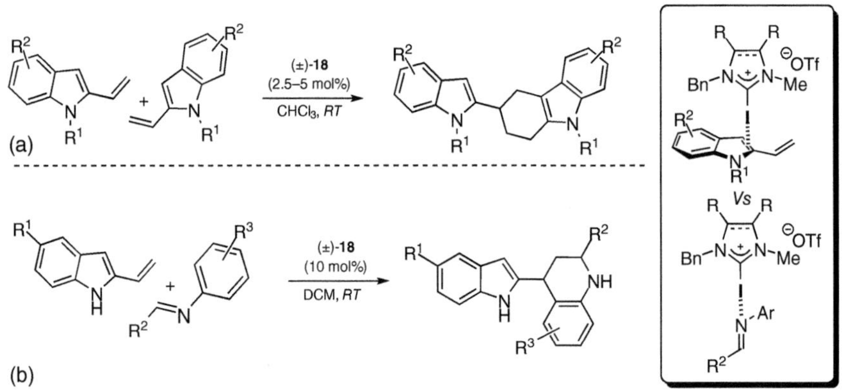

Scheme 3.16 Catalysis by XB-induced π-activation. (a) [4+2] Cycloaddition, (b) Povarov reaction.

biomolecular environments, π-systems often form halogen bonds with XB donors [51]. For example, such coordination is involved in the selective substrate recognition during the nitrilase-catalyzed synthesis of chlorophenylacetic acids [52a]. For the first time, Arai *et al.* used this phenomenon to perform the catalysis of homo- and cross-[4+2] cycloadditions of 2-alkenyl indoles using monodentate XB donors ((±)-**20**) (Scheme 3.16a) [52b]. The postulated electrophilic activation of 2-alkenyl indoles by C–I···π halogen bonds was supported by experimental indications and quantum chemical calculations. In another study, the authors used a combination of imine and olefin and found that *n*-electrons dominate over π-bonding electrons in coordination to the XB donor (**20**) [52c] (Scheme 3.16b). Thus, a Povarov reaction was preferentially observed via the C–I···N XB interaction and diverse indolyl-tetrahydroquinoline derivatives were obtained in good yields.

3.4 Asymmetric Catalysis Through Halogen Bonding

Contrary to enantioselective recognition through halogen bonding [21a, b, 53], there are only very few reports on asymmetric catalysis through this interaction.

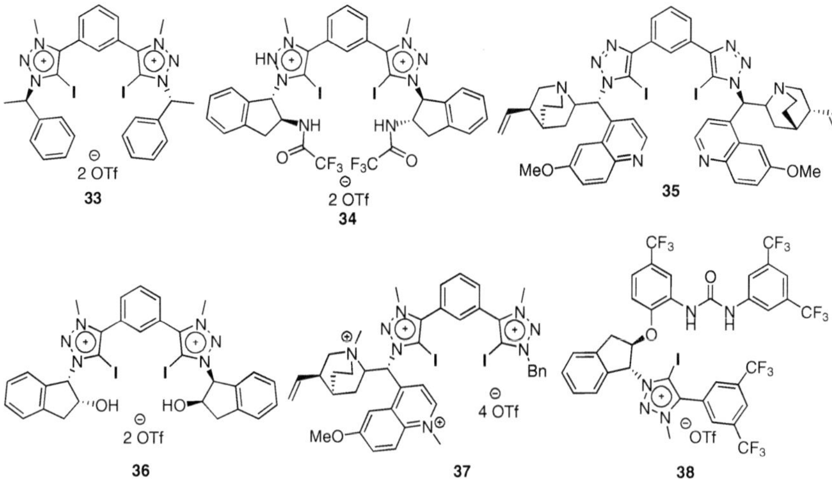

Figure 3.6 Triazolium-based chiral XB donors.

Considering the ease of synthesis of chiral triazoles by the alkyne-azide click reaction of chiral azides, initially, XB donors based on these structures were attempted as chiral catalysts. The first chiral bidentate XB catalyst of this kind (**33**) was prepared by Huber and coworkers but, unfortunately, it did not yield any enantioselectivity [32a]. Later, Kanger's group synthesized and tested several more sophisticated chiral variants of this core structure (**34**–**37**) as well as a multifunctional iodotriazolium-based XB donor containing an HB-donating urea unit (**38**; Figure. 3.6) [21a, c, 22c]. Most of them showed good catalytic activities. However, still asymmetric induction was not achieved. As already mentioned in the sections above, several other attempts involving the use of monodentate chiral cationic XB donors based on imidazolinium (**20**) [36], 1,2,3-triazolium (**24**) [37d], 1,2,4-triazolium (**26**, **27**) [20a] and bidentate imidazolinium (**21**) groups [36] were also unsuccessful.

All the above efforts indicate that it is challenging to design a system that could transfer chiral information from the XB-based catalyst to prochiral substrates in the transition state. This is mainly rooted in the relatively weak strength and high directionality of this interaction along with the relatively large size of the interacting halogens and the long length of the R–X bond. In a recent study, Huber's group used preorganization for the rigidification of chiral bidentate XB donors and showed that indeed asymmetric catalysis is possible solely through halogen bonding. Thus, a chiral analogue of their bidentate imidazolium-based XB donors (*syn*-**39**) gave moderate enantioselectivity in a Mukaiyama aldol reaction when a glyoxal derivative was used as the electrophile (Scheme 3.17) [27e]. However, no asymmetric induction was observed when simple aldehydes (monocarbonyl systems) were used. In addition, a clear enantio-discrimination of *trans*-1,2-diphenyl ethylene diamine was achieved through spatial differentiation of its enantiomers in the XB adducts, which is in dcorrelation with the asymmetric induction observed in the Mukaiyama aldol reaction of glyoxal-derived substrates.

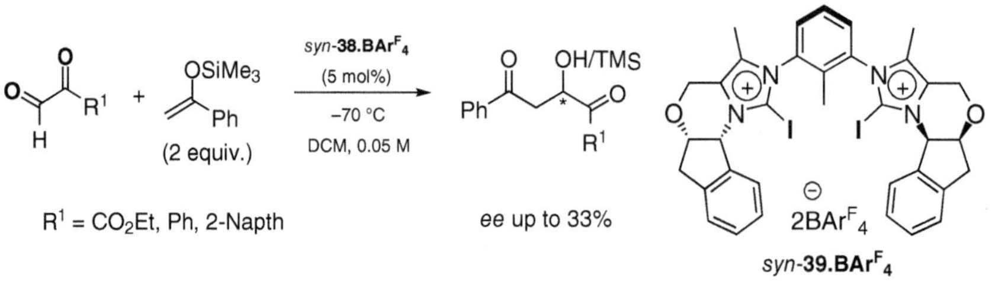

Scheme 3.17 A bis(imidazolium)-based bidentate chiral XB donor in asymmetric catalysis.

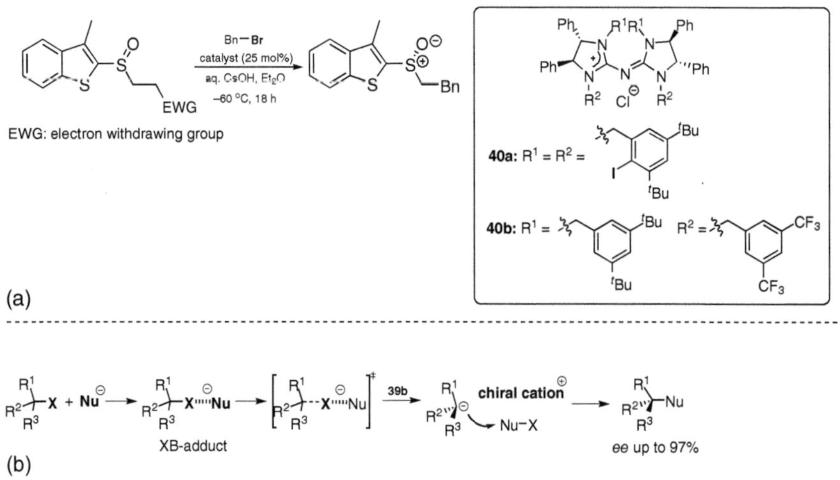

Scheme 3.18 Chirality transfer by pentanidium salts. (a) Alkylation of sulfinate anions, (b) SN^2 (X) reaction. *Source*: Arai e al. [56].

In another approach, chiral cationic species were used to induce the chirality in the enantiodifferentiation steps. For example, chiral phase-transfer catalysts consisting of halogenated pentanidium cations (**40a**) have been used in the enantioselective alkylation of sulfenate anions (Scheme 3.18a) [54]. This bifunctional catalytic system involves multiple noncovalent interactions including halogen bonding and yields various chiral sulfoxides with high *ee*. However, the contribution of XB in the induction of chirality or catalytic activity is difficult to determine.

In a further example, ion pairing of a chiral cation along with transient XB formation was used for chirality transfer [55]. Through this analogue of asymmetric counterion-directed catalysis (ACDC), an enantioconvergent substitution reaction of activated tertiary bromides with thiocarboxylates (or azides) was achieved in the presence of chiral pentanidium salts (**40b**) (Scheme 3.18b). This is a less common variant of S_N2 reactions referred to as halogenophilic $S_N2(X)$. The typical mechanism involves activation of the tertiary bromide substrate via transient halogen bonding with the thiocarboxylate salts. This generates a carbanionic intermediate which then forms an ion pair with the chiral cation of the catalyst. The chiral information is then transferred from the cation to the prochiral substrate in the enantiodifferentiation step.

In a few examples, chiral neutral halogen bond donors equipped with additional functional groups were also used to achieve asymmetry. One of the more ambiguous cases uses bis(imidazolidine)-based compound **41** (I-Bidine) for catalyzing the Michael/Henry reaction of thiosalicyl aldehydes with nitroalkenes (Scheme 3.19a) [56]. Although moderate enantioselectivity is reported by the authors, the role of XB is not certain, as non-fluorinated iodobenzene derivatives hardly form XBs in solution.

The same group subsequently used quinidine-based chiral Brønsted base bearing a halogen-bonding substituent (**42a**) for the highly enantioselective Mannich reaction between malononitrile and *N*-Boc imines (Scheme 3.19b) [57a]. Although the involvement of XB seems to be important, other factors may also have a pivotal role in chirality induction. The scope of this catalytic system was further expanded by using the pseudo-enantiomer of the catalyst (quinidine-derivative, **42b**) for enantioselective addition of malononitrile to *N*-Boc α-ketiminoesters [57b]. However, the hydrogen analogue of **42b** was also quite active, at least when *N*-Boc α-ketiminoesters were used as substrate, which complicates the mechanistic analysis. Changing the halogen substituent did not alter the yields much, although an improvement in asymmetric induction was observed from 37% for fluorine to 82% for iodine, and thus other factors such as sterics may have some influence.

All the above examples use XB as one of the primary interactions for catalyzing the desired reaction. There are also several cases in which XB is involved as secondary interaction, mostly to control the orientation of substrates.

3.5 Halogen Bonding as Supporting Interaction in Catalysis

Metrangolo and Resnati first demonstrated the combined use of XB and π stacking interactions for controlling the orientation (and thus the reactivity) of substrates in the solid state [58]. There are also a few examples where XB formation can be used as one of the noncovalent interactions for the rigidification of (enantiopure) ligand backbones of metal complexes [59]. In addition, there are several cases in which the formation of a halogen bond has a beneficial effect in improving the solubility and stability of reagents (Figure 3.7) [60, 61].

A catalytically useful version of this approach uses complementary XB ligands to form chelated Rh(I) complexes [62]. Monodentate pyridyl- and iodo(tetrafluoro)aryl-substituted phosphines act both as ligands to an Rh(I) center and as XB acceptor/donor respectively, to generate *in situ* the corresponding halogen-bonded complex (Scheme 3.20). Generally, for the efficient behavior as a halogen bond donor, the presence of electron-withdrawing substituents like fluorine on the iodoaryl moiety is required, but in this case, these substituents are not needed on the iodoaryl-phosphine ligand for XB formation with the pyridine-phosphine ligand. One of the most plausible reasons for this seems to be the template effect of the metal center. This XBphos-Rh complex (**43**) showed impressive performance and high selectivity for branched products in the hydroboration of terminal alkynes

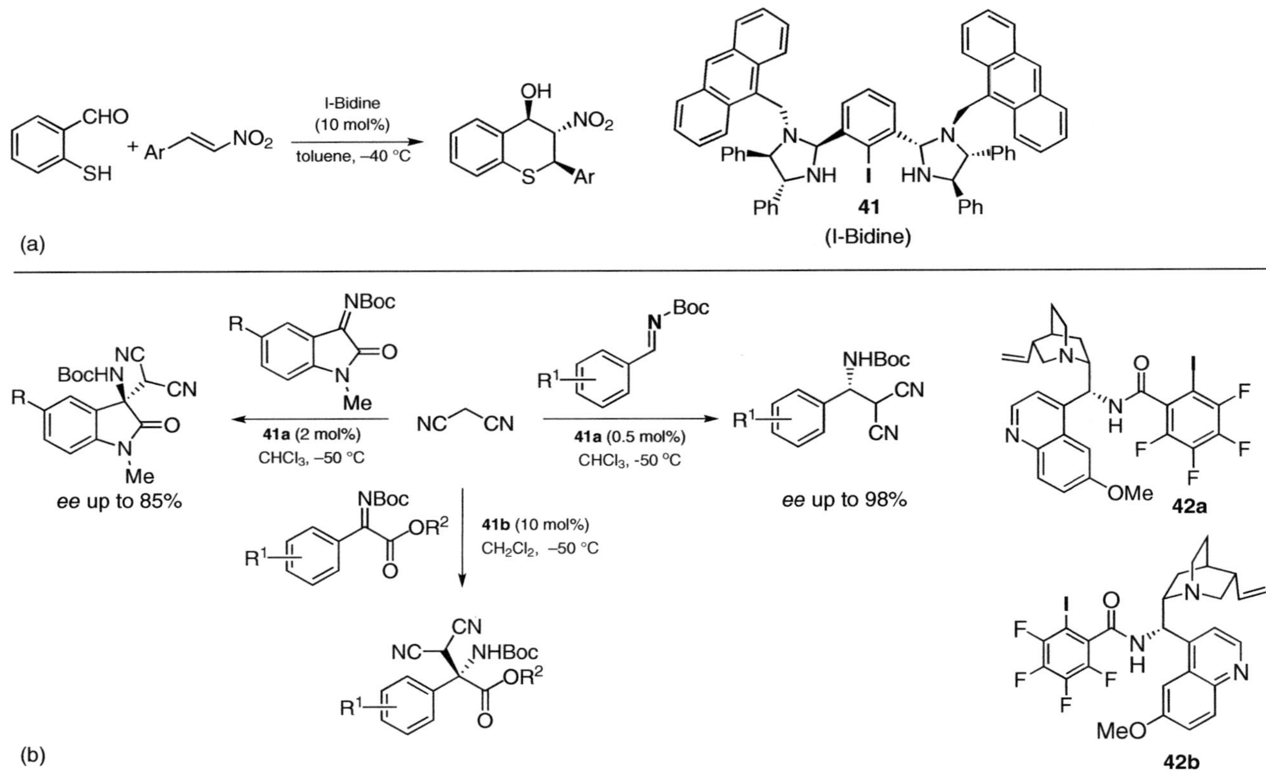

Scheme 3.19 Neutral XB-catalyzed enantioselective reactions. (a) Michael/Henry reaction, *Source*: Kuwano et al. [57a]. (b) Mannich reactions.

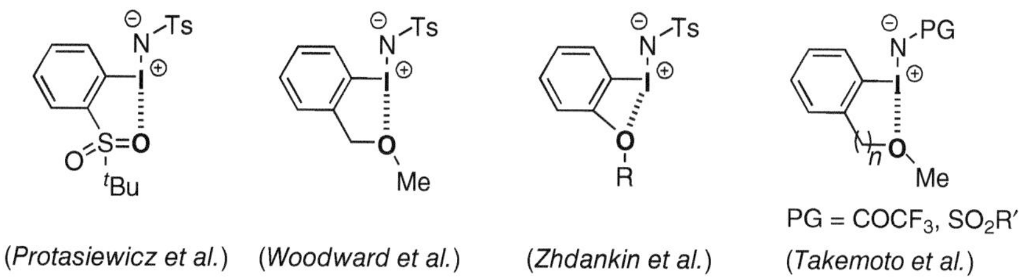

Figure 3.7 Stabilization of reaction precursors through XB. *Sources*: Zhdankin and Protasiewicz [60], Boucher et al. [61a], Macikenas et al. [61b], Meprathu and Protasiewicz [61c], Blake et al. [61d], Yoshimura et al. [61e], Kobayashi et al. [61f].

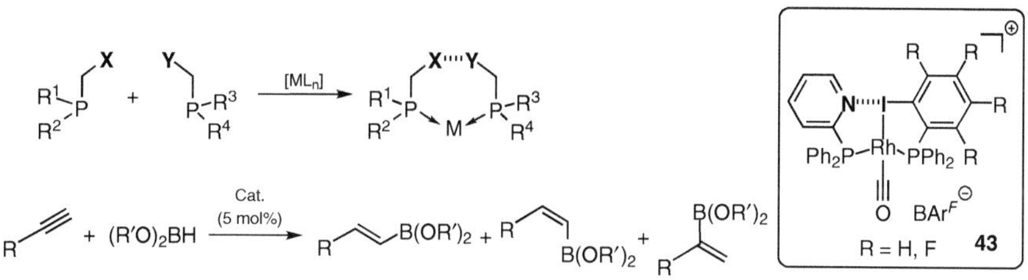

Scheme 3.20 XB-assembled bidentate phosphine-complexed Rh(I) catalyst.

[62a]. Later, it was shown that the formation of such halogen-bonded chelates around a metal center also favors oxidative addition of C_{Ar}–I bonds and generates cyclometalated products [62b].

In another approach, the XB-induced template formation between bisallyl-functionalized pyridine derivatives (as XB acceptor) and halogen bond donor **45** has been used to properly hold the olefin substrates for the expected metathesis reaction leading to macrocyclization (Scheme 3.21) [63]. The fast exchange between the XB donor and acceptor was observed on the ^{1}H and ^{19}F NMR time scales. However, DFT calculations strongly supported the involvement of halogen bonding. Subsequent hydrogenation of the intermediate olefins was achieved by using sodium borohydride, thus providing a tandem protocol for the synthesis of macrocycles.

In many cases mentioned above, the XB donor is superior to an HB donor. Nevertheless, the binding of an XB donor to the Lewis basic centers of hydrogen-bond donors sometimes enhances the Lewis acidity of the latter [64]. This strategy has been employed for enhancing the catalytic activity of Schreiner's thiourea in a glycofunctionalization reaction using a 2-iodobenzimidazolium salt as XB donor [64a]. Various amides (including the asparagine residues of several peptides) were thus coupled with glycosyl trichloroacetimidates in DCM. Several *N*-acylorthoamides along with the corresponding *β*-*N*-glycosides were selectively prepared by this method in good yields (Scheme 3.22). A reversal of anomeric selectivity (from *β* to *α*) of the *N*-glycoside product was observed on changing the protecting group at the 2-position from Ac to Bn and the solvent from DCM to ether (preferably n-butyl methyl ether) [64b].

In addition, recently the XB-induced abstraction of chloride anions from the metal catalyst has been shown as the main reason for gold-catalyzed cycloisomerization

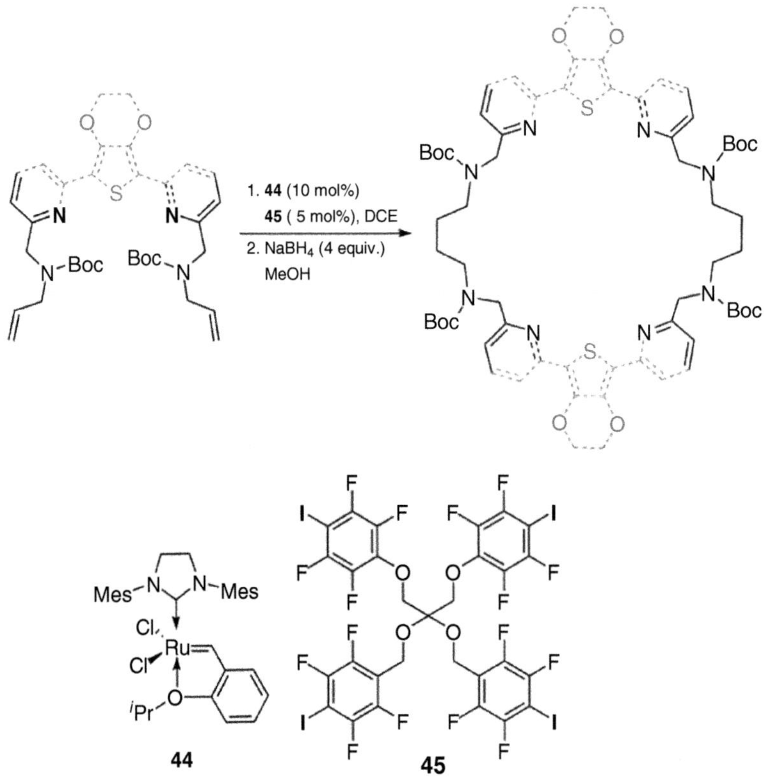

Scheme 3.21 XB template-induced macrocyclization.

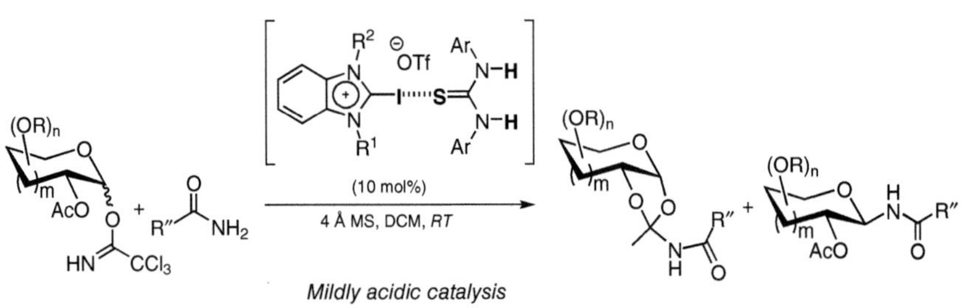

Scheme 3.22 Thiourea/XB donor co-catalysis for direct *N*-glycofunctionalization of amides.

reactions (Scheme 3.23) [65]. The role of XB was noted based on the trends in the rate enhancement of the cyclizations and ^{31}P NMR measurements along with the DFT calculations. Even though this constitutes the first activation of a metal-ligand bond by organohalogen XB donors, it still represents mostly a proof-of-principle case, as the more easily accessible $NaBAr^F{}_4$ provides superior performance as an activator.

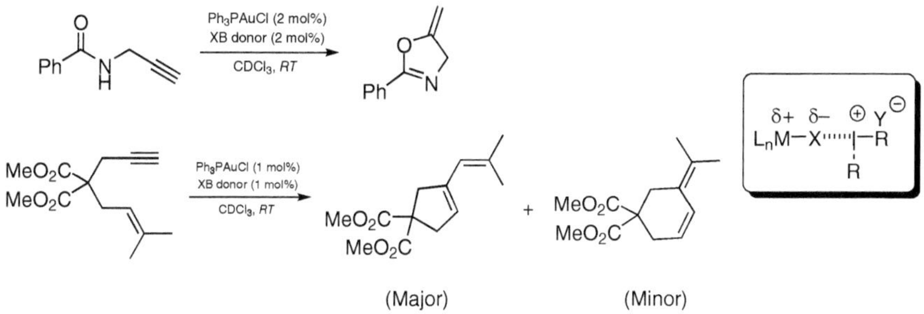

Scheme 3.23 XB-induced cleavage of an Au-Cl bond to generate an activated metal catalyst.

3.6 Conclusion

Based on all the examples presented above, it is very clear that halogen bonding can be utilized to activate a variety of Lewis basic organic functional groups including organohalides, carbonyls, imines, and even olefins. Although, in many instances, the mechanisms of these reactions are rigorously supported by experimental and theoretical means, there are also numerous cases in which the contribution of XB remains somewhat ambiguous and thus further mechanistic studies are needed. Encouragingly, the field is continuously growing and, in many examples, XB donors have shown sometimes even better performance than well-established catalysts such as Lewis acids, Brønsted acids, and hydrogen-bonding catalysts. In addition, there are already several cases in which the synthetic applicability of XB-catalyzed methods or the commercial importance of products has been noted – meaning that XB catalysis is steadily leaving a proof-of-principle phase.

Nevertheless, many challenges remain, not the least being the further application of XB catalysis to synthetically relevant reactions and reaction environments. Current limitations like the sometimes weak interaction with oxyanions and the occasionally low stability of cationic XB donors should be overcome. All three aspects mentioned here demand for a proper balance between the stability of an XB donor and its halogen-bonding strength. Key developments to further broaden this field of research are the realization of additional methods for asymmetric induction based predominantly on XB and the implementation of novel and simple XB donor motifs as well as of innovative modes of activation. A solid ground for these endeavors is surely provided by now, and many opportunities remain.

References

1 Hibbert, H. (1915). Use of iodine as dehydrating and condensing agent. *J Am. Chem. Soc.* 37: 1748–1763.

2 a) Banerjee, A.K., Vera, W., Mora, H. et al. (2006). Iodine in organic synthesis. *J. Sci. Ind. Res.* 65: 299–308. b) Togo, H. and Iida, S. (2006). Synthetic use of molecular iodine for organic synthesis. *Synlett* 2006: 2159–2175. c) Das, S.,

Borah, R., Devi, R.R., and Thakur, A.J. (2008). Molecular iodine in protection and deprotection chemistry. *Synlett* 2008: 2741–2762. d) Jereb, M., Vražič, D., and Zupan, M. (2011). Iodine-catalyzed transformation of molecules containing oxygen functional groups. *Tetrahedron* 67: 1355–1387. e) Parvatkar, P.T., Parameswaran, P.S., and Tilve, S.G. (2012). Recent developments in the synthesis of five- and six-membered heterocycles using molecular iodine. *Chem. Eur. J.* 18: 5460–5489. f) Ren, Y.-M., Cai, C., and Yang, R.-C. (2013). Molecular iodine-catalyzed multicomponent reactions: an efficient catalyst for organic synthesis. *RSC Adv.* 3: 7182–7204. g) Finkbeiner, P. and Nachtsheim, B.J. (2013). Iodine in modern oxidative catalysis. *Synthesis* 45: 979–999. h) von der Heiden, D., Bozkus, S., Klussmann, M., and Breugst, M. (2017). Reaction mechanism of iodine-catalyzed Michael additions. *J. Org. Chem.* 82: 4037–4043. i) Breugst, M. and von der Heiden, D. (2018). Mechanism in iodine catalysis. *Chem. Eur. J.* 24: 9187–9199.

3 Colin, J.J. (1814). Note Sur quelques combinaisons de l'Iode. *Ann. Chim.* 91: 252–272.

4 a) Cametti, M., Raatikainen, K., Metrangolo, P. et al. (2012). 2-Iodo-imidazolium receptor binds oxoanions*via* charge-assisted halogen bonding. *Org. Biomol. Chem.* 10: 1329–1333. b) Schulz, N., Sokkar, P., Engelage, E. et al. (2018). The interaction modes of haloimidazolium salts in solution. *Chem. Eur. J.* 24: 3464–3473.

5 a) Laurence, C. and Gal, J.-F. (eds.) (2010). *Lewis Basicity and Affinity Scales: Data and Measurement*, 1e. Chichester, UK: John Wiley & Sons. b) Laurence, C., Graton, J., Berthelot, M., and El Ghomari, M.J. (2011). The diiodine basicity scale: toward a general halogen-bond basicity scale. *Chem. Eur. J.* 17: 10431–10444.

6 a) Cox, B.G. and Maskill, H. (1983). Catalysis by iodine in the solvolysis of tertiary alkyl halides. *J. Chem. Soc. Perkin Trans.* 2: 1901–1905. b) Jereb, M. and Vražič, D. (2017). Iodine-catalyzed transformation of aryl-substituted alcohols under solvent-free and highly concentrated reaction conditions. *Acta Chim. Slov.* 64: 747–762. c) Jereb, M. and Vražič, D. (2013). Iodine-catalyzed disproportionation of aryl-substituted ethers under solvent-free reaction conditions. *Org. Biomol. Chem.* 11: 1978–1999.

7 a) Jereb, M., Vražič, D., and Zupan, M. (2009). Dual behavior of alcohols in iodine-catalyzed esterification under solvent-free reaction conditions. *Tetrahedron Lett.* 50: 2347–2352. b) Jereb, M., Vražič, D., and Zupan, M. (2009). Chemistry of organo halogenic molecules. Part 229. The role of iodine in acetyl group transfer to oxygen-containing molecules under solvent-free reaction conditions. *Acta Chim. Slov.* 56: 652–658.

8 a) aus dem Kahmen, M. and Sch-fer, H.J. (1998, 1998). Conversion of unsaturated fatty acids – cycloadditions with unsaturated fatty acids [1]. *Fett/Lipid* 100: 227–235. b) Xia, M. and Lu, Y.-d. (2005). Molecular iodine-catalyzed imino-Diels-Alder reactions: efficient one-pot synthesis of pyrano[3,2-*c*]quinolones. *Synlett* 2005: 2357–2361. c) Li, Y.-C., Zhang, J.-M., Dong, L.-T., and Yan, M. (2006). Imino Diels-Alder reaction catalyzed by iodine: efficient synthesis of tetrahydroquinolines. *Chin. J. Chem.* 24: 929–932. d) Shen,

S.-S. and Ji, S.-J. (2008). Molecular iodine catalyzed one-pot aza-Diels-Alder reaction under solvent-free condition. *Chin. J. Chem.* 26: 935–940. e) Wang, J., Xu, F.-X., Lin, X.-F., and Wang, Y.-G. (2008). Molecular iodine-catalyzed diastereoselective synthesis of cis-fused pyranobenzopyrans and furanobenzopyrans. *Tetrahedron Lett.* 49: 5208–5210. f) Wang, W., Jiang, H., Zhang, M.-M., and Wang, X.-S. (2014). Iodine-catalyzed synthesis of cyclopenta[*c*]quinoline derivatives via imino Diels–Alder reaction. *J. Heterocycl. Chem.* 51: 830–834.

9 Chuang, T.-H., Chang, W.-Y., Li, C.-F. et al. (2011). Substituent effects on the iodine-catalyzed thermal cyclization of 3,4-diphenylbuta-1,3-dienyl isocyanates: mechanistic studies. *J. Org. Chem.* 76: 9678–9686.

10 Deuri, S. and Phukan, P. (2011). Catalytic effect of molecular iodine in Diels–Alder reaction: a density functional theory study. *Struct. Chem.* 22: 1075–1085.

11 a) Wang, S.-Y., Ji, S.-J., and Loh, T.-P. (2003). The Michael addition of indole to α,β-unsaturated ketones catalyzed by iodine at room temperature. *Synlett* 2003: 2377–2379. b) Banik, B.K., Fernandez, M., and Alvarez, C. (2005). Iodine-catalyzed highly efficient Michael reaction of indoles under solvent-free condition. *Tetrahedron Lett.* 46: 2479–2482. c) Lin, C., Hsu, J., Sastry, M. N. V. et al. (2006). I2-catalyzed Michael addition of indole and pyrrole to nitroolefins. *Tetrahedron* 61: 11751–11757. d) Das, B., Chowdhury, N., and Damodar, K. (2007). Iodine-catalyzed efficient conjugate addition of pyrroles to α,β-unsaturated ketones. *Tetrahedron Lett.* 48: 2867–2870. e) Breugst, M., Detmar, E., and von der Heiden, D. (2016). Origin of the catalytic effects of molecular iodine: a computational analysis. *ACS Catal.* 6: 3203–3212.

12 Gao, S., Tzeng, T., Sastry, M.N.V. et al. (2006). Iodine catalyzed conjugate addition of mercaptans to α,β-unsaturated carboxylic acids under solvent-free condition. *Tetrahedron Lett.* 47: 1889–1893.

13 a) Phukan, P. (2004). Mukaiyama aldol reactions of silyl enolates catalyzed by iodine. *Synth. Commun.* 34: 1065–1070. b) Phukan, P. (2004). Iodine as a very powerful catalyst for three-component synthesis of protected homoallylic amines. *J. Org. Chem.* 69: 4005–4006. c) Wang, L. and Wong, M.W. (2008). Mechanism of halogen-catalyzed Mukaiyama aldol reactions: concerted or stepwise? *Tetrahedron Lett.* 49: 3916–3920. d) Deuri, S. and Phukan, P. (2012). Iodine catalyzed Mukaiyama–Michael reaction: experimental evaluation of catalytic effect in conjunction with computational study of the reaction mechanism. *J. Phys. Org. Chem.* 25: 1228–1235.

14 a) Kartha, K.P.R. (1986). Iodine, a novel catalyst in carbohydrate reactions I. O-isopropylidination of carbohydrates. *Tetrahedron Lett.* 27: 3415–3416. b) Kalita, D.J., Borah, R., and Sarma, J.C. (1998). A new selective catalytic acetalization method promoted by microwave irradiation. *Tetrahedron Lett.* 39: 4573–4574. c) Kumar, H.M.S., Reddy, B.V.S., Reddy, E.J., and Yadav, J.S. (1999). Iodine-catalyzed mild and efficient tetrahydropyranylation / depyranylation of alcohols. *Chem. Lett.* 28: 857–858. d) Basu, M.K., Samajdar, S., Becker, F.F., and Banik, B.K. (2002). A new molecular iodine-catalyzed acetalization of carbonyl compounds. *Synlett* 2002: 319–321. e) Karimi, B. and Golshani, B.

(2002). Iodine-catalyzed, efficient and mild procedure for highly chemoselective acetalization of carbonyl compounds under neutral aprotic conditions. *Synthesis* 2002: 784–788. f) Sun, J., Dong, Y., Cao, L. et al. (2004). Highly efficient chemoselective deprotection of *O,O*-acetals and *O,O*-ketals catalyzed by molecular iodine in acetone. *J. Org. Chem.* 69: 8932–8934. g) Banik, B.K., Chapa, M., Marquez, J., and Cardona, M. (2005). A remarkable iodine-catalyzed protection of carbonyl compounds. *Tetrahedron Lett.* 46: 2341–2343. h) Bailey, A.D., Cherney, S.M., Anzalone, P.W. et al. (2006). A convenient method for in situ generation of I_2 using $CuSO_4$/NaI and its applications to the deprotection of acetals, etherifications and iodolactonizations. *Synlett* 2006: 215–218.

15 a) Firouzabadi, H., Iranpoor, N., and Hazarkhani, H. (2001). Iodine catalyzes efficient and chemoselective thioacetalization of carbonyl functions, transthioacetalization of *O,O*- and *S,O*-acetals and acylals. *J. Org. Chem.* 66: 7527–7529. b) Zahouily, M., Mezdar, A., Rakik, J. et al. (2005). A mild and efficient method for the protection of carbonyl compounds as dithioacetals, dithiolanes and dithianes catalysed by iodine supported on natural phosphate. *J. Mol. Catal. A: Chem.* 233: 43–47.

16 a) Žmitek, K., Zupan, M., Stavber, S., and Iskra, J. (2006). Iodine as a catalyst for efficient conversion of ketones to *gem*-dihydroperoxides by aqueous hydrogen peroxide. *Org. Lett.* 8: 2491–2494. b) Žmitek, K., Zupan, M., Stavber, S., and Iskra, J. (2012). The effect of iodine on the peroxidation of carbonyl compounds. *J. Org. Chem.* 72: 6534–6540.

17 a) Yadav, J.S., Reddy, B.V.S., Hara Gopal, A.V. et al. (2008). Iodine as a mild and versatile reagent for the synthesis of 1,3-dioxane derivatives via the Prins reaction. *Tetrahedron Lett.* 49: 4420–4423. b) Silva, L.F. Jr, and Quintiliano, S.A. (2009). An expeditious synthesis of hexahydrobenzo[*f*]isochromenes and of hexahydrobenzo[*f*]isoquinoline via iodine-catalyzed Prins and aza-Prins cyclization. *Tetrahedron Lett.* 50: 2256–2260. c) Reddy, K.R.K.K., Rosa, I.M.L., Doriguetto, A.C. et al. (2013). Iodine-catalyzed prins cyclization of homoallylic alcohols and aldehydes. *Molecules* 18: 11100–11130. d) Figueiredo, C.A.M., Reddy, K.R.K.K., Monteiro, P.A. et al. (2013). Iodine-catalyzed aza-Prins cyclization: metal-free synthesis and antiproliferative activity of hexahydrobenzo[*f*]isoquinolines. *Synthesis* 45: 1076–1082. e) Harnying, W., Neudçrfl, J.-M., and Berkessel, A. (2017). Catalytic Prins reaction effected by molecular iodine in the presence of bis(trifluoromethanesulfonyl)imide salts. *Synthesis* 49: 269–274.

18 a) Marsili, L.A., Pergomet, J.L., Gandon, V., and Riveira, M.J. (2018). Iodine-catalyzed iso-Nazarov cyclization of conjugated dienals for the synthesis of 2-cyclopentenones. *Org. Lett.* 20: 7298–7303. b) Koenig, J.J., Arndt, T., Gildemeister, N. et al. (2019). Iodine-catalyzed Nazarov cyclizations. *J. Org. Chem.* 84: 7587–7605.

19 For reviews on the application of halogen bonding in catalysis see a) Schindler, S. and Huber, S. M. (2014). Halogen bonds in organic synthesis and organocatalysis. In *Halogen Bonding II. Topics in Current Chemistry* (eds. P. Metrangolo and G. Resnati) 359: 167–203; b) Bulfield, D. and Huber, S.M. (2016). Halogen bonding in organic synthesis and organocatalysis. *Chem. Eur. J.* 22:

14434–14450. c) Guha, S., Kazi, I., Nandy, A., and Sekar, G. (2017). Role of Lewis base-coordinated halogen(I) intermediates in organic synthesis: the journey from unstable intermediates to versatile reagents. *Eur. J. Org. Chem.* 2017: 5497–5518. d) Breugst, M., von der Heiden, D., and Schmauck, J. (2017). Novel noncovalent interactions in catalysis: a focus on halogen, chalcogen, and anion-π bonding. *Synthesis* 49: 3224–3236. e) Bamberger, J., Ostler, F., and Mancheño, O.G. (2019). Frontiers in halogen and chalcogen-bond donor organocatalysis. *ChemCatChem.* 11: 1–15. f) Sutar, R.L. and Huber, S.M. (2019). Catalysis of organic reactions through halogen bonding. *ACS Catal.* 9: 9622–9639.

20 a) Squitieri, R.A., Fitzpatrick, K.P., Jaworski, A.A., and Scheidt, K.A. (2019). Synthesis and and evaluation of evaluation of azaolium-based halogen bond-donors. *Chem. Eur. J.* 25: 10069–10073. b) Ser, C.T., Yang, H., and Wong, M.W. (2019). Iodoimidazolinium-catalyzed reduction of quinoline by Hantzsch ester: Halogen bond or Brønsted acid catalysis. *J. Org. Chem.* 84: 10338–10348.

21 a) Kaasik, M., Kaabel, S., Kriis, K. et al. (2017). Synthesis and characterisation of chiral triazole-based halogen-bond donors: Halogen bonds in the solid state and in solution. *Chem. Eur. J.* 23: 7337–7344. b) Borissov, A., Lim, J.Y.C., Brown, A. et al. (2017). Neutral iodotriazole foldamers as tetradentate halogen bonding anion receptors. *Chem. Commun.* 53: 2483–2486. c) Kaasik, M., Kaabel, S., Kriis, K. et al. (2019). Synthesis of chiral triazole-based halogen bond donors. *Synthesis* 51: 2128–2135.

22 a) Erdélyi, M. (2012). Halogen bonding in solution. *Chem. Soc. Rev.* 41: 3547–3557. b) Beale, T.M., Chudzinski, M.G., Sarwar, M.G., and Taylor, M.S. (2013). Halogen bonding in solution: thermodynamics and applications. *Chem. Soc. Rev.* 42: 1667–1680. c) Peterson, A., Kaasik, M., Metsala, A. et al. (2019). Tunable chiral triazole-based halogen bond donors: assessment of donor strength in solution with nitrogen-containing acceptors. *RSC Adv.* 9: 11718–11721.

23 Chang, Y.-P., Tang, T., Jagannathan, J.R. et al. (2020). NMR quantification of halogen-bonding ability to evaluate catalyst activity. *Org. Lett.* 22: 6647–6652.

24 a) Cabot, R. and Hunter, C.A. (2009). Non-covalent interactions between iodo-perfluorocarbons and hydrogen bond acceptors. *Chem. Commun.* 2009: 2005–2007. b) Huber, S.M., Jimenez-Izal, E., Ugaldeb, J.M., and Infante, I. (2012). Unexpected trends in halogen-bond based noncovalent adducts. *Chem. Commun.* 48: 7708–7710. c) Nepal, B. and Scheiner, S. (2015). Substituent effects on the binding of halides by neutral and dicationic bis(triazolium) receptors. *J. Phys. Chem. A* 119: 13064–13073. d) Han, N., Zeng, Y., Sun, C. et al. (2014). N···I Halogen bonding interactions: influence of Lewis bases on their strength and characters. *J. Phys. Chem. A* 118: 7058–7065. e) Scheiner, S. (2017). Halogen bonds formed between substituted imidazoliums and N bases of varying *N*-hybridization. *Molecules* 22: 1634. c) Engelage, E., Reinhard, D., and Huber, S.M. (2020). Is there a single ideal parameter for halogen-bonding-based Lewis acidity? *Chem. Eur. J.* 26: 3843–3861.

25 a) Holthoff, J.M., Engelage, E., Weiss, R., and Huber, S.M. (2020). "Anti-electrostatic" halogen bonding. *Angew. Chem. Int. Ed.* 59: 11150–11157. *Angew. Chem.* 132: 11244–11251. b) Maxson, T., Jalilov, A.S., Zeller, M., and

Rosokha, S.V. (2020). Halogen bonding between anions: association of anion radicals of tetraiodo-*p*-benzoquinone with iodide anions. *Angew. Chem. Int. Ed.* 59: 17197–17201. *Angew. Chem.* 132: 17350–17354. c) Heinen, F., Engelage, E., Cramer, C.J., and Huber, S.M. (2020). Hypervalent iodine(III) compounds as biaxial halogen bond donors. *J. Am. Chem. Soc.* 142: 8633–8640.

26 Walter, S.M., Jungbauer, S.H., Kniep, F. et al. (2013). Polyfluorinated versus cationic multidentate halogen-bond donors: a direct comparison. *J. Fluor. Chem.* 2013: 14–20.

27 a) Zapata, F., Caballero, A., Molina, P. et al. (2014). Open bis(triazolium) structural motifs as a benchmark to study combined hydrogen- and halogen-bonding interactions in oxoanion recognition processes. *J. Org. Chem.* 79: 6959–6969. b) Jungbauer, S.H. and Huber, S.M. (2015). Cationic multidentate halogen-bond donors in halide abstraction organocatalysis: catalyst optimization by preorganization. *J. Am. Soc.* 137: 12110–12120. c) Tepper, R., Schulze, B., Görls, H. et al. (2015). Preorganization in a cleft-type anion receptor featuring iodo-1,2,3-triazoles as halogen bond donors. *Org. Lett.* 17: 5740–5743. d) Riel, A.M.S., Decato, D.A., Sun, J. et al. (2018). The intramolecular hydrogen bonded–halogen bond: a new strategy for preorganization and enhanced binding. *Chem. Sci.* 9: 5828–5836. e) Sutar, R.L., Engelage, E., Stoll, R., and Huber, S.M. (2020). Bidentate chiral bis(imidazolium)-based halogen-bond donors: synthesis and applications in enantioselective recognition and catalysis. *Angew. Chem. Int. Ed.* 59: 6806–6810. *Angew. Chem.* 132: 6872–6877. f) Voelkel, M.H.H., Wonner, P., and Huber, S.M. (2020). Preorganization: a powerful tool in intermolecular halogen bonding in solution. *ChemistryOpen* 9: 214–224.

28 Bruckmann, A., Pena, M.A., and Bolm, C. (2008). Organocatalysis through halogen-bond activation. *Synlett* 2008: 900–902.

29 For some important reviews, see: a) Denmark, S. E., Kuester, W. E., and Burk, M. T. (2012). Catalytic, asymmetric halofunctionalization of alkenes—a critical perspective. *Angew. Chem. Int. Ed.* 51: 10938–10953; b) Cheng, Y.A., Yu, W.Z., and Yeung, Y.Y. (2014). Recent advances in asymmetric intra- and intermolecular halofunctionalizations of alkenes. *Org. Biomol. Chem.* 12: 2333–2343.

30 a) Nakatsuji, H., Sawamura, Y., Sakakura, A., and Ishihara, K. (2014). Cooperative activation with chiral nucleophilic catalysts and *N*-haloimides: enantioselective iodolactonization of 4-arylmethyl-4-pentenoic acids. *Angew. Chem. Int. Ed.* 53: 6974–6977. *Angew. Chem.* 126: 7094–7097. b) Lu, Y., Nakatsuji, H., Okumura, Y. et al. (2018). Enantioselective halo-oxy- and halo-azacyclizations induced by chiral amidophosphate catalysts and halo-Lewis acids. *J. Am. Chem. Soc.* 140: 6039–6043. c) Horibe, T., Tsuji, Y., and Ishihara, K. (2020). Halogen-bonding interaction between I_2 and *N*-iodosuccinimide in Lewis base-catalyzed iodolactonization. *Org. Lett.* 22: 4888–4892. d) Horibe, T., Tsuji, Y., and Ishihara, K. (2018). Thiourea–I_2 as Lewis base–Lewis acid cooperative catalysts for iodochlorination of alkene with *in situ*-generated I–Cl. *ACS Catal.* 8: 6362–6366. e) Mizar, P., Burrelli, A., Günther, E. et al. (2014). Organocatalytic stereoselective iodoamination of alkenes. *Chem. Eur. J.* 20: 13113–13116. f)

Veitch, G.E. and Jacobsen, E.N. (2010). Tertiary aminourea-catalyzed enantioselective iodolactonization. *Angew. Chem. Int. Ed.* 49: 7332–7335. *Angew. Chem.* 122: 7490–7493. g) Arai, T., Horigane, K., Watanabe, O. et al. (2019). Association of halogen bonding and hydrogen bonding in metal acetate-catalyzed asymmetric halolactonization. *iScience* 2019: 280–292.

31 a) Walter, S.M., Kniep, F., Herdtweck, E., and Huber, S.M. (2011). Halogen-bond-induced activation of a carbon–heteroatom bond. *Angew. Chem. Int. Ed.* 50: 7187–7191. *Angew. Chem.* 123: 7325–7329. b) Heinen, F., Engelage, E., Dreger, A. et al. (2018). Iodine(III) derivatives as halogen bonding organocatalysts. *Angew. Chem. Int. Ed.* 57: 3830–3833. *Angew. Chem.* 130: 3892–3896. c) Wang, Y. and Su, P. (2020). Why can cationic halogen bond donors activate the Ritter-type solvolysis of benzhydryl bromide but cationic hydrogen bond donors can not? *ACS Omega* 5 (34): 21862–21872.

32 a) Kniep, F., Rout, L., Walter, S.M. et al. (2012). 5-Iodo-1,2,3-triazolium-based multidentate halogen-bond donors as activating reagents. *Chem. Commun* 48: 9299–9301. b) Kniep, F., Walter, S.M., Herdtweck, E., and Huber, S.M. (2012). 4,4′-Azobis-(halopyridinium) derivatives: strong multidentate halogen-bond donors with a redox-active core. *Chem. Eur. J.* 18: 1306–1310. c) Castelli, R., Schindler, S., Walter, S.M. et al. (2014). Activation of glycosyl halides by halogen bonding. *Chem. Asian J.* 9: 2095–2098. d) Dreger, A., Engelage, E., Mallick, B. et al. (2018). The role of charge in 1,2,3-triazol(ium)-based halogen bonding activators. *Chem. Commun.* 54: 4013–4016. e) Perera, M.D. and Aakeröy, C.B. (2019). Organocatalysis by a multidentate halogen-bond donor: an alternative to hydrogen-bond based catalysis. *New J. Chem.* 43: 8311–8314.

33 Tsuji, N., Kobayashi, Y., and Takemoto, Y. (2014). Electrophilic iodine(I) compounds induced semipinacol rearrangement via C–X bond cleavage. *Chem. Commun.* 50: 13691–13694.

34 a) Kniep, F., Jungbauer, S.H., Zhang, Q. et al. (2013). Organocatalysis by neutral multidentate halogen-bond donors. *Angew. Chem. Int. Ed.* 52: 7028–7032. *Angew. Chem.* 125: 7166–7170. b) Saito, M., Tsuji, N., Kobayashi, Y., and Takemoto, Y. (2015). Direct dehydroxylative coupling reaction of alcohols with organosilanes through Si-X bond activation by halogen bonding. *Org. Lett.* 17: 3000–3003. c) Chan, Y.-C. and Yeung, Y.-Y. (2019). Halogen-bond-catalyzed addition of carbon-based nucleophiles to *N*-acylimminium ions. *Org. Lett.* 21: 5665–5669.

35 a) Coulembier, O., Meyer, F., and Dubois, P. (2010). Controlled room temperature ROP of L-lactide by ICl_3: a simple halogen-bonding catalyst. *Polym. Chem.* 1: 434–437. b) Takagi, K., Yamauchi, K., and Murakata, H. (2017). Halogen-bonding-mediated and controlled cationic polymerization of isobutyl vinyl ether: expanding the catalytic scope of 2-iodoimidazolium salts. *Chem. Eur. J.* 23: 9495–9500. c) Haraguchi, R., Nishikawa, T., Kanazawa, A., and Aoshima, S. (2020). Metal-free living cationic polymerization using diaryliodonium salts as organic Lewis acid catalysts. *Macromolecules* 53: 4185–4192.

36 He, W., Ge, Y.C., and Tan, C.H. (2014). Halogen-bonding-induced hydrogen transfer to C = N bond with Hantzsch ester. *Org. Lett.* 16: 3244–3247. (2019) Additions and Corrections *Org. Lett.* 21: 7684.

37 a) Takeda, Y., Hisakuni, D., Lin, C.-H., and Minakata, S. (2015). Halogenoimidazolium salt catalyzed aza-Diels-Alder reaction through halogen-bond formation. *Org. Lett.* 17: 318–321. b) Nziko, V.d.P.N. and Scheiner, S. (2016). Catalysis of the aza-Diels-Alder reaction by hydrogen and halogen bonds. *J. Org. Chem.* 81: 2589–2597. c) Haraguchi, R., Hoshino, S., Sakai, M. et al. (2018). Bulky iodotriazolium tetrafluoroborates as highly active halogen-bonding-donor catalysts. *Chem. Commun.* 54: 10320–10323. d) Kaasik, M., Metsala, A., Kaabel, S. et al. (2019). Halo-1,2,3-triazolium salts as halogen bond donors for the activation of imines in dihydropyridinone synthesis. *J. Org. Chem.* 84: 4294–4303.

38 a) Jungbauer, S.H., Walter, S.M., Schindler, S. et al. (2014). Activation of a carbonyl compound by halogen bonding. *Chem. Commun.* 50: 6281–6284. b) Kee, C.W. and Wong, M.W. (2016). In silico design of halogen-bonding-based organocatalyst for Diels–Alder reaction, Claisen rearrangement, and Cope-type hydroamination. *J. Org. Chem.* 81: 7459–7470.

39 Heinen, F., Reinhard, D., Engelage, E., and Huber, S.M. (2020). A bidentate iodine(III)-based halogen bond donor as powerful organocatalyst. *Angew. Chem. Int. Ed.* 60: 5069–5073 https://doi.org/10.1002/anie.202013172.

40 a) Gliese, J.P., Jungbauer, S.H., and Huber, S.M. (2017). A halogen-bonding-catalyzed Michael addition reaction. *Chem. Commun.* 53: 12052–12055. b) von der Heiden, D., Detmar, E., Kuchta, R., and Breugst, M. (2018). Activation of Michael acceptors by halogen-bond donors. *Synlett* 29: 1307–1313.

41 a) Ge, Y.C., Yang, H., Heusler, A. et al. (2019). Halogen bonding induced conjugate addition of thiophenes to enones and enals. *Chem. Asian J.* 14: 2656–2661. b) Zhang, H. and Toya, P.H. (2020). Halogen-bond-catalyzed Friedel–Crafts reactions of furans using a 2,2′-bipyridine-based catalyst. *Adv. Synth. Catal.* 363: 215–221.

42 a) Koenig, J.J., Arndt, T., Gildemeister, N. et al. (2019). Iodine-catalyzed Nazarov cyclizations. *J. Org. Chem.* 84: 7587–7605. b) Dreger, A., Wonner, P., Engelage, E. et al. (2019). A Halogen-bonding-catalysed Nazarov cyclisation reaction. *Chem. Commun.* 55: 8262–8265.

43 Liu, X., Ma, S., and Toy, P.H. (2019). Halogen bond-catalyzed Friedel–Crafts reactions of aldehydes and ketones using a bidentate halogen bond donor catalyst: synthesis of symmetrical bis(indolyl)methanes. *Org. Lett.* 21: 9212–9216.

44 Zhang, Y., Han, J., and Liu, Z.J. (2015). Diaryliodonium salts as efficient Lewis acid catalysts for direct three component Mannich reactions. *RSC Adv.* 5: 25485–25488.

45 Matsuzawa, A., Takeuchi, S., and Sugita, K. (2016). Iodoalkyne-based catalyst-mediated activation of thioamides through halogen bonding. *Chem. Asian J.* 11: 2863–2866.

46 a) Matsuzaki, K., Uno, H., Tokunaga, E., and Shibata, N. (2018). Fluorobissulfonylmethyl iodides: an efficient scaffold for halogen bonding catalysts with an sp^3-hybridized carbon–iodine moiety. *ACS Catal.* 8: 6601–6605. b) Sutar, R.L., Erochok, N., and Huber, S.M. (2020). Mukaiyama aldol reaction catalyzed by

(benz)imidazolium-based halogen bond donors. *Org. Biomol. Chem.* 19: 770–774 https://doi.org/10.26434/chemrxiv. 12820652.v1.
47 a) Bergamaschi, G., Lascialfari, L., Pizzi, A. et al. (2018). A halogen bond-donor amino acid for organocatalysis in water. *Chem. Commun.* 54: 10718–10721. b) Chakraborty, S., Maji, S., Ghosh, R. et al. (2019). Aryl-platform-based tetrapodal 2-iodo-imidazolium as an excellent halogen bond receptor in Aqueous Medium. *Chem. Commun.* 55: 1506–1509.
48 a) Xu, C. and Loh, C.C.J. (2019). A multistage halogen bond catalyzed strain-release glycosylation unravels new hedgehog signaling inhibitors. *J. Am. Chem. Soc.* 141: 5381–5391. b) Xu, C., Bhaskara Rao, V.U., Weigen, J., and Loh, C.C.J. (2020). A robust and tunable halogen bond organocatalyzed 2-deoxyglycosylation involving quantum tunneling. *Nat. Commun.* 2020: 4911.
49 Chan, Y.C. and Yeung, Y.Y. (2018). Halogen bond catalyzed bromocarbocyclization. *Angew. Chem. Int. Ed.* 57: 3483–3487. *Angew. Chem.* 133: 3541–3545.
50 Saito, M., Kobayashi, Y., Tsuzuki, S., and Takemoto, Y. (2017). Electrophilic activation of iodonium ylides by halogen-bond-donor catalysis for cross-enolate coupling. *Angew. Chem. Int. Ed.* 56: 7653–7657. *Angew. Chem.* 129: 7761–7765.
51 a) Mulliken, R.S. (1950). Structures of complexes formed by halogen molecules with aromatic and with oxygenated solvents. *J. Am. Chem. Soc.* 72: 600–608. b) Laurence, C., Queignec-Cabanetos, M., Dziembowska, T. et al. (1981). 1-Iodoacetylenes. 1. Spectroscopic evidence of their complexes with Lewis bases. A spectroscopic scale of soft basicity. *J. Am. Chem. Soc.* 103: 2567–2573. c) Xu, Z., Yang, Z., Liu, Y. et al. (2014). Halogen bond: its role beyond drug–target binding affinity for drug discovery and development. *J. Chem. Inf. Model.* 54: 69–78. d) Cao, J., Yan, X., He, W. et al. (2017). C–I···π Halogen bonding driven supramolecular helix of bilateral *N*-amidothioureas bearing β-turns. *J. Am. Chem. Soc.* 139: 6605–6610.
52 a) Jiang, S., Zhang, L., Cui, D. et al. (2016). The important role of halogen bond in substrate selectivity of enzymatic catalysis. *Sci. Rep.* 2016: 34750. b) Kuwano, S., Suzuki, T., Yamanaka, M. et al. (2019). Catalysis based on C–I···π halogen bonds: Electrophilic activation of 2-alkenylindoles by cationic halogen-bond donors for [4+2] cycloadditions. *Angew. Chem. Int. Ed.* 58: 10220–10224. *Angew. Chem.* 131: 10326–10330. c) Suzuki, T., Kuwano, S., and Arai, T. (2020). Non-bonding electron pair versus π-electrons in solution phase halogen bond catalysis: Povarov reaction of 2-vinylindoles and imines. *Adv. Synth. Catal.* 362: 3208–3212.
53 a) Farina, A., Meille, S.V., Messina, M.T. et al. (1999). Resolution of racemic 1,2-dibromohexafluoropropane through halogen-bonded supramolecular helices. *Angew. Chem. Int. Ed.* 38: 2433–2436. *Angew. Chem.* 111: 2585–2588. b) Lim, J.Y.C., Marques, I., Ferreira, L. et al. (2016). Enhancing the enantioselective recognition and sensing of chiral anions by halogen bonding. *Chem. Commun.* 52: 5527–5530. c) Lim, J.Y.C., Marques, I., Félix, V., and Beer, P.D. (2017). Enantioselective anion recognition by chiral halogen-bonding [2]rotaxanes. *J. Am. Chem. Soc.* 139: 12228–12239. d) Lim, J.Y.C., Marques, I., Félix, V., and Beer,

P.D. (2018). A chiral halogen-bonding [3]rotaxane for the recognition and sensing of biologically relevant dicarboxylate anions. *Angew. Chem. Int. Ed.* 57: 584–588. *Angew. Chem.* 130: 593–597. e) Peluso, P., Mamane, V., Dallocchio, R. et al. (2018). Polysaccharide-based chiral stationary phases as halogen bond acceptors: a novel strategy for detection of stereoselective σ-hole bonds in solution. *J. Sep. Sci.* 41: 1247–1256. f) Nicolas, I., Jeannin, O., Pichon, D., and Fourmigue, M. (2016). Dibromohydantoins as halogen bond (XB) donors: a route toward the introduction of chirality in halogen bonded systems. *CrystEngCom* 18: 9325–9333. g) Lim, J.Y.C., Marques, I., Félix, V., and Beer, P.D. (2018). Chiral halogen and chalcogen bonding receptors for discrimination of stereo- and geometric dicarboxylate isomers in aqueous media. *Chem. Commun.* 54: 10851–10854.

54 Zong, L., Ban, X., Kee, C.W., and Tan, C.-H. (2014). Catalytic enantioselective alkylation of sulfenate anions to chiral heterocyclic sulfoxides using halogenated pentanidium salts. *Angew. Chem. Int. Ed.* 53: 11849–11853. *Angew. Chem.* 126: 12043–12047.

55 Zhang, X., Ren, J., Tan, S.M. et al. (2019). An enantioconvergent halogenophilic nucleophilic substitution (S_N2X) reaction. *Science* 363: 400–404.

56 Arai, T., Suzuki, T., Inoue, T., and Kuwano, S. (2017). Chiral Bis(imidazolidine)iodobenzene (I-Bidine) organocatalyst for thiochromane synthesis using an asymmetric Michael/Henry reaction. *Synlett* 28: 122–127.

57 a) Kuwano, S., Suzuki, T., Hosaka, Y., and Arai, T. (2018). A chiral organic base catalyst with halogen-bonding-donor functionality: asymmetric Mannich reactions of malononitrile with *N*-Boc aldimines and ketimines. *Chem. Commun.* 54: 3847–3850. b) Kuwano, S., Nishida, Y., Suzuki, T., and Arai, T. (2020). Catalytic asymmetric Mannich-type reaction of malononitrile with *N*-Boc α-ketiminoesters using chiral organic base catalyst with halogen bond donor functionality. *Adv. Synth. Catal.* 362: 1674–1678.

58 Caronna, T., Liantonio, R., Logothetis, T.A. et al. (2004). Halogen bonding and π···π stacking control reactivity in the solid state. *J. Am. Chem. Soc.* 126: 4500–4501.

59 For selected reviews see: a) Breit, B. (2005). Supramolecular approaches to generate libraries of chelating bidentate ligands for homogeneous catalysis. *Angew. Chem. Int. Ed.* 44: 6816–6825; b) Sandee, A.J. and Reek, J.N.H. (2006). Bidentate ligands by supramolecular chemistry—the future for catalysis? *Dalton Trans.* 2006: 3385–3391. c) Breit, B. (2008). Catalysts through self-assembly for combinatorial homogeneous catalysis. *Pure Appl. Chem.* 80: 855–860. d) Carboni, S., Gennari, C., Pignataro, L., and Piarulli, U. (2011). Supramolecular ligand–ligand and ligand–substrate interactions for highly selective transition metal catalysis. *Dalton Trans.* 40: 4355–4373. e) Bellini, R., van der Vlugt, J.I., and Reek, J.N.H. (2012). Supramolecular self-assembled ligands in asymmetric transition metal catalysis. *Israel J. Chem.* 52: 613–629. f) Raynal, M., Ballester, P., Vidal-Ferran, A., and van Leeuwen, P.W.N.M. (2014). Supramolecular catalysis. Part 1: non-covalent interactions as a tool for building and modifying homogeneous catalysts. *Chem. Soc. Rev.* 43: 1660–1733. g) Raynal, M., Ballester, P.,

Vidal-Ferran, A., and van Leeuwen, P.W.N.M. (2014). Supramolecular catalysis. Part 2: artificial enzyme mimics. *Chem. Soc. Rev.* 43: 1734–1787.

60 For a review on hypervalent iodine reagents with ortho coordinating groups, see: Zhdankin, V. V. and Protasiewicz, J. D. (2014). Development of new hypervalent iodine reagents with improved properties and reactivity by redirecting secondary bonds at iodine center. *Coord. Chem. Rev.* 2014: 54–62.

61 a) Boucher, M., Macikenas, D., Ren, T., and Protasiewicz, J.D. (1997). Secondary bonding as a force dictating structure and solid-state aggregation of the primary sitrene Sources (arylsulfonylimino)iodoarenes ($ArINSO_2Ar'$). *J. Am. Chem. Soc.* 119: 9366–9376. b) Macikenas, D., Skrzypczak-Jankun, E., and Protasiewicz, J.D. (1999). A New Class of Iodonium Ylides Engineered as Soluble Primary Oxo and Nitrene Sources. *J. Am. Chem. Soc.* 121: 7164–7165. (2011) Additions and Corrections *J. Am. Chem. Soc.* 133: 4151. c) Meprathu, B.V. and Protasiewicz, J.D. (2003). Synthesis and characterization of novel polyvalent organoiodine compounds. *ARKIVOC* 2003: 83–90. d) Blake, A.J., Novak, A., Davies, M. et al. (2009). Preparation of 1,1′-oxy-bis(3,3-bis(trifluoromethyl)-3(1*H*)-1,2-benziodoxole) and 2-(*N*-(*p*-toluenesulfonyl)imino)iodobenzylmethyl ether. *Synth. Commun.* 39: 1065–1075. e) Yoshimura, A., Nemykin, V.N., and Zhdankin, V.V. (2011). *o*-Alkoxyphenyliminoiodanes: highly efficient reagents for the catalytic aziridination of alkenes and the metal-free amination of organic substrates. *Chem. Eur. J.* 17: 10538–10541. f) Kobayashi, Y., Masakado, S., and Takemoto, Y. (2018). Photoactivated *N*-acyliminoiodinanes applied to amination: an *ortho*-methoxymethyl group stabilizes reactive precursors. *Angew. Chem. Int. Ed.* 57: 693–697. *Angew. Chem.* 130: 701–705.

62 a) Carreras, L., Serrano-Torne, M., van Leeuwen, P.W.N.M., and Vidal-Ferran, A. (2018). XBphos-Rh: a halogen-bond assembled supramolecular catalyst. *Chem. Sci.* 9: 3644–3648. b) Carreras, L., Benet-Buchholz, J., Franconetti, A. et al. (2019). Halogen bonding effects on the outcome of reactions at metal centres. *Chem. Commun.* 55: 2380–2383.

63 Guillier, K., Caytan, E., Dorcet, V. et al. (2019). A halogen-bond donor catalyst for templated macrocyclization. *Angew. Chem. Int. Ed.* 58: 14940–14943. *Angew. Chem.* 131: 15082–15085.

64 a) Kobayashi, Y., Nakatsuji, Y., Li, S. et al. (2018). Direct *N*-glycofunctionalization of amides with glycosyl trichloroacetimidate by thiourea/halogen bond donor co-catalysis. *Angew. Chem. Int. Ed.* 57: 3646–3650. *Angew. Chem.* 130: 3708–3712. b) Li, S., Kobayashi, Y., and Takemoto, Y. (2018). Organocatalytic direct α-selective *N*-glycosylation of amide with glycosyl trichloroacetimidate. *Chem. Pharm. Bull.* 66: 768–770.

65 Wolf, J., Huber, F., Erochok, N. et al. (2020). Activation of a metal-halogen bond through halogen bonds. *Angew. Chem. Int. Ed.* 59: 16496–16500. *Angew. Chem.* 132: 16638–16643.

4

Catalytic Transformations with R_4NI Catalyst Precursors

Xiao-Xian Li and Yun-Fei Du

School of Pharmaceutical Science and Technology (SPST), Tianjin University, Tianjin, P. R. China

4.1 Introduction

Quaternary ammonium iodides (R_4NI), due to their unique ionic structure, have not only been widely applied in emulsification, disinfection, and polarographic analysis as cationic surfactants but also found extensive applications in various organic catalytic transformations. Based on different aliphatic substituents (R), quaternary ammonium iodides contain many species. However, the most common ones that have been applied in organic catalytic transformations include tetraethylammonium iodide (Et_4NI) and tetrabutylammonium iodide (n-Bu_4NI) as they have been proven to be nontoxic, readily available, neutral, and mild chemicals [1]. In this chapter, the organic catalytic transformations enabled by R_4NI catalyst precursors were summarized and discussed, based on the bond-forming types catalyzed by R_4NI. It is worthy to note that the organic transformations using R_4NI as phase-transfer catalysts are not covered in this chapter [2].

4.2 The Construction of C–N Bond

The formation of C–N bond is considered to be an important synthetic strategy in organic synthesis, especially a predominant approach for the synthesis of nitrogen-containing compounds. The C–N bond formation reaction catalyzed by quaternary ammonium iodides (R_4NI) have received extensive attention from synthetic chemists as it possesses the features of mild reaction conditions and good reaction selectivity.

4.2.1 The Synthesis of Amides

The combing use of quaternary ammonium iodide (R_4NI) with an oxidant has proven to be an effective system for the construction of C–N bond, leading to the synthesis of variously substituted amides.

Iodine Catalysis in Organic Synthesis, First Edition. Edited by Kazuaki Ishihara and Kilian Muñiz.

Mediated by n-Bu_4NI and *tert*-butyl hydroperoxide (TBHP), α-ketoamides **3** could be readily synthesized through the radical coupling reaction between aryl methyl ketones **1** and dialkylformamides **2** [3]. The reaction was postulated that the *tert*-butoxyl radical **4** and *tert*-butylperoxy radical **5** were formed initially through the interaction of iodide anion and TBHP. Then with the abstraction of hydrogen radical by *tert*-butoxyl radical **4**, aryl methyl ketone and DMF were converted to phenacyl radical **6** and radical **7**, respectively. Next, radical **7** was converted to dimethylamino radical **8**, which coupled with phenacyl radical **6** to produce 2-(dimethylamino)-1-phenylethanone **9**. Then, intermediate **9** was oxidized by TBHP to afford carbon radical **10**, iminium salt **11**, which captured hydroxide ion to give alcohol intermediate **12**. Finally, in the presence of TBHP, alcohol **12** was oxidized into the final product **3** (Scheme 4.1).

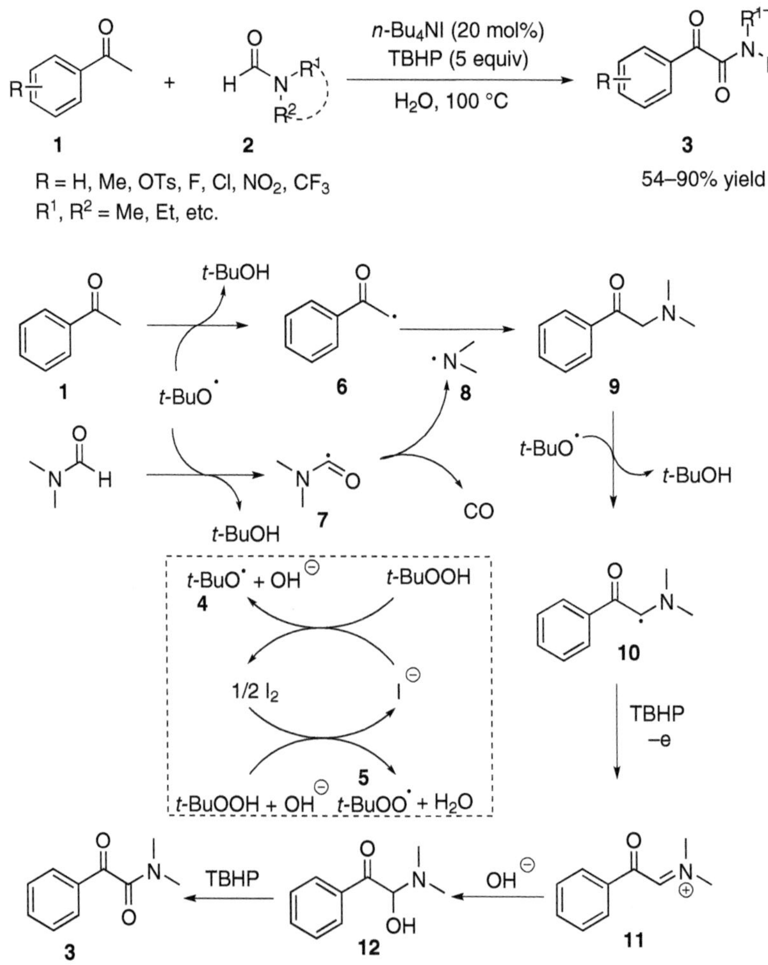

Scheme 4.1 n-Bu_4NI-catalyzed synthesis of α-ketoamides from aryl methyl ketones and dialkylformamides.

In 2014, Sun and colleagues [4] discovered that mediated by n-Bu_4NI and TBHP, ethylarenes **13** could also react with *N,N*-dialkylformamides **14** to afford α-ketoamides **15** *via* sequential C–O and C–N bond formation (Scheme 4.2). It was proposed that ethylarenes first reacted with the *tert*-butoxyl radical **4**, generated from the reaction of n-Bu_4NI and TBHP, to give the benzylic radical **16**. Then radical **16** underwent further oxidations with TBHP to give acetophenone. Following the same mechanistic pathway mentioned above, the reaction of acetophenone with *N,N*-dialkylformamide **14** mediated by n-Bu_4NI and TBHP furnished the desired α-ketoamide **15**.

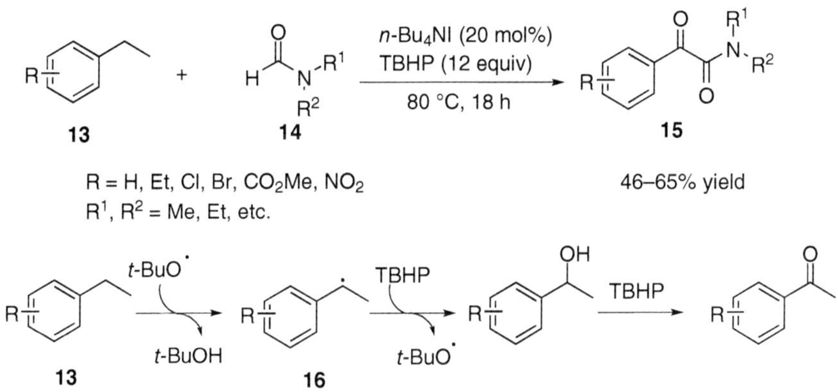

Scheme 4.2 n-Bu_4NI-catalyzed synthesis of α-ketoamides from ethylarenes and dialkylformamides.

Under oxygen atmosphere, 2′-aminoacetophenones **17** could undergo n-Bu_4NI-catalyzed oxidative annulation intramolecularly to give isatin compounds **18** [5]. It was postulated that the reaction was enabled by a n-Bu_4NI-catalyzed electrochemical oxidation, which resulted in C–O and C–N bond formation (Scheme 4.3).

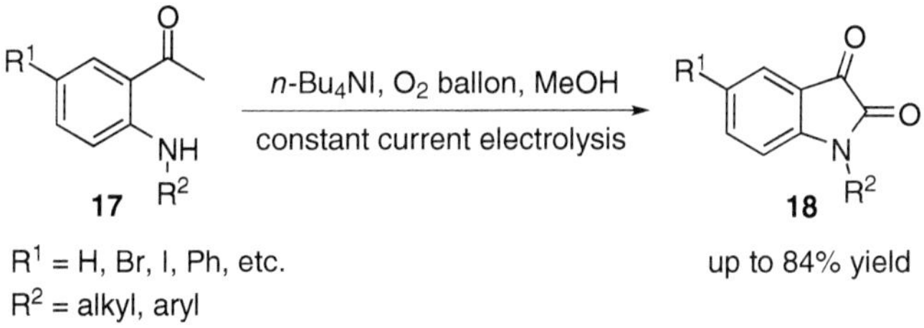

Scheme 4.3 n-Bu_4NI-catalyzed synthesis of isatins from 2′-aminoacetophenones.

Mediated by n-Bu_4NI and TBHP, aldehydes could react with tertiary amines to afford amide compounds. In 2013, Yu et al. [6] reported that the reaction of aldehydes **19** and tertiary amines **20** in the presence of n-Bu_4NI and TBHP underwent oxidative coupling to afford amides **21**. The proposed mechanism indicated that the key step

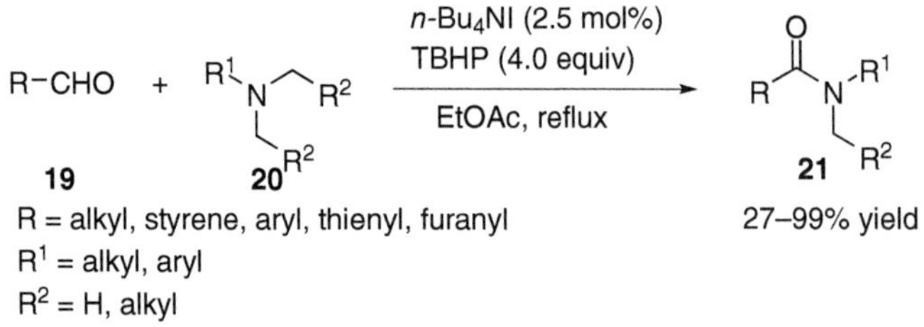

Scheme 4.4 *n*-Bu_4NI-catalyzed synthesis of amides from aldehydes and tertiary amines.

for this transformation is the *n*-Bu_4NI/TBHP-mediated conversion of tertiary amine to a secondary amine *via* the removal of one molecule of aldehyde (Scheme 4.4).

Amide compounds **24** could also be synthesized from the *n*-Bu_4NI/TBHP-mediated reaction between aldehydes **22** and formamides **23**. It was postulated that aldehydes were oxidized to acyl radicals *via* oxidative dehydrogenation, and formamides were oxidized to aminyl radicals *via* oxidative dehydrogenation followed by the removal of carbon monoxide. Finally, acyl radicals and aminyl radicals underwent cross-coupling to give amides [7] (Scheme 4.5). The research conducted by Zhu and colleagues [8] showed that when various alcohols were adjusted to similar conditions, amide derivatives could be obtained in moderate-to-high yield.

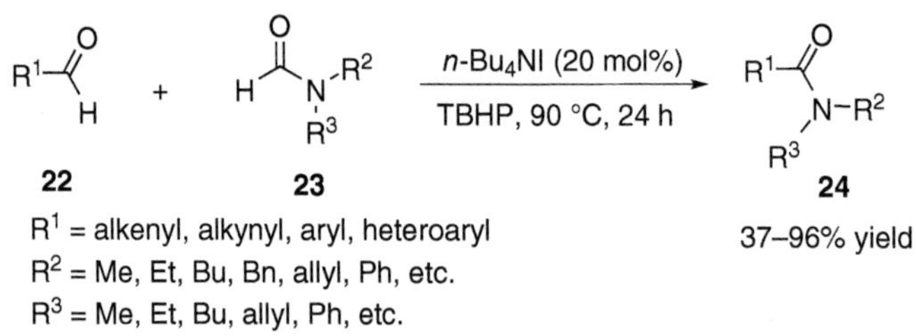

Scheme 4.5 *n*-Bu_4NI-catalyzed synthesis of amides from aldehydes and formamides.

A one-pot procedure for the synthesis of amides **27** has been developed by Yu and Chen *via* treating benzyl alcohols **25** with *N*-hydroxysuccinimide (NHS), TBHP, and a catalytic amount of *n*-Bu_4NI, followed by the adding of amines **26**. This *n*-Bu_4NI-catalyzed oxidative transformation was postulated to proceed *via* active esters **28**, which undergoes aminolysis with amines to afford the corresponding products [9] (Scheme 4.6).

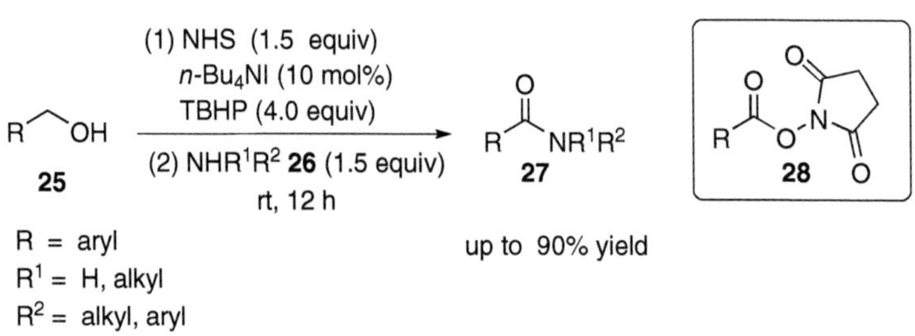

Scheme 4.6 *n*-Bu_4NI-catalyzed synthesis of amides from benzyl alcohols and amines.

In separate work, Yu and Chen [10] realized the oxidative amination of benzylic aldehydes **29** for the synthesis of primary amides **30** by using ammonium salt as the amine source, Et_4NI as the catalyst, TBHP as the oxidant (Scheme 4.7). It is presumed that the reaction of alcohols with TBHP in the presence of Et_4NI can afford aldehydes **31**. Then aldehydes **31** reacted with ammonia to generate hemiaminals **32**, which could be converted into amides *via* oxidative dehydrogenation.

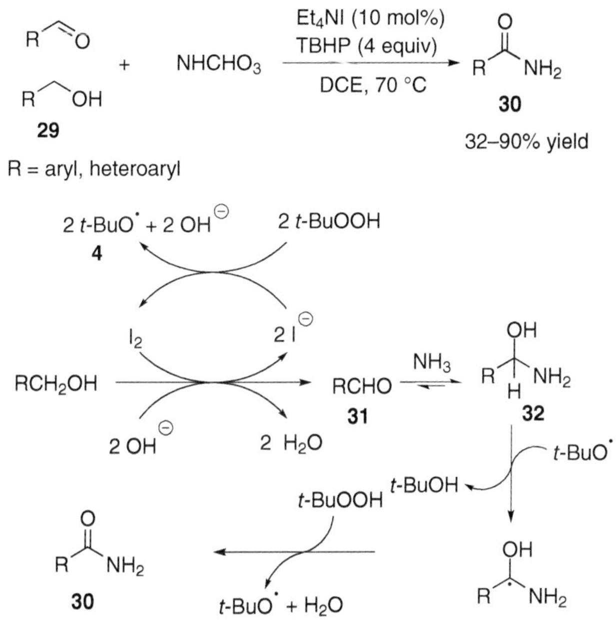

Scheme 4.7 Et_4NI-catalyzed synthesis of primary amides from benzylic aldehydes and ammonium salt.

n-Bu_4NI has been used as a catalyst to enable the oxygen-centered radical addition between acyl peroxides **33** and isocyanides **34**, leading to the synthesis of a diverse arylcarboxyamides **35** [11]. According to the mechanism, benzoyl peroxide was initially decomposed to benzoyl radical **36**, which reacted with isocyanide *via* radical addition to give C-radical intermediate **37**. The subsequent iodine oxidation of intermediate **37** led to the formation of nitrilium intermediate **38**, which captured one molecule of water to give amide **39**. Finally, amide **39** was converted to the four-member intermediate **40**, which undergoes carbon dioxide elimination to give the title product **35** (Scheme 4.8).

4.2.2 The Synthesis of Hemiaminal Ethers

The combining use of *n*-Bu_4NI as catalyst and TBHP as oxidant is an efficient oxidative system for the synthesis of hemiaminal ether, *via* oxidative cross-coupling between an amine and an ether. In 2014, Zhao and Du [12] realized the TBHP/*n*-Bu_4NI oxidative system-mediated construction of hemiaminal ether

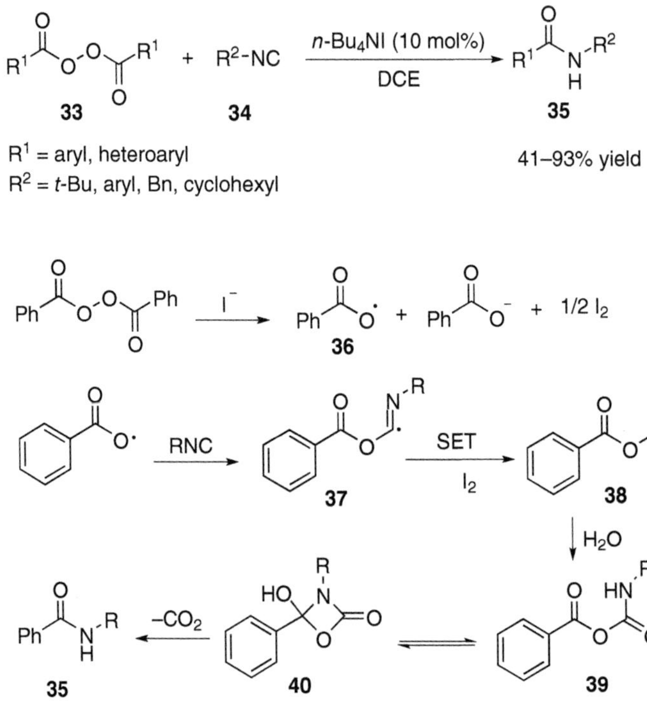

Scheme 4.8 n-Bu_4NI-catalyzed synthesis of arylcarboxyamides from acyl peroxides and isocyanides.

framework **43** through intermolecular oxidative Csp^3–N bond formation between alkyl ethers **42** and different amines **41** (Scheme 4.9). The mechanism process was proposed as follows: initially, active iodine species (ammonium hypoiodite **44** and iodite **45**) were generated *via* the n-Bu_4NI/TBHP oxidative system. Next, through the iodine species **44**/**45** mediated homolytic cleavage of the alkyl C–H bond in **42**, the carbon-centered radical **46** was formed with the concomitant formation of n-Bu_4NI or ammonium hypoiodite **44**, as well as hydroxide radical. Furthermore, the abstraction of one electron from **46** by this released hydroxide radical furnished oxonium ion **47** and a hydroxide ion (Note: in the original paper, it was described that alkyl radical **46** was generated through iodine species **44** or **45** induced homolytic cleavage of the alkyl C–H bond. Then the alkyl radical **46** was further oxidized by active iodine species **44** or **45** to form the oxonium **47**, with a hydroxide ion being released). In the final step, the amine was deprotonated by the hydroxide ion to generate the anionic species **48**, which nucleophilically attacked the oxonium **47** to give product **43**.

Intermolecular oxidative C–N bond formation *via* cross-dehydrogenative coupling (CDC) of aryl ethers **49** and tetrazoles **50** could be realized by n-Bu_4NI/TBHP for the construction of hemiaminal ether skeletons **51**. This strategy showed a high level of regioselectivity for substrates possessing multiple Csp^3–H bonds adjacent

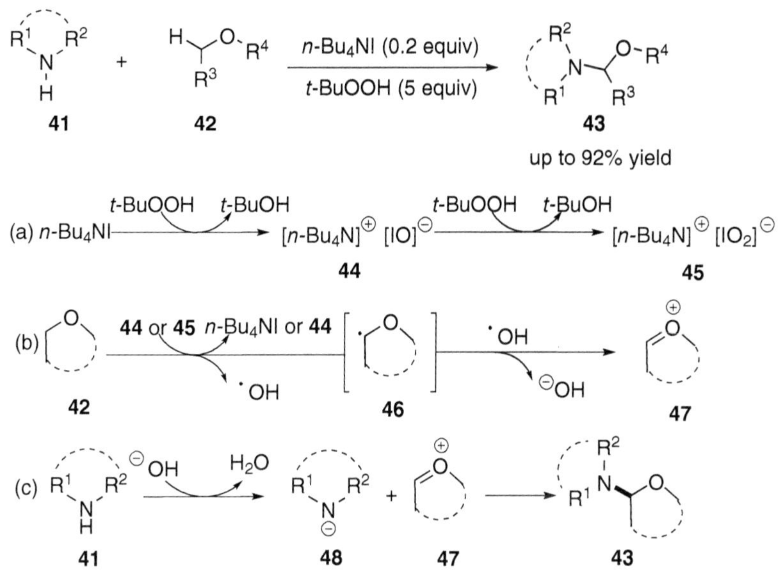

Scheme 4.9 *n*-Bu_4NI-catalyzed synthesis of hemiaminal ether framework from alkyl ethers and amines.

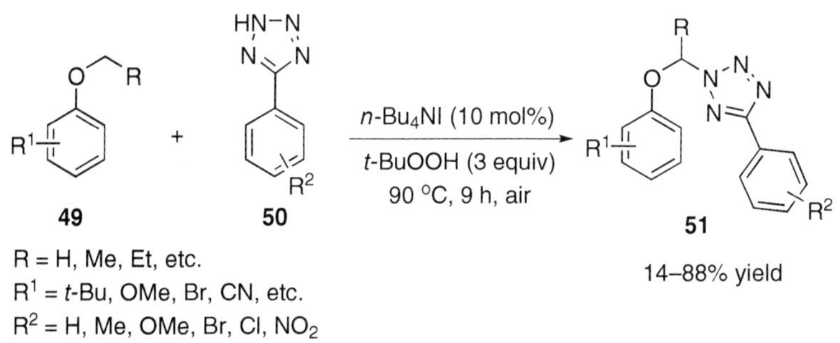

Scheme 4.10 *n*-Bu_4NI-catalyzed synthesis of hemiaminal ether skeletons from aryl ethers and tetrazoles.

to the ethereal oxygen [13] (Scheme 4.10). Later, Wang and He reported a related approach, which realized the synthesis of hemiaminal ether derivatives, through *n*-Bu_4NI-mediated amination of ethers with aryl tetrazoles and triazoles [14].

In 2017, Battula [15] realized the *n*-Bu_4NI/TBHP-mediated synthesis of hemiaminal ethers **54** by linking ethers **53** with 1*H*-1,2,4-triazoles **52** (Scheme 4.11). This method was applicable to a series of ethers and 1*H*-1,2,4-triazoles, resulting in the synthesis of the corresponding hemiaminal ethers in moderate-to-good yields.

In 2018, Lin [16] reported an oxidative coupling reaction between purines **55** and alkyl ethers **56** for synthesizing a series of 9-alkylated purine derivatives **57** *via* the *n*-Bu_4NI/TBHP-catalyzed cross-dehydrogenative coupling (Scheme 4.12).

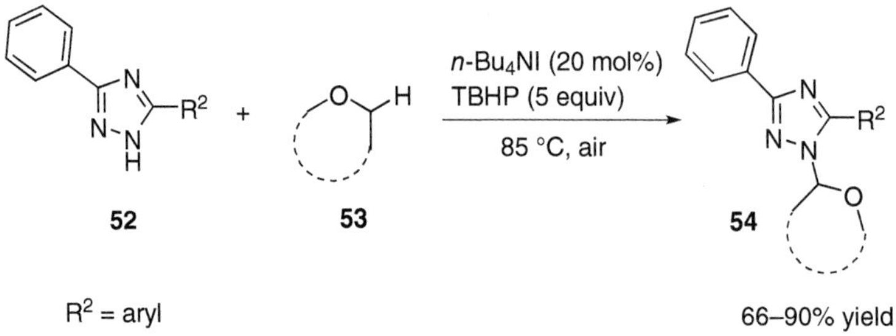

Scheme 4.11 *n*-Bu_4NI-catalyzed synthesis of hemiaminal ethers from ethers and 1*H*-1,2,4-triazoles.

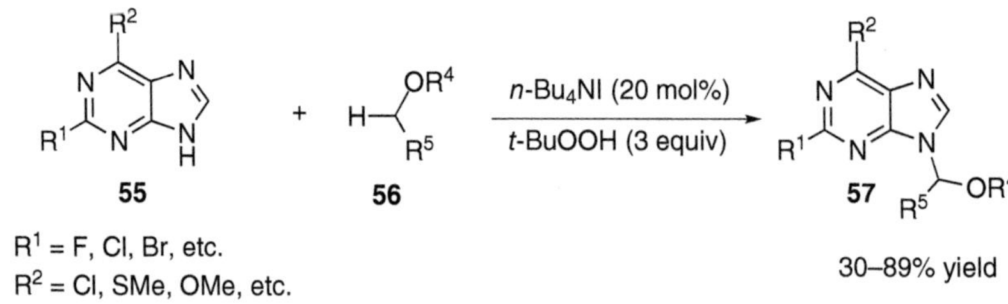

Scheme 4.12 *n*-Bu_4NI-catalyzed synthesis of 9-alkylated purine derivatives from purines and alkyl ethers.

4.2.3 The C–H Bond Amination

The combining use of *n*-Bu_4NI and catalyst and TBHP as oxidant is an efficient oxidative system for the amination of C–H bond. In 2014, He and Wang reported [17] that by employing *n*-Bu_4NI as the catalyst and *t*-BuOOH as the oxidant, the alkylated aryl tetrazoles **60** could be efficiently accessed from benzylic C–H substrates **58** and aryl tetrazoles **59** (Scheme 4.13a). The mechanism showed that *tert*-butoxyl radical **4**, iodine, and a hydroxyl anion were initially generated through the oxidation of *n*-Bu_4NI and TBHP. Subsequently, tetrazole was deprotonated by hydroxyl anion giving anionic species **65**. At the same time, the benzyl C–H bond underwent oxidation mediated by *tert*-butoxyl radical **4** and iodine furnishing benzyl radical **66** and cation **67**, respectively. Target product **60** was acquired through the reaction of aryl tetrazole anion **65** and benzyl cation **67**. The mechanistic process indicated that the I_2/I^- redox system plays a vital role in this reaction.

Battula and coworkers [18] discovered an *n*-Bu_4NI/TBHP-catalyzed C–N bond formation protocol for the synthesis of 1-benzyl-1,2,4-triazoles **62** *via* cross-dehydrogenative coupling of 1*H*-1,2,4-triazoles **61** and methylarenes **58** (Scheme 4.13b). In addition, the oxidative coupling reaction between purines **63** and benzyl compounds **58** mediated *by* *n*-Bu_4NI and *t*-BuOOH could lead to the synthesis of a series of 9-alkylated purine derivatives **64** [16] (Scheme 4.13c).

In 2020, Patel et al. [19] realized the *n*-Bu_4NI-catalyzed methylation, alkylation, and arylation of tetrazoles **68** by using TBHP as the methyl source, alkyl diacyl peroxides **69** as the primary alkyl source, alkyl peresters **71** as the secondary and tertiary

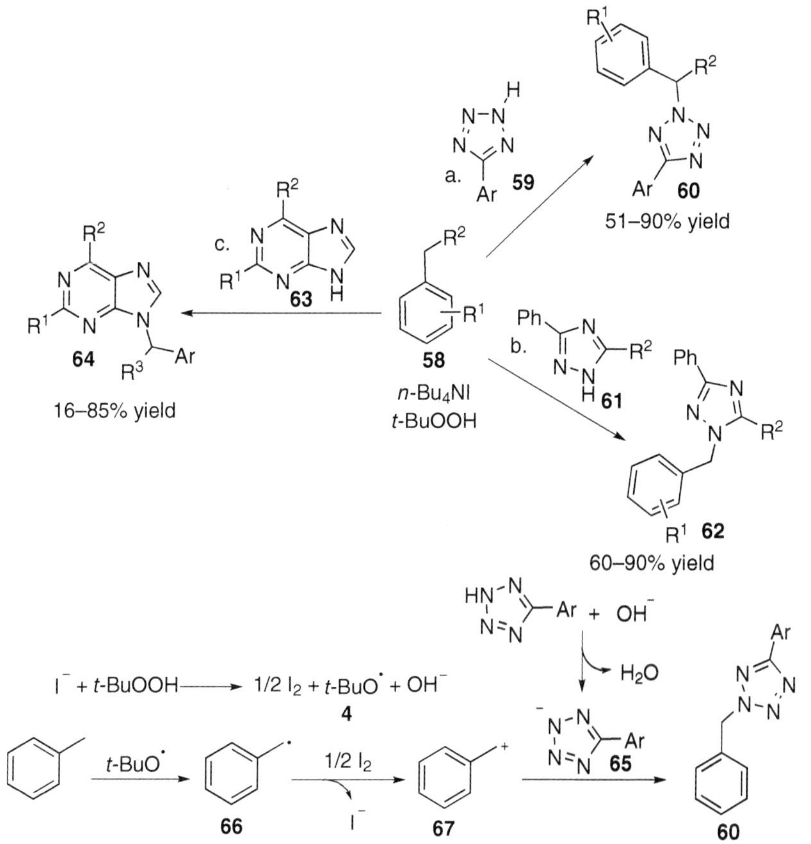

Scheme 4.13 *n*-Bu_4NI-catalyzed synthesis of alkylated aryl tetrazoles from methylarenes.

alkyl sources, and aryl diacyl peroxides as the arylating source. A series of 2-alkylated tetrazoles **70**/**72**/**73** could be achieved in satisfactory-to-good yield by this method (Scheme 4.14).

In 2014, Wang and colleagues [20] reported the *n*-Bu_4NI/TBHP-mediated coupling of aminyl radical and benzoxazole radical, generated from formamides **75** and benzoxazoles **74**, respectively, leading to the synthesis of 2-aminobenzoxazole derivatives **76** in moderate-to-good yields (Scheme 4.15). Later, Qu and Yuan [21] discovered that the tertiary amines also could be used as the nitrogen source for direct C–H amination of benzoxazoles under microwave irradiation condition.

An *n*-Bu_4NI/TBHP-catalyzed direct amination of allylic and benzylic Csp^3–H **78** with anilines **77** to form *N*-substituted anilines **79** had been developed by Wang's group [22]. The allylic cation intermediate **80** could be obtained through consecutive oxidation of compound **78** by *n*-Bu_4NI/TBHP system. Then the nucleophilic attack of amine onto the allylic cation **80** furnished the desired product (Scheme 4.16).

In 2015, Wan and coworkers developed an *n*-Bu_4NI/TBHP-catalyzed protocol for constructing aminonitrile skeletons **83** from cyanoacetic acids **82** and amines **81** [23]. The key intermediate 2-iodoacetonitrile **84** could be obtained through

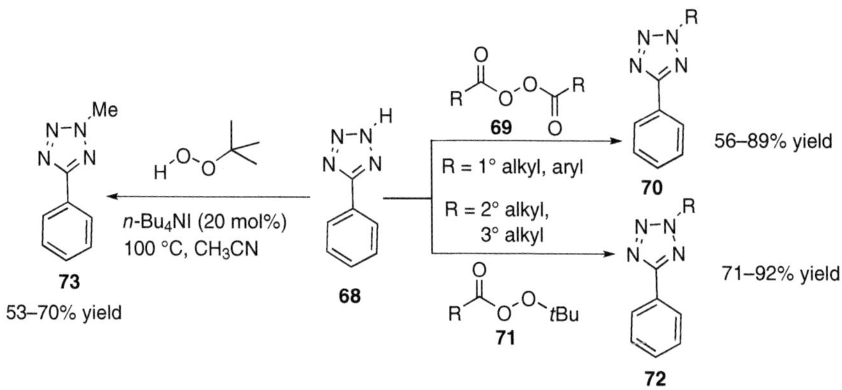

Scheme 4.14 *n*-Bu_4NI-catalyzed synthesis of 2-alkylated tetrazoles from tetrazoles and peroxides.

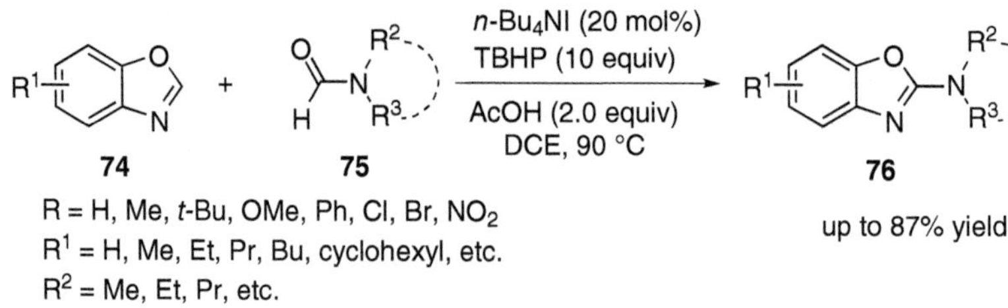

Scheme 4.15 *n*-Bu_4NI-catalyzed synthesis of aminonitrile 2-aminobenzoxazole derivatives from formamides and benzoxazoles.

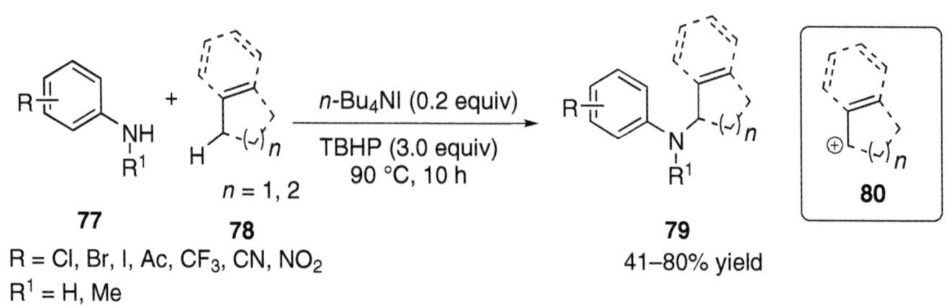

Scheme 4.16 *n*-Bu_4NI-catalyzed synthesis of *N*-substituted anilines from anilines and methylarenes.

the iodination and decarboxylation of 2-cyanoacetic acid. Then the reaction of 2-iodoacetonitrile with amines triggered the nucleophilic substitution to give compound **83** (Scheme 4.17).

The synthesis of 3,2′-pyrrolidinyl spirooxindoles **86** was achieved by Liu and Cheng [24] *via* *n*-Bu_4NI/H_2O_2-mediated intramolecular oxidative amination of 3-aminoalkyl-2-oxindoles **85** (Scheme 4.18). This transformation process was started with the formation of active iodine species IO^-/IO_2^- **44**/**45** under the interaction of H_2O_2 and *n*-Bu_4NI. Subsequently, the active iodine species reacted

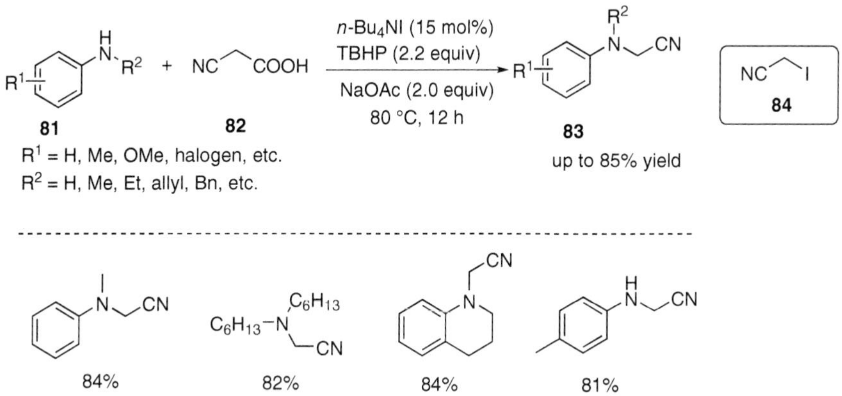

Scheme 4.17 *n*-Bu_4NI-catalyzed synthesis of aminonitrile from cyanoacetic acids and amines.

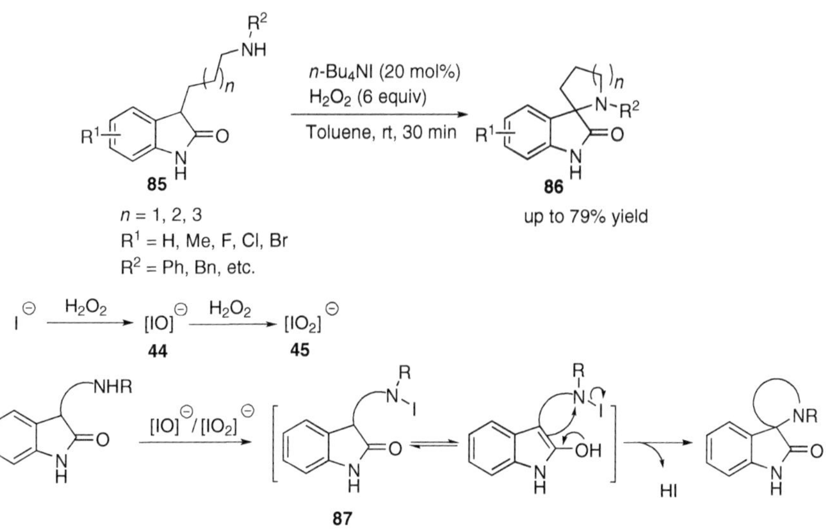

Scheme 4.18 *n*-Bu_4NI-catalyzed synthesis of 3,2′-pyrrolidinyln spirooxindoles from 3-aminoalkyl-2-oxindoles.

with aminoalkyl-2-oxindole to form the iodoamino intermediate **87**. Then the enolate form of iodoamino intermediate underwent the intramolecular cyclization to afford the title compound **86**.

In 2016, Prabhu and colleagues [25] described an azidation reaction of 1,3-dicarbonyl compounds **88** for furnishing tertiary azides **89** by using *n*-Bu_4NI as a catalyst and TBHP as an oxidant. Intermediate α-iodo adduct **90** could be obtained through the reaction of 1,3-dicarbonyl compound and hypervalent iodine species, which was generated by *n*-Bu_4NI and TBHP *in situ* (Scheme 4.19). Later, Waser and coworkers [26] disclosed room-temperature oxidative azidation protocols using hydrogen peroxide as oxidant and achiral bifunctional ammonium iodide as the

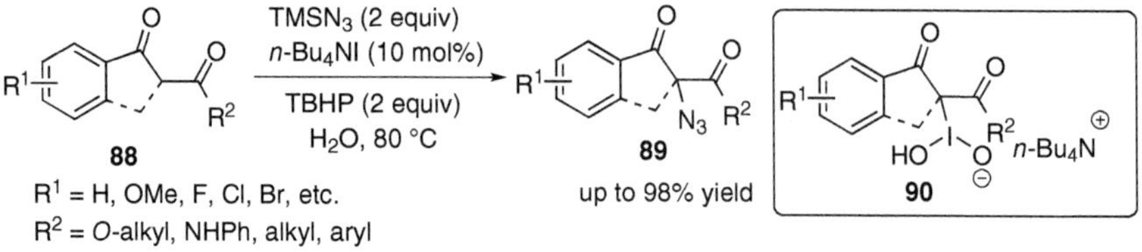

Scheme 4.19 n-Bu_4NI-catalyzed synthesis of tertiary azides from 1,3-dicarbonyl compounds and azide source.

catalyst. Both these two methods required $TMSN_3$ as an azide source. Most recently, Ishihara [27] reported an efficient protocol for the synthesis of 1,3-dicarbonyl compounds *via* enantioselective oxidative α-azidation of carbonyl compounds by using n-Bu_4NI as catalyst and H_2O_2 as oxidant. It is worth mentioning that in this transformation, the much cheaper and more commercially available NaN_3 acts as an azide source.

In 2013, Minakata et al. [28] found that organic trivalent iodine oxidants (PhI = NTs) **92** could react with iodine to form active species **94**, which could easily react with styrene compounds **91**, leading to aziridines **93** in medium yield (Scheme 4.20). This transformation was catalyzed by n-Bu_4NI_3 **95** which was generated *in situ* by n-Bu_4NI and I_2. Triiodide (I_3^-) **95** could be activated by the visible light to give iodine radical ($I^\bullet$) and iodine radical anion ($I_2^{-\bullet}$). Iodine radical anion ($I_2^{-\bullet}$) could promote the homolytic cleavage of the N–I bond of **94** to form the

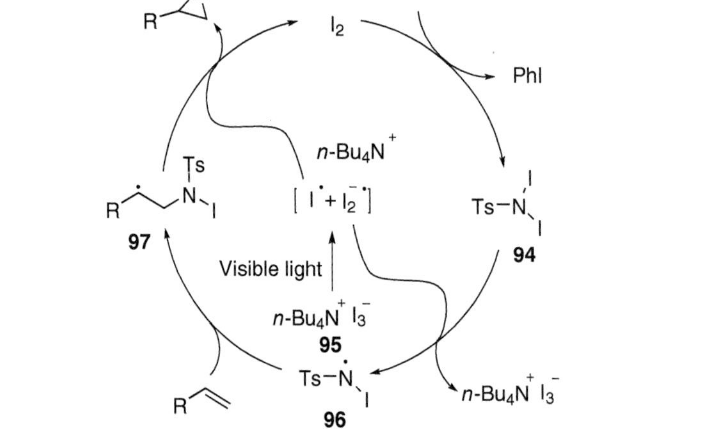

Scheme 4.20 n-Bu_4NI-catalyzed synthesis of aziridines from styrenes.

amidyl radical **96**. Then the amidyl radical **96** was trapped by the alkene affording the alkyl radical **97**, which could convert to aziridine *via* the intramolecular cyclization. Later, a similar *n*-Bu_4NI mediate electrocatalytic aziridination of alkenes **99** with *N*-aminophthalimide **98** were reported by Little and Zeng [29]. Through this protocol, various alkenes could be tolerated, and aziridines **100** were obtained in good yield (Scheme 4.21).

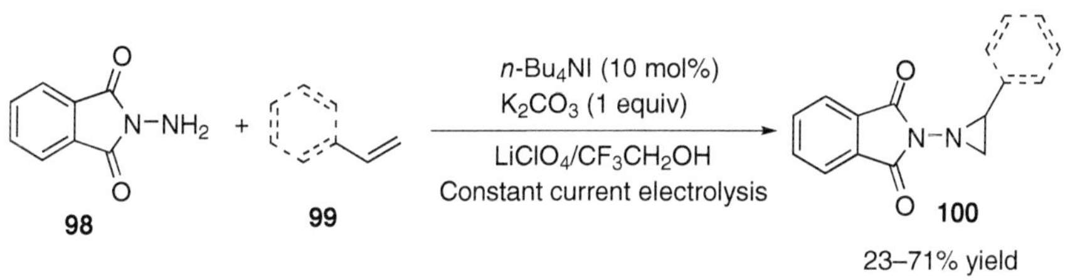

Scheme 4.21 *n*-Bu_4NI-catalyzed synthesis of aziridines from alkenes and *N*-aminophthalimide.

The *n*-Bu_4NI/TBHP-catalyzed strategy for synthesizing azaspirocyclohexadienones **102** *via* the spirocyclization of (*N*-aryl-carbamoyl)-2-iminoacetate compounds **101** had been disclosed by Yu and coworkers [30] (Scheme 4.22). This protocol underwent the dehydrogenation of **101** mediated by the co-effect of *n*-Bu_4NI and TBHP to furnish the key intermediate **103** which could convert to the corresponding spiro product in the presence of oxygen.

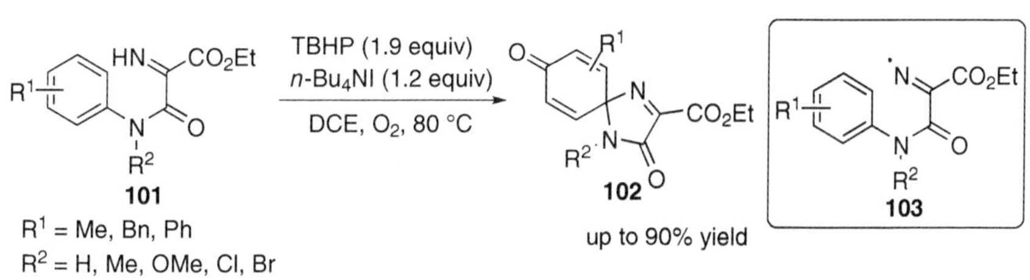

Scheme 4.22 *n*-Bu_4NI-catalyzed synthesis of azaspirocyclohexadienones from (*N*-aryl-carbamoyl)-2-iminoacetates.

In 2010, Fan and colleagues [31] reported a PhIO/*n*-Bu_4NI-mediated oxidative cyclization protocol for synthesizing highly functionalized azetidines **105** by employing amides **104** as substrates (Scheme 4.23). A new iodine(III) species **106** was generated through depolymerization of iodosobenzene and *n*-Bu_4NI, which has a higher reactivity in the ligand exchange reaction due to its basicity and the good leaving ability of iodide. By using this active hypervalent iodine reagent, professor Fan and coworkers [32] also realized the synthesis of *N*-benzoyl aziridines or oxazolines by employing the amidoalkylation adducts of activated methylene compounds as substrates. Later, a PIDA/*n*-Bu_4NI catalyzed strategy for constructing oxa-aza spirobicycles *via* a tandem carbon-hydrogen bond oxidation was developed by their group [33].

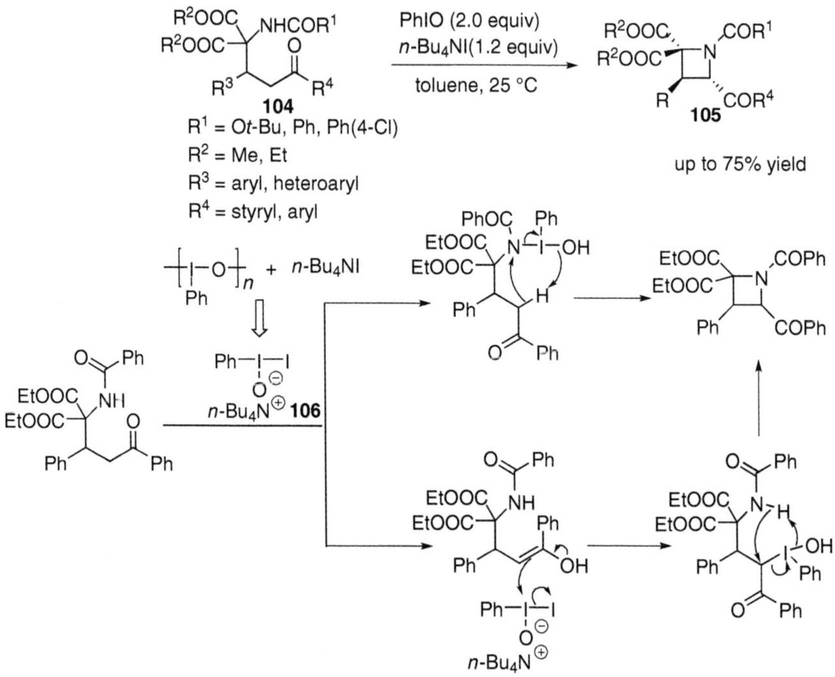

Scheme 4.23 n-Bu_4NI-catalyzed synthesis of functionalized azetidines from amides.

4.3 The Construction of C–O Bond

4.3.1 The C–O Bond Formation *via* Alkylarenes

n-Bu_4NI has been utilized as an efficient catalyst for the C–O bond formation between alkylarenes with carboxylic acids, aromatic aldehydes, alcohols, allyl alcohol, alkylbenzenes, *N*-hydroxyphthalimide (NHPI), and phosphorus nucleophiles.

In early 2012, Zhang and Yu [34] disclosed an n-Bu_4NI/TBHP-mediated oxidative esterification of benzyl C–H bonds **107** with various carboxylic acids **108** for the synthesis of benzyl esters **109**. This method was also applicable for the *O*-protection of *N*-Boc amino acids (Scheme 4.24a). It was postulated in the mechanistic pathway that n-Bu_4NI was oxidized by TBHP to form the active iodine species ($[n\text{-}Bu_4N]^+[IO]^-$ **44** or $[n\text{-}Bu_4N]^+[IO_2]^-$ **45**), which induced the hemolytic cleavage of benzyl C–H bond to give the benzyl radical. Subsequently, the benzyl radical was oxidized to benzyl cation mediated by hypoiodite species. In this redox process, the excess oxygen atom of hypoiodite captured the proton of benzoic acid to give the benzoate anion. Finally, benzoate anion coupled with benzyl cation to furnish ester product **109**. Later, Wang and coworkers reported an n-Bu_4NI-catalyzed benzylic Csp^3–H acyloxylation of alkylarenes **107** with aromatic aldehydes **110** [35]. These esters **111** were obtained from the reaction of benzyl cations and *tert*-butyl peroxybenzoate (TBPB), which were generated *in situ* from the reaction of

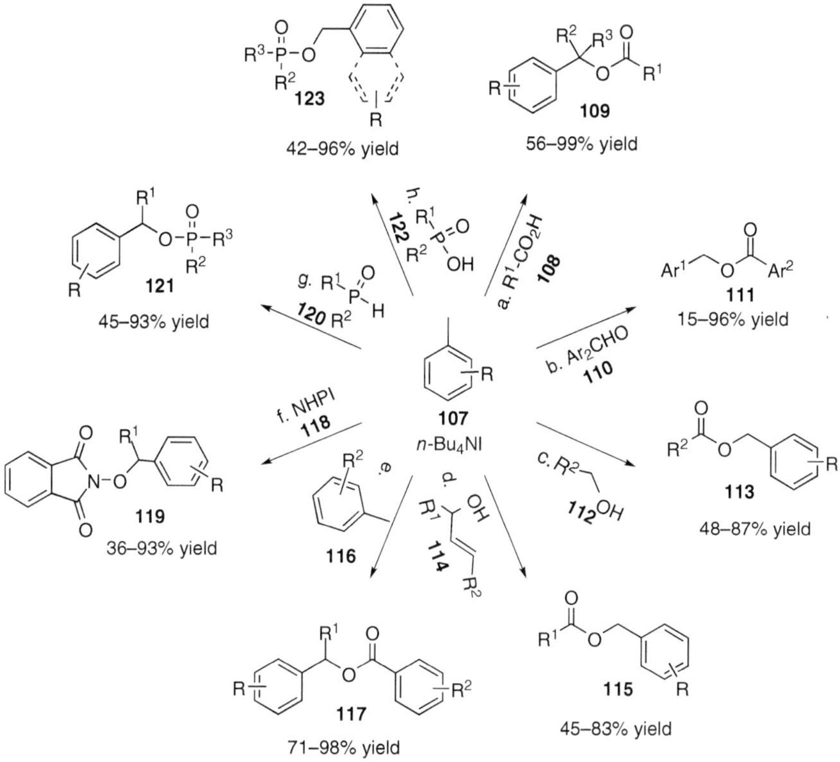

Scheme 4.24 *n*-Bu_4NI-catalyzed C–O bond formation *via* alkylarenes.

benzaldehydes and TBHP in the presence of *n*-Bu_4NI (Scheme 4.24b). Benzyl esters **113** could also be synthesized through *n*-Bu_4NI/TBHP-catalyzed oxidative coupling of alcohols **112** and toluene derivatives **107** [36]. The sequential oxidations of alcohols to aldehydes, carboxylic acids, and benzyl esters were accomplished in one pot (Scheme 4.24c). Later, Li and Jia [37] developed an *n*-Bu_4NI/TBHP-mediated oxidative approach for synthesizing benzyl benzoates **115** from toluene derivatives **107** by employing allyl alcohol **114** as the substrates. This protocol represents a new transformation of allyl alcohol involving both C–C bond cleavage and C–O bond formation (Scheme 4.24d).

In 2013, an *n*-Bu_4NI-catalyzed protocol was developed by Patel and colleagues for the synthesis of benzylic esters **117** *via* a CDC between Ar–CO_2H oxidized from toluene derivatives **107** and benzylic carbocations generated from alkylbenzenes **116** [38] (Scheme 4.24e). Later, Lv and coworkers reported an *n*-Bu_4NI-catalyzed C–O bond construction for the synthesis of alkyloxyamines **119** from the reaction of NHPI **118** and benzylic or allylic hydrocarbons **107** [39]. This transformation underwent a coupling reaction between benzyl radical and PINO radical to construct the alkyloxyamine skeleton (Scheme 4.24f).

An *n*-Bu_4NI/TBHP-catalyzed approach for the synthesis of phosphate esters **121** through Csp^3–H activation was established by Tang and coworkers [40]. In the presence of *n*-Bu_4NI/TBHP, various toluene derivatives **107** and phosphorus nucleophiles **120** were converted to the corresponding products in moderate-to-good yields (Scheme 4.24g). Phosphorus nucleophiles could be replaced by diaryl phosphinic acids **122** for furnishing organophosphorus compounds **123** *via* the dehydrogenative coupling reaction [41] (Scheme 4.24h).

Δ^2-Isoxazolines **125** could be synthesized through Selectfluor/*n*-Bu_4NI-catalyzed Csp^3–H oxidation of oximes **124** [42]. The control experiments indicated that in the presence of Selectfluor and *n*-Bu_4NI, the homolysis of the *in situ*-formed O–I bond generated an iminoxyl radical **126**, which facilitated subsequent 1,5-H transfer to give the *C*-centered radical intermediate **127**. Next, intermediate **127** was intercepted by the released iodine atom to give *β*-iodo oxime **128** or oxidized to form carbon cation **128′**, following nucleophilic substitution or addition to afford Δ^2-isoxazoline. (Scheme 4.25).

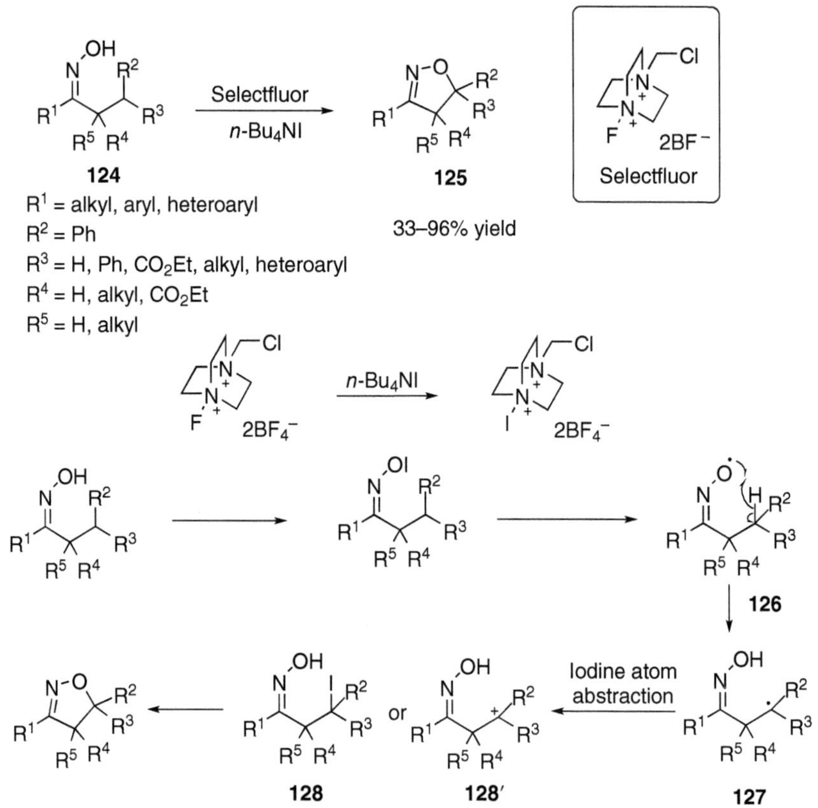

Scheme 4.25 *n*-Bu_4NI-catalyzed synthesis of Δ^2-isoxazolines from oximes.

4.3.2 Synthesis of *α*-Acyloxy Ketones

α-Acyloxy ketones and their derivatives are vital synthetic intermediates. Due to their biological importance, new methods for their synthesis are being continuously developed and documented. The useful α-acyloxy ketone-building blocks have been readily accessed by the n-Bu_4NI-catalyzed oxidative C–O bond formation.

α-Acyloxy ketones **131** could be obtained through an n-Bu_4NI/TBHP-mediated CDC reaction between ketones **129** and carboxylic acids **130** [43]. It has been well documented that benzylic alcohols [44], aromatic aldehydes [45], aryl alkenes [46], acylperoxides [47], oxo-carboxylic acids [48] and toluene derivatives [49] can be used as "acyloxy" sources for acyloxylation reaction to synthesize α-acyloxy ketones. Taking toluene derivatives as an example, the reaction is postulated to undergo this pathway: propiophenone was first oxidized to α-carbonyl radical **132** by TBHP and n-Bu_4NI. Simultaneously, toluene was oxidized by TBHP/n-Bu_4NI to perester **134** *via* acyl radical intermediate **133**. Coupling reaction between radical **132** and perester **134** led to the formation of title compound **131**. An alternative pathway involved the n-Bu_4NI/TBHP-mediated formation of the reactive cation intermediate **135** from radical **132**, and TBHP-mediated formation of carboxyl anion **136** from radical intermediate **133**. Finally, nucleophilic addition of carboxyl anion **136** to cation intermediate **135** furnished the corresponding product (Scheme 4.26).

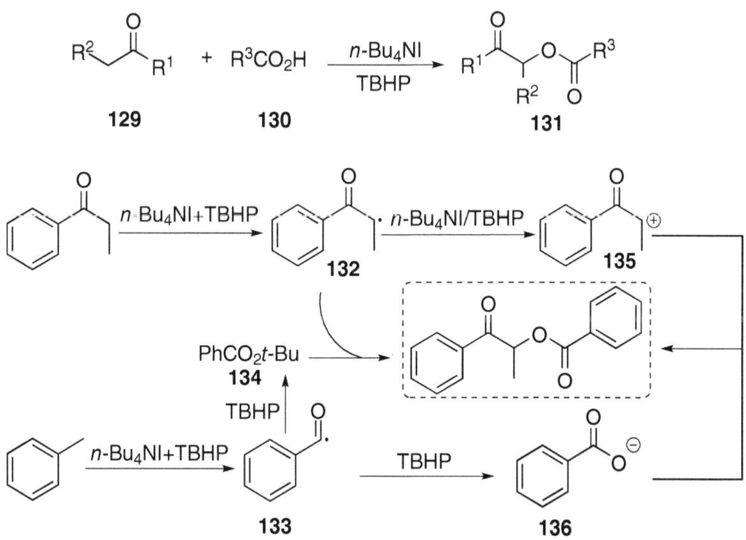

Scheme 4.26 n-Bu_4NI-catalyzed synthesis of α-acyloxy ketones from ketones and carboxylic acids.

In 2017, Xu and Zhang [50] developed an intramolecular electrocatalytic dehydrogenative esterification of aliphatic carboxylic acids **137** for the synthesis of lactones **138**. In this transformation, n-Bu_4NI had been proved to play an important catalytic role (Scheme 4.27).

In addition to the direct α-oxyacylation of ketones, the CDC reaction can also introduce the peroxy and aminoxyl groups into the α site of esters and ketones.

Synthesis of peroxides **140** could be achieved by the n-Bu_4NI/TBHP-catalyzed Csp^3–H activation of malonates, β-ketoesters, and cyanoacetic esters **139** [51]. In

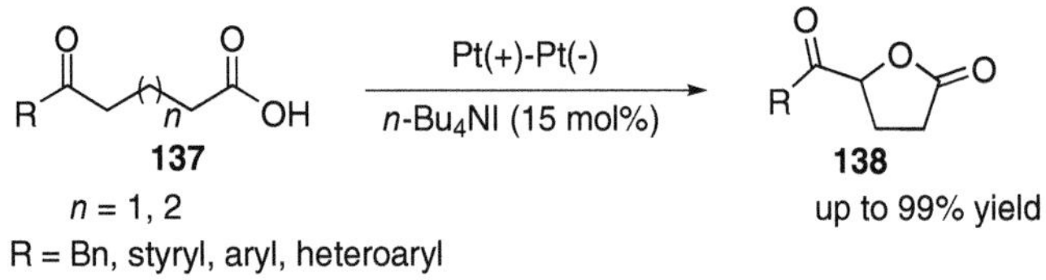

Scheme 4.27 *n*-Bu_4NI-catalyzed synthesis of lactones from aliphatic carboxylic acids.

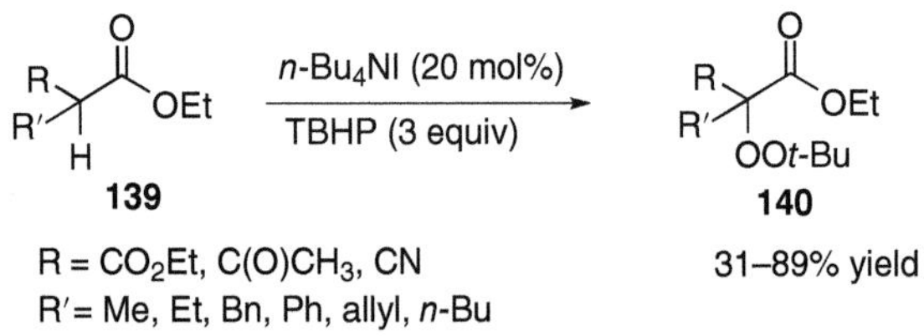

Scheme 4.28 *n*-Bu_4NI-catalyzed synthesis of peroxides from esters.

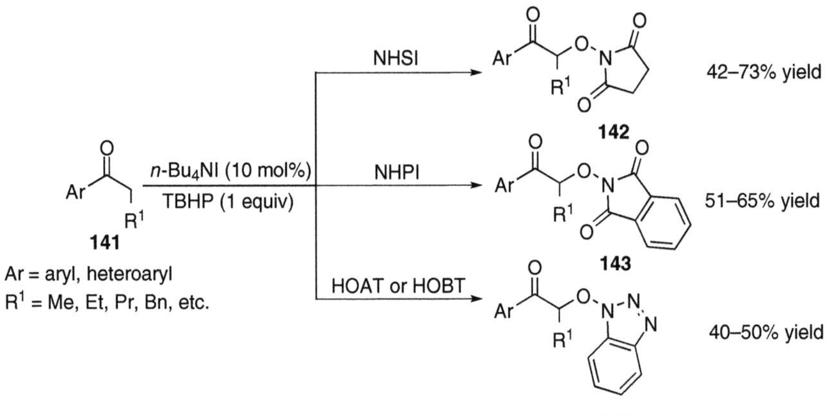

Scheme 4.29 *n*-Bu_4NI-catalyzed α-aminoxylation of ketones from ketones and *N*-hydroxyimidates.

the reaction process, TBHP plays dual roles by acting as both the oxidant and the O-reagent for the C–O coupling to occur (Scheme 4.28).

In 2015, Prabhu and coworkers [52] accomplished the *n*-Bu_4NI-catalyzed α-aminoxylation of ketones **141**, using TBHP as the oxidant. The CDC reactions of ketones with *N*-hydroxyimidates including NHSI, NHPI, *N*-hydroxybenzotriazole (HOBT) and 1-hydroxy-7-azabenzotriazole (HOAT), *via* the radial-coupling strategy, afforded the corresponding oxygenated products **142**/**143**/**144** in good-to-moderate yield (Scheme 4.29).

4.3.3 Synthesis of α-Acyloxy Ethers

α-Acyloxy ethers play an important role as key building blocks in organic synthesis. Synthesis of α-acyloxy ethers *via* an *n*-Bu_4NI/TBHP-mediated acyloxylation of ethers by carboxylic acids through a cross-dehydrogenative-coupling has been efficiently developed.

In 2011, Wan and coworkers [53] developed an n-Bu_4NI/TBHP-mediated cross-dehydrogenative-coupling reaction for synthesizing a variety of α-acyloxy ethers **147** from carboxylic acid **145** and ethers **146** (Scheme 4.30). The mechanism showed that *tert*-butoxyl radical and OH^- were first generated *in situ* from TBHP and n-Bu_4NI. Then the proton of carboxylic acid was captured by OH^-, which gave the anionic species **148**. Next, the C–H bond adjacent to an oxygen atom in ether underwent hemolytic cleavage to furnish radical intermediate **149**, which was converted to oxonium ion **150** *via* iodine-mediated oxidation. Finally, nucleophilic addition of carboxylate **148** to the electrophilic oxonium ion **150** gave the desired product **147**. The I_2/I^- redox process plays a key role in this transformation, by promoting the reductive cleavage of the O–O bond in TBHP and in the oxidation of the carbon radical to oxonium ion.

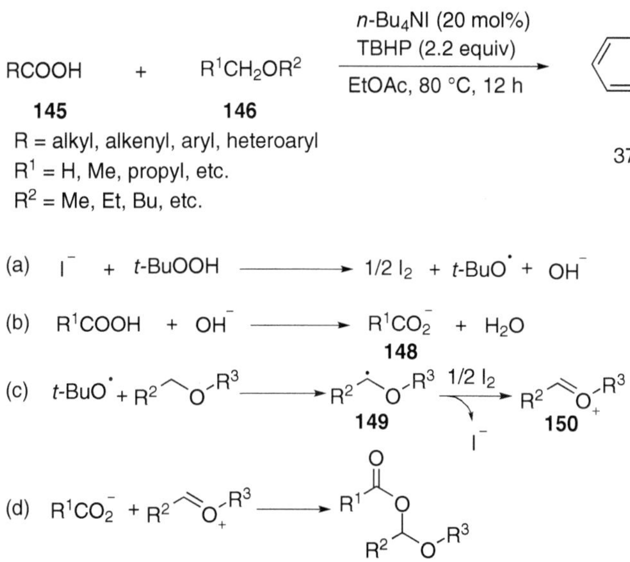

Scheme 4.30 n-Bu_4NI-catalyzed synthesis of α-acyloxy ethers from carboxylic acid and ethers.

A metal-free decarboxylative acyloxylation of a Csp^3–H bond in formamides **152** and ethers **154** with α-oxocarboxylic acids **151** for the synthesis of N-acyloxymethylamides **153** and α-acyloxy ethers **155** has been established [54] (Scheme 4.31). The mechanistic process of this reaction was similar to the cross-dehydrogenative coupling of ethers with benzoic acids. Subsequent exploration proved that the acyl oxidation of ethers also can be achieved through aldehyde [45] and alcohols [55].

4.3.4 Synthesis of Esters and Ethers

n-Bu_4NI has been utilized as an efficient catalyst for the C–O bond formation for the synthesis of esters and ethers from alkenes, aldehydes, alcohol, and epoxides.

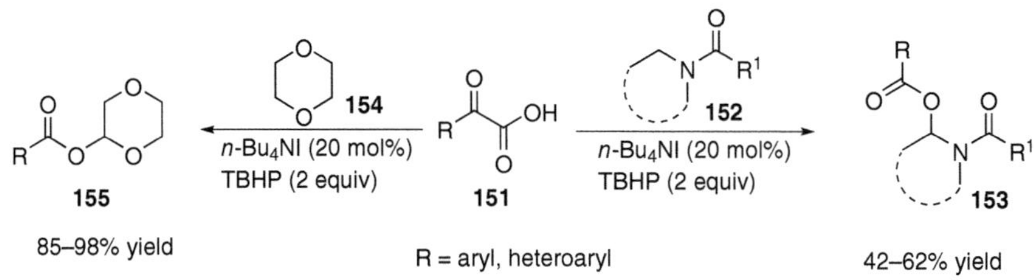

Scheme 4.31 *n*-Bu_4NI-catalyzed synthesis of *N*-acyloxymethylamides and *α*-acyloxy ethers from ethers and *α*-oxocarboxylic acids.

In 2013, Zhu and coworkers [56] developed an *n*-Bu_4NI-catalyzed synthetic approach for the synthesis of dioxygenated products **158** from carboxylic acids **157** and alkenes **156**. It is worthy to mention that carboxylic acids can be replaced by amines to furnish oxyamidation products (Scheme 4.32). Through the oxidation of *tert*-butoxyl or *tert*-butylperoxy radical generated *in situ* from *n*-Bu_4NI and TBHP, carboxylic acids **157** were converted to radial intermediates **159**. Then intermediate **159** reacted with alkenes **156** to furnish the carbon-centered radicals **160**, which were transformed into benzyl cations **161** *via* the oxidation of TBHP/*n*-Bu_4NI. Finally, the benzyl cation **161** reacted with water to form the desired product.

A coupling reaction between 2-vinylphenols **162** and carboxylic acids **163** for synthesizing 3-acyloxy-2,3-dihydrobenzofurans **164** by utilizing Bu_4NI as catalyst and TBHP as oxidant had been realized by Wan et al. [57] (Scheme 4.33). The mechanistic process indicated that the reaction between *n*-Bu_4NI and TBHP

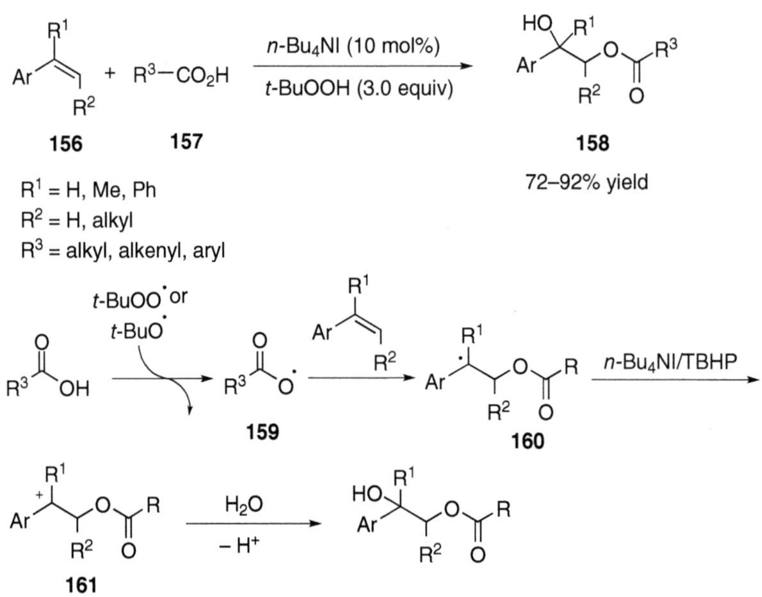

Scheme 4.32 *n*-Bu_4NI-catalyzed synthesis of deoxygenation products from carboxylic acids and alkenes.

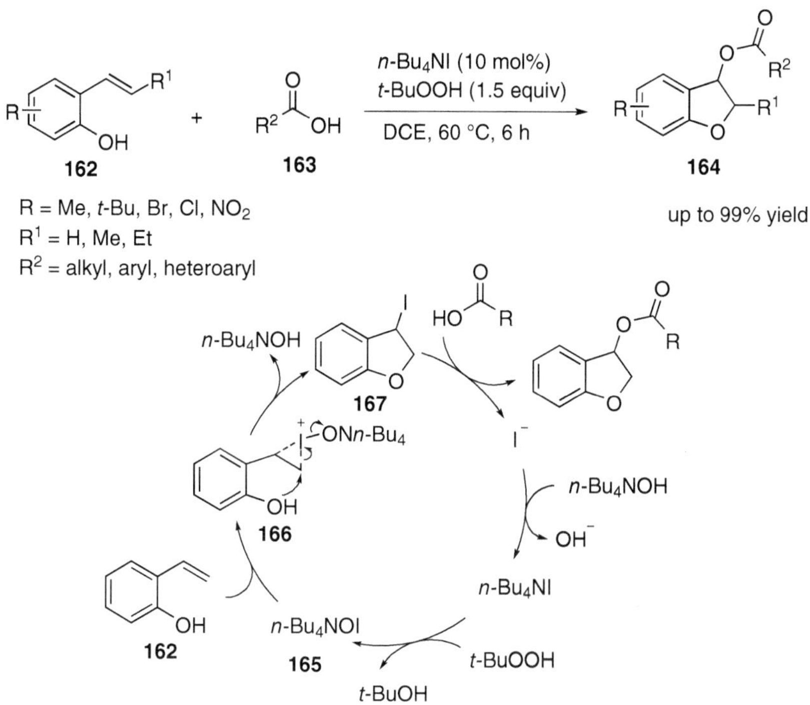

Scheme 4.33 *n*-Bu_4NI-catalyzed synthesis of 3-acyloxy-2,3-dihydrobenzofurans from 2-vinylphenols and carboxylic acids.

could afford the active species (*n*-Bu_4NOI) **165**, which enable the conversion of 2-vinylphenols to iodonium ion **166**. Finally, intramolecular cyclization of **166** gave the 3-iodo-2,3-dihydrobenzofuran **167**, which underwent nucleophilic substitution by carboxylic acid to furnish the title 3-acyloxy-2,3-dihydrobenzofurans.

tert-Butyl peresters **170** can be synthesized directly from aldehydes [58] and alcohols [59] **168** through an *n*-Bu_4NI/TBHP-catalyzed radical cross-coupling reaction. Furthermore, allylic esters **169** could also be obtained by combing this method with the Kharasch–Sosnovsky reaction *via* a two-step one-pot procedure (Scheme 4.34). Catalyzed by *n*-Bu_4NI/TBHP oxidative system, Wan and coworkers also realized the synthesis of allylic esters *via* C–H oxidation of cyclohexene with a variety of carboxylic acids [60].

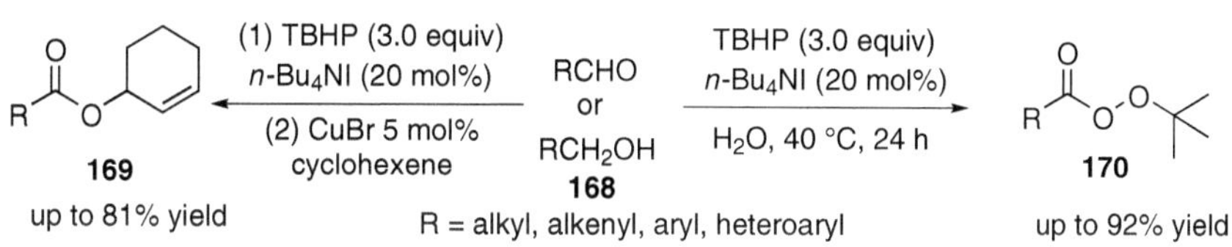

Scheme 4.34 *n*-Bu_4NI-catalyzed synthesis of *tert*-butyl peresters and allylic esters from aldehydes or alcohols.

In 2016, using TBHP as an oxidant and *n*-Bu_4NI as a catalyst, Fan and Zhang reported oxidative esterification of aldehydes **171** with alkyl halides **172** [61]. The proposed mechanism is postulated as follows: Initially, *tert*-butoxyl radical **4** and *tert*-butylperoxy radical **5** were generated from the interaction of *t*-BuOOH and *n*-Bu_4NI. Then, the hydrogen of aldehyde was abstracted by radical **4**, leading to an acyl radical **174**. The following radical coupling of **174** with **5** affords a perester intermediate **175**. Subsequently, intermediate **175** reacted with *t*-BuOOH in the presence of *n*-Bu_4NI to generate acid intermediate **176**, which could be immediately converted to the corresponding carboxylate anion **177**. Title compound **173** could be efficiently obtained through a nucleophilic substitution reaction between carboxylate **177** and alkyl bromide **172**. Notably, this esterification protocol is suitable for both aromatic and aliphatic aldehydes, and primary and secondary alkyl halides to give esters with broad structural diversity (Scheme 4.35).

Using *n*-Bu_4NI as the catalyst TBHP as terminal oxidant, Bhar and colleagues [62] discovered an efficient method for synthesis of aryl esters **179** from benzylic primary

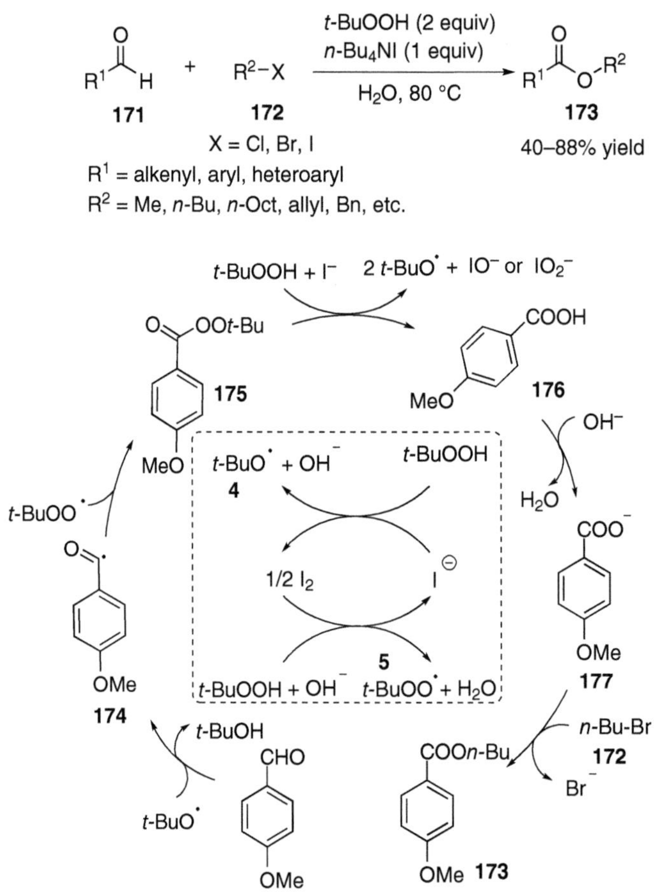

Scheme 4.35 *n*-Bu_4NI-catalyzed synthesis of esters from aldehydes and alkyl halides.

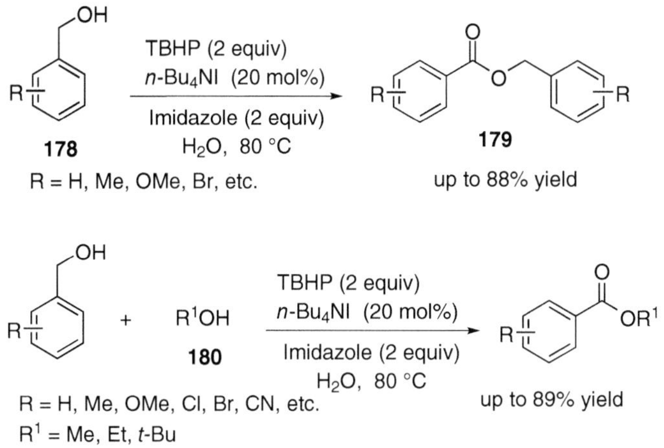

Scheme 4.36 *n*-Bu_4NI-catalyzed synthesis of aryl esters from benzylic primary alcohols and alkyl alcohols.

alcohols **178** and alkyl alcohols **180**. These reactions are highly chemoselective with wide applicability accommodating a wide range of substituents (Scheme 4.36).

An *n*-Bu_4NI-mediated electrochemical approach for the synthesis of carboxylic esters **182** from *N*-alkoxyamides **181** had been reported by Bhanage's group [63]. In this protocol, *n*-Bu_4NI acts not only as the redox catalyst but also as the supporting electrolyte (Scheme 4.37). Various esters could be obtained *via* the rearrangement of homodimerized intermediate **183** generated from *N*-alkoxyamides. Later, the same group discovered a one-pot strategy for the synthesis of esters *via* the selective electrodimerization of the amide **184** using *n*-Bu_4NI as an electrocatalyst [64]. In the absence of current, the reaction proceeds further *via* Zn catalyzed C–N bond activation of the amide dimer **183** followed by its coupling with phenol to form ester **186**.

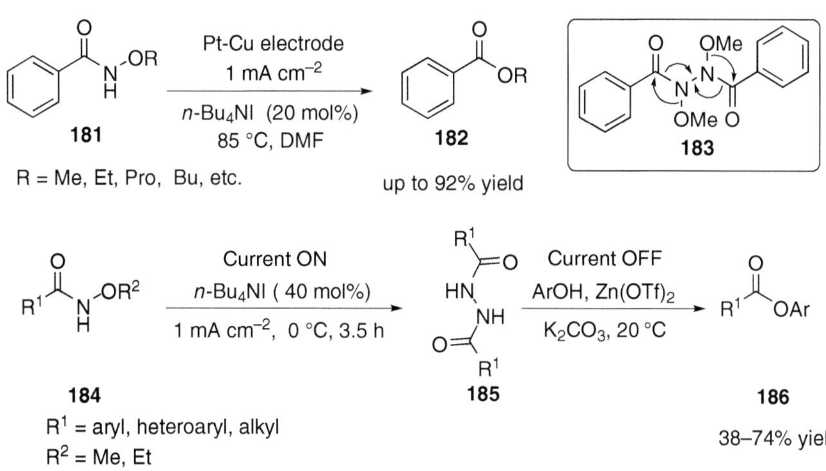

Scheme 4.37 *n*-Bu_4NI-catalyzed synthesis of carboxylic esters from *N*-alkoxyamides.

Quaternary ammonium (hypo)iodites also can be used in enantioselective oxidative reactions. In 2010, Ishihara and coworkers [65] reported a highly efficient chiral quaternary ammonium iodide **188**-catalyzed cross-coupling reaction for synthesizing of 2-acyl-2,3-dihydrobenzofuran derivatives **189**. The reaction proceeded through enantioselective oxidative cycloetherification, by employing the ketophenols **187** as the starting materials (Scheme 4.38).

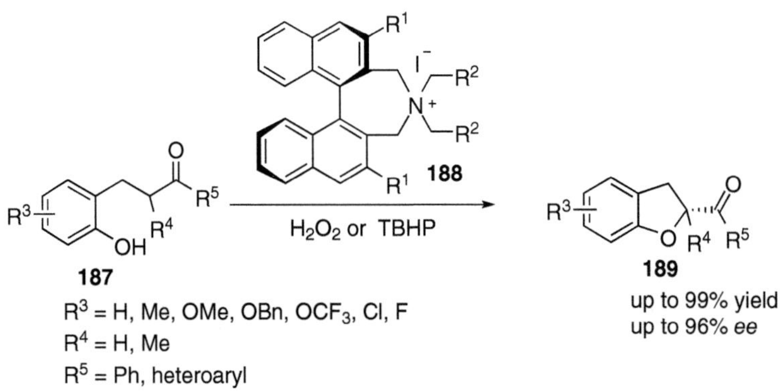

Scheme 4.38 *n*-Bu_4NI-catalyzed synthesis of 2-acyl-2,3-dihydrobenzofuran derivatives from ketophenols.

In addition to combining use with various oxidants for the construction of C–O bonds, *n*-Bu_4NI can also be used together with metal reagents to achieve the best catalytic effect.

Using $PhI(OAc)_2$ as the oxidant, the synthesis of diacetoxylated compounds **191** from the ring-opening reaction of methylenecyclopropanes **190** was realized by Shi and colleagues [66], through a trimeric palladium complex, which was generated *in situ* by $Pd(OAc)_2$ and *n*-Bu_4NI. This palladium-catalyzed process can provide an alternative approach to access functionalized diacetoxylation products through C–C bond cleavage and C–O bond formation using highly strained small rings as substrates (Scheme 4.39).

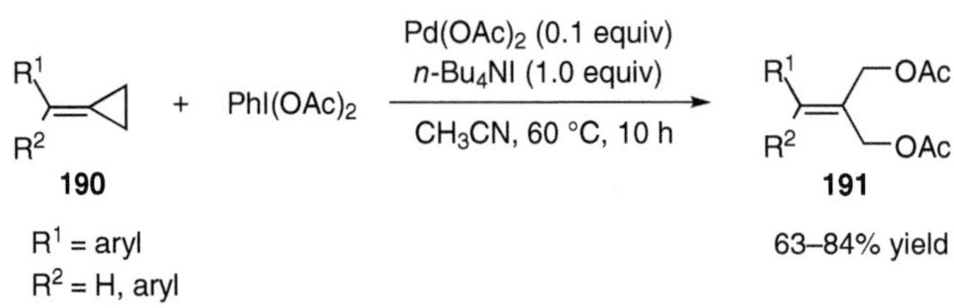

Scheme 4.39 *n*-Bu_4NI-catalyzed synthesis of functionalized diacetoxylation products from methylenecyclopropanes.

In 2012, Bai and coworkers [67] realized the synthesis of cyclic carbonates **193** from epoxides **192** and carbon dioxide catalyzed by cyanocobalamin and *n*-Bu_4NI at 140 °C without organic solvents (Scheme 4.40). The cyanocobalamin (VB_{12}) could be recycled with water, with the retention of moderate catalytic activity (Scheme 4.40).

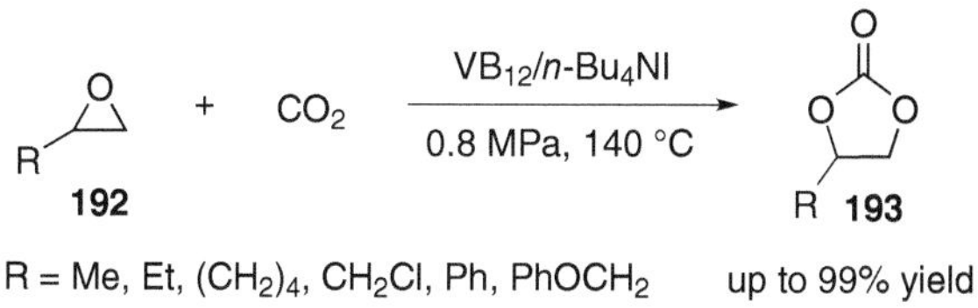

Scheme 4.40 *n*-Bu_4NI-catalyzed synthesis of cyclic carbonates from epoxides and carbon dioxide.

Later, various metal/*n*-Bu_4NI-catalytic systems including $ZnBr_2$/*n*-Bu_4NI [68], lanthanide complexes/*n*-Bu_4NI [69], and aluminum complex/*n*-Bu_4NI [70] were reported for synthesizing cyclic carbonates from epoxides *via* the insertion of CO_2. In all of these catalytic systems, *n*-Bu_4NI plays a role as a cocatalyst.

4.4 The Construction of C–C Bond

In addition to the construction of C–N and C–O bonds, *n*-Bu_4NI has also found application in the construction of C–C bond.

Promoted by TBHP/*n*-Bu_4NI oxidative system, Shia group [71] reported a radical cascade protocol for the synthesis of functionalized [6.6.5] tricyclic frameworks **195** from the R-cyano R-TMS/aryl-capped alkynyl aryl ketones **194**, in 2012 (Scheme 4.41). The sequential C–C bond formation was proposed to be initiated with the abstraction of the proton H-atom from substrate **194**, mediated by the *t*-BuO˙ or *t*-BuOO˙ generated *in situ* by TBHP and *n*-Bu_4NI to form the radical intermediate **196**. Through the *5-exo-dig* cyclization, radical intermediate **196** was transformed into vinyl radical intermediate **197**. Cyclic pentadienyl radical **198** was then obtained *via* the *6-endo-trig* intramolecular addition. Finally, rearomatization occurred *via* the C–H bond homolytic cleavage to give title product **195**.

An *n*-Bu_4NI-mediated oxidative method for synthesizing methylene-bridged bis-1,3-dicarbonyl compounds **201** by utilizing tertiary aliphatic amines **200** and 1,3-dicarbonyl compounds **199** has been reported by Wang and coworkers [72]. The *β*-carbon of disopropylethylamine (DIPEA) was transformed into the methylene carbon through an oxidative C–C bond cleavage (Scheme 4.42).

In 2014, Sun and labmates reported an Et_4NI-catalyzed method for synthesizing diaryl ketones **203** from the reaction of stilbenes **202** and $NaIO_4$. Mechanistically, stilbene **202** was first converted to epoxide **204** in the presence of Et_4NI and $NaIO_4$. Next, epoxide **204** underwent a ring-opening reaction to give intermediate **205**. The loss of iodide from **205** gave a carbocation intermediate **206**, which underwent rearrangement to provide aldehyde **207**. Subsequently, mediated by $NaIO_4$, aldehyde deformylation was occurred in aldehyde **207** to provide the desired diarylketone **203**, along with the release of formic acid (Scheme 4.43).

In 2015, Wu and Yan [73] reported an efficient *n*-Bu_4NI-catalyzed approach for the synthesis of 1,3-diketones **210** from aldehydes **208** and ketones **209**. Di-tert-butyl peroxide (DTBP) can promote homolysis of TBHP to generate hydroxyl radicals. Mediated by TBHP, *tert*-butylperoxyl radical **5** was first generated *via* the iodide/iodine

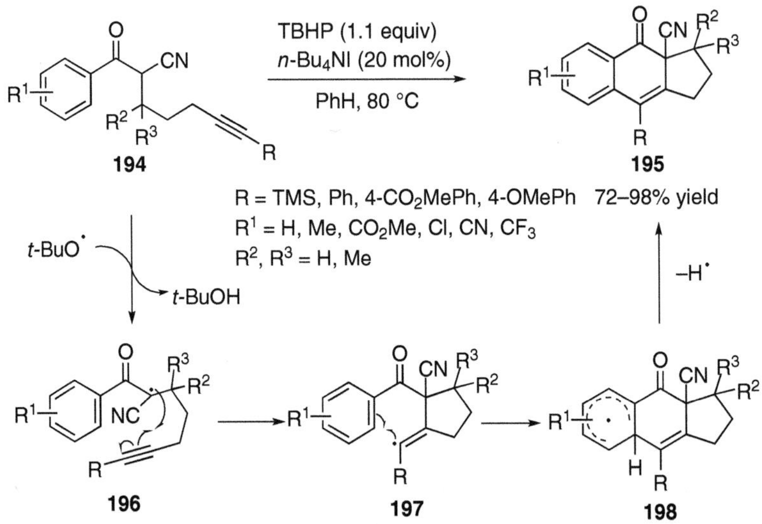

Scheme 4.41 *n*-Bu_4NI-catalyzed synthesis of functionalized [6.6.5] tricyclic frameworks from alkynyl aryl ketones.

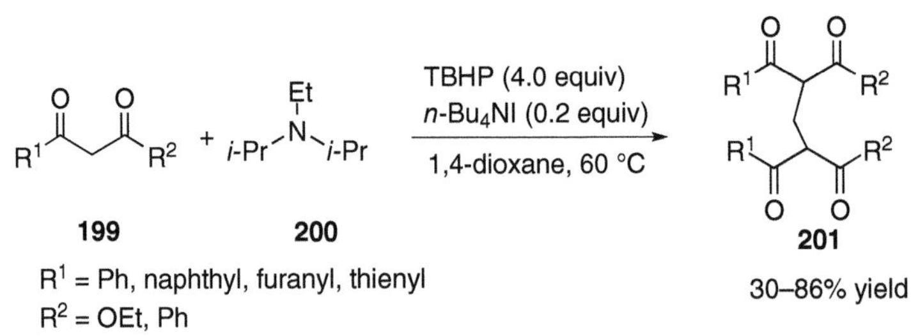

Scheme 4.42 *n*-Bu_4NI-catalyzed synthesis of methylene-bridged bis-1,3-dicarbonyl compounds from tertiary aliphatic amines and 1,3-dicarbonyl compounds.

catalytic redox process. Next, acyl radical **211** can be obtained through the abstraction of a hydrogen atom from the aldehyde. Under the collective effect of TBHP and *p*-TsOH, ketone radical **212**, which coupled with acyl radical **211** to afford the corresponding compound 1,3-diketone **210**. Although this pathway can tolerate various substituents on the aromatic ring of aldehydes, this approach is limited by the scope of ketones (Scheme 4.44).

Studer and colleagues [74] developed an *n*-Bu_4NI-mediated metal-free approach for the synthesis of phenanthridines **215** and quinoxalin-2-ones **213** through fluoroalkylation of vinyl azides **215**. This operational transformation employed *n*-Bu_4NI as an initiator, Togni reagent **214** as the CF_3 source (Scheme 4.45).

Using *n*-Bu_4NI as a redox catalyst and electrolyte under constant current electrolysis in an undivided cell, Xu and coworkers [75] realized an efficient electrochemical method for synthesis of enaminones **218** *via* a decarboxylative coupling reaction of α-keto acids **217** and vinyl azides **216**. A broad vinyl azide substrate scope and high functional group tolerance were observed (Scheme 4.46).

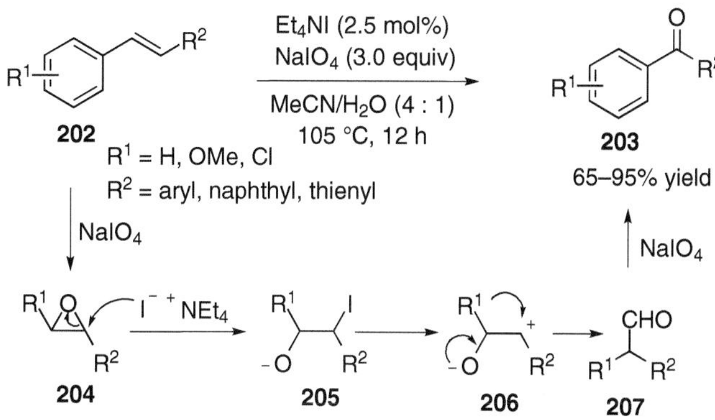

Scheme 4.43 *n*-Bu_4NI-catalyzed synthesis of diaryl ketones from stilbenes.

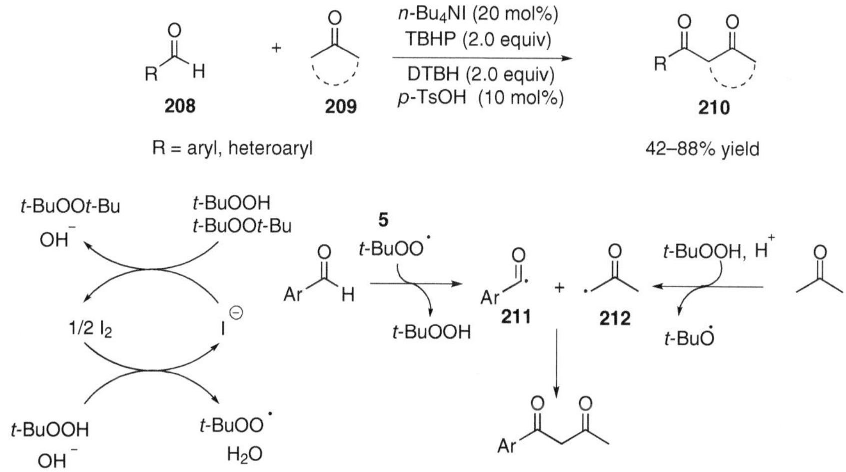

Scheme 4.44 *n*-Bu_4NI-catalyzed synthesis of 1,3-diketones from aldehydes and ketones.

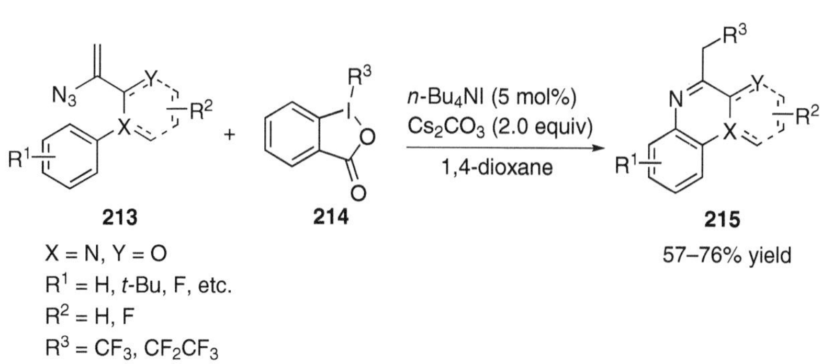

Scheme 4.45 *n*-Bu_4NI-catalyzed synthesis of phenanthridines and quinoxalin-2-ones from vinyl azides.

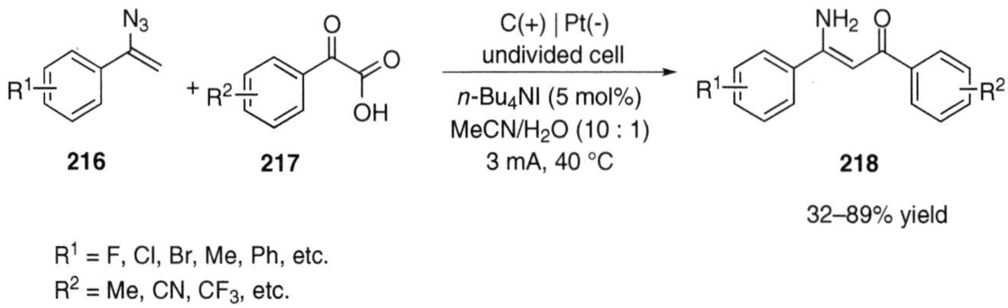

Scheme 4.46 *n*-Bu_4NI-catalyzed synthesis of enaminones from *α*-keto acids and vinyl azides.

Mediated by *n*-Bu_4NI and $PhI(OAc)_2$, Murphy and partners realized the efficient synthesis of cyclopropanes **221** through intermolecular cyclopropanation between iodonium ylides **219** and alkenes **220** [76]. This transformation underwent a radical-mediated reaction pathway, which requires the presence of both *n*-Bu_4NI and $PhI(OAc)_2$. First, $[I(OAc)_2]^-$ **222** transferred an electron to ylide **219**, generating a carbonyl-stabilized radical anion **223**. When treated with styrene, the benzylic radical intermediate **224** was generated. Subsequently, single-electron oxidation occurred in intermediate **224** to furnish **225**. After further radical recombination and reductive elimination, the title cyclopropane **221** was obtained (Scheme 4.47).

In 2017, Boens and coworkers [77] reported an *n*-Bu_4NI/*t*-BuOOH-catalyzed alkylation reaction of BODIPYs **226** with allylic alkenes **227** for furnishing *α*-alkylated BODIPYs **228** (Scheme 4.48a). The cyclohexenyl allylic radical **229** generated through the reaction between active hypoiodite $[n\text{-}Bu_4N]^+[IO]^-$ or iodite $[n\text{-}Bu_4N]^+[IO_2]^-$ species and allylic alkenes is the key reactive species for this regioselective transformation. Later, an *n*-Bu_4NI-catalyzed regioselective Csp^2–Csp^3 cross-dehydrogenative coupling was realized by Chaudhary and coworkers [78] for synthesizing the C-3 benzylated 2*H*-indazoles **232** from 2*H*-indazoles **230** and methylated arenes **231**. This protocol exhibits good functional group tolerance, and various 2-substituted-2*H*-indazoles could be benzylated with methylated arenes to furnish title products in high yield (Scheme 4.48b).

Zaki et al. [79] reported a TBHP/an *n*-Bu_4NI-mediated protocol for the synthesis of 4*H*-chromen-4-ones **234** from chroman-4-ones **233** *via* oxidative C–C bond formation. In the presence of TBHP and *n*-Bu_4NI, the *tert*-butoxy radicals were initially generated. The desired product was further obtained through oxidative single-electron transformation and elimination (Scheme 4.49).

4.5 The Construction of C–S Bond

In addition to the C–O bond formation, *n*-Bu_4NI has also been demonstrated to be an efficient catalyst for the construction of C–S bond, leading to the synthesis of a series of S–containing building blocks such as allylic sulfones, *β*-carbonyl sulfones, 3-sulfenylindoles, and difluoromethylthiolated aromatics.

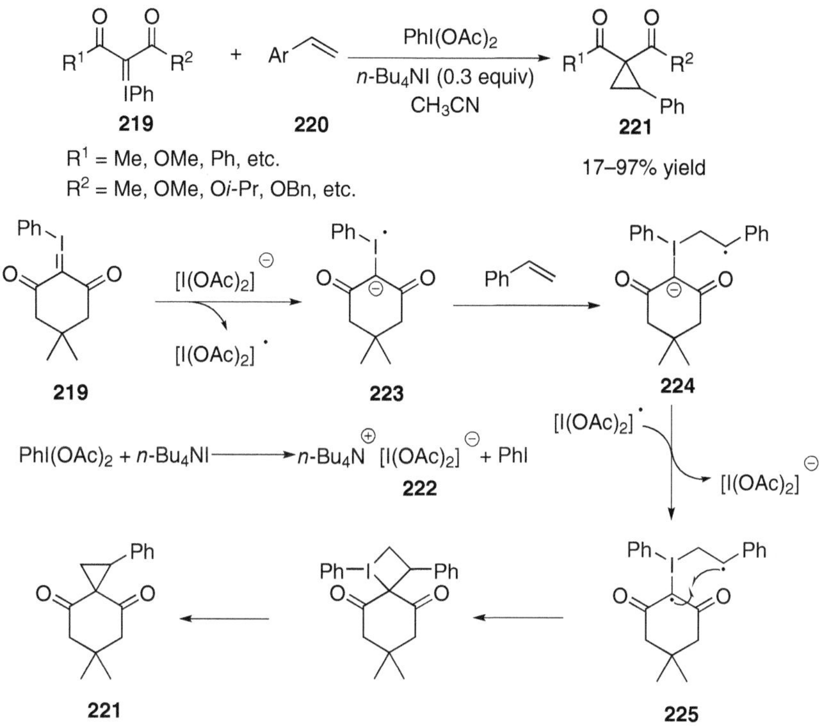

Scheme 4.47 *n*-Bu$_4$NI-catalyzed synthesis of cyclopropanes from iodonium ylides and alkenes.

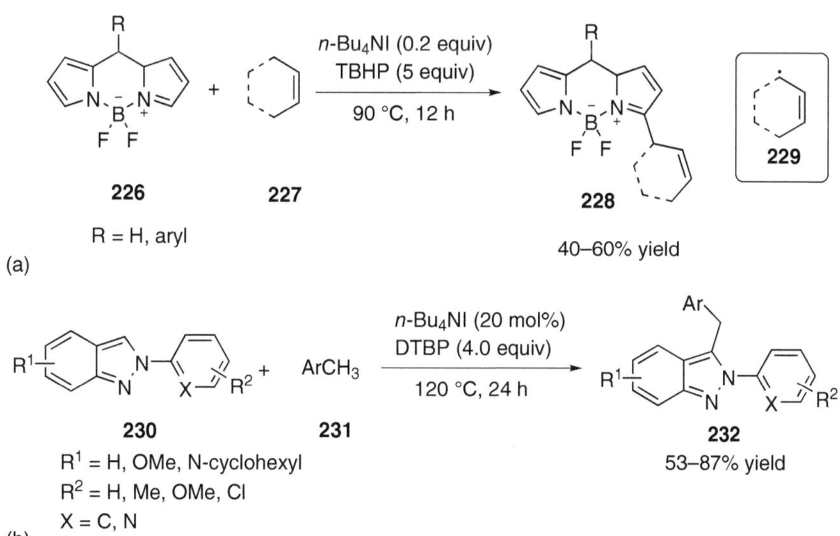

Scheme 4.48 *n*-Bu$_4$NI-catalyzed synthesis of alkylated BODIPYs from BODIPYs and allylic alkenes.

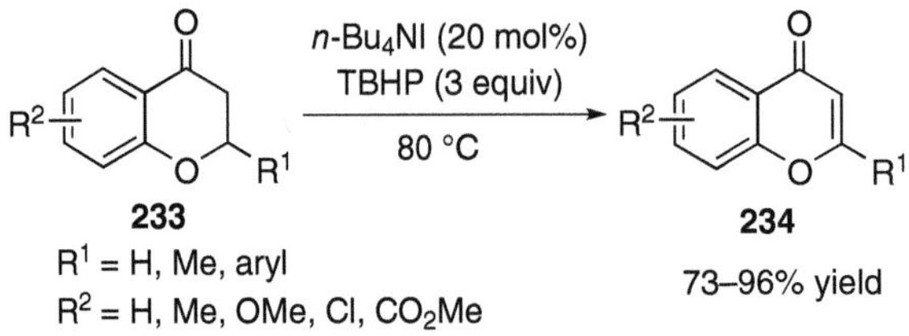

Scheme 4.49 *n*-Bu_4NI-catalyzed synthesis of 4*H*-chromen-4-ones from chroman-4-ones.

In 2012, by using an *n*-Bu_4NI/TBHP catalytic system, Li and coworkers [80] realized the synthesis of allylic sulfones **237** *via* a radical process from sulfonyl hydrazides **236** and *α*-methyl styrene derivatives **235** (Scheme 4.50). The key step for this transformation process is the generation of sulfonyl radicals **238**, which was realized through the oxidation of sulfonyl hydrazides by *tert*-butoxy radicals, with the release of molecular nitrogen. In 2019, Liu and Wei reported an *n*-Bu_4NI/TBHP-catalyzed radical reaction between styrenes and sulfonyl hydrazides, which also realized the synthesis of vinyl sulfones [81].

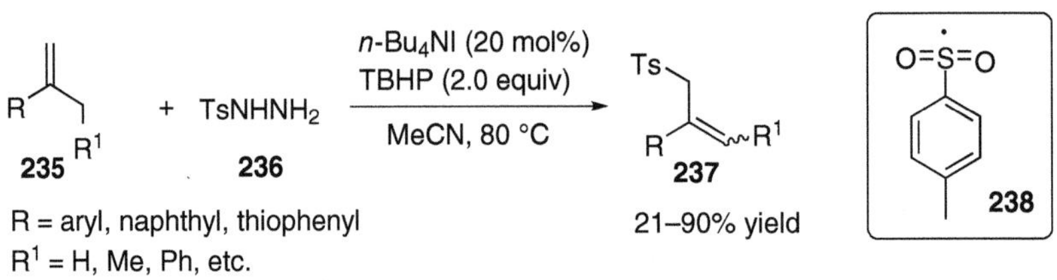

Scheme 4.50 *n*-Bu_4NI-catalyzed synthesis of allylic sulfones from sulfonyl hydrazides and *α*-methyl styrenes.

Wan and coworkers in 2016 described an *n*-Bu_4NI-mediated cross-coupling reaction between sulfonyl hydrazides **239** and diazo derivatives **240** for synthesizing *β*-carbonyl sulfones **241** by using molecular oxygen as the oxidant [82] (Scheme 4.51). The mechanistic process indicated that two molecules of nitrogen were released during this radical coupling strategy.

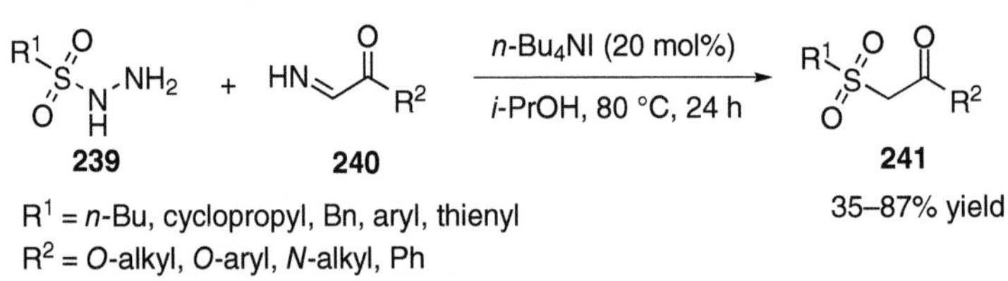

Scheme 4.51 *n*-Bu_4NI-catalyzed synthesis of *β*-carbonyl sulfones from sulfonyl hydrazides and diazo derivatives.

3-Sulfenylindoles **244** could be obtained through an *n*-Bu_4NI-mediated radical addition strategy from indole derivatives **242** and sulfinic acids **243** [83]. The reaction selectively afforded structurally diverse indole thioethers in good-to-excellent yields

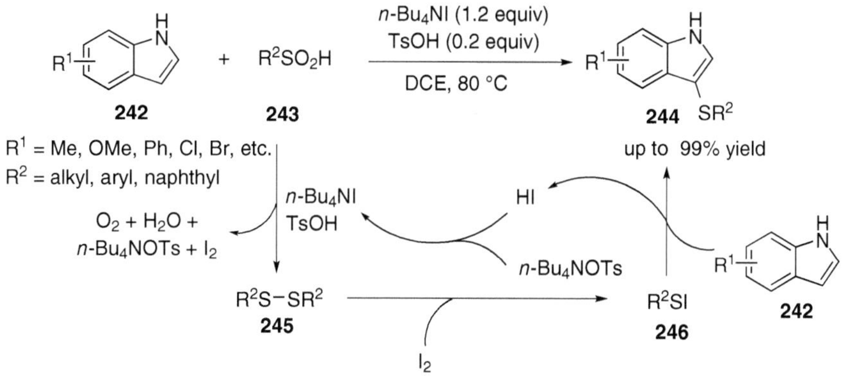

Scheme 4.52 *n*-Bu_4NI-catalyzed synthesis of 3-sulfenylindoles from indoles and sulfinic acids.

(Scheme 4.52). The key diphenyldisulfane **245** and iodine were generated through the reaction of sulfinic acid TsOH and *n*-Bu_4NI. Then disulfide **245** reacted with I_2 to give sulfenyl iodide **246**, which could be attacked by indole-furnishing thioether **246** and HI. Then byproduct HI reacted with sulfinic acid to give disulfide and iodine, which can promote the reaction continually.

Mediated by phosphine/*n*-Bu_4NI oxidative system, difluoromethylthiolation of indole derivatives and other electron-rich aromatics **247** with difluoromethanesulfonyl chloride **248** could be achieved for the synthesis of difluoromethylthiolated aromatics **249** [84]. It has been demonstrated that the presence of *n*-Bu_4NI could promote this transformation by generating iodine *in situ*. Initially, difluoromethanesulfonyl chloride **248** reacts with PPh_3 and iodide anion to form iodine, which reacted with PPh_3 to give iodotriphenylphosphonium iodide **250**. Then intermediate **250** reacted with difluoromethanesulfonyl chloride **248** to form the transition state difluoromethanesulfinic chloride **251**. Through a similar sequence, difluoromethanesulfenyl chloride **252** was generated. Electrophilic difluoromethylthiolation then occurred, giving the corresponding indole derivatives **249** (Scheme 4.53). The same research group also reported the synthesis of trifluoromethyl disulfides and difluoromethyl disulfides by using CF_3SO_2Cl and CHF_2SO_2Cl, under similar reaction conditions [85].

4.6 Cascade Bond Formation

Cascade bond formation is a quite powerful approach in organic synthesis. *n*-Bu_4NI has also been applied in reactions involving various cascade bond formations. Various heterocyclic compounds can be obtained through the *n*-Bu_4NI-catalyzed cascade C–C/C–hetero bond formation strategies.

In 2011, Wei and coworkers [86] realized the synthesis of benzimidazole derivatives **255** from diamines **253** and aldehydes **254**, mediated by $[n\text{-}Bu_4N]^+[IO]^-$ (or $[n\text{-}Bu_4N]^+[IO_2]^-$) generated *in situ* with H_2O_2 and *n*-Bu_4NI (Scheme 4.54). This protocol can also be used for the synthesis of benzoxazole and benzothiazole derivatives.

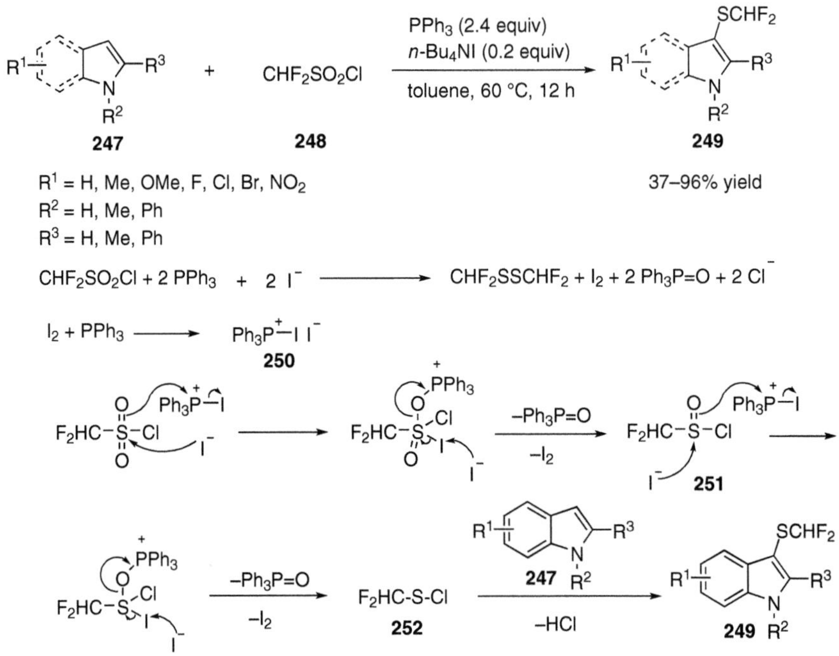

Scheme 4.53 *n*-Bu_4NI-catalyzed synthesis of difluoromethylthiolated aromatics from indoles and difluoromethanesulfonyl chloride.

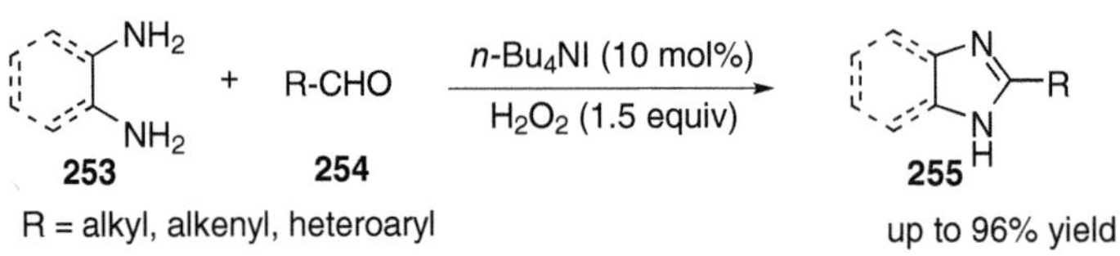

Scheme 4.54 *n*-Bu_4NI-catalyzed synthesis of benzimidazole derivatives from diamines and aldehydes.

n-Bu_4NI has been reported to possess the ability to catalyze the direct oxidative C–N coupling of 2-aminopyridines **256** with *β*-keto esters and 1,3-diones **257** to give imidazo[1,2-*a*]pyridines **258** [87] (Scheme 4.55). Mechanistically, *n*-Bu_4NI first reacted with TBHP to furnish the active iodide species, namely $[n\text{-}Bu_4N]^+[IO]^-$ **44** or $[n\text{-}Bu_4N]^+[IO_2]^-$ **45**. Then the active oxidant **44** reacted with **257** to generate iodide intermediate **259**, which can be converted to intermediate **260** through the nucleophilic attack by **256**. After an intramolecular condensation happened in intermediate **260**, the title compound **258** was obtained. Although the effect of $BF_3{\cdot}Et_2O$ was not clarified in this pathway, it might facilitate the leaving of iodide anion as a Lewis acid. An alternative pathway proposed that in the presence of $BF_3{\cdot}Et_2O$, the reaction of iodide species **45** and compound **257** gave intermediate **261**. Then nucleophilic attack of **261** by **256** afforded **260**, from which product **258** was generated. Later, an *n*-Bu_4NI-catalyzed method for the synthesis of imidazo[1,2-*a*]pyridines **264** through the oxidative coupling of aminopyridines

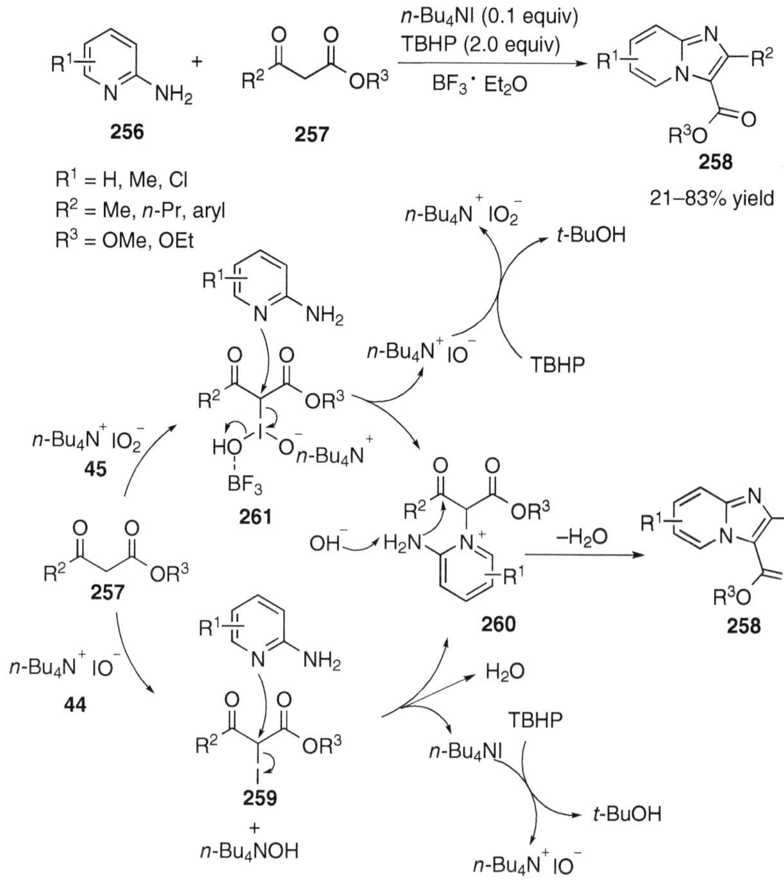

Scheme 4.55 *n*-Bu_4NI-catalyzed synthesis of imidazo[1,2-*a*]pyridines from 2-aminopyridines and 1,3-diones.

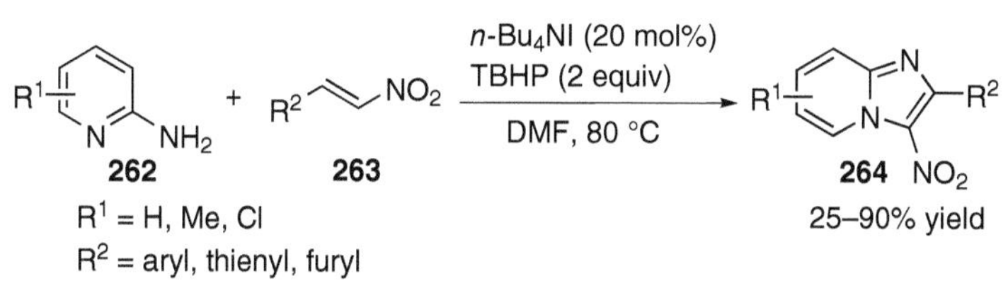

Scheme 4.56 *n*-Bu_4NI-catalyzed synthesis of imidazo[1,2-*a*]pyridines from aminopyridines and nitroolefins.

262 and nitroolefins **263** using TBHP as oxidant was realized by Xu and Li [88] (Scheme 4.56).

In 2014, Li and colleagues developed [89] an *n*-Bu_4NI-catalyzed domino reaction for the synthesis of imidazo[1,5-*c*]-quinazolines **268** *via* selective dual amination of the Csp^3–H bonds of 4-methylquinazolines **265**, using the benzylamines **266** or *α*-amino acids **267** as the nitrogen source (Scheme 4.57). Initially, acyloxy **269**

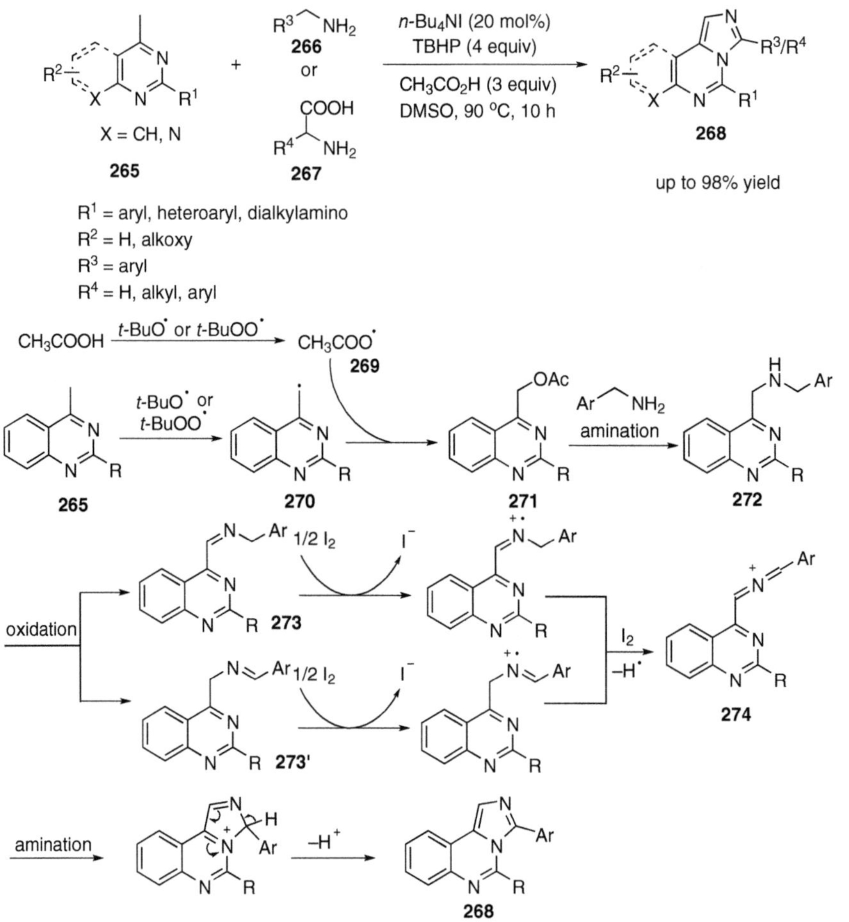

Scheme 4.57 *n*-Bu_4NI-catalyzed synthesis of imidazo[1,5-*c*]-quinazolines from 4-methylquinazolines.

and benzylic radical **270** were obtained through *tert*-butoxyl or *tert*-butylperoxy radical-mediated abstraction of hydrogen atoms in acetic acid and substrate **265**, respectively. Through the radical coupling reaction between acyloxy radical **269** and benzylic radical **270**, intermediated **271** was furnished. Next, intermediated **271** was converted to compound **272** *via* amination, followed by *n*-Bu_4NI/TBHP-mediated oxidation to give imine intermediate **273** or **273′**. Subsequently, these two intermediates were oxidized by iodine to form iminium intermediate **274**, which underwent intramolecular amination/cyclization and rearrangement sequentially to furnish the title compound.

In 2014, by applying *t*-BuOOH and *n*-Bu_4NI, Han and team workers [90] realized an efficient synthesis of isoxazoline-functionalized phenanthridines **277** *via* the reaction of β,γ-unsaturated ketoximes **276** and 2-arylphenylisonitriles **275**. The mechanism for this transformation involves the initial generation of iminoxyl radicals **278** from the reaction of β,γ-unsaturated ketoximes **276** with TBHP and *n*-Bu_4NI,

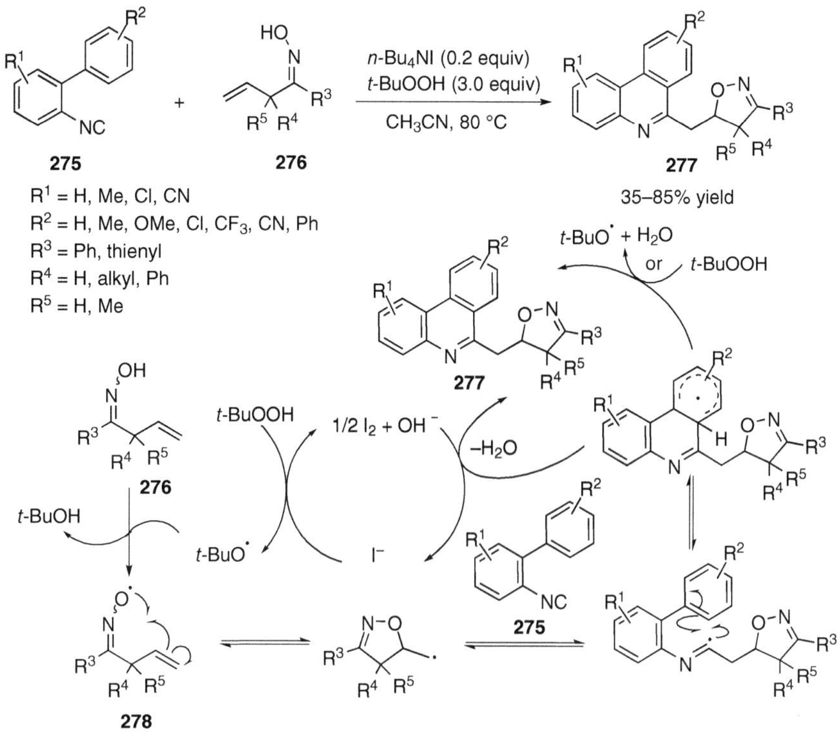

Scheme 4.58 *n*-Bu_4NI-catalyzed synthesis of isoxazoline functionalized phenanthridines from *β*,*γ*-unsaturated ketoximes and 2-arylphenylisonitriles.

followed by a cascade radical cyclization/addition/cyclization sequence to give the corresponding products (Scheme 4.58). In 2018, Shao and colleagues [91] discovered a microwave-assisted method for the synthesis of isoxazoline-functionalized isoquinolines **281** *via* the *n*-Bu_4NI-catalyzed radical cascade cyclization of vinyl isocyanides **279** with *β*,*γ*-unsaturated ketoximes **280** (Scheme 4.59).

Utilizing *n*-Bu_4NI as the catalyst and $K_2S_2O_8$ as the oxidant, aminooxylated oxindoles **284** could be obtained through oxidative aminooxyarylation of alkenes **282** with NHPI **283** [92]. This protocol underwent a radical C–H functionalization process, constructing C–O and C–C bonds in one pot. The benzylic and *α*-methylene C(sp^3)–H bonds were also aminooxylated under the standard reaction conditions (Scheme 4.60).

An *n*-Bu_4NI/TBHP-catalyzed cascade bonds formation protocol for synthesizing 2-aminobenzofuran-3(2*H*)-ones **287** through an oxidative coupling of *o*-acylphenols **285** and amines **286** was realized by Yu and coworkers [93] (Scheme 4.61). Benzofuran-3(2*H*)-one **288** which was generated from substrate **285** *via* an iodination/cyclization process, was recognized as the key intermediate in this transformation.

Using a simple beaker-type undivided cell with *n*-Bu_4NI serving as a redox catalyst under constant current electrolysis condition, Sun and Zeng [94] conducted an efficient electrochemical amino-oxygenation of styrenes **289** with alcohols **290** for

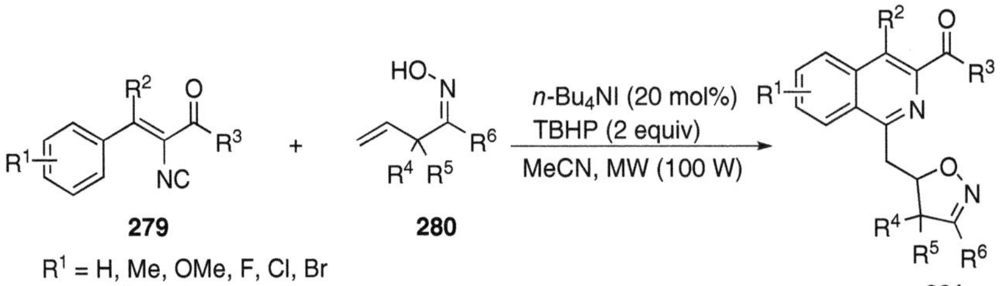

279 280 281

57–91% yield

R^1 = H, Me, OMe, F, Cl, Br
R^2 = aryl
R^3 = OMe, OEt, pyrrolidine, piperidine
R^4 = H, Me
R^5 = H, Me
R^6 = alkyl, aryl, heteroaryl

Scheme 4.59 *n*-Bu_4NI-catalyzed synthesis of isoxazoline-functionalized isoquinolines from vinyl isocyanides and *β*,*γ*-unsaturated ketoximes.

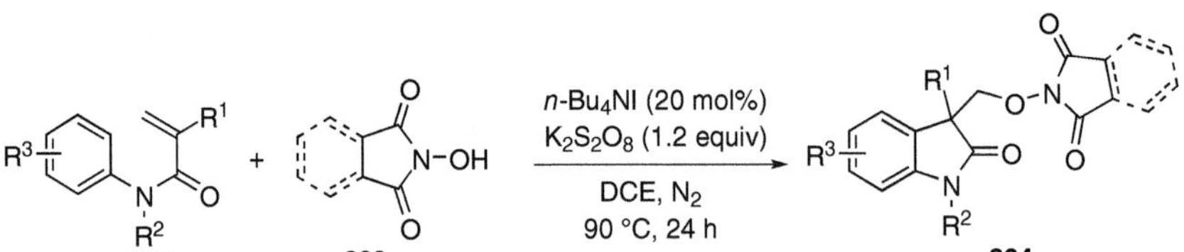

282 283 284

up to 88% yield

R^1 = Me
R^2 = Me, Et, *i*-Pr, *n*-Bu, Ph, Bn
R^3 = H, OMe, halogen, CF_3, CN, NO_2

Scheme 4.60 *n*-Bu_4NI-catalyzed synthesis of aminooxylated oxindoles from alkenes and *N*-hydroxyphthalimides.

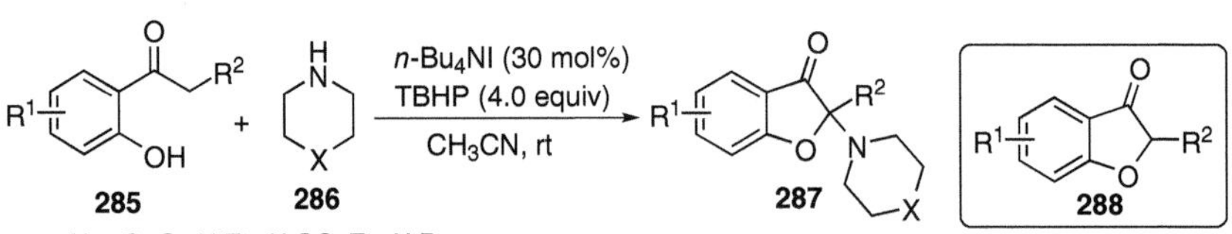

285 286 287 288

up to 90% yield

X = C, O, *N*-Et, *N*-CO_2Et, *N*-Boc
R^1 = H, Me, OEt, F
R^2 = Me, Et, Ph

Scheme 4.61 *n*-Bu_4NI-catalyzed synthesis of 2-aminobenzofuran-3(2*H*)-ones from *o*-acylphenols and amines.

the synthesis of 3-methoxyindolines and 3-ethoxyindoline **291** (Scheme 4.62). The control experiment indicated that iodonium intermediate **292** is a crucial reactive species for the reaction.

The construction of oxazolidinone skeletons **295** could be realized through a three-component reaction consisting of CO_2, epoxides **294** and amines **293** facilitated by *n*-Bu_4NI and DUB [95]. *n*-Bu_4NI was believed to be responsible for the opening of the epoxide ring. This metal-free and easily available catalytic system was applicable to a broad range of substrates (Scheme 4.63).

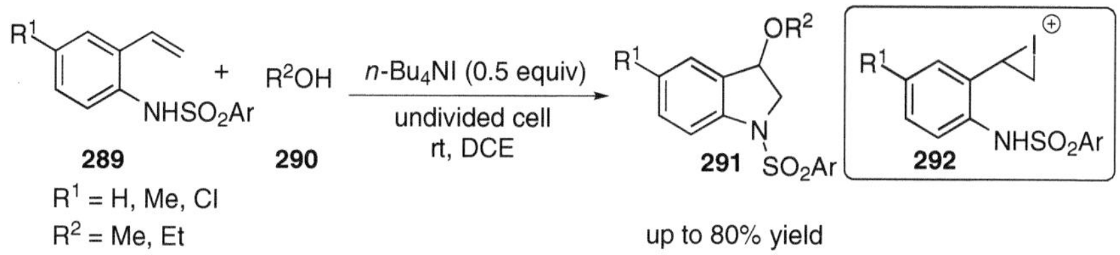

Scheme 4.62 n-Bu_4NI-catalyzed synthesis of 3-methoxyindolines and 3-ethoxyindoline from styrenes and alcohols.

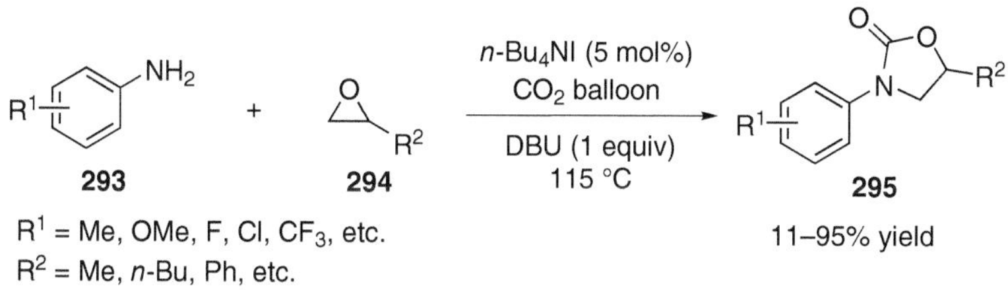

Scheme 4.63 n-Bu_4NI-catalyzed synthesis of oxazolidinone derivatives from CO_2, epoxides and amines.

In 2016, Zhu and Huang [96] developed a convenient method for synthesizing 2-aminobenzothiazoles **298**, from aryl isothiocyanates **296** and formamides **297** using n-Bu_4NI/TBHP as the radical initiator. Mechanistic studies indicated that decarbonylative aminyl radical **299** was formed initially under the effect of the n-Bu_4NI/TBHP system, followed by aminyl radical addition to isothiocyanates and intramolecular cyclization to afford product **298** (Scheme 4.64).

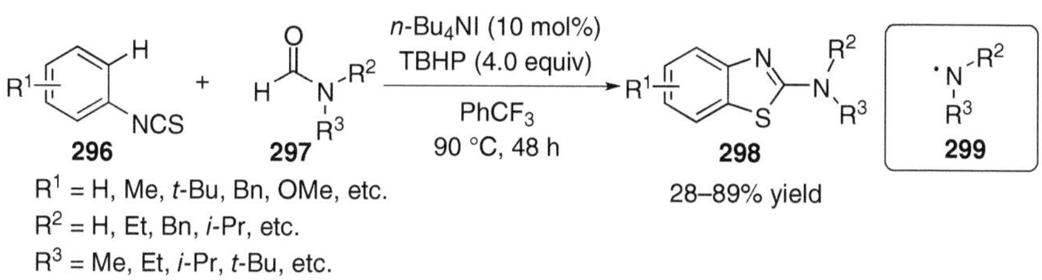

Scheme 4.64 n-Bu_4NI-catalyzed synthesis of 2-aminobenzothiazoles from aryl isothiocyanates and formamides.

In 2020, Liu and colleagues [97] accomplished the synthesis of 1,2,4-thiadiazoles and pyrido-fused 1,2,4-thiadiazoles **302** from isothiocyanates **300** and 2-aminopyridines/amidines **301**, using n-Bu_4NI as the catalyst and electrolyte (Scheme 4.65). This electrochemical intermolecular dehydrogenative coupling reaction was proposed to undergo this mechanistic pathway: initially, an iodide anion undergoes electro-oxidation in a platinum anode, generating iodine. Then 2-aminopyridine **301** tautomerizes to generate an imine form **303**, which reacts with iodine to give N-iodopyridin-2-amine **304**. The active species **304** undergoes a further [3+2]

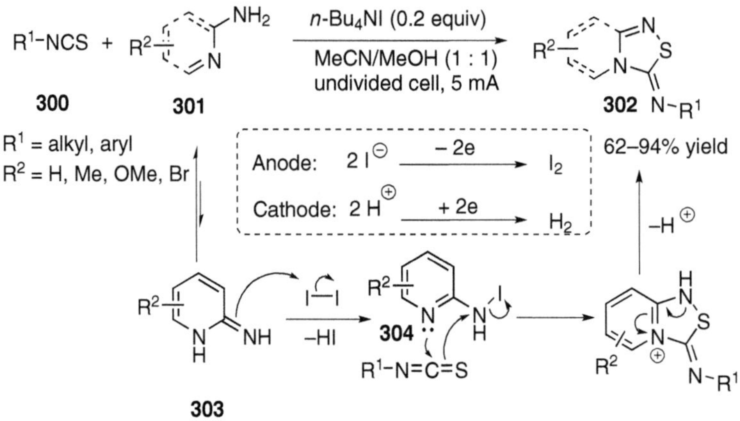

Scheme 4.65 *n*-Bu_4NI-catalyzed synthesis of 1,2,4-thiadiazoles and pyrido-fused 1,2,4-thiadiazoles from isothiocyanates and 2-aminopyridines/amidines.

annulation with an isothiocyanate **300** *via* deiodination and deprotonation to afford 1,2,4-thiadiazole **302**.

In 2017, Du and coworkers reported that substituted pyrrolin-4-ones **306** could be synthesized *via* *n*-Bu_4NI/TBHP-catalyzed oxidative cascade bond formation by employing enamino esters **305** as substrates [98]. The reaction process was postulated to involve the key steps of the coupling of two radical intermediates, an intramolecular ring closure, and an exclusive 1,2-aryl radical migration (Scheme 4.66).

Zhou and coworkers [99] explored an *n*-Bu_4NI-catalyzed reaction of 3-methylindoles **307** with primary amines **308** for the synthesis of quinazolinones **309** using TBHP as oxidant. It was postulated that *tert*-butoxyl **4** and *tert*-butylperoxy radicals **5** were first formed through the reaction of TBHP and *n*-Bu_4NI. Then the radical initiated hydrogen abstraction in indole **307** occurred to afford indolyl radical **310** and its resonance species **311**. Subsequently, *tert*-butylperoxy radical **5** coupled

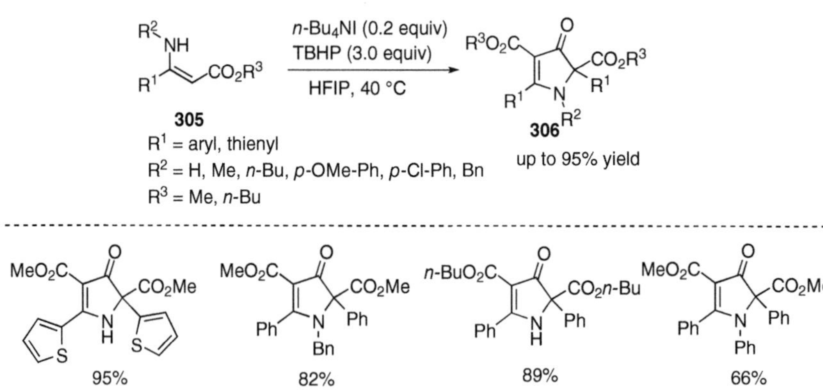

Scheme 4.66 *n*-Bu_4NI-catalyzed synthesis of substituted pyrrolin-4-ones from enamino esters.

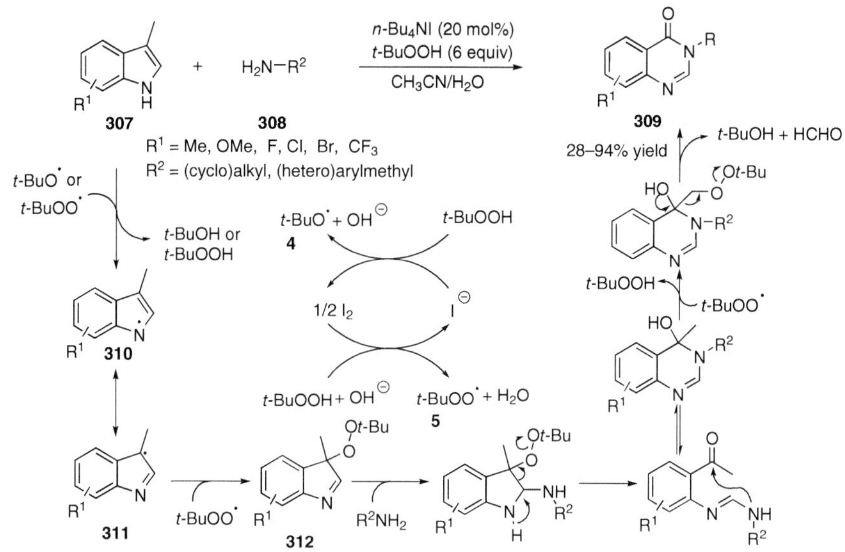

Scheme 4.67 *n*-Bu_4NI-catalyzed synthesis of quinazolinones from 3-methylindoles and primary amines.

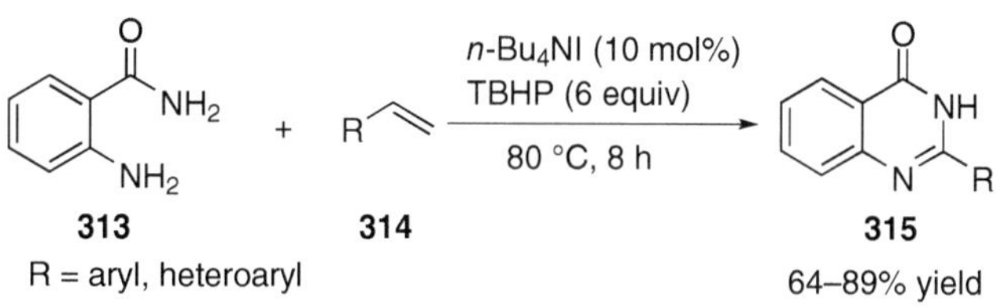

Scheme 4.68 *n*-Bu_4NI-catalyzed synthesis of quinazolin-4(3*H*)-ones from *o*-aminobenzamide and alkenes.

with radical **311** to afford intermediate **312**, which underwent nitrogenation, ring-opening, and recyclization sequentially, affording various quinazolinones in good-to-excellent yields (Scheme 4.67).

In 2020, Vidavalur and colleagues [100] developed an *n*-Bu_4NI-catalyzed protocol for synthesizing quinazolin-4(3*H*)-ones **315** from *o*-aminobenzamide **313** and alkenes **314** (Scheme 4.68). The mechanism for this transformation was proposed as follows: Firstly, the alkene was converted to aldehyde *via* an *n*-Bu_4NI/TBHP-mediated oxidative cleavage of the C=C bond. Then aldehyde underwent condensation, cyclization, and oxidative elimination to give the final product.

Wei and colleagues [101] disclosed that acid hydrazides **316** or araldehyde *N*-acylhydrazones **317** could be converted to symmetrical or unsymmetrical 2,5-disubstituted 1,3,4-oxadiazoles **318** *via* an *n*-Bu_4NI/TBHP-catalyzed pathway. This one-pot transformation was initiated by the formation of aldehydes **319** *via* *n*-Bu_4NI/TBHP-mediated radical reaction of hydrazide **316**. Next, the condensation reaction between aldehyde **319** and hydrazide **316** occurred to give *N*-acylhydrazone

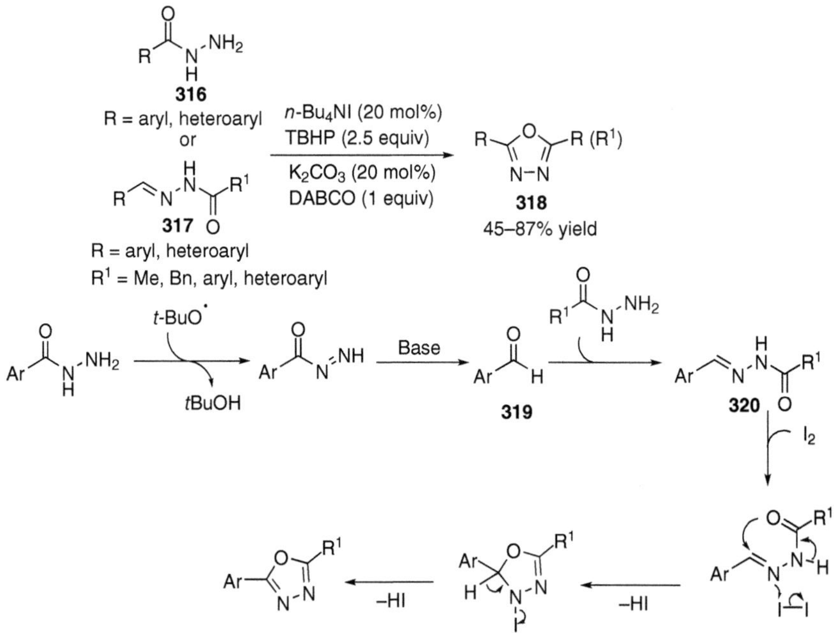

Scheme 4.69 *n*-Bu_4NI-catalyzed synthesis of 2,5-disubstituted 1,3,4-oxadiazoles from acid hydrazides or araldehyde *N*-acylhydrazones.

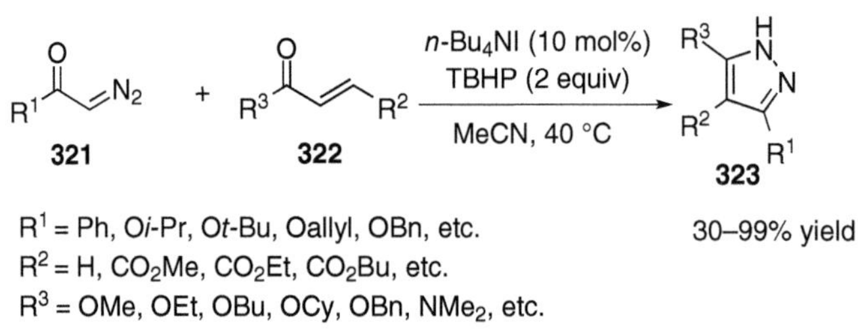

Scheme 4.70 *n*-Bu_4NI-catalyzed synthesis of pyrazole skeletons from alkenes and diazo compounds.

320. The intramolecular cyclization, triggered by iodine-mediated nucleophilic attack and the subsequent rearomatization occurred to furnish product **318** (Scheme 4.69).

In 2016, Wan and Shao developed an efficient *n*-Bu_4NI-catalyzed reaction for the construction of pyrazole skeletons **323** *via* sequential [3+2] cycloaddition and oxidative dehydrogenation reactions between alkenes **322** and diazo compounds **321** in the presence of TBHP [102] (Scheme 4.70).

Liu and coworkers [103] in 2018 developed an electro-catalyzed method for synthesis of racemic Favorskii amides **326** from the *n*-Bu_4NI-mediated reaction of 1,3-diarylacetones **324** and aliphatic amines **325**. The reaction began with the anodic oxidation of iodide to generate molecular iodine, which undergoes reaction

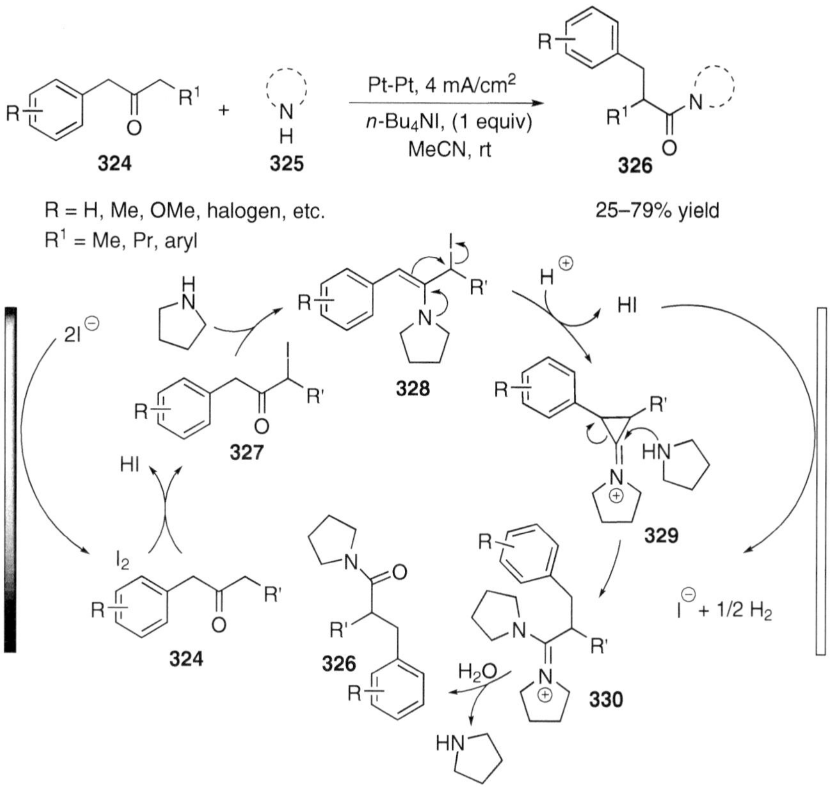

Scheme 4.71 n-Bu$_4$NI-catalyzed synthesis of Favorskii amides from 1,3-diarylacetones and aliphatic amines.

with ketone **324** to form α-iodo ketone **327**. Secondary amines **325** can promote Favorskii rearrangements by the formation of an enamine **328**. Subsequently, the cyclopropenone intermediate **329** attacked by **325** gave a more stable carbanion **330**. Finally, benzylated amides **326** are obtained by hydrolysis of **330** (Scheme 4.71).

4.7 The Formation of P–O/C, S–N Bond

n-Bu$_4$NI has also found application in the catalytic transformation involving P–O/C, S–N bond formations.

In 2016, Zeng and coworkers [104] developed an NH$_4$I-mediated approach for the synthesis of sulfonamides **333** *via* the electrochemical oxidative amination of sodium sulfinates **331**. The chemistry proceeds in a simple undivided cell employing a substoichiometric amount of NH$_4$I that serves both as a redox catalyst and a supporting electrolyte. A wide range of substrates including aliphatic or aromatic secondary and primary amines, as well as aqueous ammonia **332**, have been proved to be compatible with the protocol (Scheme 4.72).

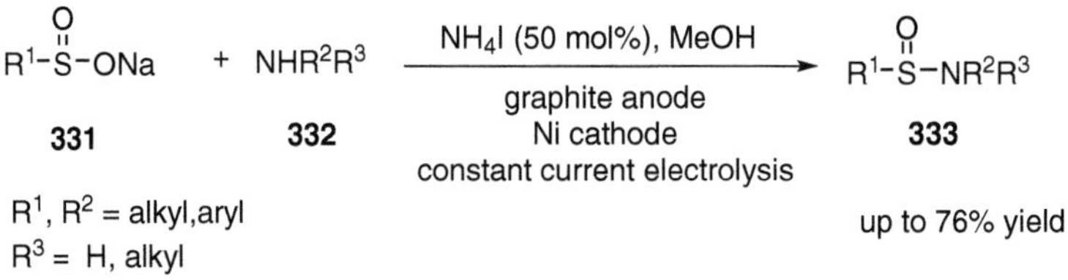

Scheme 4.72 NH_4I-catalyzed synthesis of sulfonamides from sodium sulfonates and amines.

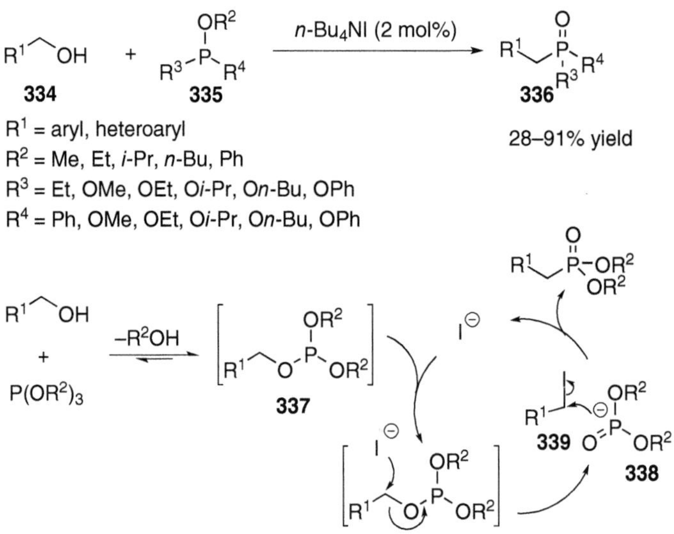

Scheme 4.73 *n*-Bu_4NI-catalyzed synthesis of phosphoryl compounds from alcohols.

In 2018, an *n*-Bu_4NI-catalyzed C–P(O) bond formation reaction was reported by Han and coworkers [105] for the synthesis of phosphoryl compounds **336** from alcohols **334** and phosphites, phosphonites, and phosphinites **335** (Scheme 4.73). The transformation was initiated by the formation of phosphate intermediate **337** through nucleophilic substitution. Next, iodide attacked the alkylenyl carbon of phosphate intermediate **337**, trigging the C–O cleavage to give the phosphoryl anion **338** and alkyl iodide **339**. Finally, through the nucleophilic substitution reaction of phosphoryl anion **338** and alkyl iodide **339**, the corresponding phosphoryl compounds were obtained.

Carbamoylphosphonates **342** could be synthesized through an *n*-Bu_4NI-catalyzed P–H oxidation and functionalization method by employing phosphoramides **340** and alcohols **341** as the substrates [106] (Scheme 4.74).

4.8 Conclusion

R_4NI, especially tetrabutylammonium iodide, have been widely used in various catalytic transformations in organic chemistry, leading to the construction of a plethora

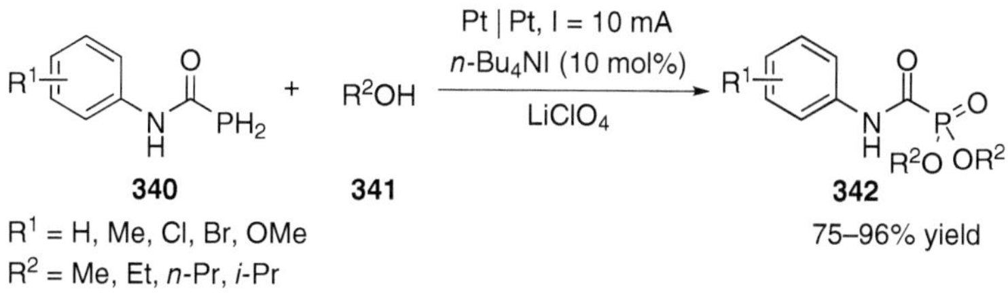

Scheme 4.74 *n*-Bu_4NI-catalyzed synthesis of carbamoylphosphonates from phosphoramides and alcohols.

of useful building blocks. It can be envisaged that more efficient R_4NI-mediated reactions, especially the chiral iodide-catalyzed asymmetric transformations, will emerge in this vigorously developing field.

References

1 (a) Chen, R., Chen, J., Zhang, J., and Wan, X. (2018). Combination of tetrabutylammonium iodide (TBAI) with *tert*-butyl hydroperoxide (TBHP): an efficient transition-metal-free system to construct various chemical bonds. *The Chemical Record* 18: 1292–1305. (b) Huang, Y.L., Bao, W.H., Zhu, W.M., and Wei, W.T. (2018). Recent advances in tetrabutylammonium iodide/*t*-BuOOH promoted C–N bond formation. *Chinese Journal of Organic Chemistry* 38: 752–759. (c) Wu, X.F., Gong, J.L., and Qi, X. (2014). A powerful combination: recent achievements on using TBAI and TBHP as oxidation system. *Organic & Biomolecular Chemistry* 12: 5807–5817. (d) Jiao, K.-J., Xing, Y.-K., Yang, Q.-L. et al. (2020). Site-selective C–H functionalization *via* synergistic use of electrochemistry and transition metal catalysis. *Accounts of Chemical Research* 53: 300–310.

2 (a) Ooi, T., Uematsu, Y., Fujimoto, J. et al. (2007). Advantage of *in situ* generation of *N*-arylsulfonyl imines from *α*-amide sulfones in the phase-transfer-catalyzed asymmetric Strecker reaction. *Tetrahedron Letters* 48: 1337–1340. (b) Kawamura, M. and Sato, K. (2006). Magnetically separable phase-transfer catalysts. *Chemical Communications*: 4718–4719. (c) Ooi, T., Uematsu, Y., and Maruoka, K. (2006). Asymmetric Strecker reaction of aldimines using aqueous potassium cyanide by phase-transfer catalysis of chiral quaternary ammonium salts with a tetranaphthyl backbone. *Journal of the American Chemical Society* 128: 2548–2549. (d) Soltani Rad, M.N., Khalafi-Nezhad, A., Behrouz, S. et al. (2009). One-pot synthesis of *N*-alkyl purine, pyrimidine and azole derivatives from alcohols using Ph_3P/CCl_4: a rapid route to carboacyclic nucleoside synthesis. *Synthesis* 2009: 3067–3076.

3 Mai, W.-P., Wang, H.-H., Li, Z.-C. et al. (2012). *n*-Bu_4NI-catalyzed direct synthesis of *α*-ketoamides from aryl methyl ketones with dialkylformamides in water using TBHP as oxidant. *Chemical Communications* 48: 10117–10119.

4 Du, B., Jin, B., and Sun, P. (2014). The syntheses of α-ketoamides *via* *n*-Bu_4NI-catalyzed multiple sp^3C–H bond oxidation of ethylarenes and sequential coupling with dialkylformamides. *Organic & Biomolecular Chemistry* 12: 4586–4589.

5 Qian, P., Su, J.-H., Wang, Y. et al. (2017). Electrocatalytic C–H/N–H coupling of 2′-aminoacetophenones for the synthesis of isatins. *The Journal of Organic Chemistry* 82: 6434–6440.

6 Wang, S., Wang, J., Guo, R. et al. (2013). *n*-Bu_4NI-catalyzed oxidative amidation of aldehydes with tertiary amines. *Tetrahedron Letters* 54: 6233–6236.

7 Liu, Z., Zhang, J., Chen, S. et al. (2012). Cross coupling of acyl and aminyl radicals: direct synthesis of amides catalyzed by Bu_4NI with TBHP as an oxidant. *Angewandte Chemie International Edition* 51: 3231–3235.

8 Li, H., Xie, J., Xue, Q. et al. (2012). Metal-free *n*-Bu_4NI-catalyzed direct synthesis of amides from alcohols and *N,N*-disubstituted formamides. *Tetrahedron Letters* 53: 6479–6482.

9 Wang, G., Yu, Q.-Y., Wang, J. et al. (2013). Iodide-catalyzed amide synthesis from alcohols and amines. *RSC Advances* 3: 21306–21310.

10 Wang, G., Yu, Q.-Y., Chen, S.-Y., and Yu, X.-Q. (2014). Et_4NI-catalyzed amidation of aldehydes and alcohols with ammonium salts. *Organic & Biomolecular Chemistry* 12: 414–417.

11 Chen, M., Li, Y., Tang, H. et al. (2017). Bu_4NI-catalyzed oxygen-centered radical addition between acyl peroxides and isocyanides. *Organic Letters* 19: 3147–3150.

12 Dian, L., Wang, S., Zhang-Negrerie, D. et al. (2014). Organocatalytic amination of alkyl ethers *via* *n*-Bu_4NI/*t*-BuOOH-mediated intermolecular oxidative C(sp^3)–N bond formation: novel synthesis of hemiaminal ethers. *Chemical Communications* 50: 11738–11741.

13 Rajamanickam, S., Majji, G., Santra, S.K., and Patel, B.K. (2015). Bu_4NI catalyzed C–N bond formation *via* cross-dehydrogenative coupling of aryl ethers (Csp^3–H) and tetrazoles (N–H). *Organic Letters* 17: 5586–5589.

14 Wang, L., Zhu, K.-Q., Wu, W.-T. et al. (2015). *n*-Bu_4NI-catalyzed direct amination of ethers with aryl tetrazoles and triazoles *via* cross-dehydrogenative coupling reaction. *Catalysis Science & Technology* 5: 2891–2896.

15 Abebe, H., Vidavalur, S., and Battula, V.R. (2017). An efficient *n*-Bu_4NI-catalyzed protocol for the synthesis of hemiaminal ethers by linking ethers with 1,2,4-triazoles and indazoles *via* cross dehydrogenative coupling reaction. *ChemistrySelect* 2: 4165–4168.

16 Luo, Z., Jiang, Z., Jiang, W., and Lin, D. (2018). C–H amination of purine derivatives *via* radical oxidative coupling. *The Journal of Organic Chemistry* 83: 3710–3718.

17 Wang, L., Zhu, K., Chen, Q., and He, M. (2014). Transition-metal-free direct alkylation of aryl tetrazoles *via* intermolecular oxidative C–N formation. *The Journal of Organic Chemistry* 79: 11780–11786.

18 Abebe, H., Vidavalur, S., and Battula, V.R. (2016). *n*-Bu_4NI/TBHP-catalyzed C–N bond formation *via* cross-dehydrogenative coupling of 1H-1,2,4-triazoles (N–H) and methylarenes (Csp^3–H). *RSC Advances* 6: 82289–82293.

19 Rajamanickam, S., Sah, C., Mir, B.A. et al. (2020). Bu_4NI-catalyzed, radical-induced regioselective *N*-alkylations and arylations of tetrazoles using organic peroxides/peresters. *The Journal of Organic Chemistry* 85: 2118–2141.

20 Wang, R., Liu, H., Yue, L. et al. (2014). Transition metal-free direct amination of benzoxazoles using formamides as nitrogen sources. *Tetrahedron Letters* 55: 2233–2237.

21 Yuan, J.-W., Jin, M., Yin, Q.-Y. et al. (2016). *n*-Bu_4NI-catalyzed direct amination of benzoxazoles with tertiary amines using TBHP as oxidant under microwave irradiation. *Zeitschrift für Naturforschung B* 71: 317–325.

22 Zhang, X., Wang, M., Li, P., and Wang, L. (2014). *n*-Bu_4NI/TBHP-catalyzed direct amination of allylic and benzylic C(sp^3)–H with anilines under metal-free conditions. *Chemical Communications* 50: 8006–8009.

23 Wang, H., Shao, Y., Zheng, H. et al. (2015). Cyanoacetic acid as a masked electrophile: transition-metal-free cyanomethylation of amines and carboxylic acids. *Chemistry – A European Journal* 21: 18333–18337.

24 Gao, Y.-T., Jin, X.-Y., Liu, Q. et al. (2018). Iodide/H_2O_2 catalyzed intramolecular oxidative amination for the synthesis of 3,2′-pyrrolidinyl spirooxindoles. *Molecules* 23: 2265.

25 Dhineshkumar, J. and Prabhu, K.R. (2016). An efficient tertiary azidation of 1,3-dicarbonyl compounds in water catalyzed by tetrabutylammonium iodide. *European Journal of Organic Chemistry* 2016: 447–452.

26 Tiffner, M., Stockhammer, L., Schörgenhumer, J. et al. (2018). Towards an asymmetric organocatalytic α-azidation of β-ketoesters. *Molecules* 23: 1142.

27 Uyanik, M., Sahara, N., Tsukahara, M. et al. (2020). Chemo- and enantioselective oxidative α-azidation of carbonyl compounds. *Angewandte Chemie International Edition* 59: 17110–17117.

28 Kiyokawa, K., Kosaka, T., and Minakata, S. (2013). Metal-free aziridination of styrene derivatives with iminoiodinane catalyzed by a combination of iodine and ammonium iodide. *Organic Letters* 15: 4858–4861.

29 Chen, J., Yan, W.-Q., Lam, C.M. et al. (2015). Electrocatalytic aziridination of alkenes mediated by *n*-Bu_4NI: a radical pathway. *Organic Letters* 17: 986–989.

30 Li, D., Yang, T., Su, H., and Yu, W. (2015). *tert*-Butyl hydroperoxide and tetrabutylammonium iodide-promoted free radical cyclization of α-imino-*N*-arylamides and α-azido-*N*-arylamides. *Advanced Synthesis & Catalysis* 357: 2529–2539.

31 Ye, Y., Wang, H., and Fan, R. (2010). Stereoselective construction of highly functionalized azetidines *via* a [2 + 2]-cycloaddition. *Organic Letters* 12: 2802–2805.

32 Fan, R., Wang, H., Ye, Y., and Gan, J. (2010). PhIO/Bu_4NI mediated oxidative cyclization of amidoalkylation adducts for the synthesis of *N*-benzoyl aziridines and oxazolines. *Tetrahedron Letters* 51: 453–456.

33 Sun, Y., Gan, J., and Fan, R. (2011). Facile construction of oxa-aza spirobicycles *via* a tandem carbon-hydrogen bond oxidation. *Advanced Synthesis & Catalysis* 353: 1735–1740.

34 Feng, J., Liang, S., Chen, S.-Y. et al. (2012). A metal-free oxidative esterification of the benzyl C–H bond. *Advanced Synthesis & Catalysis* 354: 1287–1292.

35 Huang, J., Li, L.-T., Li, H.-Y. et al. (2012). Bu_4NI-catalyzed benzylic acyloxylation of alkylarenes with aromatic aldehydes. *Chemical Communications* 48: 10204–10206.

36 Liu, L., Yun, L., Wang, Z. et al. (2013). Transition metal free oxidative esterification of alcohols with toluene. *Tetrahedron Letters* 54: 5383–5386.

37 Chen, Y., Li, C., Cui, Y. et al. (2019). Bu_4NI-catalyzed C–C bond cleavage and oxidative esterification of allyl alcohols with toluene derivatives. *Synthesis* 51: 3667–3674.

38 Majji, G., Guin, S., Gogoi, A. et al. (2013). Easy access to benzylic esters directly from alkyl benzenes under metal-free conditions. *Chemical Communications* 49: 3031–3033.

39 Lv, Y., Sun, K., Wang, T. et al. (2015). *n*-Bu_4NI-catalyzed intermolecular C–O cross-coupling reactions: synthesis of alkyloxyamines. *RSC Advances* 5: 72142–72145.

40 Xu, J., Zhang, P., Li, X. et al. (2014). Tetrabutylammonium iodide-catalyzed phosphorylation of benzyl C–H bonds *via* a cross-dehydrogenative coupling (CDC) reaction. *Advanced Synthesis & Catalysis* 356: 3331–3335.

41 Xiong, B., Wang, G., Zhou, C. et al. (2018). Bu_4NI-catalyzed dehydrogenative coupling of diaryl phosphinic acids with C(sp^3)–H bonds of arenes. *The Journal of Organic Chemistry* 83: 993–999.

42 Shi, D., Qin, H.-T., Zhu, C., and Liu, F. (2015). Selectfluor–Bu_4NI-mediated C(sp^3)–H oxidation in aqueous media: synthesis of Δ^2-isoxazolines from oximes. *European Journal of Organic Chemistry* 2015: 5084–5088.

43 Uyanik, M., Suzuki, D., Yasui, T., and Ishihara, K. (2011). *In situ* generated (hypo)iodite catalysts for the direct α-oxyacylation of carbonyl compounds with carboxylic acids. *Angewandte Chemie International Edition* 50: 5331–5334.

44 Guo, S., Yu, J.-T., Dai, Q. et al. (2014). The Bu_4NI-catalyzed α-acyloxylation of ketones with benzylic alcohols. *Chemical Communications* 50: 6240–6242.

45 Zhu, F. and Wang, Z.-X. (2014). Bu_4NI-catalyzed α-acyloxylation reaction of ethers and ketones with aldehydes and *tert*-butyl hydroperoxide. *Tetrahedron* 70: 9819–9827.

46 Mondal, B., Sahoo, S.C., and Pan, S.C. (2015). *n*-Bu_4NI-catalyzed α-benzoxylation of ketones with terminal aryl alkenes. *European Journal of Organic Chemistry* 2015: 3135–3140.

47 Zhou, Z., Cheng, J., and Yu, J.-T. (2015). Bu_4NI-catalyzed direct α-oxyacylation of diarylethanones with acyl peroxides. *Organic & Biomolecular Chemistry* 13: 9751–9754.

48 Chang, L.-M. and Yuan, G.-Q. (2016). One-pot synthesis of α-acyloxycarbonyl compounds *via* oxidative decarboxylation coupling reaction of α-oxo carboxylic acids with carbonyl compounds. *Tetrahedron* 72: 7003–7007.

49 Li, C., Jin, T., Zhang, X. et al. (2016). Bu_4NI-catalyzed α-oxyacylation of carbonyl compounds with toluene derivatives. *Organic Letters* 18: 1916–1919.

50 Zhang, S., Lian, F., Xue, M. et al. (2017). Electrocatalytic dehydrogenative esterification of aliphatic carboxylic acids: access to bioactive lactones. *Organic Letters* 19: 6622–6625.

51 Terent'ev, A.O., Zdvizhkov, A.T., Levitsky, D.O. et al. (2015). Organocatalytic peroxidation of malonates, β-ketoesters, and cyanoacetic esters using n-Bu_4NI/t-BuOOH-mediated intermolecular oxidative $C(sp^3)$–O coupling. *Tetrahedron* 71: 8985–8990.

52 Siddaraju, Y. and Prabhu, K.R. (2015). A chemoselective α-aminoxylation of aryl ketones: a cross dehydrogenative coupling reaction catalysed by Bu_4NI. *Organic & Biomolecular Chemistry* 13: 11651–11656.

53 Chen, L., Shi, E., Liu, Z. et al. (2011). Bu_4NI-catalyzed C–O bond formation by using a cross-dehydrogenative coupling (CDC) reaction. *Chemistry – A European Journal* 17: 4085–4089.

54 Zhang, S., Guo, L.-N., Wang, H., and Duan, X.-H. (2013). Bu_4NI-catalyzed decarboxylative acyloxylation of an sp^3 C–H bond adjacent to a heteroatom with α-oxocarboxylic acids. *Organic & Biomolecular Chemistry* 11: 4308–4311.

55 Wang, Q., Feng, J., Chai, W. et al. (2014). Bu_4NI-catalyzed construction of CO bonds by oxidative coupling of alcohols with ethers. *Tetrahedron Letters* 55: 4785–4789.

56 Xue, Q., Xie, J., Xu, P. et al. (2013). Metal-free, n-Bu_4NI-catalyzed regioselective difunctionalization of unactivated alkenes. *ACS Catalysis* 3: 1365–1368.

57 Chen, J., Shao, Y., Zheng, H. et al. (2015). Metal-free coupling of 2-vinylphenols and carboxylic acids: an access to 3-acyloxy-2,3-dihydrobenzofurans. *The Journal of Organic Chemistry* 80: 10734–10741.

58 Wei, W., Zhang, C., Xu, Y., and Wan, X. (2011). Synthesis of *tert*-butyl peresters from aldehydes by Bu_4NI-catalyzed metal-free oxidation and its combination with the Kharasch–Sosnovsky reaction. *Chemical Communications* 47: 10827–10829.

59 Zhang, H., Dong, D.-Q., Hao, S.-H., and Wang, Z.-L. (2016). Bu_4NI-catalyzed construction of *tert*-butyl peresters from alcohols. *RSC Advances* 6: 8465–8468.

60 Shi, E., Shao, Y., Chen, S. et al. (2012). Tetrabutylammonium iodide catalyzed synthesis of allylic ester with *tert*-butyl hydroperoxide as an oxidant. *Organic Letters* 14: 3384–3387.

61 Wang, Q.-Q., Wang, Z.-X., Xu, Y.-S. et al. (2016). Bu_4NI-catalyzed and t-BuOOH-oxidized esterification of aldehydes with alkyl halides in water. *Asian Journal of Organic Chemistry* 5: 1304–1308.

62 Nandy, S., Ghatak, A., Das, A.K., and Bhar, S. (2018). Chemoselective and metal-free synthesis of aryl esters from the corresponding benzylic alcohols in aqueous medium using TBHP/TBAI as an efficient catalytic system. *Synlett* 29: 2208–2212.

63 Subramanian, K., Yedage, S.L., and Bhanage, B.M. (2017). An electrochemical method for carboxylic ester synthesis from N-alkoxyamides. *The Journal of Organic Chemistry* 82: 10025–10032.

64 Subramanian, K., Yedage, S.L., and Bhanage, B.M. (2018). Electrodimerization of *N*-alkoxyamides for zinc(II) catalyzed phenolic ester synthesis under mild reaction conditions. *Advanced Synthesis & Catalysis* 360: 2511–2521.

65 Uyanik, M., Okamoto, H., Yasui, T., and Ishihara, K. (2010). Quaternary ammonium (hypo)iodite catalysis for enantioselective oxidative cycloetherification. *Science* 328: 1376–1379.

66 Jiang, M. and Shi, M. (2009). Palladium-catalyzed diacetoxylation of methylenecyclopropanes *via* $C(sp^3)$–$C(sp^3)$ bond breaking. *Organometallics* 28: 5600–5602.

67 Bai, D., Jing, H., and Wang, G. (2012). Cyclic carbonate synthesis from epoxides and CO_2 over cyanocobalamin/*n*-Bu_4NI. *Applied Organometallic Chemistry* 26: 600–603.

68 Sun, J., Fujita, S.-I., Zhao, F., and Arai, M. (2005). A highly efficient catalyst system of $ZnBr_2$/*n*-Bu_4NI for the synthesis of styrene carbonate from styrene oxide and supercritical carbon dioxide. *Applied Catalysis A: General* 287: 221–226.

69 Qin, J., Wang, P., Li, Q. et al. (2014). Catalytic production of cyclic carbonates mediated by lanthanide phenolates under mild conditions. *Chemical Communications* 50: 10952–10955.

70 (a) Woo, W.H., Hyun, K., Kim, Y. et al. (2017). Highly active salen-based aluminum catalyst for the coupling of carbon dioxide with epoxides at ambient temperature. *European Journal of Inorganic Chemistry* 2017: 5372–5378. (b) Whiteoak, C.J., Kielland, N., Laserna, V. et al. (2013). A powerful aluminum catalyst for the synthesis of highly functional organic carbonates. *Journal of the American Chemical Society* 135: 1228–1231.

71 Wong, Y.-C., Tseng, C.-T., Kao, T.-T. et al. (2012). Tandem cyclization of α-cyano α-alkynyl aryl ketones induced by *tert*-butyl hydroperoxide and tetrabutylammonium iodide. *Organic Letters* 14: 6024–6027.

72 Xing, L.-J., Wang, X.-M., Li, H.-Y. et al. (2014). Metal-free synthesis of methylene-bridged bis-1,3-dicarbonyl compounds *via* oxidative C–C bond cleavage of tertiary aliphatic amines. *RSC Advances* 4: 26783–26786.

73 Shen, X., Borah, A.J., Cao, X. et al. (2015). Direct synthesis of 1,3-dicarbonyl compounds *via* radical coupling of aldehydes with ketones under metal-free conditions. *Tetrahedron Letters* 56: 6484–6487.

74 Mackay, E.G. and Studer, A. (2016). Electron-catalyzed fluoroalkylation of vinyl azides. *Chemistry – A European Journal* 22: 13455–13458.

75 Kong, X., Liu, Y., Lin, L. et al. (2019). Electrochemical synthesis of enaminones *via* a decarboxylative coupling reaction. *Green Chemistry* 21: 3796–3801.

76 Tao, J., Estrada, C.D., and Murphy, G.K. (2017). Metal-free intermolecular cyclopropanation between alkenes and iodonium ylides mediated by $PhI(OAc)_2 \cdot Bu_4NI$. *Chemical Communications* 53: 9004–9007.

77 Yu, Y., Jiao, L., Wang, J. et al. (2017). Bu_4NI/*t*-BuOOH catalyzed, α-regioselective cross-dehydrogenative coupling of BODIPY with allylic alkenes and ethers. *Chemical Communications* 53: 581–584.

78 Yadav, L. and Chaudhary, S. (2020). Bu_4NI-catalyzed, oxidative $C(sp^2)–C(sp^3)$ cross dehydrogenative coupling for the regioselective direct C-3 benzylation of 2*H*-indazoles. *Organic & Biomolecular Chemistry* 18: 5927–5936.

79 Agisho, H.A., Hairat, S., and Zaki, M. (2020). An efficient TBHP/TBAI-mediated protocol for the synthesis of 4*H*-chromen-4-ones from chroman-4-ones *via* oxidative C–C bond formation. *Monatshefte für Chemie-Chemical Monthly* 151: 599–603.

80 Li, X., Xu, X., and Zhou, C. (2012). Tetrabutylammonium iodide catalyzed allylic sulfonylation of α-methyl styrene derivatives with sulfonylhydrazides. *Chemical Communications* 48: 12240–12242.

81 Bao, W.-H., Ying, W.-W., Xu, X.-D. et al. (2019). Radical heck-type reaction of styrenes with sulfonyl hydrazides on water at room temperature. *Tetrahedron Letters* 60: 55–58.

82 Wang, Y., Ma, L., Ma, M. et al. (2016). Bu_4NI-catalyzed cross-coupling between sulfonyl hydrazides and diazo compounds to construct β-carbonyl sulfones using molecular oxygen. *Organic Letters* 18: 5082–5085.

83 Liu, C.-R. and Ding, L.-H. (2015). Byproduct promoted regioselective sulfenylation of indoles with sulfinic acids. *Organic & Biomolecular Chemistry* 13: 2251–2254.

84 Zhao, X., Wei, A., Li, T. et al. (2017). Transition-metal free direct difluoromethylthiolation of electron-rich aromatics with difluoromethanesulfonyl chloride. *Organic Chemistry Frontiers* 4: 232–235.

85 Zhao, X., Li, T., Yang, B. et al. (2017). Transition-metal-free trifluoromethylthiolation and difluoromethylthiolation of thiols with trifluoromethanesulfonyl chloride and difluoromethanesulfonyl chloride. *Tetrahedron* 73: 3112–3117.

86 Zhu, C. and Wei, Y. (2011). An inorganic iodine-catalyzed oxidative system for the synthesis of benzimidazoles using hydrogen peroxide under ambient conditions. *ChemSusChem* 4: 1082–1086.

87 Ma, L., Wang, X., Yu, W., and Han, B. (2011). TBAI-catalyzed oxidative coupling of aminopyridines with β-keto esters and 1,3-diones–synthesis of imidazo[1,2-*a*]pyridines. *Chemical Communications* 47: 11333–11335.

88 Xu, X., Hu, P., Yu, W. et al. (2014). Bu_4NI-catalyzed synthesis of imidazo[1,2-*a*]pyridines *via* oxidative coupling of aminopyridines with nitroolefins. *Synlett* 25: 718–720.

89 Zhao, D., Wang, T., Shen, Q., and Li, J.-X. (2014). *n*-Bu_4NI-catalyzed selective dual amination of sp^3 C–H bonds: oxidative domino synthesis of imidazo[1,5-*c*]quinazolines on a gram-scale. *Chemical Communications* 50: 4302–4304.

90 Yang, X.-L., Chen, F., Zhou, N.-N. et al. (2014). Synthesis of isoxazoline-functionalized phenanthridines *via* iminoxyl radical-participated cascade sequence. *Organic Letters* 16: 6476–6479.

91 Xu, Y., Chen, H., Li, W. et al. (2018). Synthetic access to isoxazoline-functionalized isoquinolines *via* microwave-assisted iminoxyl radical-participated cascade cyclization of vinyl isocyanides. *Organic & Biomolecular Chemistry* 16: 4996–5005.

92 Zhang, M.-Z., Luo, N., Long, R.-Y. et al. (2018). Transition-metal-free oxidative aminooxyarylation of alkenes: annulations toward aminooxylated oxindoles. *The Journal of Organic Chemistry* 83: 2369–2375.

93 Yu, H., Huang, W., and Zhang, F. (2014). Metal-free oxidative cyclization of *o*-acylphenols for the construction of 2-aminobenzofuran-3(2*H*)-ones. *European Journal of Organic Chemistry* 2014: 3156–3162.

94 Liang, S., Zeng, C.-C., Luo, X.-G. et al. (2016). Electrochemically catalyzed amino-oxygenation of styrenes: *n*-Bu_4NI induced C–N followed by a C–O bond formation cascade for the synthesis of indolines. *Green Chemistry* 18: 2222–2230.

95 Lv, M., Wang, P., Yuan, D., and Yao, Y. (2017). Conversion of carbon dioxide into oxazolidinones mediated by quaternary ammonium salts and DBU. *ChemCatChem* 9: 4451–4455.

96 He, Y., Li, J., Luo, S. et al. (2016). A metal-free synthesis of 2-aminobenzothiazoles through aminyl radical addition to aryl isothiocyanates. *Chemical Communications* 52: 8444–8447.

97 Li, J.-S., Xie, X.-Y., Yang, P.-P. et al. (2020). Electrochemical synthesis of 1,2,4-thiadiazoles through intermolecular dehydrogenative S–N coupling. *Advanced Synthesis & Catalysis* 362: 771–775.

98 Zhao, X., Zhang, Y., Deng, J. et al. (2017). TBHP/TBAI-mediated oxidative cascade reaction consisting of dimerization, cyclization, and 1,2-aryl migration: metal-free synthesis of pyrrolin-4-ones and highly substituted pyrroles. *The Journal of Organic Chemistry* 82: 12682–12690.

99 He, J., Dong, J., Su, L. et al. (2020). Selective oxidative cleavage of 3-methylindoles with primary amines affording quinazolinones. *Organic Letters* 22: 2522–2526.

100 Karasala, B.K., Gollamudi, P., Inkollu, B., and Vidavalur, S. (2020). A metal- and ligand-free synthesis of quinazolin-4(3*H*)-ones *via* a Bu_4NI/TBHP-mediated oxidative cleavage of the olefinic C=C bond. *Russian Journal of Organic Chemistry* 56: 1446–1454.

101 Gao, P. and Wei, Y. (2013). Efficient oxidative cyclisation of acid hydrazides to 2,5-disubstituted 1,3,4-oxadiazoles catalysed by Bu_4NI with *t*-BuOOH as oxidant. *Journal of Chemical Research* 37: 506–510.

102 Shao, Y., Tong, J., Zhao, Y. et al. (2016). [3 + 2] Cycloaddition and subsequent oxidative dehydrogenation between alkenes and diazo compounds: a simple and direct approach to pyrazoles using TBAI/TBHP. *Organic & Biomolecular Chemistry* 14: 8486–8492.

103 Liu, W., Huang, W., Lan, T. et al. (2018). Synthesis of α-benzylated amides *via* electrocatalytic favorskii rearrangement of 1, 3-diarylacetones. *Tetrahedron* 74: 2298–2305.

104 Jiang, Y.-Y., Wang, Q.-Q., Liang, S. et al. (2016). Electrochemical oxidative amination of sodium sulfinates: synthesis of sulfonamides mediated by NH_4I as a redox catalyst. *The Journal of Organic Chemistry* 81: 4713–4719.

105 Ma, X., Xu, Q., Li, H. et al. (2018). Alcohol-based michaelis-arbuzov reaction: an efficient and environmentally-benign method for C–P(O) bond formation. *Green Chemistry* 20: 3408–3413.

106 Wu, Q.-L., Chen, X.-G., Huo, C.-D. et al. (2019). Electrochemically driven P–H oxidation and functionalization: synthesis of carbamoylphosphonates from phosphoramides and alcohols. *New Journal of Chemistry* 43: 1531–1535.

5

Catalytic Transformations Based on Iodine(I) Involving Radical Pathways

Alexandra E. Bosnidou[1] *and R. Martin Romero*[2]

[1] *Laboratoire de Synthese Organique, Ecole Polytechnique, CNRS, ENSTA, Institut Polytechnique de Paris, Palaiseau, France*
[2] *Université Paris-Saclay, CEA, CNRS, NIMBE, CEA Saclay, Gif-sur-Yvette cedex, France*

5.1 Introduction

Iodine and its related compounds containing iodine in higher oxidation states have found synthetic applications as key tools in various areas of organic, inorganic, and supramolecular chemistry [1]. Recent achievements involve a newly developed concept of iodine catalysis that establishes a powerful strategy toward a variety of oxidative transformations with the use of either electrophilic iodine reagents, or iodinated substrates formed during catalytic cycles, that usually accelerate catalyst turnover and allow for the formation of carbon–carbon or carbon–heteroatom bond [2]. These catalytic transformations in many cases are reminiscent of transition metal-catalyzed reactions. Nevertheless, iodine compounds are mild, stable, easy to handle, and eco-compatible reagents, advantageously competing with heavy metals. As such, it is not surprising that iodine catalysis has attracted the interest of the synthetic community [3].

Common reaction pathways for hypervalent iodine chemistry involve oxidative addition, ligand exchange, and reductive elimination. However, iodine can be found in the oxidation state of (I). Electrophilic iodine(I) reagents are now widely established as efficient synthetic tools for the selective functionalization of hydrocarbon molecules [4], which, in many cases, are employed as stable salt precursors. Alternatively, alkyliodines, that are formed during catalytic processes, are involved in numerous transformations involving radical pathways.

In this chapter, we summarize syntheses of iodine(I) complexes and catalytic applications of iodine(I) reagents or intermediates with a focus on radical pathways. Catalytic transformations involving R_4NI are not included as they are discussed in detail in another chapter of this book. Examples covered herein include the application of alkyl iodides(I) as reaction intermediates for carbon–heteroatom bond formation, among other catalytic applications of electrophilic iodine reagents.

Iodine Catalysis in Organic Synthesis, First Edition. Edited by Kazuaki Ishihara and Kilian Muñiz.

5.2 Synthesis of Iodine(I) Complexes

A useful concept that has recently emerged to attract significant attention for the synthesis of iodine(I) complexes consists of Lewis-base coordination to iodine. Such coordination not only results in stabilization of the electrophilic halogen atom but may also enable the fine-tuning of unique reactivity in subsequent synthetic transformations. Historically, Barluenga's reagent IPy_2BF_4 [5, 6] represents the most prominent of these compounds in which the electrophilic iodine(I) coordinates with two pyridine units. Subsequent modification has identified a large number of related coordination complexes based on numerous additional heterocycles that are prone to iodine coordination [7]. Moreover, the enhanced stability of these iodonium complexes has also been exploited in supramolecular chemistry [8].

Hayashi demonstrated that such nitrogen-based coordination to electrophilic iodine(I) can be extended to primary amines [9], and Yoshida introduced neutral donors such as DMSO that have also enabled the stabilization of electrochemically generated iodonium ions [10].

Additional noteworthy accomplishments include activation based on halogen bonding [11] and other cases of Lewis-base coordination [12].

In contrast to neutral ligands, charged ligands have received significantly less attention in the stabilization of electrophilic iodine reagents. In principle, coordination of two anionic ligands Y^- to the electrophilic iodine(I) in the presence of a suitable counter-cation Q^+ gives rise to complexes of the general structure $[Q][I(Y)_2]$.

Compounds of this type of molecular composition date back to the Simonini complex $AgI(OAc)_2$ derived from I_2 and 2AgOAc [13]. This compound was recognized as a source of acetyl hypoiodite (IOAc) in the Prevost deoxygenation of alkenes [14].

It is interesting to draw a comparison to the related class of usually highly reactive diarylated or dialkylated iodine-centered ate-complexes, which have been invoked as the key intermediates in halogen–metal exchange reactions [15].

5.2.1 Synthesis via Complexation of Neutral Precursors

Historically, the formation of iodide complexes was discovered by Ross, subsequently of the formation of bromides complex from preformed electrophilic halides. When investigating the electrochemical reduction of *N*-bromosuccinimide (NBS, **1**), they discovered that the addition of tetraalkylammonium succinimide $[NR_4][NSuccin]$ **2** (R = ethyl, butyl; NSuccin = succinimide) altered the behavior of NBS (Scheme 5.1). Based on IR data, the authors concluded the formation of a complex halide of the composition $[NR_4][Br(NSuccin)_2]$ **3** [16].

Subsequent work revealed that this and related compounds could be isolated and fully characterized. In an important extension, the authors demonstrated that the same reaction takes place for the corresponding *N*-iodosuccinimide (NIS, **4**) as well [17]. In addition, ammonium salts **5** with different anions than the formal ones in **4**

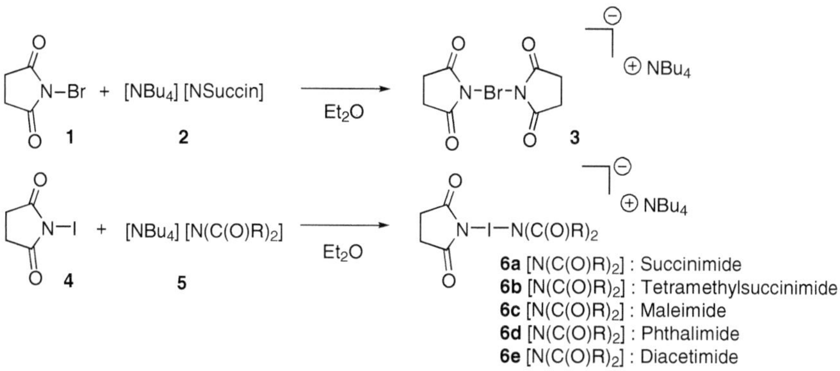

Scheme 5.1 Syntheses of bromate and iodate compounds through complexation.

can be employed leading to the formation of complex iodides **6** with nonsymmetrical coordination.

5.2.2 From Oxidation of Low-valent Iodine

The reaction between salts of iodide (I^-) and iodine(III) reagents results in an effective comproportionation reaction making use of the high oxidation potential of the unnatural iodine(III) oxidation state triggering an iodine(I) formation, which coordinates the two anionic groups from the former iodine(III) reagent. This context was first observed by Doleschall and Tóth [18]. Usually, ammonium salts of iodide are employed as iodine sources. This was demonstrated by Suárez, who reported the pioneering synthesis of $[NBu_4][I(OAc)_2]$ **8a** from the most common iodine(III) reagent (diacetoxyiodo)benzene (PIDA) [19]. While this compound has demonstrated expeditious reactivity in a number of oxidation reactions, it exhibits relatively low stability and usually is prepared *in situ*. This is a versatile strategy since the only by-product is the reduced iodobenzene, which is usually inert under the subsequent oxidative transformations. More stable iodine(I) reagents **10** can be generated using benzoate groups as ligands in the iodine(I) core. Their synthesis follows the same route and uses [bis(benzoyloxy)iodo]benzene **9** (Ar = Ph) instead of PIDA. A series of benzoate compounds were generated from the corresponding iodine(III) oxidants using this route (Scheme 5.2) [20].

In addition to ammonium salts, the corresponding phosphonium salts **12** were introduced by Kirschning following the standard oxidation procedure with PIDA as iodine(III) oxidant (Scheme 5.3a) [21]. Recently, Kashyap added the corresponding trimethylsulfonium salt **14** to this class of reagents, which was generated in an identical manner by treatment of commercial Me_3SI (**13**) with PIDA in dichloromethane solution (Scheme 5.3b) [22].

Additional aminated compounds can be generated by the use of the corresponding preformed iodine(III) reagents incorporating the transferable nitrogenated entities. For example, the saccharide derivative $[NBu_4][I(NSacc)_2]$ **16a** and the phthalimide derivative $[NBu_4][I(NPhth)_2]$ **16b** were obtained in this way (Scheme 5.4) [23, 24].

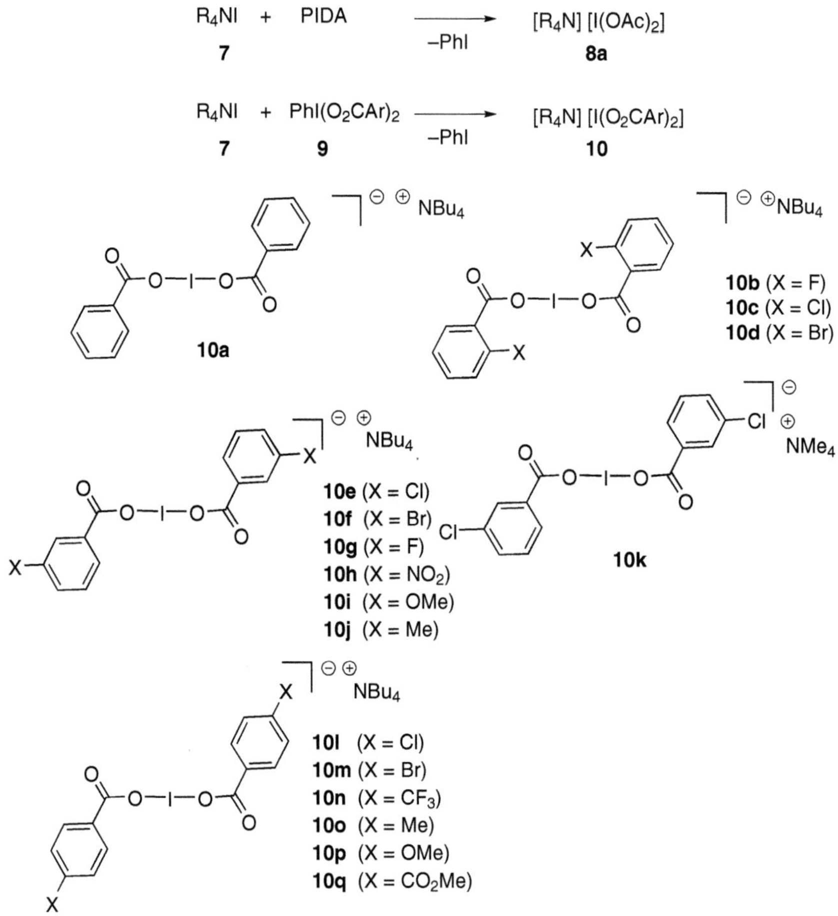

Scheme 5.2 Syntheses of iodate compounds through oxidation of iodides.

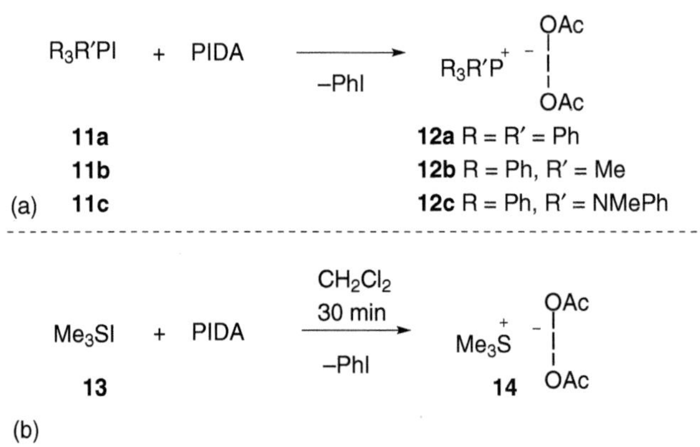

Scheme 5.3 Cation variation in iodate complexes: (a) Phosphonium salts; (b) Sulfonium salt. *Source*: Reddy et al. [22]; Rao et al. [22].

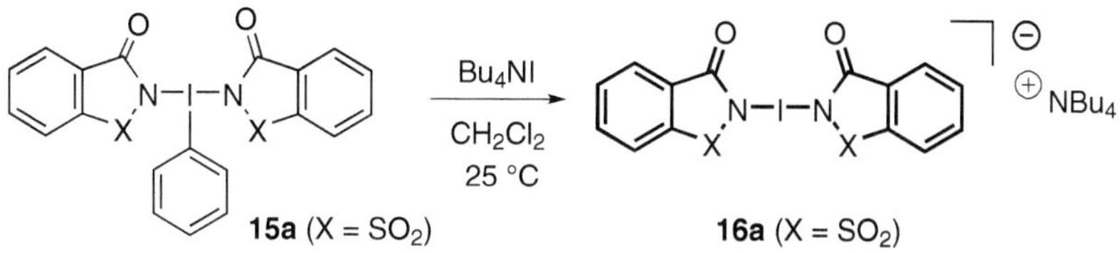

15a (X = SO_2)
15b (X = CO)

16a (X = SO_2)
16b (X = CO)

Scheme 5.4 Synthesis of bis(amidato)iodate complexes.

Comparison of the values of the bond distances between ligands and the iodine core from complexes with anionic ligands such as carboxylates, which are in the range of 2.1778–2.1962 Å [20], with the corresponding complexes with neutral ligands such as pyridine (2.255–2.261 Å) [5, 6] and DMSO adducts (2.27 Å) [10], demonstrates a stronger dative bond in the former due to shorter bonding, and, thus, stronger interaction with the iodine core for the anionic ligands.

5.2.3 Preparation of Polymer-bound Reagents

In a seminal accomplishment, Kirschning explored the possibility to incorporate the ammonium counterion into polymeric support in order to arrive at polymer-bound electrophilic iodine reagents [25]. This approach does not apply to the concept of catalysis. However, it enables a convenient recovery of the ammonium core for subsequent reuse. While the synthesis of the preformed reagents can usually be accomplished readily throughout the synthetic routes outlined above, there are several reports that rely on the *in situ* formation from the corresponding precursors. This strategy has been used preferentially for the cases employing the direct oxidative transformation of low-valent halogen precursors.

The stability of the iodine-carboxylate coordination in iodate complexes has been exploited in the construction of supramolecular assemblies. Sindelár used iodates as axles in the synthesis of three rotaxanes **18** in combination with the bambusuril macrocycle **17** as a wheel (Scheme 5.5) [26].

The authors propose that the threading process involves dissociation of the iodate to a neutral acylhypoiodite and tetrabutylammonium carboxylate. In such a way, the carboxylate **8** could interact with the bambusuril core **17** followed by re-complexation to the electrophilic iodine in a process that would resemble the reactions discussed in Scheme 5.2.

5.3 Reactivity

In iodine(I) reagents, the iodine atom is stabilized by the simultaneous coordination of two electron-donating ligands. According to molecular orbital theory, the model of three-center four-electron (3c-4e) provides an understanding of the structure of the iodine-ligand bonds. The three atoms of the system [L-I-L]* may either be in a

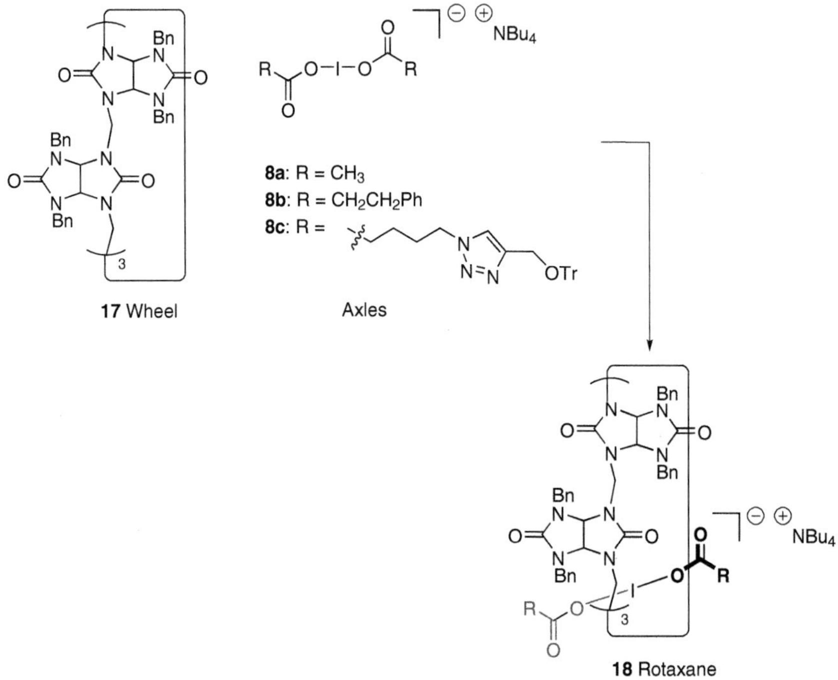

Scheme 5.5 Use of iodate complexes in the construction of rotaxanes.

static, symmetric geometry [L--I--L]*, or in fast equilibrium [L-X--L]* ↔ [L--X-L]*. In the symmetric geometric state, the iodine core forms two ligand-iodine bonds of equal distance and strength. However, for the case of nonsymmetric and dynamic systems, the iodine core forms one stronger and shorter covalent bond and a weaker and longer one, in each of the isomers.

Thus, the electrophilic iodine(I) reagents usually react through ligand exchange reactions, which is very briefly mentioned here as we are going to focus on the involvement of these types of reagents in radical pathways.

5.3.1 Ligand Exchange Reactions at Iodine

Kirschning reported the synthesis of a novel iodine(I) reagent that was used for haloazidation of alkenes via ligand exchange. Treatment of diacetoxyiodate(I) reagents **19** with TMS azide led to the formation of the corresponding isolable disazidoiodate(I) species **20** (Scheme 5.6) [27]. The same compounds can be

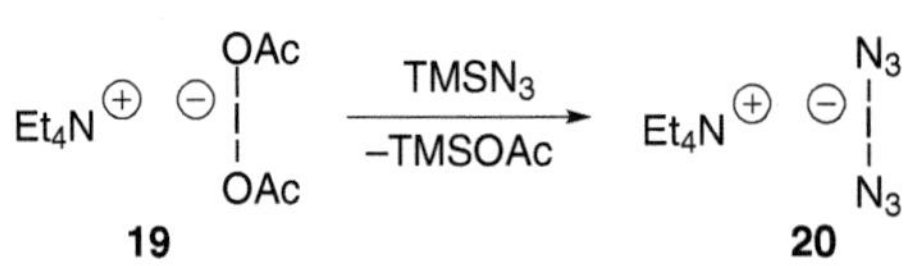

Scheme 5.6 Ligand exchange reactions at iodate complexes.

accessed through oxidation of the halide precursors with (diazidoiodo)benzene $PhI(N_3)_2$. Since the synthesis of this particular reagent [28] is cumbersome, the anion exchange is usually the preferred route. The same process was successfully conducted for a polymer-supported iodine reagent [29].

Attempts to identify anionic ligand exchange reactions in case of the benzoate complexes **10** have so far not led to conclusive results [20].

5.3.2 Radical Reactivity

The majority of the reactions involving Barluenga's reagent are stoichiometric and proceed through ionic mechanisms [6]. However, they can also be involved in radical pathways. The latter is exemplified by the synthesis of ω-iodocarbonyl compounds **22** from cyclic alcohols **21** (Scheme 5.7) [6b, 30].

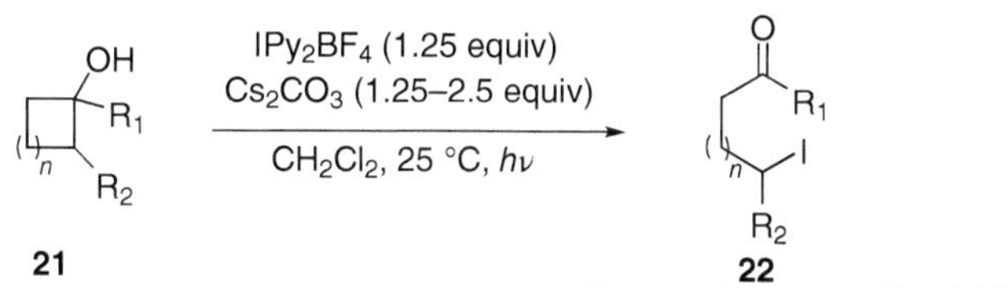

21 **22**

Scheme 5.7 Synthesis of ω-iodocarbonyl compounds with Ipy_2BF_4.

This radical pathway involves the *in situ* formation of the oxonium ion **A** that leads to the hypoiodite **B** upon deprotonation. Light-driven homolytic cleavage affords the radical **C** which, through a β-scission process, forms the carbon-centered radical **D** that finally recombines with the iodanyl radical to afford the final product **22** (Scheme 5.8).

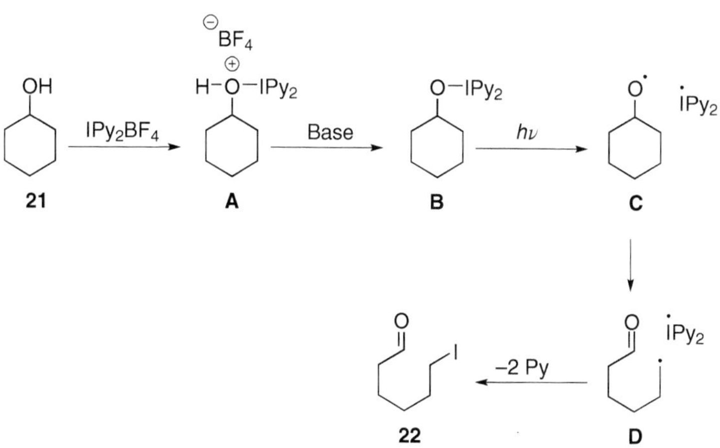

21 **A** **B** **C** **22** **D**

Scheme 5.8 Radical mechanism involving iodine(I).

The interaction between a heteroatom with the iodine core followed by a homolytic cleavage will be the basis for the iodine(I) reactivity which is involved in radical transformations.

5.4 Application in Catalysis

Most of the reactions mentioned in the previous section encompass the incorporation of the iodine atom into the product, thereby removing it from potential regeneration. However, in those cases that iodine is liberated throughout the oxidative transformation, the newly formed iodide(I^-) should be prone to reoxidation and thus the regeneration of the original iodide reagent (Figure 5.1, iodine(III) oxidant is chosen as representative reoxidant). Such a conceptual approach would enable the use of iodide complexes as catalysts in homogeneous organic transformations.

Alternatively, alkyl iodide(I) species are formed commonly throughout catalytic reactions. Such a concept envisions iodine catalysis that departs from an iodine(III) catalyst state at the outset of the functionalization of the hydrocarbon substrate (Figure 5.2 left). Prominent examples to this end are catalytic alkene difunctionalization and oxidation reactions of enolates. Here, the interaction with the organic molecule initiates the oxidation reaction, and upon subsequent manipulation of the resulting intermediary alkyl-iodine(III) single bond, reductive deiodination of an iodine(I) catalyst state occurs. A subsequent reoxidation step to the initial iodine(III) state proceeds outside the functionalization of the organic substrate leading to an overall catalytic cycle of iodine(III/I).

As to the strategic alternative, an initial functionalization of the organic substrate at the stage of an iodine(I) catalyst leads to the formation of an alkyl iodide(I) catalyst

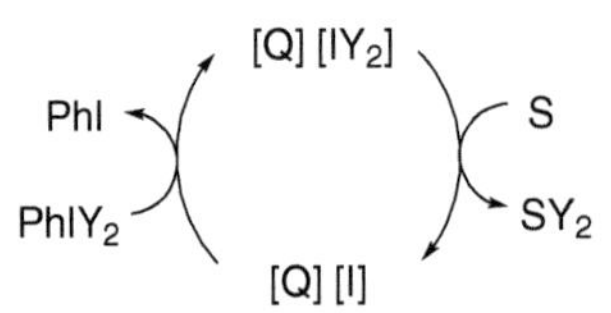

Figure 5.1 Conceptual background for iodine(I^-/I(I)) oxidation catalysis involving complex halide catalyst states.

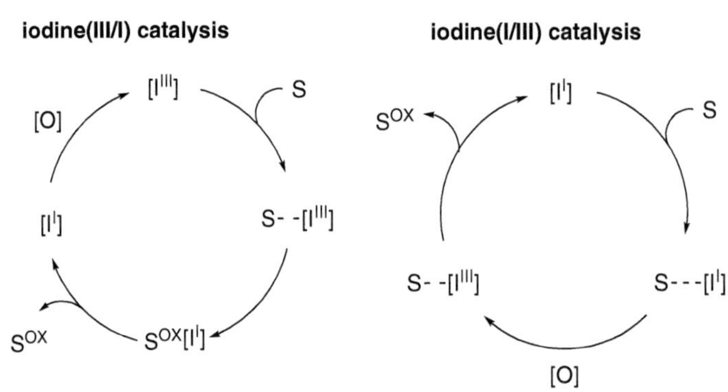

Figure 5.2 Conceptual background for iodine(I/III) and iodine(III/I) oxidation catalysis involving complex halide catalyst states.

state. In the presence of a suitable oxidant, the subsequent oxidation to a low-living iodine(III) species induces reductive displacement of iodine to furnish the oxidized substrate and regenerate the iodine(I) catalyst from the outset (Figure 5.2 right). Such a concept promotes the oxidation of the iodine catalyst within its presence in the transformable organic substrate [31].

5.4.1 Amination Reactions

The general interest in the field of intra- and intermolecular amination reactions has led organic chemists to the development of reliable methods based on iodine catalysis. Herein, we summarize the existing protocols that follow radical pathways for C(sp^3)-H and C(sp^2)-H intra- and intermolecular aminations.

Based on the pioneering stoichiometric work by Suárez [32], Muñiz was the first to employ iodine catalysis for the Hofmann–Löffler reaction, [33] providing 5-membered heterocycles, commonly pyrrolidines **24**. In this context, two different protocols were developed, which proceed within iodine(I/III) [33a] and iodine (-I/I) [33b] catalysis manifolds (Scheme 5.9, conditions **a** and **b** respectively).

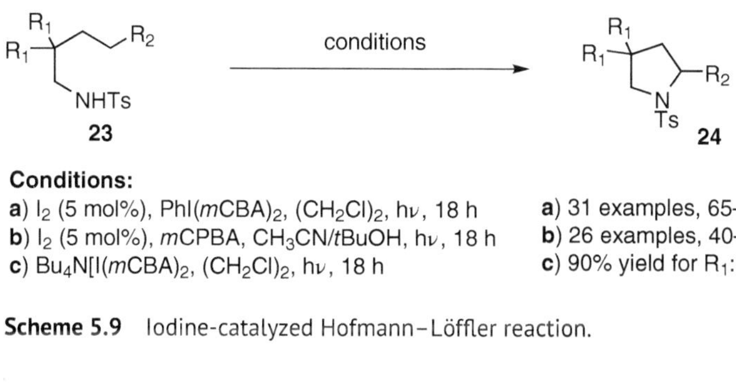

Conditions:
a) I_2 (5 mol%), PhI(*m*CBA)$_2$, $(CH_2Cl)_2$, hν, 18 h
b) I_2 (5 mol%), *m*CPBA, CH_3CN/*t*BuOH, hν, 18 h
c) Bu_4N[I(*m*CBA)$_2$, $(CH_2Cl)_2$, hν, 18 h

a) 31 examples, 65–98% yield
b) 26 examples, 40–99% yield
c) 90% yield for R_1:Me and R_2:Ph

Scheme 5.9 Iodine-catalyzed Hofmann–Löffler reaction.

The initial step of the reaction includes the comproportionation between molecular iodine and the oxidant for the formation of the iodine(I) catalyst or hypoiodite bearing a *meta*-chlorobenzoate [I(*m*CBA)]. This newly formed electrophilic catalyst promotes the iodination of the substrate **23**, toward **A**. Upon exposure to light, the N–I bond is cleaved affording the nitrogen-centered radical **B**, which engages the 1,5-hydrogen transfer (1,5-HAT) yielding the carbon-centered radical **C**. Iodination through iodine abstraction from another *N*-iodinated intermediate in a radical chain mechanism allows the formation of the benzylic iodide **D**. Depending on the chosen terminal oxidant, iodine(III) or *m*CPBA, and the degree of reactivity at the respective iodinated carbon of **D**, immediate cyclization to the pyrrolidine **24** can occur at this stage through the nitrogen nucleophile (red pathway). Alternatively, suitably engineered oxidation to an intermediary alkyliodine(III) state **E** may be pursued (green pathway). Oxidation to the alkyliodine(III) intermediate **E** triggers the final cyclization step to the corresponding pyrrolidine product **24** (Scheme 5.10).

Alkyl iodides(I) have been postulated as intermediates in iodine-catalyzed amination reactions. A common feature of these reactions is the formation of

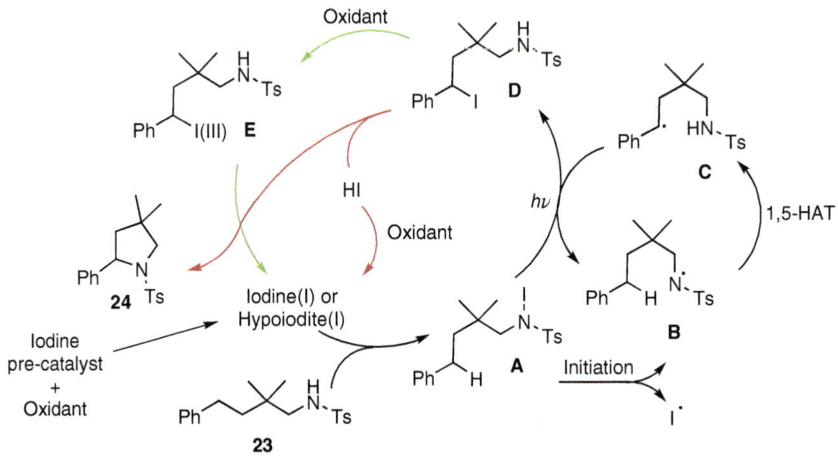

Scheme 5.10 Reaction mechanism for the iodine catalytic protocols of Hofmann–Löffler reaction.

an electrophilic iodine(I) catalyst state, which reacts with the free N-H group of the sulfonamide substituent of the substrate. In order to demonstrate that the iodine(I) is the putative catalyst state $H[I(O_2CAr)_2]$, the authors synthesized $[Bu_4N][I(O_2CAr)_2]$ (Scheme 5.2) as a more stable model compound and submitted it for reaction. They observed the product with 90% yield (Scheme 5.19, condition c).

This iodine catalysis has been designed based on the reactivity of the sulfonamidyl radical and the oxidation of alkyl iodide(I) to alkyliodine(III) that would allow for the functionalization of nonactivated positions. Worthy to note that computational chemistry has confirmed the involvement of the putative sulfonamidyl radical [33], and recently their experimental determination has been achieved [34]. The enhanced nucleofugacity of alkyliodine(III), in comparison with the parent iodine(I) species has been disclosed recently by Bosnidou and Muñiz [31].

Noteworthy, Muñiz reported on an enantioselective total synthesis of nicotine using a direct C–H amination reaction within the Hofmann–Löffler cyclization as the key step. This synthesis stands out as it tolerates free pyridine units without inhibition of the electrophilic iodine reagent. This successful approach streamlines chemoselective C–H amination promoted by iodine(I) reagents and it is expected to trigger further application of this methodology in the synthesis of heterocyclic alkaloids [35].

Interestingly, a different protocol that affords the same pyrrolidine products has also been developed, involving photoredox catalyst and molecular iodine as promoters for the amination reaction [36]. Mechanistically, the proposed reaction sequence consists of two individual light-induced catalytic reactions within several intertwined individual cycles. The authors used Raman spectroscopy to identify the active iodine(I) catalyst. Upon the formation of the iodine(I) catalyst, the reaction follows the same mechanistic scenario as described in Scheme 5.10.

Nagib, based on the studies of Muñiz, reported a catalytic β C–H amination via imidate radical relay [37]. The reaction requires catalytic amounts of molecular iodine in combination with the iodine(III) oxidant in order to generate the acyl hypoiodite, that is the active catalyst, and leads to the iodination of the imine **25**, which upon homolytic cleavage is going to generate the nitrogen-centered radical. This radical via 1,5-HAT and a radical chain mechanism is transferred to the carbon and affords the corresponding alkyl iodide(I), which upon oxidation from the terminal oxidant and intramolecular attack is going to lead to the cyclized intermediate which submitted afterward in hydrolysis conditions to afford the product **26** (Scheme 5.11).

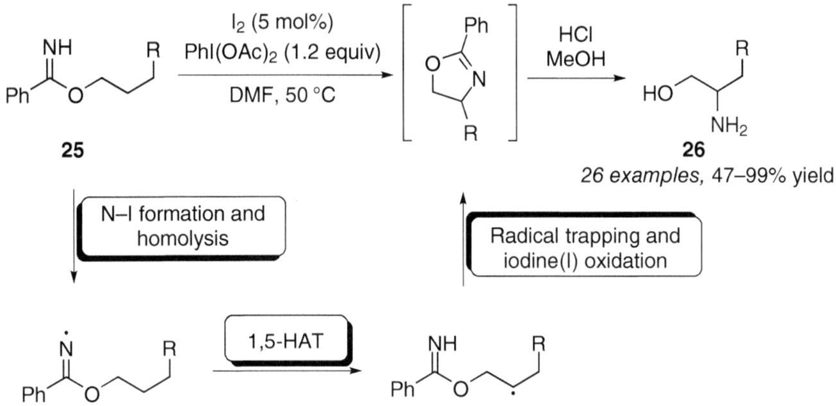

Scheme 5.11 Iodine-catalyzed β C–H amination.

Apart from the concept of five-membered ring formation based on 1,5-HAT, the exploration of conditions in order to access six-membered rings was initiated by Muñiz. In 2017, they identified mild and uniform conditions for a selective iodine-catalyzed C–H amination to obtain 2-aryl-substituted piperidines **28** (Scheme 5.12). This reaction overrides the common preference for pyrrolidine formation within the Hofmann–Löffler manifold and significantly enlarges both the scope of light-induced catalysis and position-selective C–H amination reactions [38].

The reaction is initiated with a visible-light-assisted homolytic cleavage of the N–Br bond in NBS. The N-centered succinimidyl radical then abstracts a hydrogen atom at the benzylic position of substrate **27**. The intermediary benzylic radical **A** abstracts an iodine atom from a halogen-bonded I_2–NBS adduct **B** [39] to generate IBr and the intermediary benzyl iodide **C** and regenerates the succinimide radical. The formation of the latter closes the catalytic cycle of radical C–H functionalization. The benzylic iodide undergoes nucleophilic substitution to the pyrrolidine product **28**. The liberated HI regenerates the molecular iodine catalyst upon reaction with IBr. Molecular iodine recoordinates with NBS within a halogen-bonding mode, which closes the second catalytic cycle of iodine catalysis. The postulation of a free radical mechanism outside the classical *N*-centered radical from *N*-halogenation as in the Hofmann–Löffler scenario is in agreement with the observation that

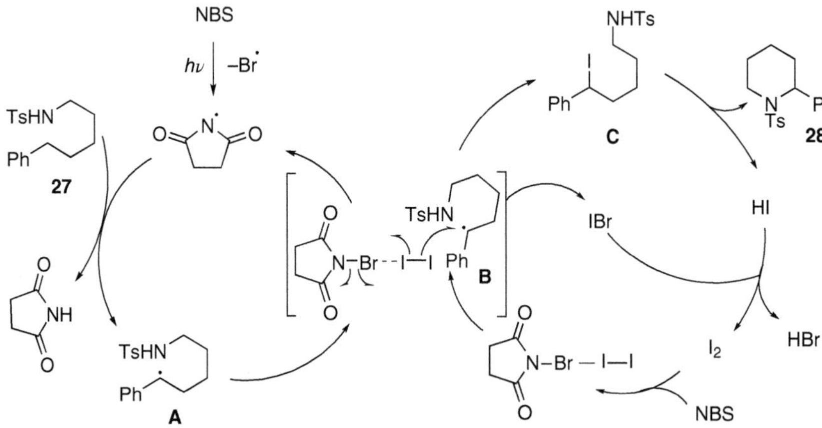

Scheme 5.12 Iodine-catalyzed C–N bond formation through radical C–H abstraction.

molecular iodine alone does not convert tosylamides into their *N*-iodinated derivatives.

An alternative route to compounds containing 6-membered rings via 1,6-HAT processes was highly desirable. Following the behavior of sulfamate esters that could drive this process [40], Minakata and Kiyokawa developed an intramolecular amination of aliphatic bonds of sulfamate esters with stoichiometric amounts of *t*-BuOI, which led to the formation of protected 1,3-aminoalcohols [41].

At exactly the same time, Muñiz reported on a route to 1,3-diamines **30** through Ritter-type amination, with the use of sulfamides as amidyl radical precursors and, at the same time, traceless removable functional entities [42] (Scheme 5.13). The idea was based on the selective halogenation reactions at tertiary C–H bonds, where the related sulfamate ester group was of unique effectiveness as it performs 1,6-HAT, and it exhibits a low tendency to undergo amination itself [40a, b, d].

The mechanistic scenario is reminiscent of one of the iodine-catalyzed Hoffmann–Loffler reactions (Scheme 5.10), but with the difference that the nucleophilic attack at the stage of the alkyliodine(III) is performed by an external nucleophile (Scheme 5.13).

The application of the concept on high-oxidation state iodine catalysis [31] recently enabled the realization of a unique intermolecular C–H amination of alkanes **31** to amines **32** under iodine catalysis (Scheme 5.14) [43]. This reaction proceeds with high regioselectivity regarding the C–H bond and enables the synthesis of alkylamines, benzylamines, α- and β-amino acids, and aminals of α-amino esters. Its scope and applicability have been demonstrated through four intermediates in the synthesis of marketed pharmaceuticals.

The underlying catalytic cycle employs an iodine(I) catalyst that is reminiscent of the intramolecular version. Its reaction with the sulfonamide generates the N–I derivative **B**, which under visible light generates an amidyl radical **C** that triggers the intermolecular abstraction of hydrogen from the alkane substrate **31**. Radical recombination with the iodine radical leads to the formation of the intermediary

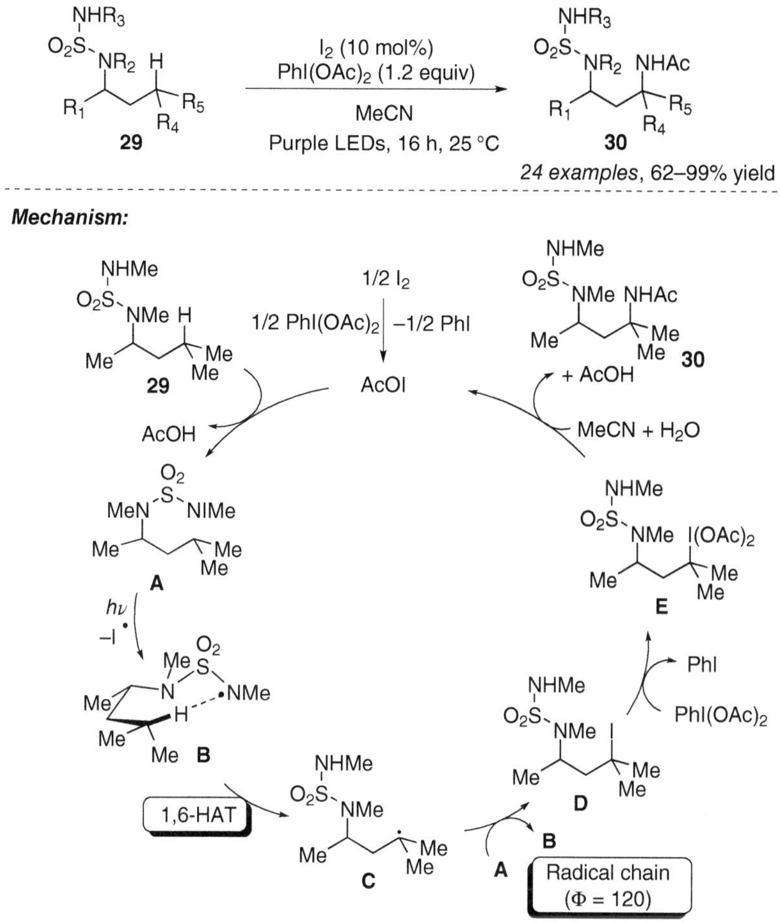

Scheme 5.13 1,3-Diamines formation through 1,6-HAT promoted by iodine(I).

alkyl iodide(I) **D**, which requires the oxidation to the corresponding high-oxidation state iodine(III) state **E** for the nucleophilic amination to proceed. This context was unambiguously demonstrated by a control experiment with 1-iodoethylbenzene **33**, which gives the C–N bond formation to **32** only in the presence of $PhI(O_2CAr)_2$ oxidant. The low nucleophilicity of the amino group is overcompensated by the high nucleophilicity of iodine(III) confirming the viability of the high-oxidation state concept as an enabling tool in catalysis.

Similarly, Minakata reported on the Ritter-type amination of tertiary aliphatic bonds under iodine catalysis with the use of *N*-hydroxy phthalimide (NHPI) and iodic acid in catalytic amounts [44].

The combination of catalytic amounts of molecular iodine and hypervalent iodine(III) as oxidant served for the oxidative C(sp^2)-H amination as well. A new iodine catalysis protocol was developed that provides conditions for a mild and selective intramolecular aryl amination reaction [45]. This chemistry demonstrated that

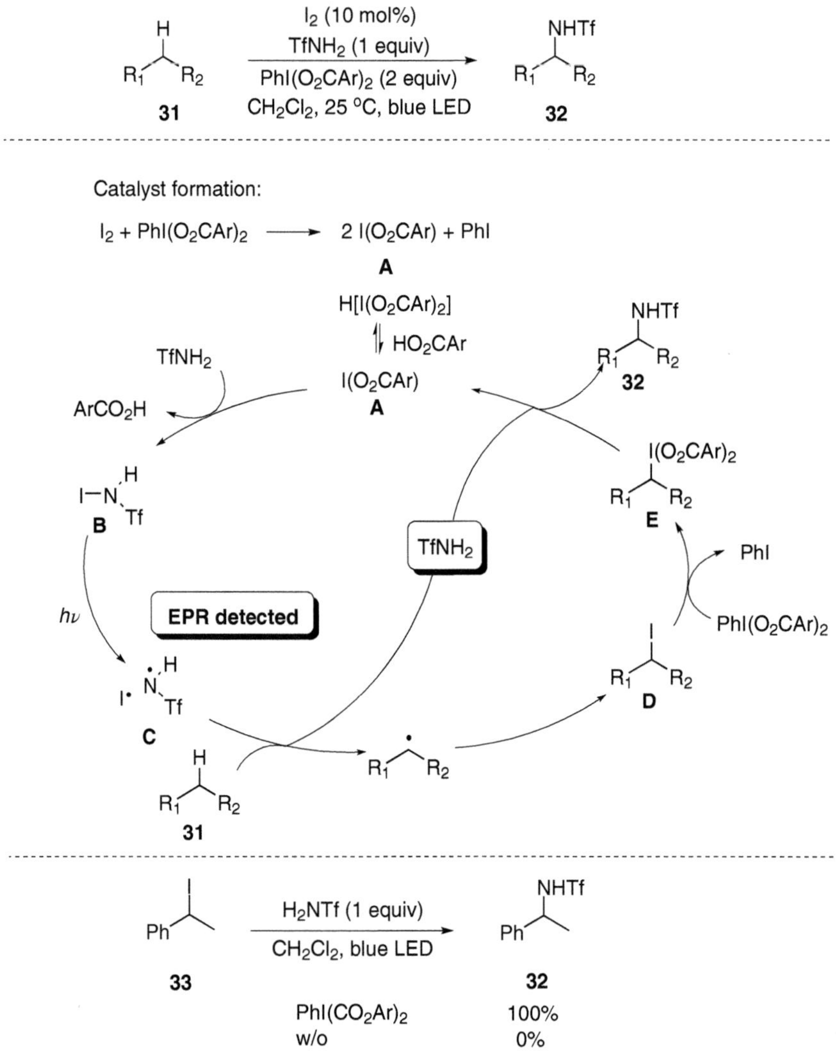

Scheme 5.14 Iodine(I/III)-catalyzed radical intermolecular amination of aliphatic C–H bonds. Ar = 4-Br-C_6H_4.

photochemical iodine catalysis can provide general oxidative amination conditions for the cases of aliphatic and aromatic hydrocarbons alike (Scheme 5.15).

Regarding the mechanism of the reaction, the authors claim a radical mechanism for the C–N bond formation (Scheme 5.16). The authors, based on the same principle as discussed before for the C(sp^3)-H amination, claimed that the electrophilic iodine(I) promotes the iodination of the substrate **34**, toward intermediate **A** which upon homolytic cleavage leads to the radical **B**. Its oxidation and aromatic substitution lead to the final product **35**. In order to prove the formation of the electrophilic iodine(I) catalyst, which has been postulated, they use ammonium iodates as suitable surrogates for the unstable putative compounds $H[IX_2]$.

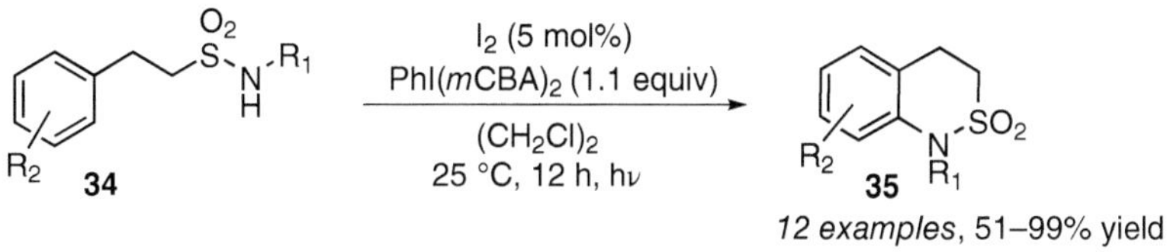

Scheme 5.15 Iodine catalyzed intramolecular amination of arenes.

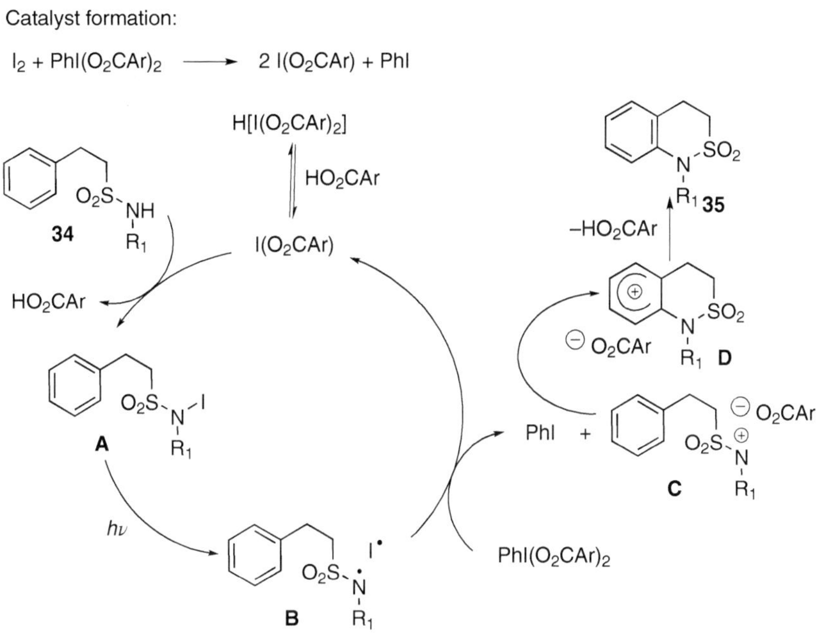

Scheme 5.16 Radical mechanism for the iodine(I)-promoted C–H amination reaction.

A related behavior of electrophilic iodine(I) reagent was observed in the intermolecular C(sp^2)-H amination of heterocycles, where sub-stoichiometric amount of the iodate **16a** was observed to be involved in the initiation of the radical chain mechanism that is operative for the amination of caffeine (**36**). [23] For this case, the bis(saccharinato)iodate **16a** from Scheme 5.4 could be used as a model compound to determine the initial formation of the crucial saccharin radical **B** (Scheme 5.17). Indeed, compound **16a** initiates the radical chain reaction with **B**, providing radical intermediate **C** which is oxidized by the iodine(III) reagent. This oxidation regenerates the self-replicating radical carrier **B** and provides reduced iodobenzene together with a saccharide, which deprotonates the oxidized cationic intermediate **D** to the final product **37**.

5.4.2 Oxygenation Reactions

Several research groups have investigated iodine catalysis for selective oxygenation reactions. In all the cases, the iodide salt is oxidized by a peracid to generate *in situ*

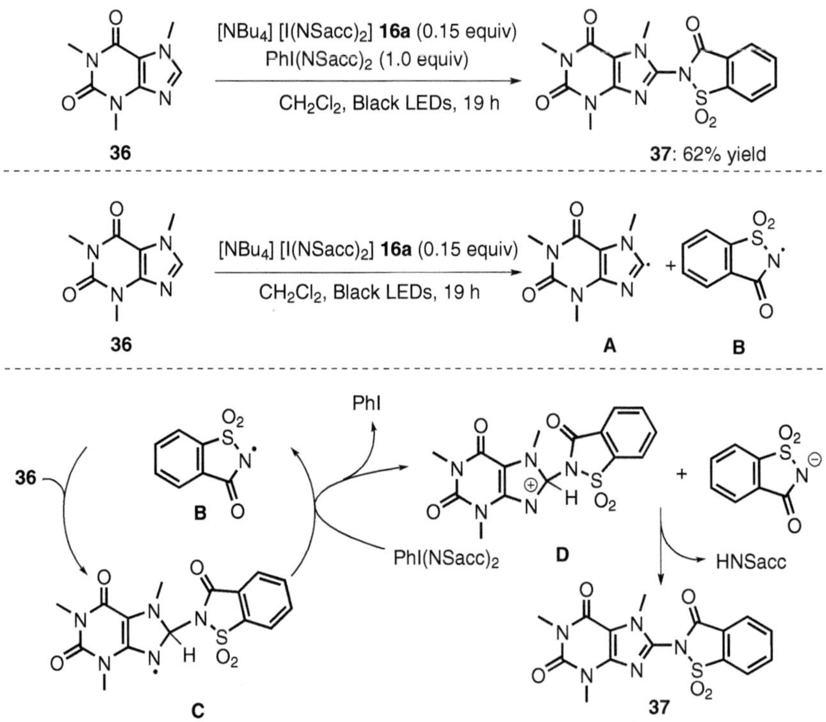

Scheme 5.17 Iodate-initiated radical C–H amination reactions.

(hypo)iodite species which is the actual catalyst for these transformations, similar to the aminations described before. Hypoiodite species are unstable under thermal or irradiation conditions providing free radicals which undergo selective hydrogen atom abstraction at the weakest C(sp^3)–H bond. Subsequent iodination and oxygenation occur to afford the desired final product.

Ishihara and coworkers established oxygenation reactions under iodide catalysis. Both intra- or intermolecular oxygenations were developed using tetrabutylammonium iodide as precatalyst and either hydrogen peroxide or *tert*-butyl hydroperoxide (TBHP) as oxidant [46]. As a result, lactonization could be accessed by cyclization of carboxylic acid derivatives and α-oxygenation of ketones could be developed using carboxylic acid as an oxygen source. Remarkably, they also designed chiral ammonium iodide salts for the enantioselective cyclization of phenols [47]. As an application of the latter, they were able to use these chiral ammonium salts to access tocopherol derivatives [48]. All these transformations were developed with an iodide salt as a catalyst. The combination of hypervalent iodine(III) oxidant and molecular iodine as catalyst enables direct oxidative cyclization of carboxylic acids as well, following a radical mechanism. The group of Yokoyama designed a methodology in which [bis(trifluoroacetoxy)iodo]benzene (PIFA) is used as a terminal oxidant [49]. 2-benzylbenzoic acid derivatives **38** were successfully cyclized using this protocol. Indeed, since the p*K*a of benzoic acids are higher than trifluoroacetic acid (TFA), ligand exchange occurs at the hypervalent iodine(III). Further reaction with molecular iodine provides the benzoyl hypoiodite(I) species **A**. At this stage, no decarboxylation reaction was observed since benzoic acid derivatives in mild conditions do not decarboxylate. A subsequent homolytic cleavage followed by a selective 1,5-HAT occurs. The newly formed C-centered radical **C** is quenched by molecular iodine to provide a benzylic iodide intermediate **D** that is prone to cyclize affording the corresponding lactones **39** (Scheme 5.18).

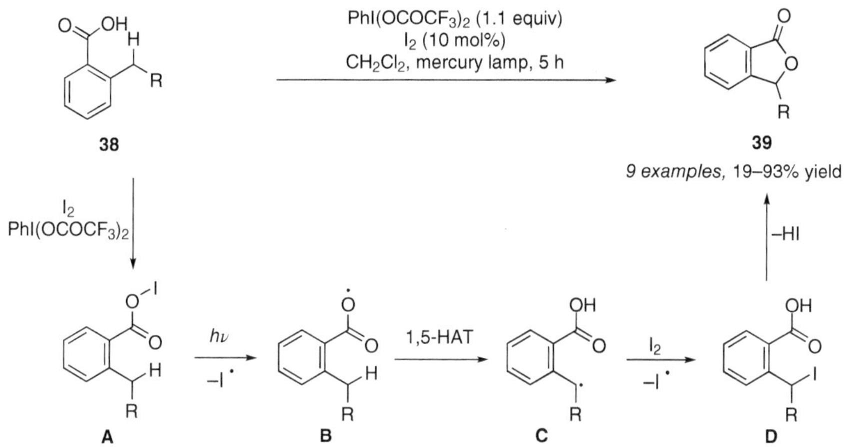

Scheme 5.18 Radical iodine catalysis for the lactonization of 2-benzylbenzoic acid derivatives.

Muñiz, after the successful application of cooperative catalysis between a photocatalyst and molecular iodine for amination reaction [36], employed the same system for the formation of γ- and δ-lactones [50]. Mechanistic investigations were carried out as well to determine with accuracy the mechanism of these transformations. Interestingly, the mechanism of the oxygenation differs from the one in the amination, in the step of the alkyliodine oxidation, where the authors claim oxidation to the iodoso compound **D**, prior to cyclization (Scheme 5.19).

Minakata based on his previously reported system for Ritter-type amination, [44] described an oxidation system with iodic acid (HIO_3) as an oxidant, in the presence of *N*-hydroxyphthalimide (NHPI) that allowed for the selective hydroxylation of tertiary C–H bonds and the lactonization of carboxylic acids containing a tertiary carbon center [51]. The mechanism involves initial oxidation of the *N*-oxo-phthalamide by iodic acid, toward the hypoiodite which, upon homolytic cleavage, leads to the formation of the carbon radical **A** and liberation of iodine and water. The resulting alkyl iodine **B** is oxidized to generate the corresponding I(III) species **C** that upon nucleophilic substitution generates the final products **43** (Scheme 5.20).

Interestingly, in a related acetoxylation of tertiary aliphatic bonds with the use of molecular iodine and PIDA as oxidants, where the reaction initiates with the formation of the hypoiodite species, the mechanistic scenario follows an ionic pathway [52].

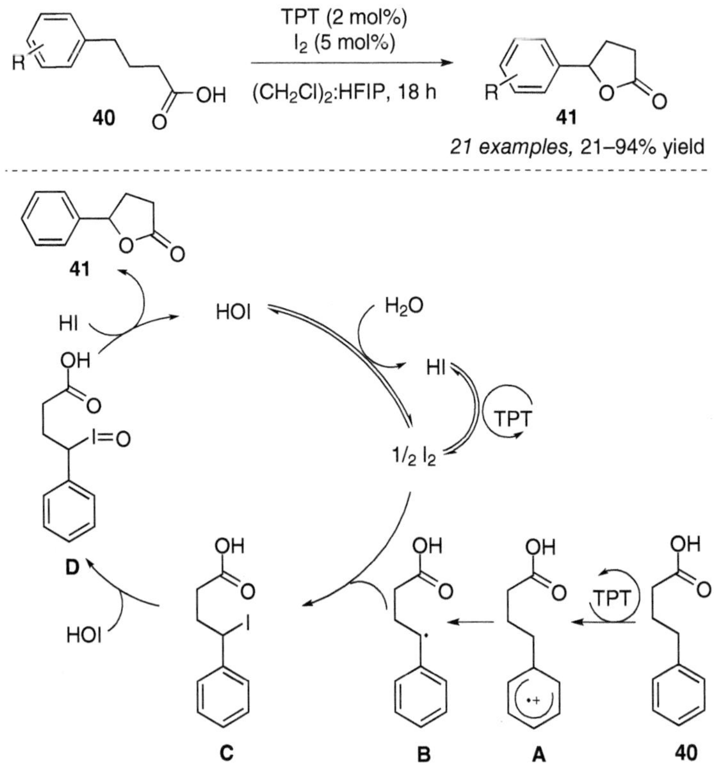

Scheme 5.19 Formation of γ-lactones via cooperative iodine catalysis.

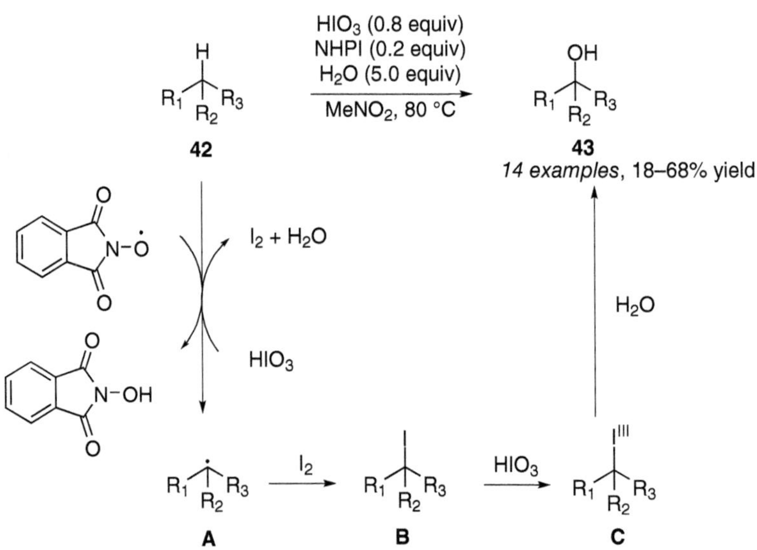

Scheme 5.20 Radical initiation for the iodine-catalyzed oxygenation of tertiary aliphatic bonds.

A detailed explanatory overview of oxygenation and amination reactions using iodine reagents has been recently disclosed by Minakata [53].

5.4.3 C–F Bond Formation

The majority of halogenation reactions that are based on the use of the electrophilic halide reagent are stoichiometric as the reagent usually reacts through ligand exchange, installing the halogen moiety on the substrate. [5, 6, 54]. An exception to that principle was published recently by Muñiz, who developed nucleophilic fluorination with the use of catalytic amounts of molecular iodine, hypervalent iodine(III) as terminal oxidant, and triethylamine trihydrofluoride as fluorine source [55]. The authors took advantage of the acquired knowledge on amidyl radicals [42] in combination with iodine catalysis to address position selectivity toward the formation of *tert*-alkyl fluorides **42** (Scheme 5.21).

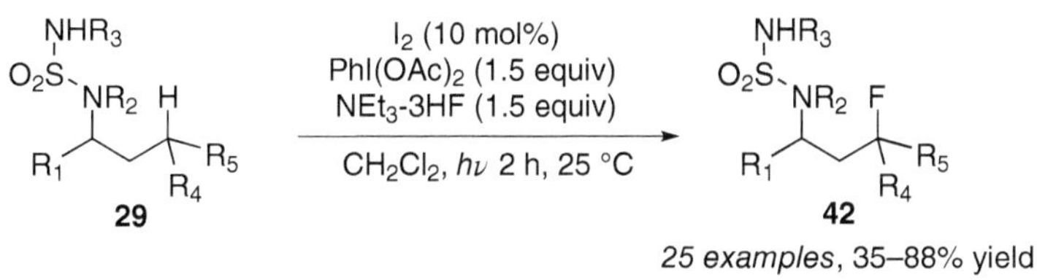

Scheme 5.21 Nucleophilic fluorination via iodine catalysis.

5.4.4 β-Functionalization of Alcohols

Apart from the concept of iodination through iodine(I) reagents and further homolytic cleavage toward the formation of two radicals, different approaches for other transformations have been developed. In 2016, Lei reported an oxidative β-Csp^3-H functionalization of *tert*-butanol (**44**) for the construction of C–S bonds through iodine catalyzed Csp^3-H/S-H coupling [53]. Notably, iodine catalysts might exhibit weaker coordination than transition metals with the mercaptans.

The authors demonstrated the radical process by trapping intermediates with TEMPO and detecting the primary alkyl iodide, by GC-MS. They claim that the reaction is initiated with the α-iodination of *tert*-butanol followed by homolytic cleavage leading to the carbon-radical **B**, which will recombine with the sulfur radical **D** produced by the aryl disulfide **C** to afford the product **45**. At the same time, oxidation of succinimide by iodine and di-*tert*-butyl peroxide (DTBP) regenerates the NIS as the iodine(I) catalyst (Scheme 5.22).

In a similar manner, Sekar in 2020 reported on the synthesis of α,β-diphenylthio enones from secondary alcohols [56]. The reaction between secondary, benzylic alcohols **46** and thiophenols gave a wide scope of the corresponding enones **47** in good yields (Scheme 5.23). Worthy to note that the authors do not consider a radical pathway for the reaction mechanism, but an ionic one instead.

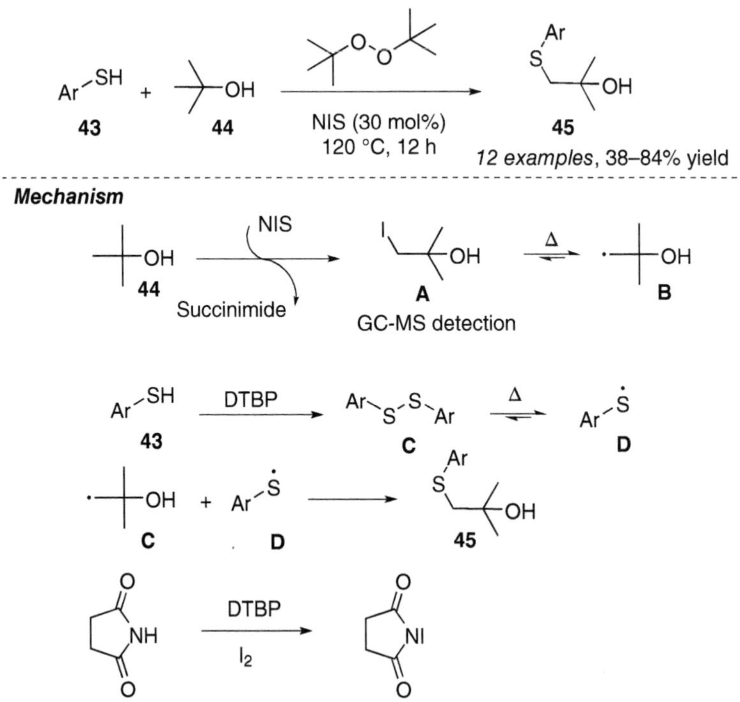

Scheme 5.22 Synthesis of β-hydroxy thioethers via alkyliodine(I) catalysis.

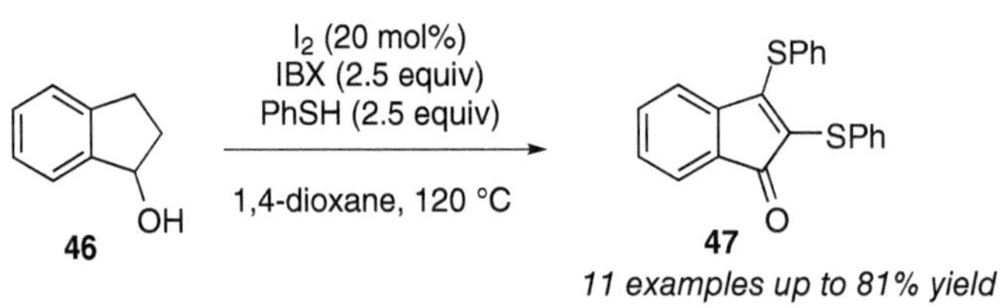

Scheme 5.23 Synthesis of diphenylthio enones.

5.4.5 Alkene Difunctionalization

In the field of alkene difunctionalization, an extended field of study also with iodine reagents, the majority of proceedings relies on ionic pathways. However, Itoh in 2014, reported on unprecedented iodine-catalyzed difuctionalization of styrene derivatives **48** toward β-hydroxy sulfones **49** (Scheme 5.24) [57].

5.4.6 α-Functionalization of Carbonyl Compounds

In this context, Murphy developed an elegant cyclopropanation reaction based on iodonium ylides **50** using 30 mol% of tetrabutylammonium iodide in combination with an equimolar amount of PIDA (Scheme 5.25) [58]. The initial transformations exploring different iodonium ylides employed an excess of styrene as the alkene

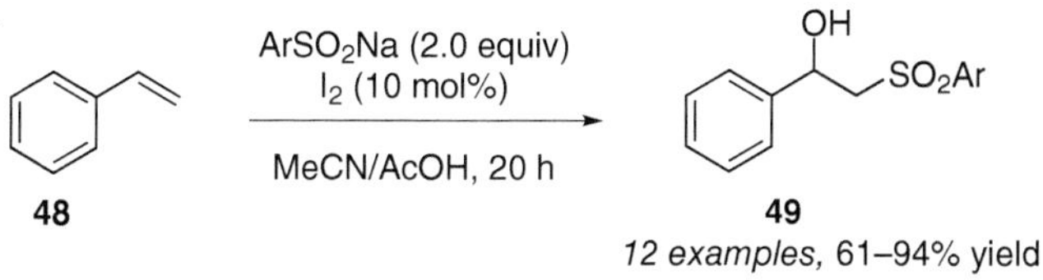

Scheme 5.24 Iodine catalyzed alkene difunctionalization.

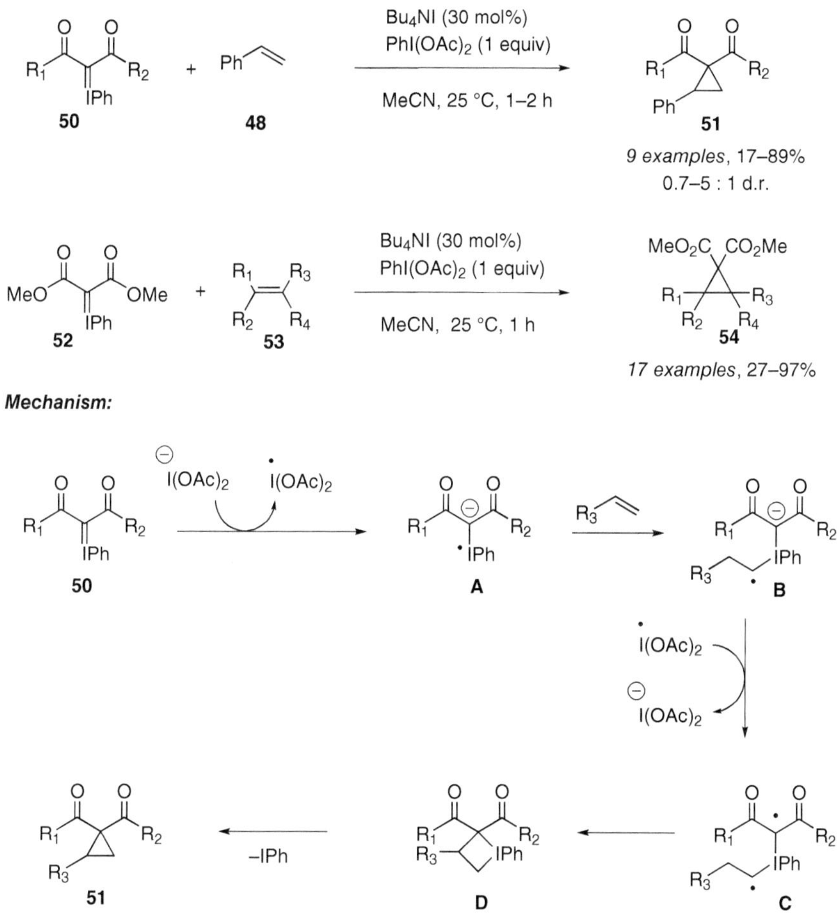

Scheme 5.25 Iodate-initiated radical cyclopropanation reactions.

component. Subsequent exploration used a limiting amount of alkenes demonstrating the economy of the process. Mechanistic investigation revealed that PIDA can also be employed in catalytic amounts, but that both Bu_4NI and PIDA are essential for the cyclopropanation to proceed. Diacetoxyiodate should be formed under the present conditions, which was confirmed by a control experiment employing 30 mol% of the preformed $[Bu_4N][I(OAc)_2]$. To account for the overall reactions, the authors propose the role of the iodate to consist as a redox mediator. In this context, single-electron transfer from the iodate to the iodonium ylide **50** will generate a carbonyl-stabilized anionic radical hypervalent iodine **A** [59], which will add the alkene. The resulting carbon-centered radical **B** can undergo a second SET reaction to regenerate the iodate furnishing a biradical species **C**. Intramolecular radical recombination will give ioda(III)cyclobutene **D**, which should engage in reductive C–C bond formation to generate the cyclopropane product **51** upon loss of iodobenzene.

In addition to the photolytic cleavage of heteroatom-iodine bonds, certain carbon-iodine bonds can undergo thermal homolytic cleavage. Lei demonstrate this possibility for 2-iodo-1,3-dicarbonyl compounds due to its relative low bond-dissociation energy [60]. In this work, catalytic amounts of iodine are employed to obtain functionalized furanes **56** or indolizines **58**, respectively with moderate to good yields (Scheme 5.26).

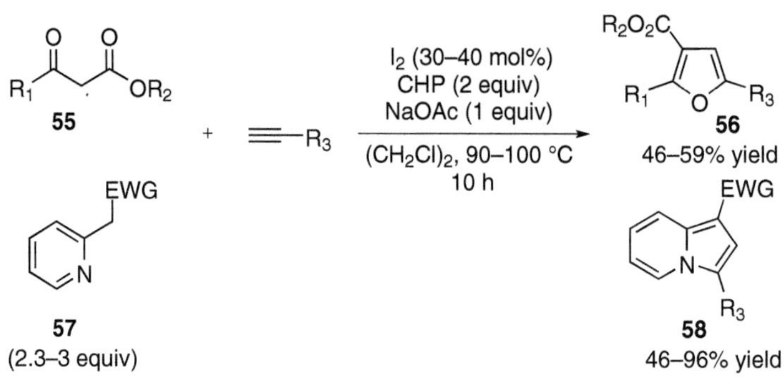

Scheme 5.26 Synthesis of furans and indolizines via iodine catalysis.

Again in this transformation, molecular iodine is not able to trigger the reaction itself, but when it is oxidized by the action of peroxide to hypoiodite **A**, α-carbonyl iodination takes place toward **B**. Thermal homolysis generates iodine radical **C** and carbon radical **E**. The latter reacts with the alkyne **F** to obtain intermediate **G**. Subsequent intramolecular radical cyclization **H** and oxidation lead to the final product **56**. After the homolysis step, peroxide oxidizes the generated radical iodine **C** to hypoiodite **A**, closing the catalytic circle (Scheme 5.27).

5.5 Conclusions

The application of iodine(I) reagents either as preformed reagents or reactive intermediates in catalytic cycles has served as a useful tool in organic synthesis. Catalytic transformations involving radical pathways are relatively rare for electrophilic iodine(I) compounds. However, its homolytic cleavage and alkyliodine synthesis constitute the basis for these radical pathways. These transformations comprise direct functionalization of sp^2, and sp^3-hybridized carbon centers and proceed without the requirement of additional metal promoters. A considerable

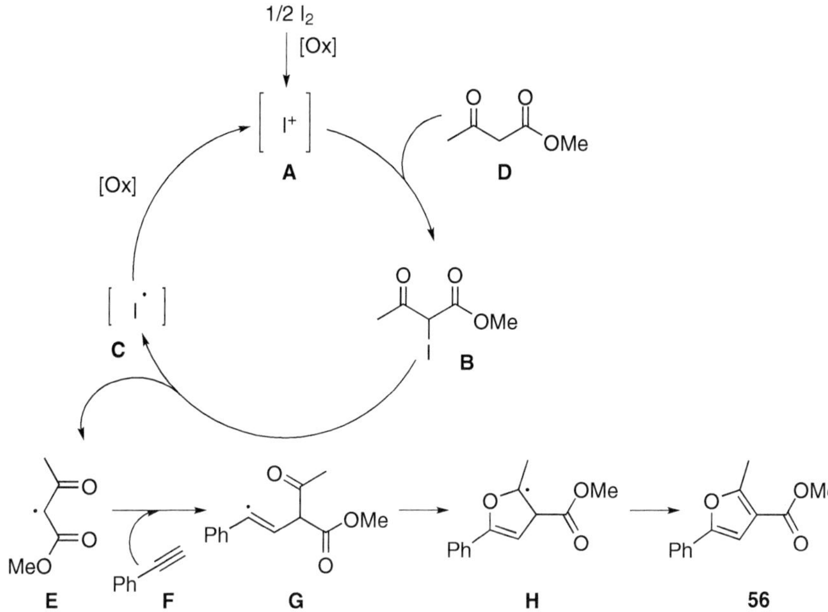

Scheme 5.27 Mechanistic scenario for the synthesis of furans.

number of iodine(I)-based transformations are now available, which can compete with the traditional metal-mediated reactions, offering huge possibilities for the development of efficient synthesis covering the demands of modern synthetic reactions.

References

1 Selected Monographs and reviews: a) Marek, I. and Olofsson, B. (eds.) (2018). *The Chemistry of Hypervalent Halogen Compounds*. PATAI's Chemistry of Functional Groups, Wiley, New York; b) Zhdankin, V.V. (2014). *Hypervalent Iodine Chemistry: Preparation, Structure, and Synthetic Applications of Polyvalent Iodine Compounds*. Chichester, UK: Wiley. c) Wirth, T. (ed.) (2016). *Hypervalent Iodine Chemistry, Topics in Current Chemistry*, vol. 373. Berlin: Springer. d) Yoshimura, A. and Zhdankin, V.V. (2016). Advances in synthetic applications of hypervalent iodine compounds. *Chem. Rev.* 116: 3328–3435. e) Zhdankin, V.V. and Muñiz, K. (2017). Editorial for the special issue on hypervalent iodine reagents. *J. Org. Chem.* 82: 11667–11668.

2 a) Dohi, T. and Kita, Y. (2009). Hypervalent iodine reagents as a new entrance to organocatalysts. *Chem. Commun.* 16: 2073–2085. b) Yusubov, M.S. and Zhdankin, V.V. (2015). Iodine Catalysis: a green alternative to transition metals in organic chemistry and technology. *Resource-Efficient Technol.* 1: 49–67. c) Claraz, A. and Masson, G. (2018). Asymmetric iodine catalysis-mediated enantioselective oxidative transformations. *Org. Biomol. Chem.* 16: 5386–5402.

3 a) Wirth, T. (2005). Hypervalent iodine chemistry in synthesis: scope and new directions. *Angew. Chem. Int. Ed.* 44: 3656–3665. b) Stang, P.J. and Zhdankin, V.V. (1996). Organic polyvalent iodine compounds. *Chem. Rev.* 96: 1123–1178. c) Grelier, G., Darses, B., and Dauban, P. (2018). Hypervalent organoiodine compounds: from reagents to valuable building blocks in synthesis. *Belstein J. Org. Chem.* 14: 1508–1528. d) Boelke, A., Finkbeiner, P., and Nachtsheim, B.J. (2018). Atom-economical group-transfer reactions with hypervalent iodine compounds. *Belstein J. Org. Chem.* 14: 1263–1280.

4 a) Kaiho, T. (2015). *Iodine Chemistry and Applications*. New York: Wiley. b) Kaiho, T. (2017). *Iodine Made Simple*. Boca Raton: CRC Press. c) Küpper, F.C., Feiters, M.C., Olofsson, B. et al. (2011). Commemorating two centuries of iodine research: an interdisciplinary overview of current research. *Angew. Chem. Int. Ed.* 50: 11598–11620. d) Bergström, M., Suresh, G., Naidu, V.R., and Unelius, C.R. (2017). *N*-iodosuccinimide in direct aromatic iodination. *Eur. J. Org. Chem.*: 3234–3239.

5 Barluenga, J. (1999). Transferring iodine: more than a simple functional group exchange in organic synthesis. *Pure Appl. Chem.* 71: 431–436.

6 a) Barluenga, J., González, J.M., Campos, P.J., and Asensio, G. (1985). $I(py)_2$, a new reagent in organic synthesis: general method for the 1,2-iodofunctionalization of olefins. *Angew. Chem. Int. Ed.* 24: 319–320. b) Barluenga, J., González-Bobes, F., Merguía, M.C. et al. (2004). Bis(pyridine)iodonium tetrafluoroborate (IPy_2BF_4): a versatile oxidizing reagent. *Chem. Eur. J.* 10: 4206–4213. c) Espuña, G., Ananthoju, S.R., Andreu, D. et al. (2006). Iodination of proteins by IPy_2BF_4, a new tool in protein chemistry. *Biochem.* 45: 5957–5963. d) Barluenga, J., González, J.M., García-Martín, M.A. et al. (1993). Acid-mediated reaction of bis(pyridine)iodonium(I) tetrafluoroborate with aromatic compounds. A selective and general iodination method. *J. Org. Chem.* 58: 2058–2060. e) Barluenga, J., Campos-Gómez, E., Minatti, A. et al. (2009). Iodoarylation reactions of allenes: inter- and intramolecular processes. *Chem. Eur. J.* 15: 8946–8950. f) Barluenga, J., Vázquez-Villa, H., Merino, I. et al. (2006). The reaction of *o*-alkynylarene and heteroarene carboxaldehyde derivatives with iodonium ions and nucleophiles: a versatile and regioselective synthesis of 1H-isochromene, naphthalene, indole, benzofuran, and benzothiophene compounds. *Chem. Eur. J.* 12: 5790–5805. g) Barluenga, J., Vázquez-Villa, H., Ballesteros, A., and González, J.M. (2005). Synthesis of indoles upon sequential reaction of 3-alkynylpyrrole-2-carboxaldehyde with iodonium ions and alkenes. Preparation of related benzofuran and benzothiophene derivatives. *Adv. Synth. Catal.* 347: 526–530. h) Barluenga, J., Trincado, M., Marco-Arias, M. et al. (2005). Intramolecular iodoarylation reaction of alkynes: easy access to derivatives of benzofused heterocycles. *Chem. Commun.*: 2008–2010. i) Barluenga, J., Trincado, M., Rubio, E., and González, J.M. (2006). Direct intramolecular arylation of aldehydes promoted by reaction with IPy_2BF_4/HBF_4: synthesis of benzocyclic ketones. *Angew. Chem. Int. Ed.* 45: 3140–3143.

7 a) Vlasenko, Y.A., Postnikov, P.S., Trusova, M.E. et al. (2018). Synthesis of five-membered iodine–nitrogen heterocycles from benzimidazole-based iodonium salts. *J. Org. Chem.* 83: 12056–12070. b) Boelke, A., Lork, E., and Nachtsheim, B.J. (2018). *N*-heterocycle-stabilized iodanes: from structure to reactivity. *Chem. Eur. J.* 24: 18653–18657.

8 Yusubov, M.S. and Zhdankin, V.V. (2012). Hypervalent iodine reagents and green chemistry. *Curr. Org. Synth.* 9: 247–272.

9 a) Li, J., Lear, M.J., Kawamoto, Y. et al. (2015). Oxidative amination of nitroalkanes with amine nucleophiles using molecular oxygen and iodine. *Angew. Chem. Int. Ed.* 127: 12178–13182. b) Li, J., Lear, M.J., Kwon, E., and Hayashi, Y. (2016). Mechanism of oxidative amidation of nitroalkanes with oxygen and amine nucleophiles by using electrophilic iodine. *Chem. Eur. J.* 22: 5538–3342.

10 a) Ashikari, Y., Shimizy, A., Nokami, T., and Yoshida, J.-i. (2013). Halogen and chalcogen cation pools stabilized by DMSO. Versatile reagents for alkene difunctionalization. *J. Am. Chem. Soc.* 135: 16070–16073. b) Shimizu, A., Hayashi, R., Ashikan, Y. et al. (2015). Switching the reaction pathways of electrochemically generated β-haloalkoxysulfonium ions - synthesis of halohydrins and epoxides. *Beilstein J. Org. Chem.* 11: 242–248.

11 a) Catalano, L., Cavallo, G., Metrangolo, P., Resnati, G. and Terraneo, G. (2016). Halogen bonding in hypervalent iodine compounds. In: *Hypervalent Iodine Chemistry. Topics in Current Chemistry* (ed. Wirth T.) 373: 289-310. Springer, Cham; b) Heinen, F., Engelage, E., Cramer, C.J., and Huber, S.M. (2020). Hypervalent iodine(III) compounds as biaxial halogen bond donors. *J. Am. Chem. Soc.* 142: 8633–8640. c) Heinen, F., Engelage, E., Dreger, A. et al. (2018). Iodine(III) derivatives as halogen bonding organocatalysts. *Angew. Chem. Int. Ed.* 57: 3830–3833.

12 Guha, S., Kazi, I., Nandy, A., and Sekar, G. (2017). Role of Lewis-base-coordinated halogen(I) intermediates in organic synthesis: the journey from unstable intermediates to versatile reagents. *Eur. J. Org. Chem.*: 5497–5518.

13 Simonini, A. (1982). Über den Abbau der fetten Säuren zu kohlenstoffärmeren Alkoholen. *Monatsh. Chem.* 13: 320–325.

14 Wilson, C.V. (1957). The reaction of halogens with silver salts of carboxylic acids. *Org. React.* 9: 332–387.

15 a) Wittig, G. and Schöllkopf, U. (1958). Zum Chemismus der Halogen-Lithium-Austauschreaktion. *Tetrahedron* 3: 91–93. b) Reich, H.J., Philips, N.H., and Reich, I.L. (1985). Lithium-metalloid exchange reactions. Kinetic evidence for an intermediate in the lithium-iodine exchange. *J. Am. Chem. Soc.* 107: 4101–4103. c) Reich, H.J., Green, D.P., and Philips, N.H. (1989). The lithium-metalloid exchange reaction. NMR studies of the phenyllithium-iodobenzene exchange. *J. Am. Chem. Soc.* 111: 3444–3445. d) Reich, H.J., Green, D.P., and Philips, N.H. (1991). Lithium-metalloid exchange. Dynamics and equilibrium in the Li/I and Li/Te exchange in tetrahydrofuran: iodine, tellurium, and mercury ate complexes. *J. Am. Chem. Soc.* 113: 1414–1416. e) Farnham, W.B. and Calabrese, J.C. (1986). Novel hypervalent (10-I-2) iodine structures. *J. Am. Chem. Soc.* 108: 2449–2451. f) Schulze, V., Brönstrup, M., Böhm, V.P.W. et al. (1988). α-iodoalkyl-iodine-ate complexes as observable intermediates in the iodine-magnesium exchange reaction. *Angew. Chem. Int. Ed.* 37: 824–826.

16 Barry, J.E., Finkelstein, M., Moore, W.M. et al. (1982). Electrochemical reduction of *N*-bromosuccinimide. Reaction mechanism for formation of the succinimidyl radical. *J. Org. Chem.* 47: 1292–1298.

17 Barry, J.E., Finkelstein, M., Moore, W.M. et al. (1984). The *N*-bromosuccinimide-succinimide anion complex, an intermediate in the electron transfer mediated reduction of *N*-bromosuccinimide by succinimide anion. *Tetrahedron Lett.* 25: 2847–2850.

18 Doleschall, G. and Tóth, G. (1980). The degradation of carboxylic acids into aldehydes: regioselective α-acetoxylation of 1,2,4-triazolium salts with diacetoxyiodate(I) anion. *Tetrahedron* 36: 1649–1665.

19 de Armas, P., Concepción, J.I., Francisco, C.G. et al. (1989). Intramolecular hydrogen abstraction. Hypervalent organoiodine compounds, convenient reagents for alkoxy radical generation. *J. Chem. Soc. Perkin Trans. I*: 405–410.

20 Muñiz, K., García, B., Martínez, C., and Piccinelli, A. (2016). Dioxoiodane compounds as versatile sources for Iodine(I) chemistry. *Chem. Eur. J.* 22: 1539–1545.

21 Kirschning, A., Plumeier, C., and Rose, L. (1988). Phosphonium salts of diacetoxyiodine(I) anions, new reagents for the iodoacetoxylation of alkenes. *Chem. Commun.*: 33–34.

22 a) Reddy, T.R., Rao, D.S., Babachary, K., and Kashyap, S. (2016). Sulfonium salts of iodine(I) species as efficient reagents for the regioselective bisfunctionalisation of glycals and enol ethers. *Eur. J. Org. Chem.*: 291–301. b) Rao, D.S., Reddy, T.R., Babachary, K., and Kashyap, S. (2016). Regioselective vicinal functionalization of unactivated alkenes with sulfonium iodate(I) reagents under metal-free conditions. *Org. Biomol. Chem.* 14: 7529–7543.

23 Lucchetti, N., Tkatcheva, A., Fantasia, S., and Muñiz, K. (2018). Radical C–H-amination of heteroarenes using dual initiation by visible light and iodine. *Adv. Synth. Catal.* 360: 3889–3893.

24 Bergès, J., Bosnidou, A.E., García, B., and Muñiz, K. (2020). A defined molecular iodine(I) reagent for aliphatic C–H amination alpha to indoles using phthalimide as amine source. *Manuscript submitted.*

25 Kirschning, A., Jesberger, M., and Schönberger, A. (2001). The first polymer-assisted solution-phase synthesis of deoxyglycosides. *Org. Lett.* 3: 3623–3626.

26 Kandrnálová, M., Kokan, Z., Havel, V. et al. (2019). Hypervalent iodine based reversible covalent bond in rotaxane synthesis. *Angew. Chem. Int. Ed.* 58: 18182–18185.

27 Kirschning, A., Hashem, M.A., Monenschein, H. et al. (1999). Preparation of novel haloazide equivalents by iodine(III)-promoted oxidation of halide anions. *J. Org. Chem.* 64: 6522–6526.

28 a) Manus, P., Lacour, J., Evans, P.A. et al. (1996). Hypervalent iodine chemistry: new oxidation reactions using the iodosylbenzene-trimethylsilyl azide reagent combination. Direct α- and β-Azide functionalization of triisopropylsilyl enol ethers. *J. Am. Chem. Soc.* 118: 3406–3418. b) Magnus, P. and Lacour, J. (1992).

New trialkylsilyl enol ether chemistry. Direct .beta.-azido functionalization of triisopropylsilyl enol ethers. *J. Am. Chem. Soc.* 114: 767–769. c) Magnus, P. and Lacour, J. (1992). New trialkylsilyl enol ether chemistry. Conjugate additions without the enone. *J. Am. Chem. Soc.* 114: 3993–3994. d) Magnus, P., Roe, M.B., and Hulme, C. (1995). New trialkylsilyl enol ether chemistry: direct 1,2-bis-azidonation of triisopropyslsilyl enol ethers: an azido-radical addition process promoted by TEMPO. *J. Chem. Soc. Chem. Commun.* 2: 263–265.

29 Kirschning, A., Monenschein, H., and Schmeck, C. (1999). Stable polymer-bound iodine azide. *Angew. Chem. Int. Ed.* 38: 2594–2596.

30 Barluenga, J., Gonzalez-Bobes, F., Ananthoju, S.R. et al. (2001). Oxidative opening of cycloalkanols: an efficient entry to ω-iodocarbonyl compounds. *Angew. Chem. Int. Ed.* 40: 3389–3392.

31 Bosnidou, A.E. and Muñiz, K. (2019). Alkyliodines in high oxidation states: enhanced synthetic possibilities and accelerated catalyst turn-over. *Chem. Eur. J.* 25: 13654–13664.

32 a) Carrau, R., Hernández, R., Suárez, E., and Betancor, C. (1987). Intramolecular functionalization of *N*-cyanamide radicals: synthesis of 1,4- and 1,5-*N*-cyanoepimino compounds. *J. Chem. Soc. Perkins.* 1: 937–943. b) Dorta, R.L., Francisco, C.G., and Suárez, E. (1989). Hypervalent organoiodine reagents in the transannular functionalisation of medium-sized lactams: synthesis of 1-azabicyclo compounds. *J. Chem. Soc. Chem. Commun.*: 1168–1169. c) Francisco, C.G., Herrera, A.J., and Suárez, E. (1989). Intramolecular hydrogen abstraction reaction promoted by *N*-radicals in carbohydrates. Synthesis of chiral 7-Ox-2-azabicyclo[2.2.1]heptanes and 8-Oxa-6-azabicyclo[3.2.1]octane ring systems. *J. Org. Chem.* 68: 1012–1017.

33 a) Martínez, C. and Muñiz, K. (2015). An iodine-catalyzed Hofmann-Löffler reaction. *Angew. Chem. Int. Ed.* 54: 8287–8291. b) Duhamel, T., Stein, C.J., Martínez, C. et al. (2018). Engineering molecular iodine catalysis for alkyl-nitrogen bond formation. *ACS Catal.* 8: 3918–3925.

34 Bosnidou, A.E., Duhamel, T., and Muñiz, K. (2020). Detection of the elusive nitrogen-centered radicals from catalytic Hofmann-Löffler reactions. *Eur. J. Org. Chem.* 2020 (40): 6361–6365.

35 del Castillo, E. and Muñiz, K. (2019). Enantioselective synthesis of nicotine via an iodine-mediated Hofmann-Löffler reaction. *Org. Lett.* 21: 705–708.

36 Becker, P., Duhamel, T., Stein, J. et al. (2017). Cooperative light-activated iodine and photoredox catalysis for the amination of Csp_3-H bonds. *Angew. Chem. Int. Ed.* 56: 8004–8008.

37 Stateman, L.M., Wappes, E.A., Nakafuku, K.M. et al. (2019). Catalytic β C–H amination *via* an imidate radical relay. *Chem. Sci.* 10: 2693–2699.

38 Zhang, H. and Muñiz, K. (2017). Selective piperidine synthesis exploiting iodine-catalyzed $C_{SP}{}^3$–H amination under visible light. *ACS Catal.* 7: 4122–4125.

39 a) Cavallo, G., Metrangolo, P., Mmilani, R. et al. (2016). The halogen bond. *Chem. Rev.* 116: 2478–2601. b) Metrangolo, P. and Resnati, G. (2015). *Halogen Bonding I: Impact on Material Chemistry and Life Sciences, Topics in Current Chemistry*, vol. 358. Berlin: Springer. c) Metrangolo, P. and Resnati, G. (2015).

Halogen Bonding II: Impact on Materials Chemistry and Life Sciences, Topics in Current Chemistry, vol. 359. Berlin: Springer. d) Breugst, M., Derman, E., and vonder Heiden, D. (2016). Origin of the catalytic effect of molecular iodine: a computational analysis. *ACS Catal.* 6: 3203–3212.

40 a) Short, M.A., Blackburn, J.M., and Roizen, J.L. (2018). Sulfamate esters guide selective radical-mediated chlorination of aliphatic C-H bonds. *Angew. Chem. Int. Ed.* 57: 296–299. b) Sathyamoorthi, S., Banerjee, S., Du Bois, J. et al. (2018). Site-selective bromination of sp3-C–H bonds. *Chem. Sci.* 9: 100–104. c) Zalatan, D.N. and Du Bois, J. (2009). Oxidative cyclization of sulfamate esters using NaOCl - a metal-mediated Hofmann-Löffler freytag reaction. *Synlett.*: 143–146. d) del Castillo, E., Martínez, M.D., Bosnidou, A.E. et al. (2018). Multiple halogenation of aliphatic C–H bonds within the Hofmann-Löffler manifold. *Chem. Eur. J.* 24: 17225–17229.

41 Kiyakawa, K., Nakamura, S., Jou, K. et al. (2019). Transition-metal-free intramolecular C–H- amination of sulfamate ester and *N*-alkylsulfamides. *Chem. Commun.* 55: 11782–11785.

42 Duhamel, T., Martinez, M.D., Sideri, I.K., and Muñiz, K. (2019). Engineering molecular iodine catalysis of alkyl-nitrogen bond formation. *ACS Catal.* 9: 7741–7745.

43 Bosnidou, A.E. and Muñiz, K. (2019). Intermolecular radical C(sp^3)–H amination under iodine catalysis. *Angew. Chem. Int. Ed.* 58: 7485–7489.

44 Kiyokawa, K., Takemoto, K., and Minakata, S. (2016). Ritter-type amination of C–H bonds at tertiary carbon centers using iodic acid as an oxidant. *Chem. Commun.* 52: 13082–11785.

45 Martínez, C., Bosnidou, A.E., Allmendinger, S., and Muñiz, K. (2016). Towards uniform iodine catalysis: intramolecular C–H amination of arenes under visible light. *Chem. Eur. J.* 22: 9929–9932.

46 Uyanik, M., Suzuki, D., Yasui, T., and Ishihara, K. (2011). *In situ* generated (hypo)iodite catalysts for the direct α-oxyacylation of carbonyl compounds with carboxylic acids. *Angew. Chem. Int. Ed.* 50: 5331–5334.

47 Uyanik, M., Suzuki, D., Watanabe, M. et al. (2015). High-performance hypoiodite/hydrogen peroxide catalytic system for the oxylactonization of aliphatic γ-oxocarboxylic acids. *Chem. Lett.* 44: 387–389.

48 Uyanik, M., Okamoto, H., Yasui, T., and Ishihara, K. (2010). Quaternary ammonium (Hypo)iodite catalysis for enantioselective oxidative cycloetherification. *Science* 328: 1376–1379.

49 Togo, H., Muraki, T., Hoshina, Y. et al. (1997). Formation and synthetic use of oxygen-centred radicals with (Diacetoxyiodo)arenes. *J. Chem. Soc., Perkin Trans.* 1: 787–793.

50 Duhamel, T. and Muñiz, K. (2019). Cooperative iodine and photoredox catalysis for direct oxidative lactonization of carboxylic acids. *Chem. Commun.* 55: 933–936.

51 Kiyokawa, K., Ito, R., Takemoto, K., and Minakata, S. (2018). C–H oxygenation at tertiary carbon centers using iodine oxidant. *Chem. Commun.* 54: 7609–7612.

52 Kiyokawa, K., Okumatsu, D., and Minakata, S. (2018). Hypervalent iodine(III)-mediated decarboxylative acetoxylation at tertiary and benzylic carbon centers. *Beilstein J. Org. Chem.* 14: 1046–1050.

53 Kiyokawa, K. and Minakata, S. (2020). Iodine-based reagents in oxidative amination and oxygenation. *Synlett* 31: 845–855.

54 a) Lal, G.S., Pez, G.P., and Syvet, R.G. (1996). Electrophilic N–F fluorinating agents. *Chem. Rev.* 96: 1737–1756. b) Timofeeva, D.S., Ofial, A.R., and Mayr, H. (2018). Kinetics of electrophilic fluorinations of enamines and carbanions: comparison of the fluorinating power of N–F reagents. *J. Am. Chem. Soc.* 140: 11474–11486.

55 Bafaluy, D., Georgieva, Z., and Muñiz, K. (2020). Iodine catalysis for $C(sp^3)$–H fluorination with a nucleophilic fluorine source. *Angew. Chem. Int. Ed.* 59: 14241–14245.

56 Li, Y., Liu, D., Liu, C., and Lei, A. (2016). Oxidative β-Csp^3-H functionalization of tBuOH: a selective radical/radical cross-coupling access to β-hydroxy thioethers. *Chem. Asian J.* 11: 2246–2249.

57 Sundaravelu, N., Guha, S., and Sekar, G. (2020). Iodonium ion-catalyzed domino synthesis of *Z*-selective α,β-diphenylthio enones from easily accessible secondary alcohols. *J. Org. Chem.* 85: 5895–5906.

58 Tao, J., Estrada, C.D., and Murphy, G.K. (2017). Metal-free intermolecular cyclopropanation between alkenes and iodonium ylides mediated by $PhI(OAc)_2{\cdot}Bu_4NI$. *Chem. Commun.* 53: 9004–9007.

59 Wang, X. and Studer, A. (2017). Iodine(III) reagents in radical chemistry. *Acc. Chem. Res.* 50: 1712–1724.

60 Kariya, A., Yamaguchi, T., Nobuta, T. et al. (2014). Molecular-iodine-catalyzed aerobic oxidative synthesis of β-hydroxy sulfones from alkenes. *RCS Adv.* 4: 13191–13194.

6

Aromatic C–H Functionalization

Mahesh Puthanveedu[1,2] and Andrey P. Antonchick[1,2,3,]*

[1] Department of Chemical Biology, Max Planck Institute for Molecular Physiology, Dortmund, Germany
[2] Department of Chemistry and Chemical Biology, Technische Universität Dortmund, Dortmund, Germany
[3] Department of Chemistry and Forensics, School of Science and Technology, Nottingham Trent University, Nottingham, United Kingdom

6.1 Introduction

The direct functionalization of C–H bonds without the requirement of prefunctionalization of starting materials is considered the holy grail of organic chemistry. Hypervalent iodine reagents are an important class of reagents used in C–H functionalization chemistry owing to their nontoxic and environmentally benign nature. Because of their easy accessibility, diverse reactivity, and broad functional group tolerance, polyvalent iodanes are considered viable alternatives to expensive transition metals in several organic transformations. Various λ^3 and λ^5 iodanes, iodonium salts, iodonium ylides, etc., are used for oxidative carbon–carbon and carbon–heteroatom bond formation [1]. Additionally, rapid progress in the area of hypervalent iodine-catalyzed asymmetric transformations, and photoredox and electrocatalytic reactions indicates the current trend toward the development of green and sustainable chemical methods [2, 3]. Iodine(III) reagents represent an attractive alternative for conventional transition metal-catalyzed/mediated coupling reactions as well as for the synthesis of various natural products and other heterocyclic compounds. In this chapter, we showcase recent significant advances made toward the synthesis and functionalization of (hetero)aromatic compounds involving stoichiometric or catalytic amounts of hypervalent iodine(III) reagents, based on the type of reactions and patterns of bond formation. The first part is devoted to the synthesis and functionalization of heterocycles through oxidative intra- and intermolecular bond-forming reactions using stoichiometric amounts of hypervalent iodine(III) reagents, and the second part describes the direct C–H bond functionalization leading to the formation of new C–C and C–heteroatom bonds in the presence of catalytic amounts of iodane compounds.

Iodine Catalysis in Organic Synthesis, First Edition. Edited by Kazuaki Ishihara and Kilian Muñiz.

6.2 Reactions with Stoichiometric Hypervalent Iodine Reagents

6.2.1 Synthesis of (Hetero)aromatic Compounds

The synthesis of *N*-heterocycles involving an electrophilic nitrogen intermediate produced by heterolytic cleavage of reagent-substrate intermediate remains the most important class of hypervalent iodine reactions in the synthesis of heteroaromatic compounds. Nitrogen heterocycles are an important class of organic compounds because of their wide occurrence in natural products and synthetic drugs. The possibility to synthesize this class of molecules in a transition metal-free manner is of great significance in the pharmaceutical industry as it eliminates any chances of trace metal contamination [4, 5].

6.2.1.1 Intramolecular C–N Bond Formation via Oxidative Cyclization

Kikugawa and coworkers reported a robust C(sp^2)–H amidation method to yield corresponding *N*-heterocycles *via* cationic cyclization of stabilized nitrenium ions (Scheme 6.1, eq. 1). The electron-donating methoxy group attached to the nitrogen in compound **1** stabilizes the cationic intermediate **1b** formed during the reaction. Hypervalent iodine reagents like (bis(trifluoroacetoxy)iodo)benzene (PIFA) or 4-chloro-1-(bis(trifluoroacetoxy)iodo)benzene (4-ClPIFA) were used to

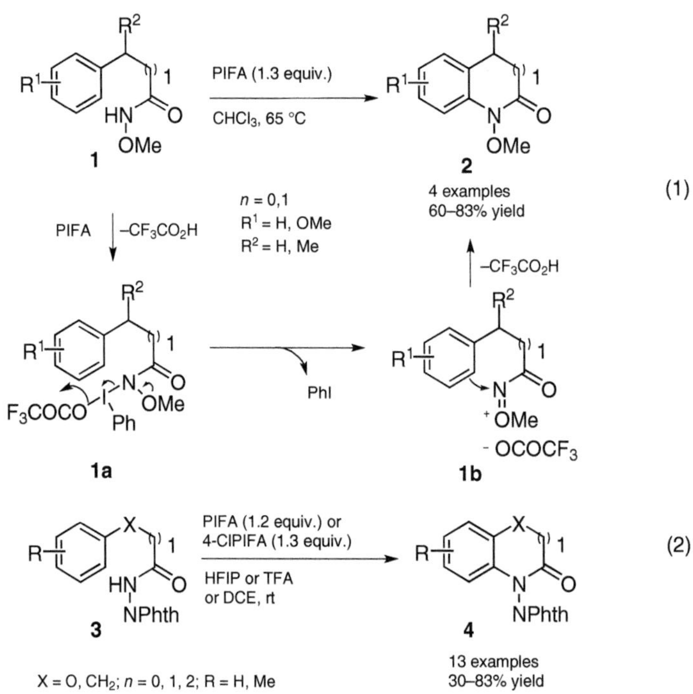

Scheme 6.1 Electrophilic arylation of *N*-substituted amides. *Source*: Kikugawa and Kawase [6], Kikugawa et al. [7].

generate nitrenium ions. Mechanistically, an initial ligand exchange of **1** with PIFA results in the formation of intermediate **1a**. Following a heterolytic cleavage and cyclization, the desired *N*-aryl amide **2** is formed. Replacement of the methoxy group in **1** with phthalimide (NPhth) enables the synthesis of a number of five-, six-, and seven-membered lactams **4** (Scheme 6.1, eq. 2) [6, 7]. This chemistry has attracted the attention of the synthetic community and significant developments were reported in the following years for the synthesis of a large number of nitrogen-containing aromatic compounds in a transition metal-free manner.

Romero's group and Fu's group independently developed the hypervalent iodine-mediated synthesis of benzoimidazolones **6** *via* oxidative cyclization of acyclic ureas **5** with hypervalent iodine reagents. The methodology shows high functional group tolerance and proceeds smoothly without any additional catalysts, ligands, or inert conditions (Scheme 6.2) [8, 9].

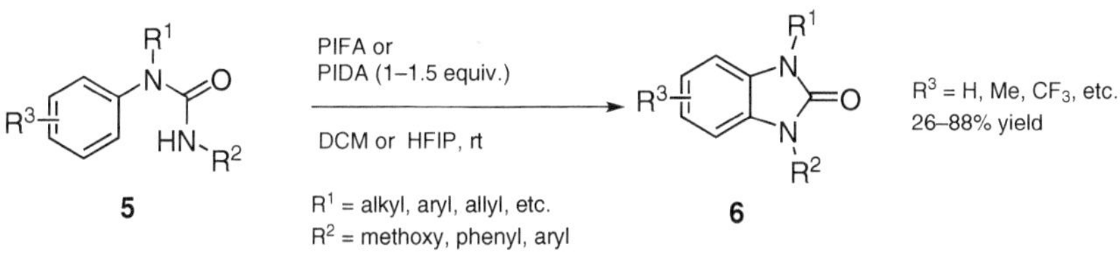

Scheme 6.2 Oxidative cyclization of acyclic ureas. *Source*: Romero et al. [8], Yu et al. [9].

A big leap forward in the research of iodane-mediated heterocycle synthesis was the development of synthetic methods for carbazole formation which is an important scaffold of many naturally occurring alkaloids and synthetic drugs. In 2011, Chang's group observed the intramolecular cyclization of 2-aminobiphenyls **7** which occurs even in the absence of copper catalyst albeit in lower yields [10]. This protocol was further extended to the synthesis of various heterocycles such as carbazolones **9** and indoloquinolinones **10** by Zhao, Du, and coworkers (Scheme 6.3) [11, 12].

In addition, Du and Zhao also developed reactions to synthesize biologically relevant heterocycles like 1,4-benzodiazepines **11**, dibenzoxazepinone **12**, and indole derivatives **13** *via* hypervalent iodine-mediated intramolecular C–N bond formation (Scheme 6.4) [13–15].

In 2018, Shibata and coworkers demonstrated the synthesis of dihydroquinoxalines using metal-free hypervalent iodine chemistry. An intramolecular C–H/N–H coupling induced by PIDA took place with the complete transfer of chirality from starting materials (Scheme 6.5) [16]. In terms of mechanism, the reaction is initiated by a ligand exchange reaction of iodine(III) reagent with **14** to form the N–I intermediate **14b**. Elimination of iodobenzene and an acetate anion results in the formation of nitrenium ion intermediate **14c**. Aromatic nucleophilic attack by the aryl ring to the nitrenium ion followed by aromatization yields the desired product **15** (Scheme 6.6).

Ding, Wu, and coworkers presented a synthesis of biologically relevant benzosultams **16** and **17** using PIDA-I_2 through intramolecular cyclization of *ortho*-arylbenzenesulfonamides (Scheme 6.7, eq. 1) [17]. Yang's group revealed an

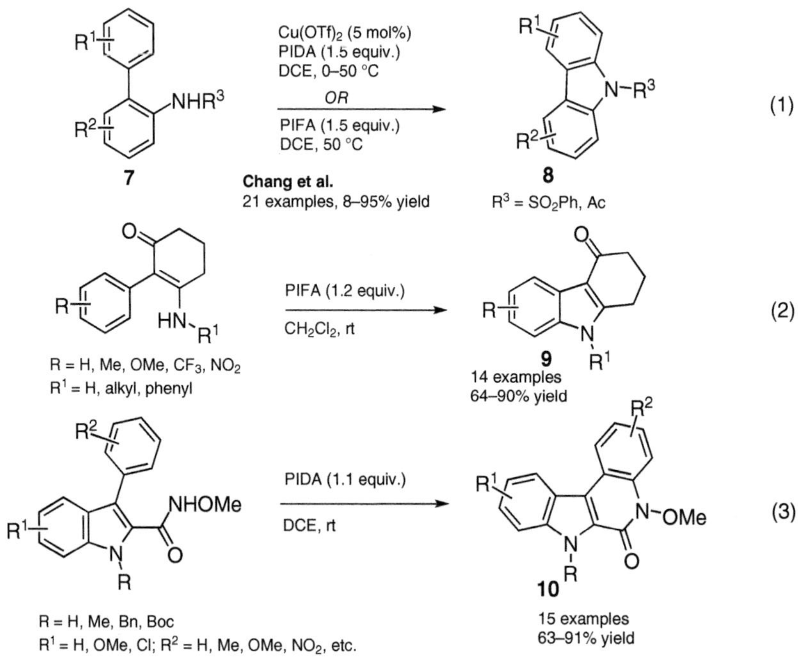

Scheme 6.3 Synthesis of carbazoles, carbazolones, and indoloquinolones *via* oxidative cyclization. *Source*: Ban et al. [11], Li et al. [12].

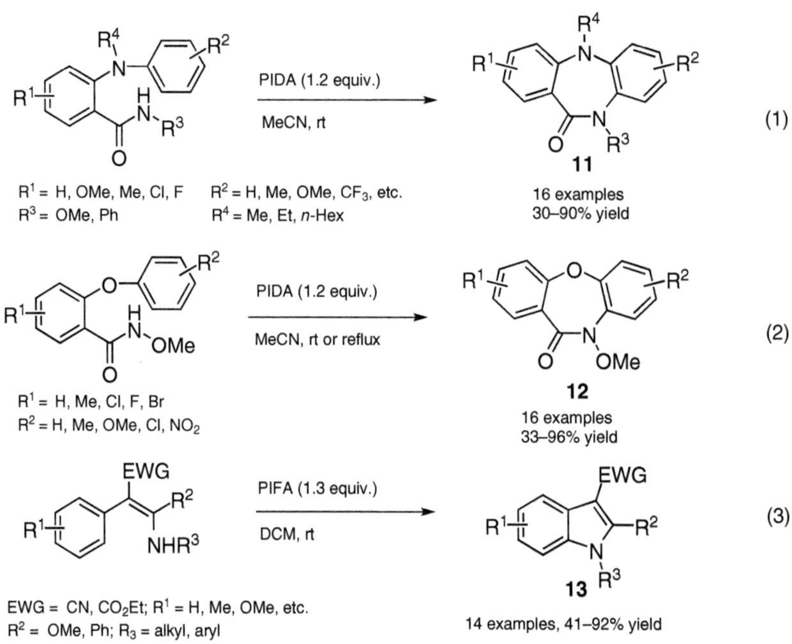

Scheme 6.4 Iodine(III)-mediated synthesis of biologically relevant heterocycles. *Source*: Li et al. [13], Guo et al. [14], Du et al. [15].

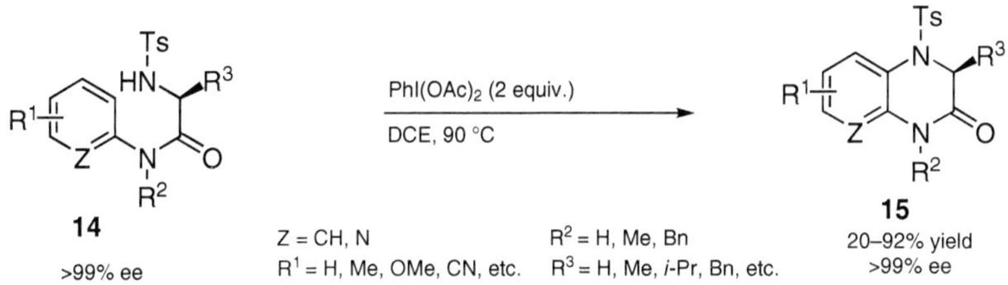

Scheme 6.5 Synthesis of dihydroquinoxalinones with retention of chirality. *Source*: Kanyiva et al. [16].

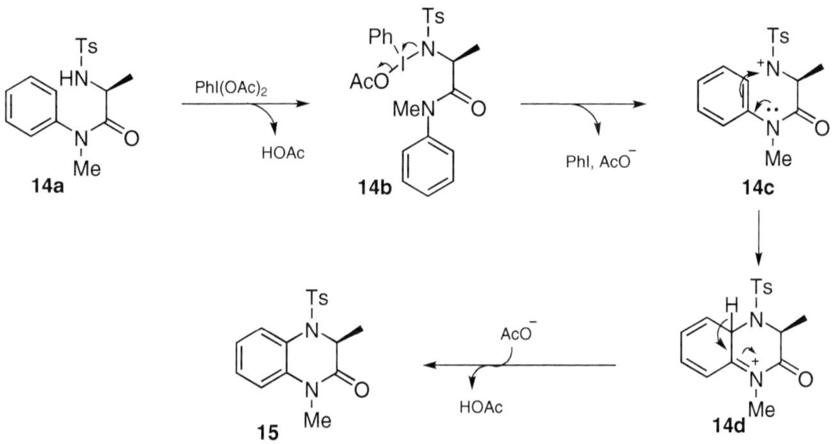

Scheme 6.6 Plausible mechanism for dihydroquinoxalinone synthesis.

appealing strategy for the synthesis of atropoisomeric phosphinamide derivative **18**. The reaction proceeded *via* tandem intramolecular oxidative C–H amination and iodination (Scheme 6.7, eq. 2) [18]. In another report, dibenzothiazines **19** were synthesized from sulphides by a one-pot *N, O*-transfer followed by an intramolecular amination of the sp^2 C–H bond (Scheme 6.7, eq. 3) [19]. Zhao, Du, and coworkers succeeded in developing a PIDA-mediated intramolecular C–N bond formation for the synthesis of chromenone-indole-fused heterocycles **20** (Scheme 6.8) [20].

Xu's group described PIFA promoted direct synthesis of imidazopyrimidines **21**. Antonchick's group developed methods to synthesize benzimidazole-fused heterocycle **22**. Afterward, Zhu's group published an interesting method to synthesize pyridobenzimidazoles **24** from *N*-benzyl-2-aminopyridines mediated by $PhI(OPiv)_2$. The authors proposed a plausible mechanism for this reaction. The initial reaction of hypervalent iodine(III) with the amino group is followed by a characteristic *ipso*-S_EAr reaction to form intermediate **23c**. Solvent attack on this intermediate and further reaction with another molecule of iodine(III) finally results in the formation of **23g**. This intermediate undergoes rearomatization to yield the desired product **24** (Scheme 6.9) [21–23]. Wu's group investigated the substituent electronic effects governing the formation of benzimidazoles and carbazoles from the reaction of *N*-(biphenyl)pyridin-2-amines with hypervalent iodine(III) reagents. The authors

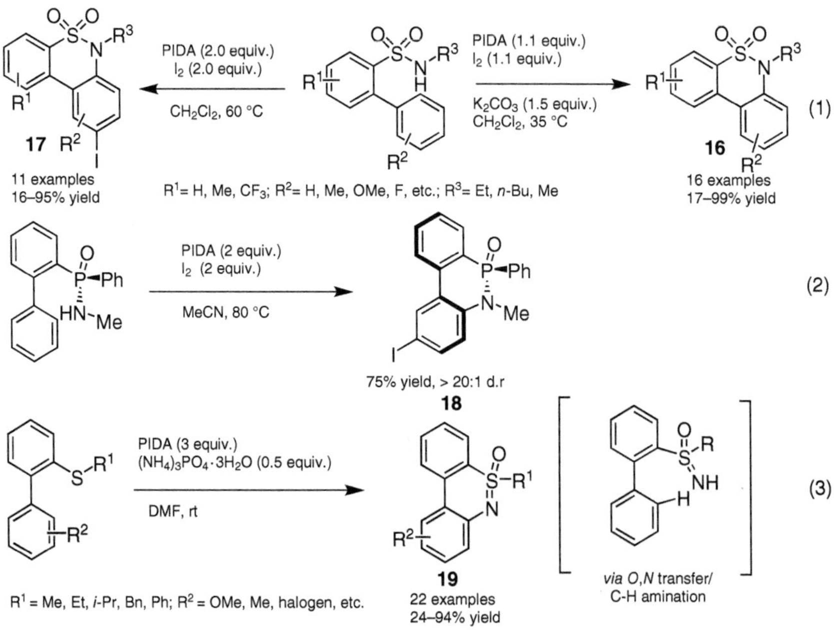

Scheme 6.7 C–H amination of biaryl compounds. *Source*: Li et al. [17].

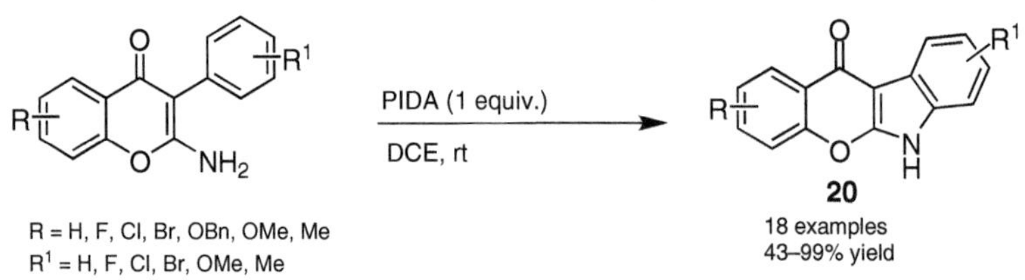

Scheme 6.8 Synthesis of chromenone-indole-fused heterocycles. *Source*: Sun et al. [20].

studied a wide number of substitutions on different positions of substrate **25** to get insights into the reaction mechanism. The optimization studies showed that PIFA shows superior selectivity for benzimidazoles **27** over PIDA. The reaction in the presence of PIDA also resulted in the acetoxylation of the formed heterocycles in some cases (Scheme 6.10) [24].

A recent report presented the synthesis of benzimidazopurine nucleosides **28** from N^6-aryl adenosine derivatives by PIDA-mediated intramolecular C–N bond formation (Scheme 6.11, eq. 1) [25]. Otani, Shibata and coworkers presented a facile two-step synthesis of aza[7]helicene **29** involving hypervalent iodine-mediated double NH/CH coupling (Scheme 6.11, eq. 2) [26].

6.2.1.2 Intramolecular C–C Bond Formation via Oxidative Cyclization

Hypervalent iodine reagent-mediated C–C bond-forming reactions are also an important class of metal-free methodologies for the synthesis of biologically relevant

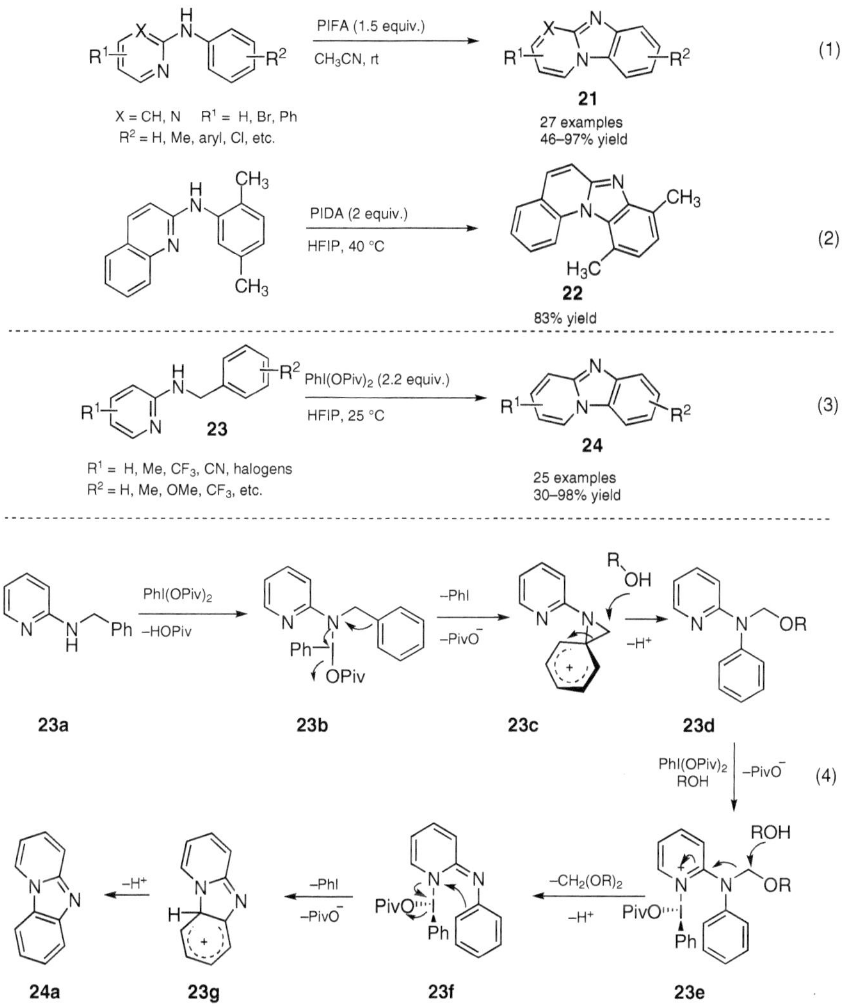

Scheme 6.9 Synthesis of imidazopyrimidines and benzimidazoles fused heterocycles. *Source*: Manna et al. [21], Qian et al. [22], Liang et al. [23].

molecules. In this context, Tellitu, Domínguez, and coworkers explored a novel method for the synthesis of heterocycle-fused quinolinone derivatives **30** *via* biaryl coupling. However, the method was too limited in terms of scope and the authors had to switch the strategy to known intramolecular C–H amination methods (Scheme 6.12) [27].

Shang *et al.* presented an unusual rearrangement reaction of 2-acylamino-*N*-phenyl benzamides or 2-hydroxy-*N*-phenyl benzamides mediated by PIDA. The method enabled the synthesis of the dibenzodihydro-1,3-diazepinone skeleton **32** through an aryl C(sp^2)–C(sp^2) coupling followed by an intramolecular condensation reaction [28]. The reaction of PIDA with substrate **31** and an *ipso*-cyclization results in the formation of intermediate **31c**. However, the lone pair of nitrogen quickly

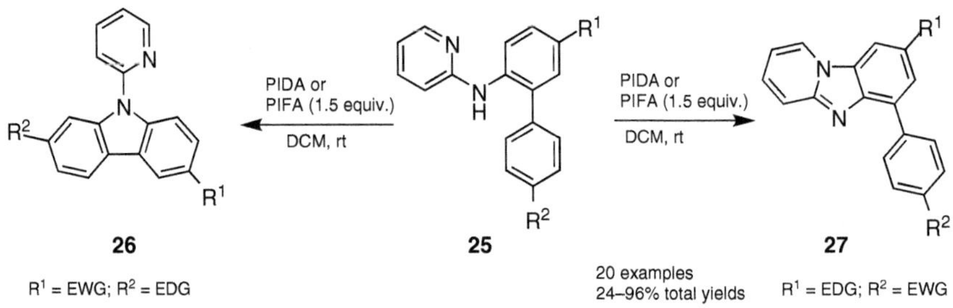

Scheme 6.10 Effect of substituents on oxidative C–N cyclization. *Source*: Chu et al. [24].

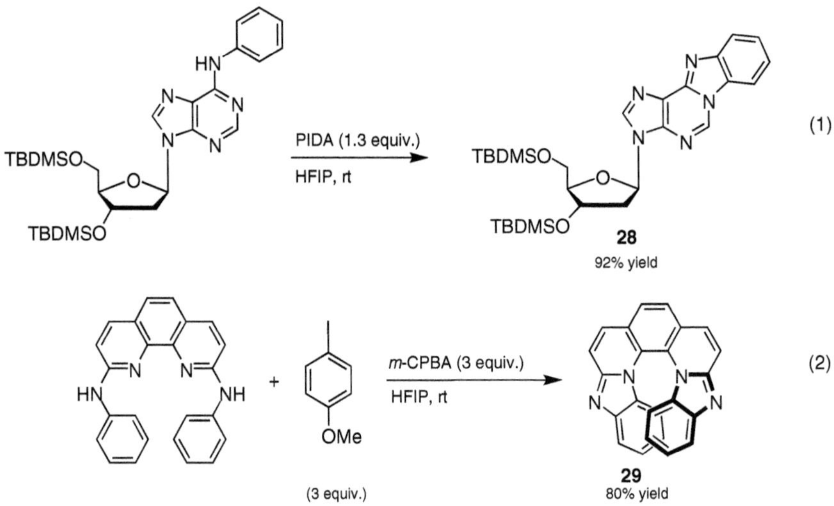

Scheme 6.11 PIDA-mediated nucleoside and helicene synthesis. *Source*: Satishkumar and Lakshman [25], Otani et al. [26].

Scheme 6.12 Biaryl coupling approach for quinolinone derivatives. *Source*: Herrero et al. [27].

initiates a ring-opening reaction to give intermediate **31d** which then undergoes lactamization to yield the final rearrangement product **32a** (Scheme 6.13).

Hypervalent iodine-mediated oxidative functionalization of enolizable carbonyl compounds are an important approach for metal-free C–C bond formation. Du, Zhao, and coworkers demonstrated a convenient route for the synthesis of spiroxindoles **33** and 3-hydroxy-2-oxindoles **34** from anilides (Scheme 6.14, eq.

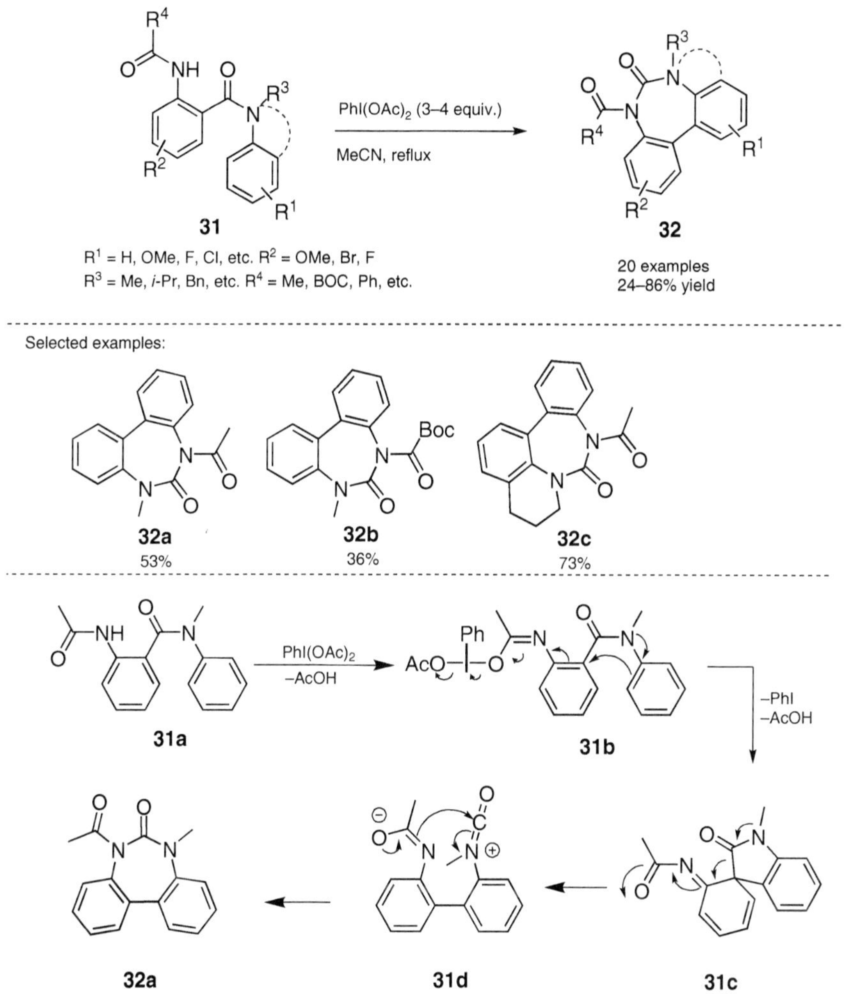

Scheme 6.13 Rearrangement reaction of *N*-phenylbenzamides and proposed mechanism.

1). Most recently, Du's group also developed a method for the synthesis of C_3-unsymmetric spirooxindoles starting from the diphenylmalonamides *via* intramolecular cyclization in the presence of PIDA. The cascade reaction proceeded by intramolecular oxidative C–C bond formation followed by C–N bond formation (Scheme 6.14, eq. 2). The authors proposed a mechanism where iodine undergoes a ligand exchange reaction with substrate **35**. Deprotonation of **35b** intermediate, followed by nucleophilic attack of the phenyl ring provides **35d** which rearranges to intermediate **35e** and then eventually yields **36a** [29, 30].

A PIFA-mediated hydroxylation followed by an acid-promoted intramolecular condensation reaction of readily available *N*-phenylacetamides enabled the synthesis of biologically important 3-hydroxyquinolinones **37** in one-pot (Scheme 6.15, eq. 1) [31]. In analogous manner, Du and Zhao along with their group developed

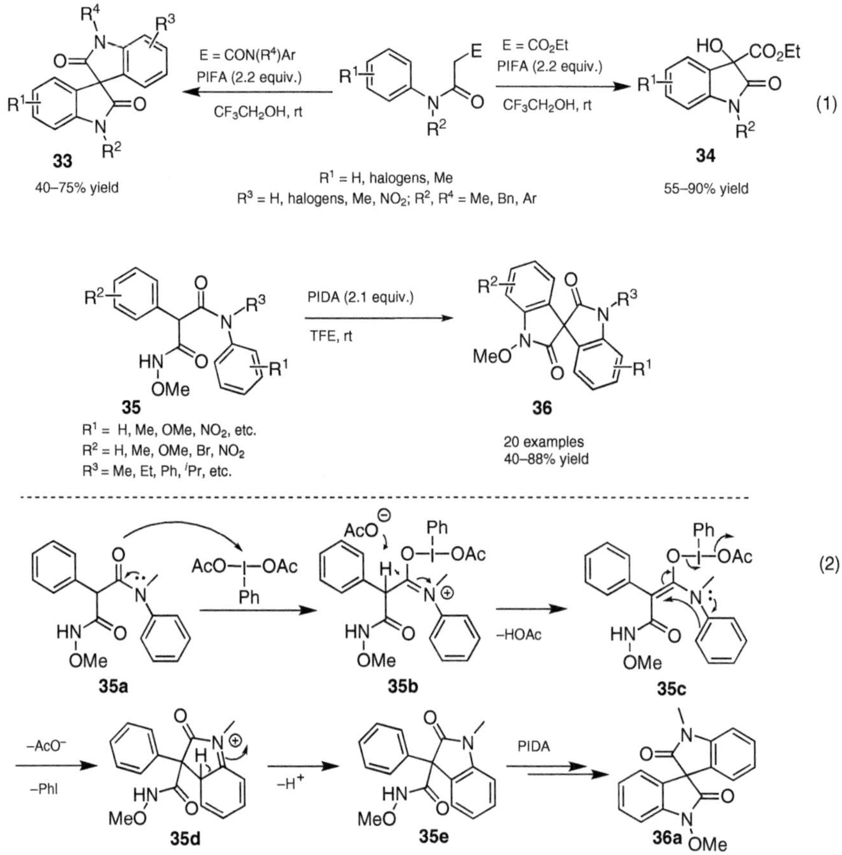

Scheme 6.14 Hypervalent iodine-mediated synthesis of oxindoles and spiroxindoles.

a synthetic strategy to access 3-arylquinoline-2-ones **38** involving a characteristic 1, 2-aryl migration (Scheme 6.15, eq. 2) [32]. Kita's group has made immense contributions to the area of I(III) reagent's oxidative coupling chemistry. They developed a very comprehensive intramolecular biaryl synthesis mediated by a PIDA–BF_3•Et_2O combination as oxidant [33]. The application of this system was most distinct in the biaryl coupling of benzyl tetrahydroisoquinoline for the synthesis of aporphine scaffold **39** (Scheme 6.16) [34].

6.2.1.3 Radical-Mediated Intramolecular Cyclization

In recent years, hypervalent iodine(III) reagents have received wide application in radical chemistry. They allow the generation of several reactive radicals, such as alkyl, azide, trifluoromethyl, acyloxy, etc., under metal-free conditions which finally leads to oxidative cyclization to afford diverse heterocyclic scaffolds. Antonchick's group developed a broadly applicable radical azidoarylation of alkenes in the presence of an azide-hypervalent iodine reagent combination (Scheme 6.17, eq. 1) [35]. Wang *et al.* were able to accomplish a novel route to 6-(trifluoromethyl)phenanthridines **41** from 2-isocyanylbiphenyls. The reaction was

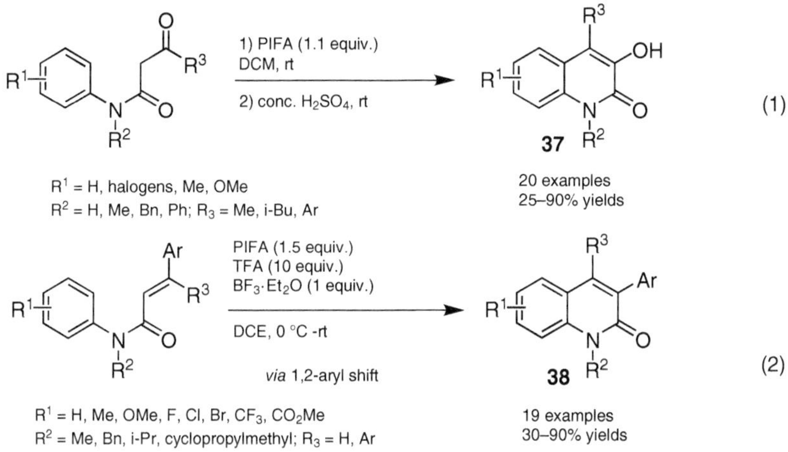

Scheme 6.15 Synthesis of quinolinone derivatives. *Source*: Liu et al. [32].

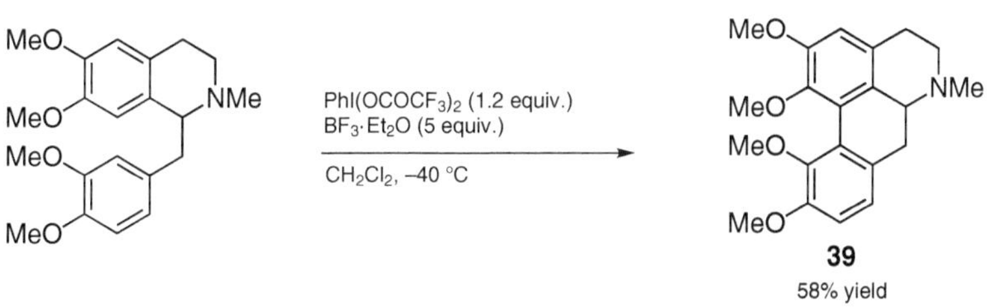

Scheme 6.16 Oxidative synthesis of aporphine scaffold. *Source*: Hamamoto et al. [34].

mediated by PIDA in the presence of a substoichiometric amount of benzoquinone (BQ) (Scheme 6.17, eq. 2). This was followed by a report from Hilton's group using the Togni reagent for the synthesis and modification of the natural product marinoquinoline (Scheme 6.17, eq. 3) [36, 37]. In the meantime, the first example of a metal-free direct trifluoromethylation of alkenes was reported by Tan, Liu, and coworkers. The method allows access to functionalized trifluoromethyl aza-heterocycles with high selectivity and broad scope [38]. Du's and Zhao's group have successfully developed a radical pathway to acridone derivatives **44** starting from *N*-arylamino aldehydes. The reaction is mediated by a PIDA-benzoyl peroxide (BPO) combination and involves a cross-dehydrogenative coupling (CDC) between an unactivated aldehyde and a C(sp^2)–H bond (Scheme 6.18) [39].

6.2.2 C–H Functionalization of (Hetero)aromatic Compounds

In the past decade, there were quite notable developments in metal-free oxidative coupling reactions. Among them, hypervalent iodine-mediated C(sp^2)–H functionalization is one of the most studied methods. This involves intramolecular and intermolecular metal-free dehydrogenative homo- and cross-couplings as well as functionalization of unactivated aromatic C–H bonds to C–C and C–heteroatom bonds [40, 41].

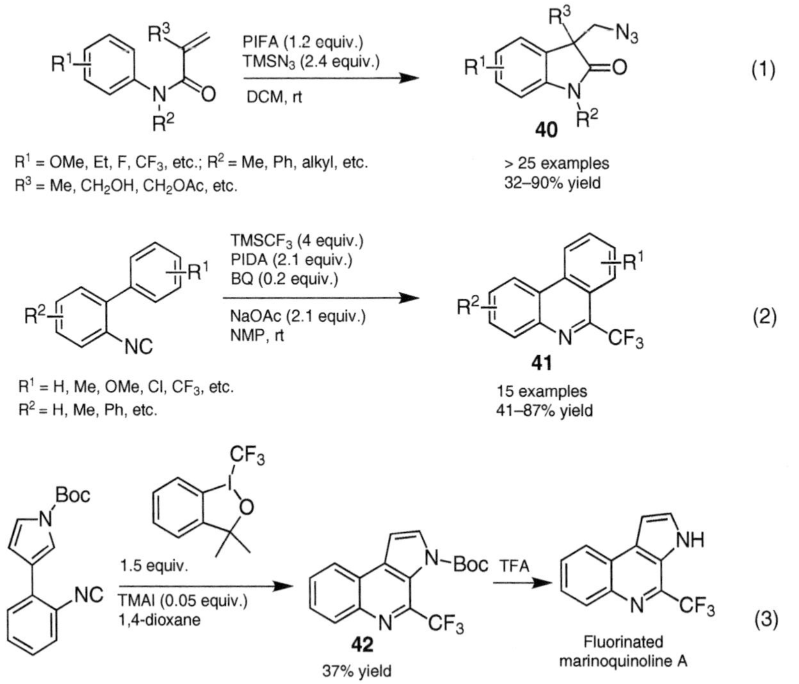

Scheme 6.17 Azidoarylation and trifluoromethylation-arylation of unsaturated bonds. *Source*: Matcha et al. [35].

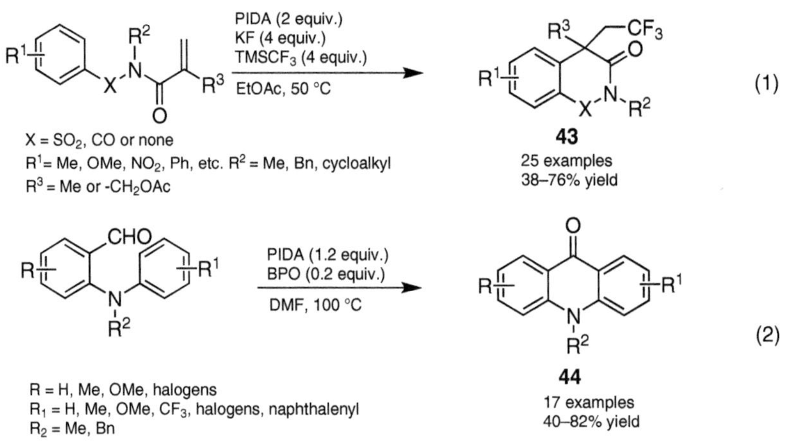

Scheme 6.18 Synthesis of aza-heterocycles and acridones. *Source*: Zheng et al. [39].

6.2.2.1 Hypervalent Iodine-Mediated C–H (Hetero)arylation

Intermolecular C–H functionalization produces diverse molecules and is practical for the synthesis of complex products. Considering these advantages, a number of metal-free routes were developed for C–C and C–heteroatom formation using hypervalent iodine chemistry. In 2002, Kita and coworkers reported the first metal-free intermolecular oxidative coupling of alkylarenes for the synthesis of alkylbiaryls **45** [42]. Along this line, homocoupling of various substituted thiophenes in the presence of different hypervalent iodine reagents was also studied by the same group. Various dithiophenes **46** were synthesized by hypervalent iodine-mediated oxidative coupling in the presence of PIDA–BF_3•Et_2O [43]. Another strategy was developed by the same group to obtain unsymmetrical heterobiaryls **47** in a unique head-to-tail-coupling fashion. This reaction involves the formation of an iodonium intermediate at the C_2 position of 3-substituted thiophenes in the presence of PhI(OH)OTs (HTIB) as iodine(III) reagent (Scheme 6.19) [44]. Such methods have also been used for the synthesis of bipyrroles **48** and homodimers of indoles [45–47].

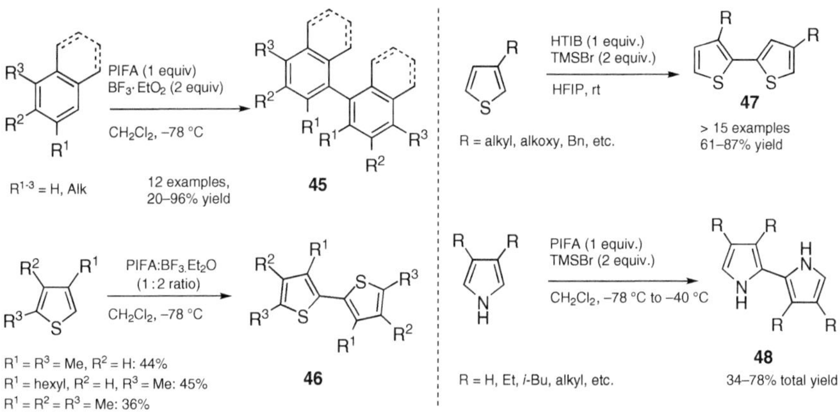

Scheme 6.19 Homocoupling of alkylarenes, thiophenes, and pyrroles.

Hypervalent iodine-mediated aryl-aryl, aryl-heteroaryl, and heteroaryl-heteroaryl cross-coupling reactions have also been studied. Kita and coworkers discovered that naphthalenes undergo oxidative-coupling reactions with electron-rich arenes *via* generation of a cation radical intermediate in the presence of ($C_6F_5I(OCOCF_3)_2$) otherwise known as FPIFA–BF_3•Et_2O system to provide cross-coupled products **49** in high yields. This work was followed by the development of oxidative cross-coupling of aromatic ethers for accessing highly oxygenated biaryls **50** [48, 49]. Similarly, a trimerization of catechol for the metal-free synthesis of hexahydroxytriphenylene (HHTP) was also realized at ambient temperature [50]. Efforts to develop cross-coupling reactions involving heteroaromatics compounds were also successful. For instance, thiophene derivatives could be effectively coupled with 1-methoxynaphthalene in the presence of HTIB and trimethylsilylbromide in HFIP [51]. Recently, Lee and coworkers demonstrated an iodosobenzene (PhIO)-mediated 3-arylation of quinoxalin-2(H)ones using arylhydrazines to give the corresponding arylated products **52** (Scheme 6.20) [52]. Canesi observed oxidative coupling of various methanesulfonamide-protected anilines with thiophenes in moderate-to-good yields. The authors proposed an interesting mechanism for this reaction based on the side-product observed. The mechanism involves formation of a sulfenium ion intermediate **53b**. This intermediate either undergoes direct

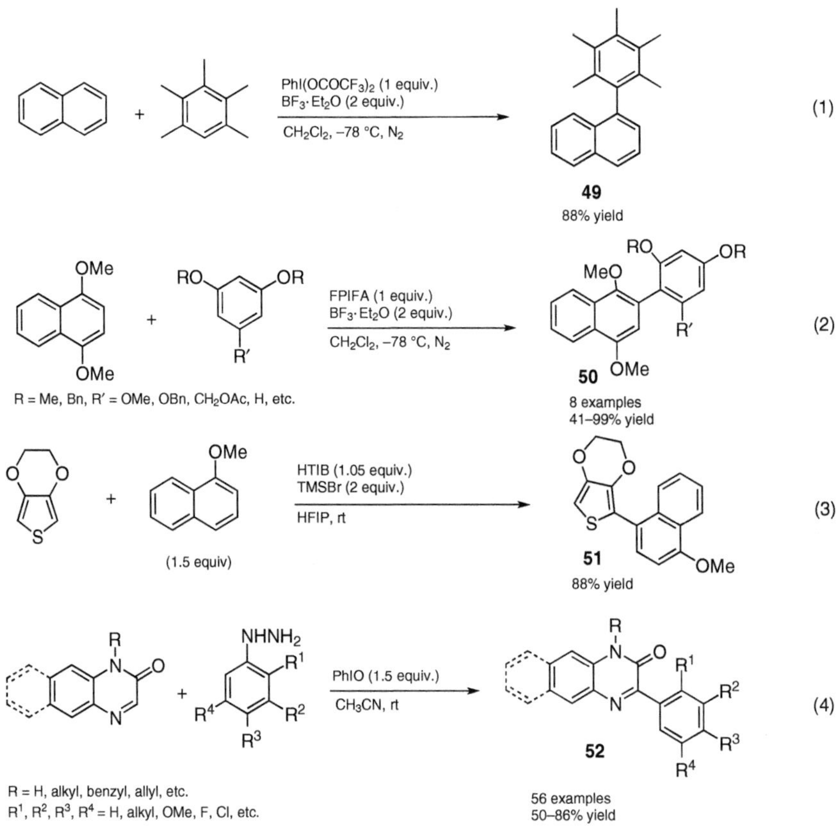

Scheme 6.20 Oxidative cross-coupling reactions mediated by hypervalent iodine. *Source:* Paul et al. [52].

conversion to final product **54** or goes through a formal [4 + 3] cycloaddition to form **53c** and subsequent aromatization to **53d**. This cycloadduct can then be converted into the final product **54** under acidic conditions using trifluoroacetic acid (TFA) or camphorsulfonic acid (CSA) (Scheme 6.21) [53].

Acyclic diaryliodonium salts are an important class of hypervalent iodine reagents. They are widely used in C–H arylation reactions of heteroaromatic compounds. Even though the transition metal-catalyzed examples of such reactions are plenty, examples of metal-free methods can also be spotted in literature. For instance, direct mono-arylation of nitrogen heterocycles like pyrrole, pyridine, pyrazine, and indole motifs was realized using diaryliodonium salts in the absence of any transition metals (Scheme 6.22) [54, 55].

6.2.2.2 Hypervalent Iodine-Mediated C–H Amination Reactions

A highly appealing method for intermolecular C–N bond formation was reported by Chang, Cho, and coworkers. In this method, arenes were treated with imide derivatives to enable the desired C(sp^2)–N bond formation (Scheme 6.23, eq. 1). The authors also described that changing the nitrogen source will render the

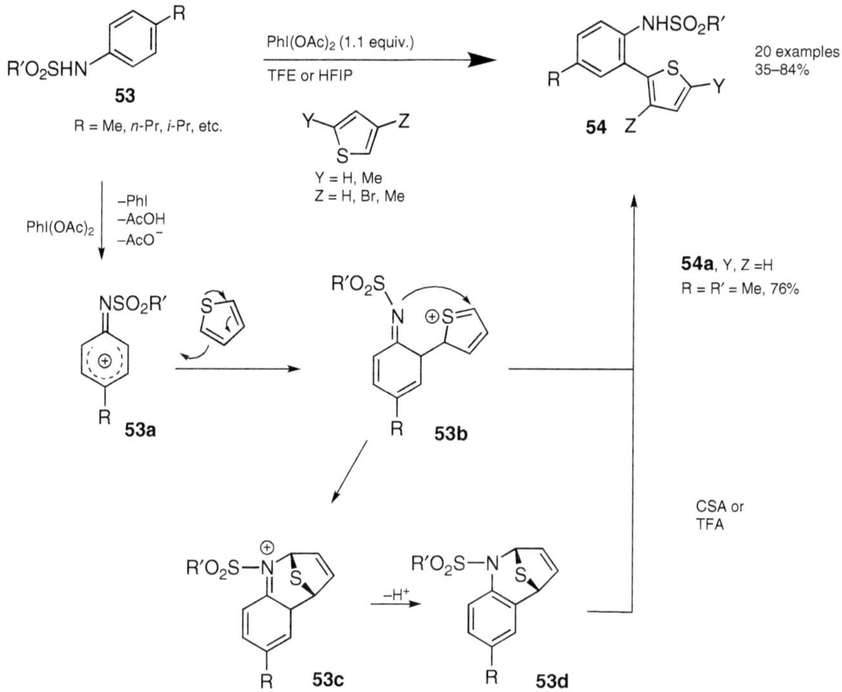

Scheme 6.21 Oxidative coupling of protected anilines with thiophenes. *Source*: Jean et al. [53].

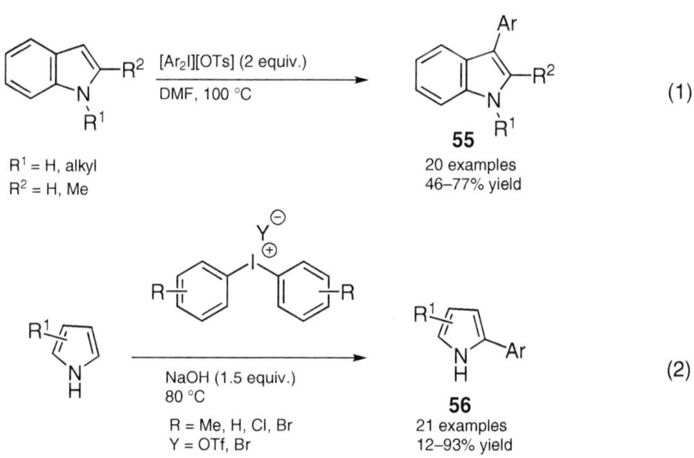

Scheme 6.22 C–H arylations mediated by diaryliodonium salts. *Source*: Wen et al. [54], Ackermann et al. [55].

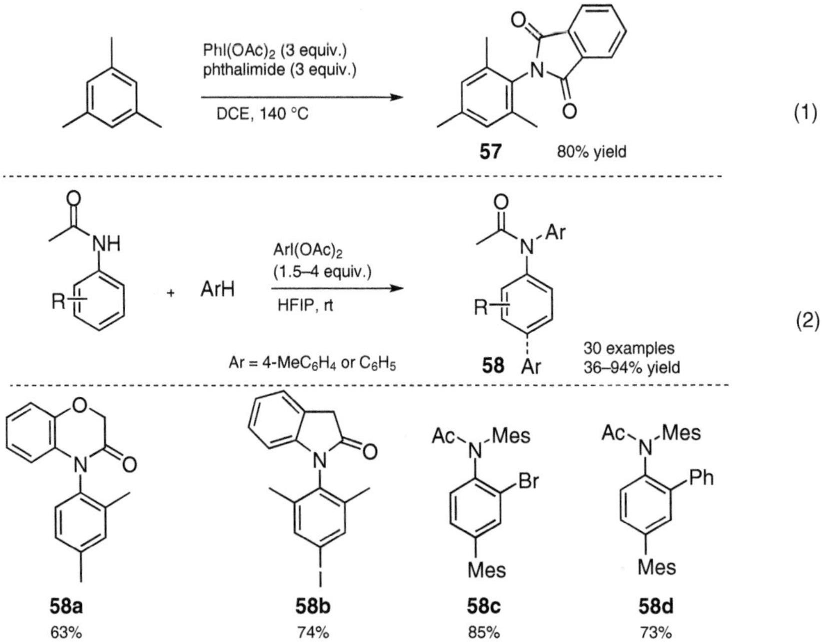

Scheme 6.23 C–H amination of arenes mediated by iodine(III) reagents. *Source*: Samanta et al. [57], Antonchick et al. [58].

reaction selective to benzylic functionalization to form C(sp^3)–N bond [56]. Meanwhile, an intermolecular, metal-free, and direct oxidative method for C–N bond formation through cross-coupling of non-prefunctionalized arenes mediated by a hypervalent iodine reagent was developed by Antonchick *et al.* Both *N*-arylation and *para*-selective diarylation of anilides are possible by this method (Scheme 6.23, eq. 1) [57, 58]. A complementary method for room temperature CDC of imidazopyridines with aliphatic cycloamines was reported by Hajra and coworkers [59]. Mild and ambient reaction conditions and broad substrate scopes are the notable advantages of this method. In an interesting report from Kita *et al.*, an N^1 selective method to couple azoles with pyrroles and related compounds was revealed (Scheme 6.24) [60].

6.2.2.3 Other Direct C–H Functionalization Reactions

In 1994, Kita *et al.* reported that the formation of diaryliodonium salts from unsubstituted phenol ethers can be blocked by substitution at the *para*-position [61]. The authors demonstrated the selective introduction of various nucleophiles such as azides, acetates, and β-dicarbonyl compounds at the *ortho*-position. The roles of radical intermediates in the reaction with hypervalent iodine reagents were also elucidated based on extensive UV/Vis and ESR studies. They also proposed that in the first reaction step, $C_6H_5I(OCOCF_3)_2$ (PIFA) reacted with the aromatic ring of the phenol ether to form a charge-transfer (CT) π-complex **61a**. Afterward, a single electron transfer (SET) occurs from the aromatic system to PIFA to generate

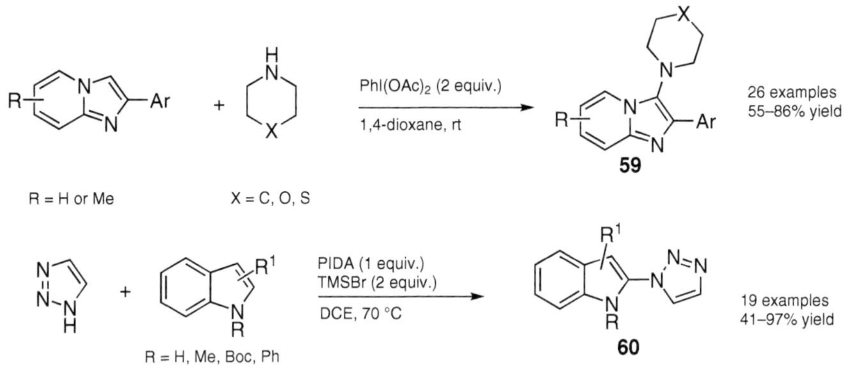

Scheme 6.24 C–H amination of heterocycles mediated by iodine(III) reagents. *Source*: Morimoto et al. [60].

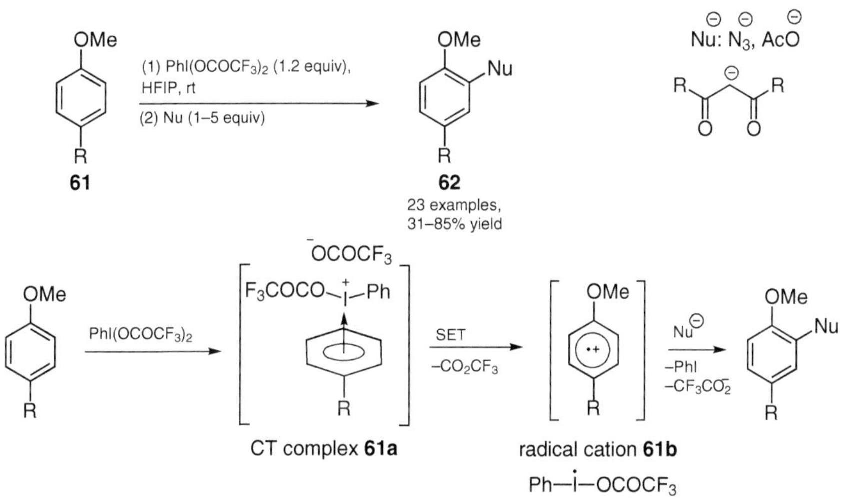

Scheme 6.25 Direct C–H functionalization at *ortho*-position of 4-substituted arenes.

a radical cation **61b**, which is promptly attacked by the nucleophile to generate the final product **62** (Scheme 6.25).

Direct oxidation of aromatic rings can be achieved using hypervalent iodine reagents in combination with Lewis acids as in the case of *N*-aryl acetamides [62–64]. A variety of *O*-nucleophiles were successfully coupled selectively at *ortho*-position of *N*-aryl acetamides to yield compounds **63-66**. In addition, the cross-coupling reaction of thiophenes and nucleophiles as well as the reaction of 3-hexylthiophene with a PIFA-TMSX (X = Cl, Br, NCS or CN) combination (Scheme 6.26) provide the corresponding coupling products **67–69** [47, 65, 66]. In recent years, hypervalent iodine-mediated C–H functionalization of heteroaryls has emerged as the most prominent alternative to metal-catalyzed methods. Various methodologies for C–C and C–heteroatom bond formation have been reported. Among them, the CDC reactions have found profound importance in the synthesis

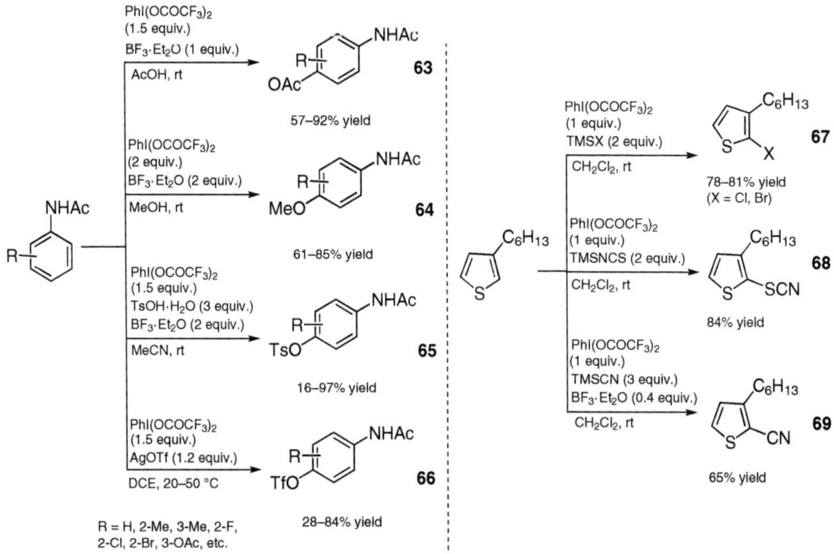

Scheme 6.26 Direct oxidative C–H functionalization of aniline and thiophene derivatives.

and functionalization of biologically relevant molecules. Antonchick *et al.* initially reported an effective acylation of nitrogen heterocycles using non-prefunctionalized starting materials in the presence of the PIFA-$TMSN_3$ system (Scheme 6.27, eq. 1) [67]. Numerous examples were illustrated in the substrate scope. The reaction is highly relevant because of generality, efficiency, and scalability. Additionally, alkylation of *N*-heterocycles was also realized by PIFA-mediated coupling between heteroaromatic compounds and simple alkanes in the presence of sodium azide (Scheme 6.27, eq. 2) [68].

Subsequently, an unprecedented C_2-alkylation of quinoline *N*-oxide derivatives *via* C–C bond activation of *tert*- and *sec*-alkyl alcohol was described using hypervalent iodine(III) reagent. The reaction proceeds efficiently with a broad range of substrates including quinoline, isoquinoline, and pyridine *N*-oxides using a variety of alcohols [69]. In an analogous manner, C_2-selective azidation of quinoline *N*-oxides was achieved by Li and coworkers (Scheme 6.28) [70].

Togni *et al.* exploited the electrophilic group transferability of the highly reactive but sufficiently stable I(III) reagent dimethyl benziodoxole **75** for the direct metal-free trifluoromethylation of various heterocycles. In the case of nitrogen heterocycles, a pronounced tendency for the incorporation of the trifluoromethyl group at the position adjacent to nitrogen was observed (Scheme 6.29) [71].

The first transition metal-free photocatalytic C–H trifluoromethylation of electron-rich heterocycles was reported by Scaiano and coworkers. This radical reaction was carried out using low concentrations of methylene blue (MB) as photosensitizer and Togni's reagent as CF_3 source. The methodology allows selective trifluoromethylation of 3-methyl indole, pyrrole, and thiophene derivatives. It is proposed that upon visible light irradiation, the triplet state of MB can be quenched by either TMEDA or DBU to form the semireduced MB radical and

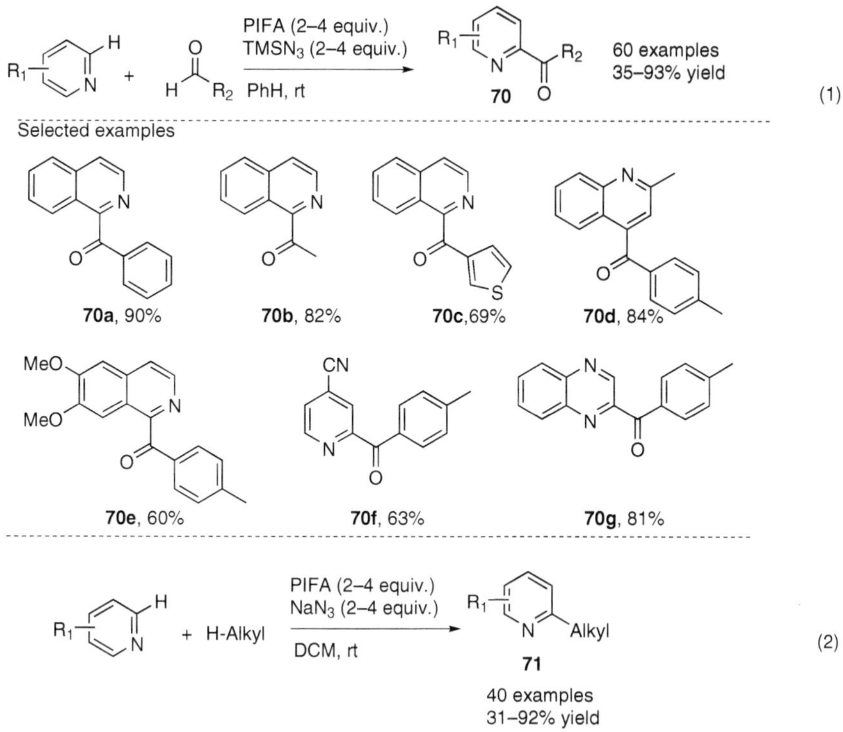

Scheme 6.27 Direct C−H functionalization of nitrogen heterocycles. *Source*: Antonchick et al. [68].

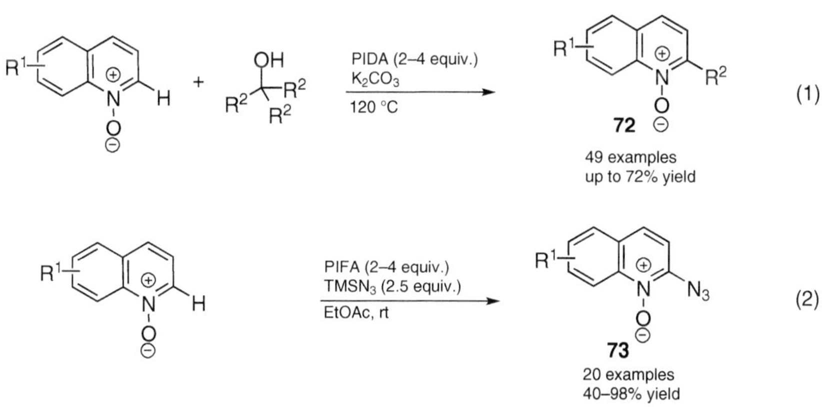

Scheme 6.28 Direct C−H functionalization of quinoline *N*-oxides. *Source*: Li et al. [70].

an α-amino radical. Both of these species can, in turn, reduce Togni's reagent, resulting in the release of a CF_3 radical which eventually reacts with heterocycles to give the desired product **77** (Scheme 6.30) [72]. An interesting method for direct alkylation of heterocycles using readily accessible carboxylic acid has been reported recently. The reaction proceeds in the presence of a hypervalent iodine reagent

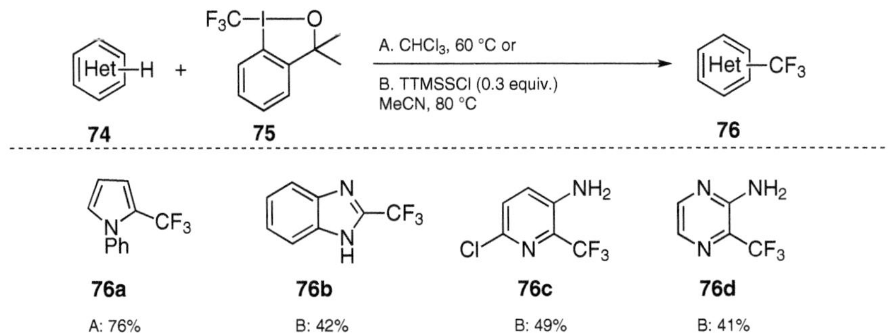

Scheme 6.29 Electrophilic trifluoromethylation of *N*-heteroarenes. *Source*: Wiehn et al. [71].

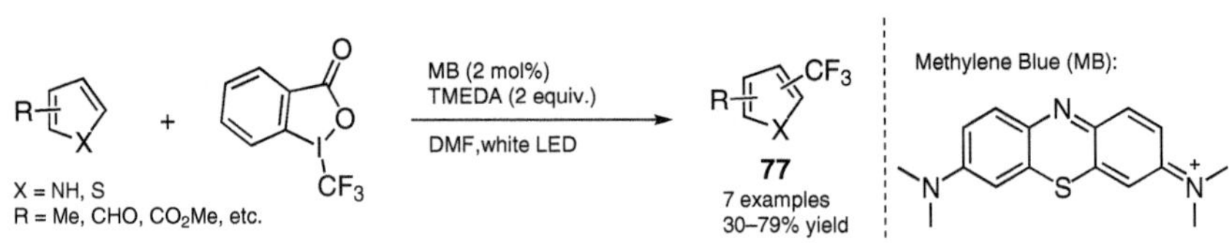

Scheme 6.30 Photocatalytic C–H trifluoromethylations. *Source*: Pitre et al. [72].

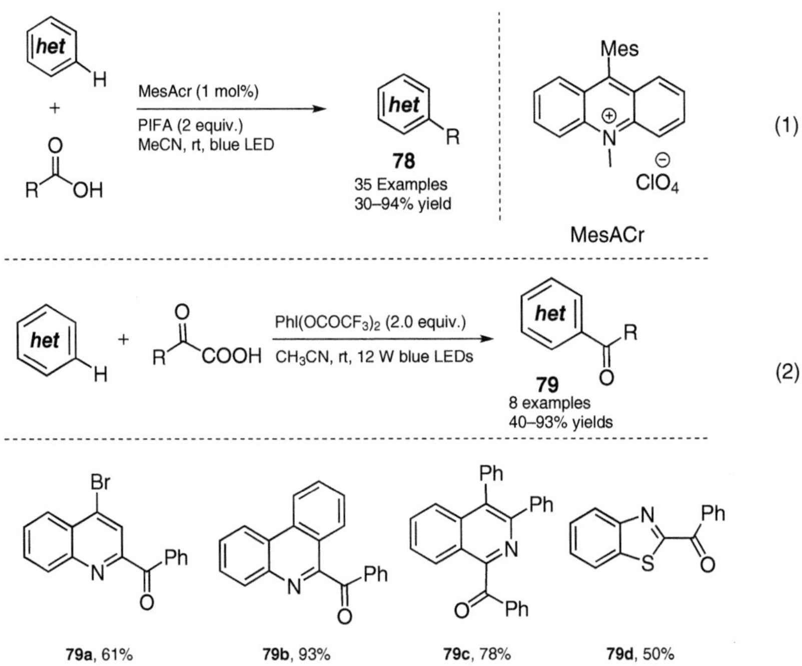

Scheme 6.31 Photocatalytic C–H alkylations and acylations. *Source*: Zhang et al. [74].

and catalytic amounts of the organo-photocatalyst MesAcr (9-mesityl-10-methyl acridinium) under mild conditions using blue LEDs as a light source. A number of heterocyclic compounds like quinoline, pyridine, benzothiazole, etc., were alkylated under metal-free conditions (Scheme 6.31, eq. 1) [73]. Zhang, Yang, and coworkers demonstrated an efficient protocol for the C–H acylation of heterocycles employing phenylglyoxalic acid as the decarboxylative acyl group donor (Scheme 6.31, eq. 2) [74].

6.3 Hypervalent Iodine-Catalyzed C–H Functionalization Reactions

Iodoarenes are frequently used in stoichiometric amounts. However, in the last couple of decades, catalytic usage of iodoarenes has been studied vastly and has become a valuable addition in the area of environmentally benign C–H functionalization. This also includes rapid progress in the area of design of new catalysts and enantioselective catalysis *via in situ* generation of active iodine(III).

6.3.1 Hypervalent Iodine-Catalyzed Intramolecular Cyclization Reactions

Numerous methods of heterocycle synthesis were illustrated using only catalytic amounts of iodoarene catalysts. In general, the *in situ* oxidation of iodoarene using oxidants such as *meta*-chloroperoxybenzoic acid (*m*-CPBA), oxone, or peracetic acid results in the formation of active iodine(III). This active iodine gets reduced while oxidizing the substrate and, hence, the iodine(I) is regenerated which continues to undergo the next oxidation cycle. In this regard, inspired by Kita's early works of iodoarene catalysis [75, 76], Antonchick and coworkers developed a catalytic version of intramolecular C–N bond cyclization for carbazole synthesis using peroxy acid as an oxidant [58]. Based on several mechanistic studies, the authors proposed the formation of a key nitrenium ion intermediate during the reaction. According to the authors, the iodine catalyst gets oxidized by peracetic acid to form the oxo-bridged hypervalent iodine species **81a**. This undergoes a ligand exchange reaction with substrate **80** and eventually results in the formation of a nitrenium species which further cyclizes followed by a nucleophilic attack of the arene ring to give the carbazole product **82a** (Scheme 6.32).

Punniyamurthy and coworkers demonstrated the successful usage of catalytic amounts of simple iodobenzene as organocatalyst in the synthesis of benzimidazole derivatives [77]. Tanimori and coworkers also studied the intramolecular C–N bond formation *via* C–H bond functionalization. They reported facile access to 1*H*-indazoles **84** in the presence of catalytic amounts of organoiodine and stoichiometric amounts of oxone as an oxidant [78]. An organocatalytic protocol for the synthesis of 2-substituted benzoxazoles and benzothiazoles **85** was

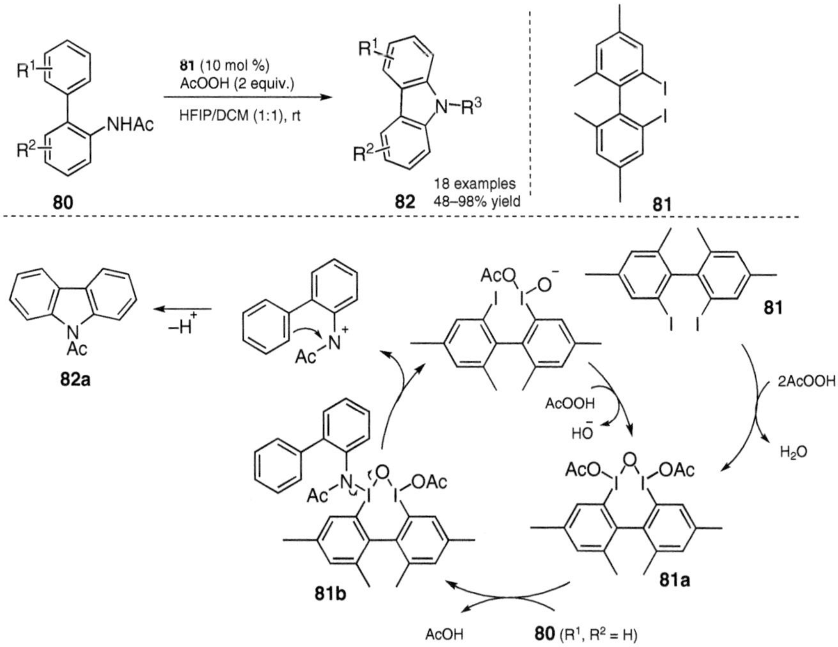

Scheme 6.32 Carbazole synthesis *via* oxidative amidation.

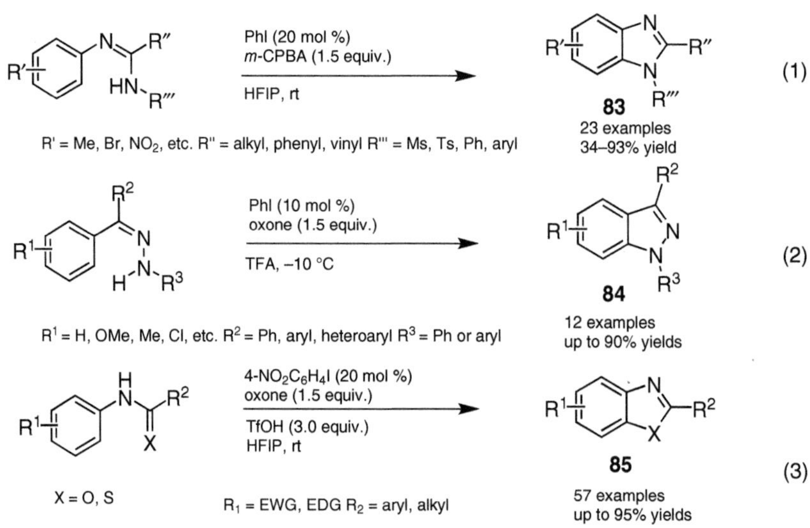

Scheme 6.33 Catalytic intramolecular C–H aminations. *Source*: Alla et al. [79].

described using 1-iodo-4-nitrobenzene as a catalyst in the presence of oxone as an environmentally safe terminal oxidant at room temperature (Scheme 6.33) [79]. Interesting literature from Xue's group revealed a novel C–H functionalization approach to phenanthridinones. They used iodine(III) catalysis for the

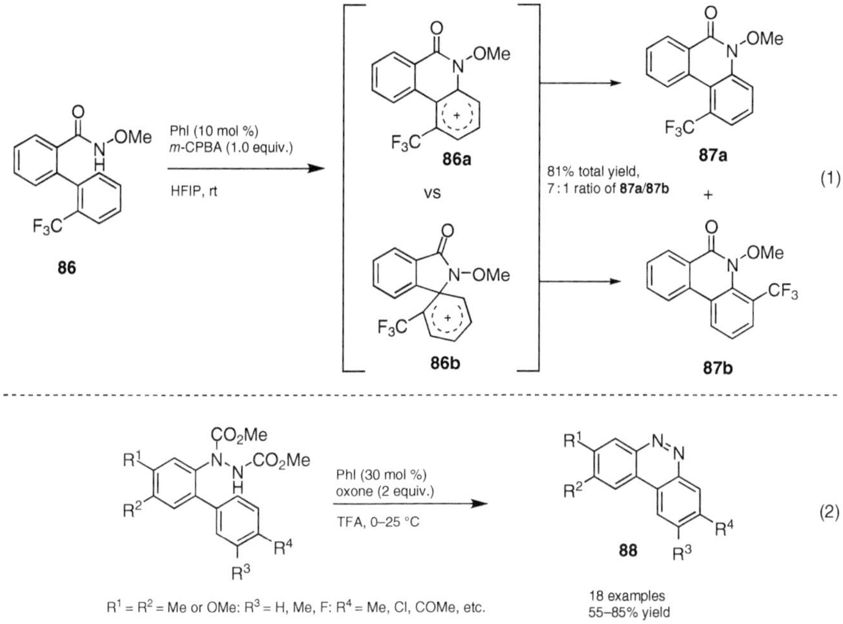

Scheme 6.34 Synthesis of phenanthridinones and cinnolines. *Source*: Reddy et al. [82].

oxidative C–H amidation of *N*-methoxybenzamides to yield the corresponding phenanthridinones. However, in the case of some substrates, a mixture of products was observed [80]. For instance, depending on the carbon attacking the electrophilic nitrenium ion, intermediate **86a** or spirocyclic intermediate **86b** could be formed. Aromatization of **86a** leads to product **87a** while rearrangement reaction of **86b** gives **87b** (Scheme 6.34, eq. 1). Meanwhile, they also observed a one-pot cyclization-halogenation of *N*-methoxybenzamides to provide 2-bromo-phenanthridinones in the presence of iodobenzene as a catalyst, tetrabutylammonium bromide as bromide source and peracetic acid as an oxidant [81]. Reddy *et al.* extended this chemistry for the synthesis of benzo[*c*]cinnoline compounds (Scheme 6.34, eq. 2) [82].

In an interesting report, Zhu's group revealed an Iodine(III) reagent-catalyzed C–H cycloamination of *N*-aryl-2-aminopyridines and *N*-arylimidines (Scheme 6.35) [83]. An iodine(III)-catalyzed bromocarbocyclization was elaborated as an efficient tool for the synthesis of C_3-disubstituted oxoindoles **93**. The ring-closing reaction occurred with complete diastereoselectivity in most cases and the reaction possesses a broad scope including substrates with chemically sensitive functionalities [84]. Das and coworkers developed a method to synthesize benzimidazole-fused heterocycles [85]. Interestingly, the reaction yield and regioselectivity of **94** is optimal using water as a green solvent for the reaction and the active hypervalent iodine reagent (Koser's reagent) is generated *in situ* using a $PhI(OAc)_2$: *m*-CPBA: PTSA.H_2O reagent combination (Scheme 6.36).

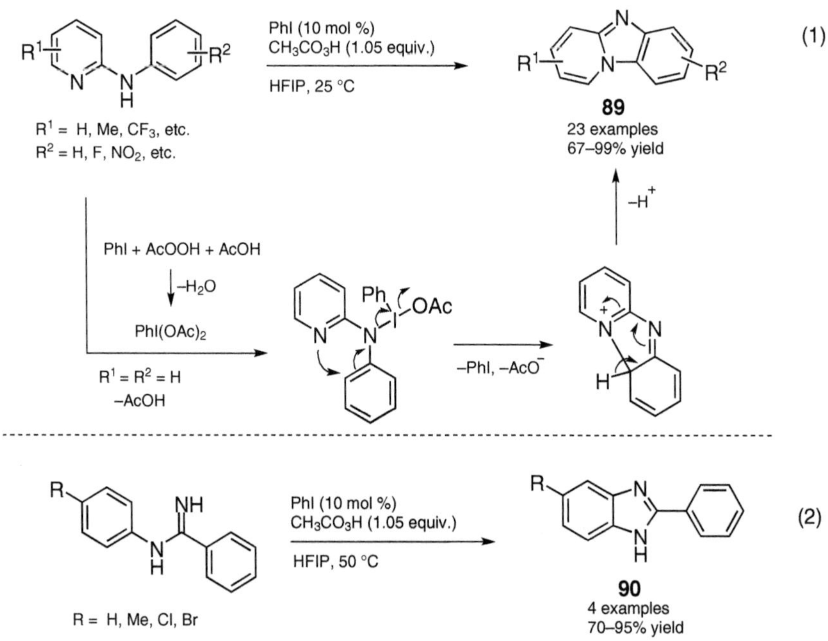

Scheme 6.35 Iodine(III) reagent-catalyzed C–H cycloamination. *Source*: He et al. [83].

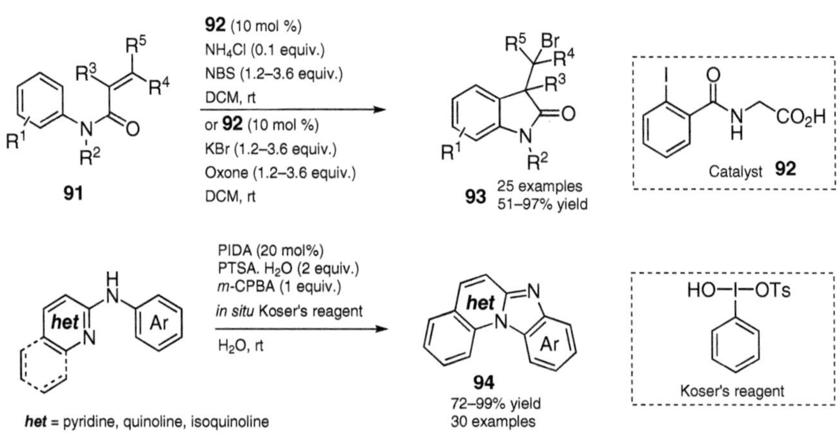

Scheme 6.36 Iodine(III)-catalyzed synthesis of oxoindoles and fused benzimidazoles.

6.3.2 Hypervalent Iodine-Catalyzed Intermolecular Reactions

The first iodoarene-catalyzed intermolecular cross-coupling reactions were reported by Kita and coworkers. The method defined specific 2,2′-diiodobiphenyls **96** as organocatalysts for the selective C–C bond formation without the use of any metal catalysts and thus providing greener access to the important biaryl motifs **97** (Scheme 6.37, eq. 1) [86]. The same group also developed a catalytic route for selective phenol-arene and phenol-phenol cross-coupling reactions in

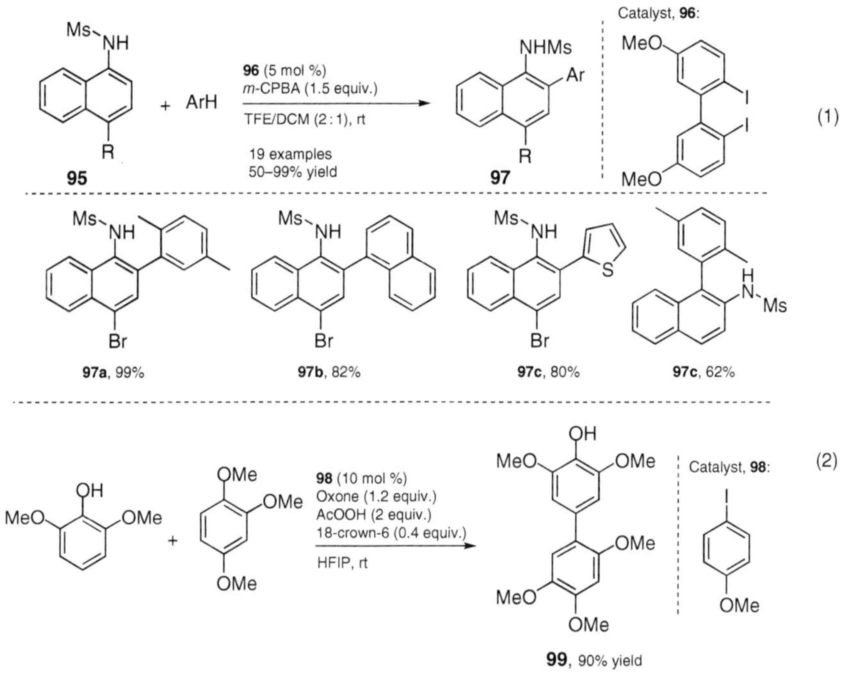

Scheme 6.37 Oxidative intermolecular C–C bond formations. *Source*: Morimoto et al. [87].

hexafluoroisopropanol. This CDC reaction uses oxone as a terminal oxidant and demonstrates easy-to-set-up access to biaryls **99** with various functional groups (Scheme 6.37, eq. 2) [87].

Following this report, Mal *et al.* studied an intermolecular tandem C–C and C–N formation between *N*-protected anilines and various arenes to yield corresponding carbazoles. They reported an appealing method for the synthesis of 1,2,4-trialkyl-substituted carbazoles **101** by a distal C–H group functionalization involving a characteristic 1,2-methyl migration using *in situ*-generated PIDA

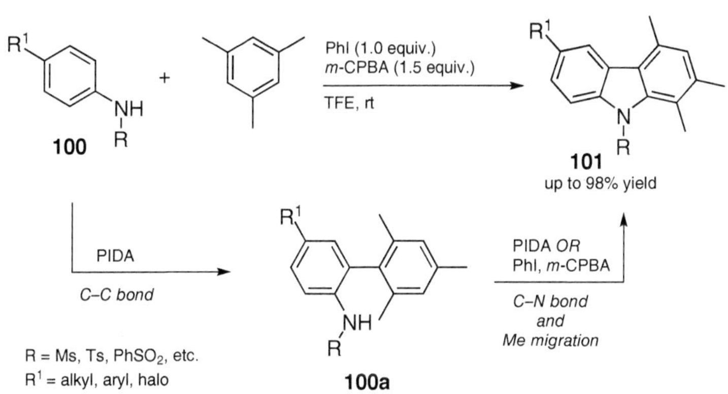

Scheme 6.38 Oxidative intermolecular C–C/C–N bond formations. *Source*: Maiti et al. [88], Maiti and Mal [89], Bal et al. [90].

from iodobenzene and *meta*-chloroperoxybenzoic acid (Scheme 6.38) [88–90]. Antonchick's group realized a tandem intermolecular C–N/C–C bond formation *via* C–H functionalization between benzamide derivatives and alkynes. The method enabled the synthesis of different isoquinolone derivatives in the presence of an *in situ*-generated hypervalent iodine catalyst [91]. The reaction starts with the formation of PIDA *via* oxidation of iodobenzene in the presence of peracetic acid. PIDA undergoes a ligand exchange reaction with the benzamide to form intermediate **102a**. Afterward, a disproportionation of this intermediate generates electronically stabilized nitrenium ion **102b**. Alkynes trap this nitrenium and a subsequent Friedel–Crafts process provides final product isoquinolone (Scheme 6.39).

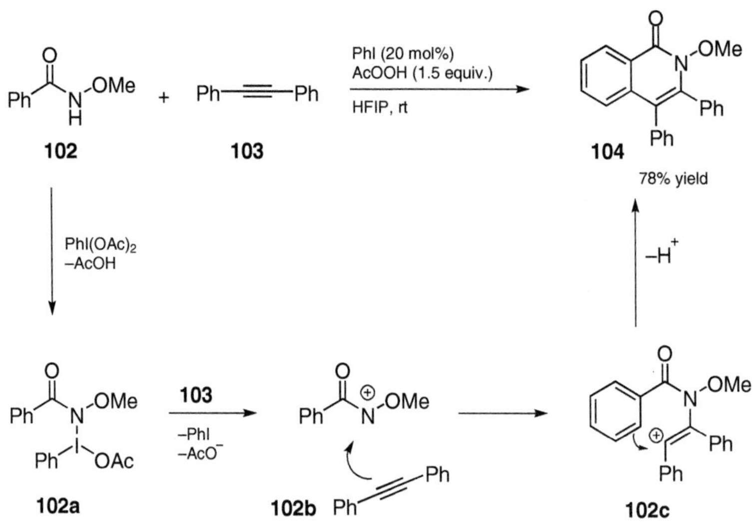

Scheme 6.39 Organocatalytic annulation of benzamides with various alkynes.

Likewise, modification of aromatic and heteroaromatic compounds through late-stage C–H functionalization is of great importance in the synthetic community. Pialat *et al.* revealed a highly *para*-selective imidation of acetanilide derivatives

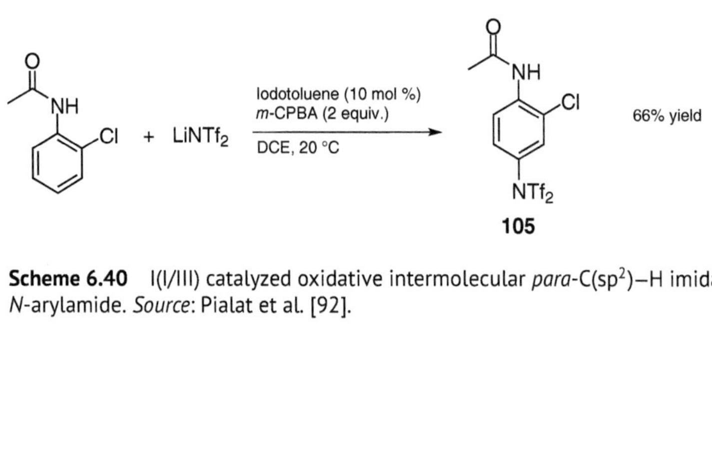

Scheme 6.40 I(I/III) catalyzed oxidative intermolecular *para*-C(sp^2)–H imidation of *N*-arylamide. *Source*: Pialat et al. [92].

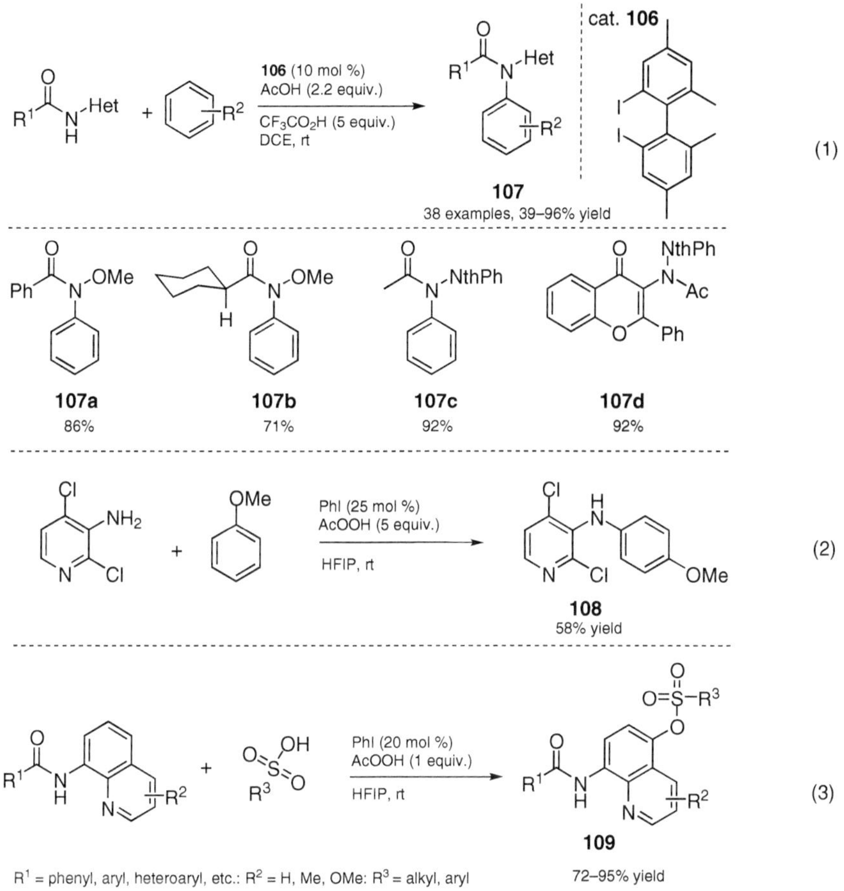

Scheme 6.41 Oxidative formation of C–N and C–O bonds *via* C–H functionalization. *Source*: Shen et al. [95].

using lithium triflimide as an imide source. The method exhibited a high functional group tolerance under iodine-catalyzed conditions (Scheme 6.40) [92]. A new atom-economical, environmentally friendly, direct oxidative intermolecular process of amination and hydrazination of non-prefunctionalized arenes was delineated by Samanta *et al.* (Scheme 6.41, eq. 1). The products were formed in a good regioselective manner under organocatalytic conditions at ambient temperature [93]. Subsequently, an arylation of 3-aminopyridines was also realized under organocatalytic conditions. The method allows a selective monoarylation of unprotected amines using electron-rich arenes (Scheme 6.41, eq. 2) [94]. An efficient protocol for the iodobenzene-catalyzed synthesis of aryl sulfonate esters from aminoquinolines *via* remote radical C–O cross-coupling was reported by Shen and coworkers. This C–O-coupling reaction proceeds under simple and mild conditions and affords the sulfonated products in good-to-excellent yields with high selectivity (Scheme 6.41, eq. 3) [95].

6.4 Conclusion

In this chapter, we focus on the developments of hypervalent iodine(III) chemistry in the synthesis and functionalization of (hetero)aromatic compounds *via* direct functionalization of one or more C(sp^2)–H bonds. This class of reagents possesses benefits like environmentally benign nature and unique oxidizing properties in addition to their easy availability and cost-effectiveness. The quest to develop sustainable chemical reactions is vital in terms of maintaining a clean and safe environment. The discovery of novel metal-free methods is an important aspect of sustainable chemistry. Iodine(III)-mediated reactions are very relevant from the sustainability point of view and will attract more attention in the coming years. For this reason, several polyvalent iodine-mediated transformations are described with a major focus on transition-metal-free methods. However, many reactions also occur in the presence of I(III) reagents as terminal oxidants, additives, or group transfer reagents of metal-catalyzed reactions. The design and application of chiral iodine catalysts is also an emerging area of research in organic chemistry.

References

1 Yoshimura, A. and Zhdankin, V.V. (2016). Advances in synthetic applications of hypervalent iodine compounds. *Chem. Rev.* 116: 3328–3435.

2 Francke, R. (2019). Electrogenerated hypervalent iodine compounds as mediators in organic synthesis. *Curr. Opin. Electrochem.* 15: 83–88.

3 Chen, J.-Y., Wu, W., Li, Q., and Wei, W.-T. (2020). Visible-light induced C(sp^3)–H functionalization for the formation of C–N bonds under metal catalyst-free conditions. *Adv. Synth. Catal.* 362: 2770–2777.

4 Maiti, S., Alam, T., Bal, A., and Mal, P. (2019). Nitrenium ions from amine-iodine(III) combinations. *Adv. Synth. Catal.* 361: 4401–4425.

5 Kandimalla, S.R., Parvathaneni, S.P., Sabitha, G., and Reddy, B.V.S. (2019). Recent advances in intramolecular metal-free oxidative C–H bond aminations using hypervalent iodine(III) reagents. *Eur. J. Org. Chem.* 2019: 1687–1714.

6 Kikugawa, Y. and Kawase, M. (1990). An electrophilic aromatic substitution by *N*-methoxyamides *via* hypervalent iodine intermediates. *Chem. Lett.* 19: 581–582.

7 Kikugawa, Y., Nagashima, A., Sakamoto, T. et al. (2003). Intramolecular cyclization with nitrenium ions generated by treatment of *N*-acylaminophthalimides with hypervalent iodine compounds: formation of lactams and spiro-fused lactams. *J. Org. Chem.* 68: 6739–6744.

8 Romero, A.G., Darlington, W.H., Jacobsen, E.J., and Mickelson, J.W. (1996). Oxidative cyclization of acyclic ureas with bis(trifluoroacetoxy)iodobenzene to generate *N*-substituted 2-benzimidazolinones. *Tetrahedron Lett.* 37: 2361–2364.

9 Yu, J., Gao, C., Song, Z. et al. (2015). Metal-free oxidative C–H amidation of N, N′-diarylureas with $PhI(OAc)_2$: synthesis of benzimidazol-2-one derivatives. *Eur. J. Org. Chem.* 2015: 5869–5875.

10 Cho, S.H., Yoon, J., and Chang, S. (2011). Intramolecular oxidative C–N bond formation for the synthesis of carbazoles: comparison of reactivity between the copper-catalyzed and metal-free conditions. *J. Am. Chem. Soc.* 133: 5996–6005.

11 Ban, X., Pan, Y., Lin, Y. et al. (2012). Synthesis of carbazolones and 3-acetylindoles *via* oxidative C–N bond formation through PIFA-mediated annulation of 2-aryl enaminones. *Org. Biomol. Chem.* 10: 3606–3609.

12 Li, Y., Zhang-Negrerie, D., Du, Y., and Zhao, K. (2015). A convenient synthesis of indoloquinolinones *via* 3-arylation of indole-2-carboxamides and PIDA-mediated C–N bond formation. *Tetrahedron* 71: 2927–2935.

13 Li, X., Yang, L., Zhang, X. et al. (2014). Construction of 1,4-benzodiazepine skeleton from 2-(arylamino)benzamides through $PhI(OAc)_2$-mediated oxidative C–N bond formation. *J. Org. Chem.* 79: 955–962.

14 Guo, X., Zhang-Negrerie, D., and Du, Y. (2015). Iodine(III)-mediated construction of the dibenzoxazepinone skeleton from 2-(aryloxy)benzamides through oxidative C–N formation. *RSC Adv.* 5: 94732–94736.

15 Du, Y.F., Liu, R.H., Linn, G., and Zhao, K. (2006). Synthesis of *N*-substituted indole derivatives *via* PIFA-mediated intramolecular cyclization. *Org. Lett.* 8: 5919–5922.

16 Kanyiva, K.S., Horiuchi, M., and Shibata, T. (2018). Metal-free N–H/C–H coupling for efficient asymmetric synthesis of chiral dihydroquinoxalinones from readily available α-amino acids. *Eur. J. Org. Chem.* 2018: 1067–1070.

17 Li, Y., Ding, Q., Qiu, G., and Wu, J. (2014). Synthesis of benzosultams *via* an intramolecular sp^2 C–H bond amination reaction of *o*-arylbenzenesulfonamides under metal-free conditions. *Org. Biomol. Chem.* 12: 149–155.

18 Ma, Y.-N., Zhang, X., and Yang, S.-d. (2017). Tandem oxidative C–H amination and iodization to synthesize difunctional atropoisomeric P-stereogenic phosphinamides. *Chem. Eur. J.* 23: 3007–3011.

19 Ma, Y.-N., Guo, C.-Y., Zhao, Q. et al. (2018). Synthesis of dibenzothiazines from sulfides by one-pot *N*, *O*-transfer and intramolecular C–H amination. *Green Chem.* 20: 2953–2958.

20 Sun, J., Zhang-Negrerie, D., Du, Y., and Zhao, K. (2015). Synthesis of chromeno[2,3-*b*]indol-11(6*H*)-one via $PhI(OAc)_2$-mediated intramolecular oxidative $C(sp^2)–N(H_2)$ bond formation. *J. Org. Chem.* 80: 1200–1206.

21 Manna, S., Matcha, K., and Antonchick, A.P. (2014). Metal-free annulation of arenes with 2-aminopyridine derivatives: the methyl group as a traceless non-chelating directing group. *Angew. Chem. Int. Ed.* 53: 9695–9695.

22 Qian, G., Liu, B., Tan, Q. et al. (2014). Hypervalent iodine(III) promoted direct synthesis of imidazo[1,2-a]pyrimidines. *Eur. J. Org. Chem.* 2014: 4837–4843.

23 Liang, D., He, Y., Liu, L., and Zhu, Q. (2013). A metal-free tandem demethylenation/$C(sp^2)$–H cycloamination process of *N*-Benzyl-2-aminopyridines via C–C and C–N bond cleavage. *Org. Lett.* 15: 3476–3479.

24 Chu, J.-H., Hsu, W.-T., Wu, Y.-H. et al. (2014). Substituent electronic effects govern direct intramolecular C–N cyclization of *N*-(Biphenyl)pyridin-2-amines induced by hypervalent iodine(III) reagents. *J. Org. Chem.* 79: 11395–11408.

25 Satishkumar, S. and Lakshman, M.K. (2017). Benzimidazopurine nucleosides from N_6-Aryl adenosine derivatives by $PhI(OAc)_2$-mediated C–N bond formation, no metal needed. *Chem. Commun.* 53: 2226–2229.

26 Otani, T., Tsuyuki, A., Iwachi, T. et al. (2017). Facile two-step synthesis of 1,10-phenanthroline-derived polyaza[7]helicenes with high fluorescence and CPL efficiency. *Angew. Chem. Int. Ed.* 56: 3906–3910.

27 Herrero, M.T., Tellitu, I., and Domínguez, E., Hernández, S., Moreno, I. and SanMartín, R. (2002). A general and efficient PIFA mediated synthesis of heterocycle-fused quinolinone derivatives. *Tetrahedron* 58: 8581–8589.

28 Shang, S., Zhang-Negrerie, D., Du, Y., and Zhao, K. (2014). Intramolecular metal-free oxidative aryl–aryl coupling: an unusual hypervalent-iodine-mediated rearrangement of 2-substituted *N*-phenylbenzamides. *Angew. Chem. Int. Ed.* 53: 6216–6219.

29 Wang, J., Yuan, Y., Xiong, R. et al. (2012). Phenyliodine bis(trifluoroacetate)-mediated oxidative C–C bond formation: synthesis of 3-hydroxy-2-oxindoles and spirooxindoles from anilides. *Org. Lett.* 14: 2210–2213.

30 Sun, J., Li, G., Zhang, G. et al. (2018). Cascade formation of C_3-unsymmetric spirooxindoles *via* $PhI(OAc)_2$-mediated oxidative C–C/C–N bond formation. *Adv. Synth. Catal.* 360: 2476–2481.

31 Yuan, Y., Yang, R., Zhang-Negrerie, D. et al. (2013). One-pot synthesis of 3-hydroxyquinolin-2(1H)-ones from *N*-phenylacetoacetamide via $PhI(OCOCF_3)_2$-mediated α-hydroxylation and H_2SO_4-promoted intramolecular cyclization. *J. Org. Chem.* 78: 5385–5392.

32 Liu, L., Lu, H., Wang, H. et al. (2013). $PhI(OCOCF_3)_2$-mediated C–C bond formation concomitant with a 1,2-aryl shift in a metal-free synthesis of 3-arylquinolin-2-ones. *Org. Lett.* 15: 2906–2909.

33 Kita, Y., Gyoten, M., Ohtsubo, M. et al. (1996). Non-phenolic oxidative coupling of phenol ether derivatives using phenyliodine(III) bis(trifluoroacetate). *Chem. Commun.* 32: 1481–1482.

34 Hamamoto, H., Shiozaki, Y., Nambu, H. et al. (2004). The efficient synthesis of morphinandienone alkaloids by using a combination of hypervalent iodine(III) reagent and heteropoly acid. *Chem. Eur. J.* 10: 4977–4982.

35 Matcha, K., Narayan, R., and Antonchick, A.P. (2013). Metal-free radical azidoarylation of alkenes: rapid access to oxindoles by cascade C–N and C–C bond-forming reactions. *Angew. Chem. Int. Ed.* 52: 7985–7989.

36 Wang, Q., Dong, X., Xiao, T., and Zhou, L. (2013). $PhI(OAc)_2$-mediated synthesis of 6-(trifluoromethyl)phenanthridines by oxidative cyclization of 2-isocyanobiphenyls with CF_3SiMe_3 under metal-free conditions. *Org. Lett.* 15: 4846–4849.

37 Patel, B. and Hilton, S.T. (2015). A radical-mediated approach to the total synthesis of fluorinated marinoquinoline A and related tricyclic and tetracyclic congeners. *Synlett* 26: 79–83.

38 Li, L., Deng, M., Zheng, S.-C. et al. (2014). Metal-free direct intramolecular carbotrifluoromethylation of alkenes to functionalized trifluoromethyl azaheterocycles. *Org. Lett.* 16: 504–507.

39 Zheng, Z., Dian, L., Yuan, Y. et al. (2014). $PhI(OAc)_2$-mediated intramolecular oxidative aryl-aldehyde Csp^2–Csp^2 bond formation: metal-free synthesis of acridone derivatives. *J. Org. Chem.* 79: 7451–7458.

40 Budhwan, R., Yadav, S., and Murarka, S. (2019). Late stage functionalization of heterocycles using hypervalent iodine(III) reagents. *Org. Biomol. Chem.* 17: 6326–6341.

41 Narayan, R., Matcha, K., and Antonchick, A.P. (2015). Metal-free oxidative C–C bond formation through C–H bond functionalization. *Chem. Eur. J.* 21: 14678–14693.

42 Tohma, H., Iwata, M., Maegawa, T., and Kita, Y. (2002). Novel and efficient oxidative biaryl coupling reaction of alkylarenes using a hypervalent iodine(III) reagent. *Tetrahedron Lett.* 43: 9241–9244.

43 Tohma, H., Iwata, M., Maegawa, T. et al. (2003). A novel and direct synthesis of alkylated 2,2′-bithiophene derivatives using a combination of hypervalent iodine(III) reagent and $BF_3{\cdot}Et_2O$. *Org. Biomol. Chem.* 1: 1647–1649.

44 Morimoto, K., Yamaoka, N., Ogawa, C. et al. (2010). Metal-free regioselective oxidative biaryl coupling leading to head-to-tail bithiophenes: reactivity switching, a concept based on the iodonium(III) intermediate. *Org. Lett.* 12: 3804–3807.

45 Dohi, T., Morimoto, K., Maruyama, A., and Kita, Y. (2006). Direct synthesis of bipyrroles using phenyliodine bis(trifluoroacetate) with bromotrimethylsilane. *Org. Lett.* 8: 2007–2010.

46 Dohi, T., Morimoto, K., Ito, M., and Kita, Y. (2007). Regioselective bipyrrole coupling of pyrroles and 3-substituted pyrroles using phenyliodine(III) bis(trifluoroacetate). *Synthesis* 2007: 2913–2919.

47 Dohi, T., Ito, M., Yamaoka, N. et al. (2009). Hypervalent iodine(III): selective and efficient single-electron-transfer (SET) oxidizing agent. *Tetrahedron* 65: 10797–10815.

48 Dohi, T., Ito, M., Morimoto, K. et al. (2008). Oxidative cross-coupling of arenes induced by single-electron transfer leading to biaryls by use of organoiodine(III) oxidants. *Angew. Chem. Int. Ed.* 47: 1301–1304.

49 Dohi, T., Ito, M., Itani, I. et al. (2011). Metal-free C–H cross-coupling toward oxygenated naphthalene-benzene linked biaryls. *Org. Lett.* 13: 6208–6211.

50 Morimoto, K., Dohi, T., and Kita, Y. (2013). Oxidative trimerization of catechol to hexahydroxytriphenylene. *Eur. J. Org. Chem.* 2013: 1659–1662.

51 Kita, Y., Morimoto, K., Ito, M. et al. (2009). Metal-free oxidative cross-coupling of unfunctionalized aromatic compounds. *J. Am. Chem. Soc.* 131: 1668–1669.

52 Paul, S., Ha, J.H., Park, G.E., and Lee, Y.R. (2017). Transition metal-free iodosobenzene-promoted direct oxidative 3-arylation of quinoxalin-2(*H*)-ones with arylhydrazines. *Adv. Synth. Catal.* 359: 1515–1521.

53 Jean, A., Cantat, J., Bérard, D. et al. (2007). Novel method of aromatic coupling between *N*-aryl methanesulfonamide and thiophene derivatives. *Org. Lett.* 9: 2553–2556.

54 Wen, J., Zhang, R.-Y., Chen, S.-Y. et al. (2012). Direct arylation of arene and *N*-heteroarenes with diaryliodonium salts without the use of transition metal catalyst. *J. Org. Chem.* 77: 766–771.

55 Ackermann, L., Dell'Acqua, M., Fenner, S. et al. (2011). Metal-free direct arylations of indoles and pyrroles with diaryliodonium salts. *Org. Lett.* 13: 2358–2360.

56 Kim, H.J., Kim, J., Cho, S.H., and Chang, S. (2011). Intermolecular oxidative C–N bond formation under metal-free conditions: control of chemoselectivity between aryl sp^2 and benzylic sp^3 C–H bond imidation. *J. Am. Chem. Soc.* 133: 16382–16385.

57 Samanta, R., Lategahn, J., and Antonchick, A.P. (2012). Metal-free direct oxidative intermolecular diarylation of anilides at ambient temperature assisted by cascade selective formation of C–C and C–N bonds. *Chem. Commun.* 48: 3194–3196.

58 Antonchick, A.P., Samanta, R., Kulikov, K., and Lategahn, J. (2011). Organocatalytic, oxidative, intramolecular C–H bond amination and metal-free cross-amination of unactivated arenes at ambient temperature. *Angew. Chem. Int. Ed.* 50: 8605–8608.

59 Mondal, S., Samanta, S., Jana, S., and Hajra, A. (2017). (Diacetoxy)iodobenzene-mediated oxidative C–H amination of imidazopyridines at ambient temperature. *J. Org. Chem.* 82: 4504–4510.

60 Morimoto, K., Ohnishi, Y., Nakamura, A. et al. (2014). N_1-selective oxidative C–N coupling of azoles with pyrroles using a hypervalent iodine reagent. *Asian J. Org. Chem.* 3: 382–386.

61 Kita, Y., Tohma, H., Hatanaka, K. et al. (1994). Hypervalent iodine-induced nucleophilic substitution of *para*-substituted phenol ethers. Generation of cation radicals as reactive intermediates. *J. Am. Chem. Soc.* 116: 3684–3691.

62 Pialat, A., Liegault, B., and Taillefer, M. (2013). Oxidative *para*-triflation of acetanilides. *Org. Lett.* 15: 1764–1767.

63 Liu, H., Xie, Y.Z., and Gu, Y.H. (2011). A novel method for tosyloxylation of anilides using phenyliodine bistrifluoroacetate (PIFA). *Tetrahedron Lett.* 52: 4324–4326.

64 Liu, H.A., Wang, X.M., and Gu, Y.H. (2011). Direct acetoxylation and etherification of anilides using phenyliodine bis(Trifluoroacetate). *Org. Biomol. Chem.* 9: 1614–1620.

65 Dohi, T., Morimoto, K., Takenaga, N. et al. (2007). Direct cyanation of heteroaromatic compounds mediated by hypervalent iodine(III) reagents: *in situ* generation of PhI(III)-CN species and their cyano transfer. *J. Org. Chem.* 72: 109–116.

66 Dohi, T., Morimoto, K., Kiyono, Y. et al. (2005). Novel and direct oxidative cyanation reactions of heteroaromatic compounds mediated by a hypervalent iodine(III) reagent. *Org. Lett.* 7: 537–540.

67 Matcha, K. and Antonchick, A.P. (2013). Metal-free cross-dehydrogenative coupling of heterocycles with aldehydes. *Angew. Chem. Int. Ed.* 52: 2082–2086.

68 Antonchick, A.P. and Burgmann, L. (2013). Direct selective oxidative cross-coupling of simple alkanes with heteroarenes. *Angew. Chem. Int. Ed.* 52: 3267–3271.

69 Sen, C. and Ghosh, S.C. (2018). Transition-metal-free regioselective alkylation of quinoline *N*-oxides via oxidative alkyl migration and C–C bond cleavage of *tert-/sec*-alcohols. *Adv. Synth. Catal.* 360: 905–910.
70 Li, P., Zhao, J., Xia, C., and Li, F. (2015). The development of carbene-stabilized N–O radical coupling strategy in metal-free regioselective C–H azidation of quinoline *N*-oxides. *Org. Chem. Front.* 2: 1313–1317.
71 Wiehn, M.S., Vinogradova, E.V., and Togni, A. (2010). Electrophilic trifluoromethylation of arenes and *N*-heteroarenes using hypervalent iodine reagents. *J. Fluor. Chem.* 131: 951–957.
72 Pitre, S.P., McTiernan, C.D., Ismaili, H., and Scaiano, J.C. (2014). Metal-free photocatalytic radical trifluoromethylation utilizing methylene blue and visible light irradiation. *ACS Catal.* 4: 2530–2535.
73 Genovino, J., Lian, Y., Zhang, Y. et al. (2018). Metal-free-visible light C–H alkylation of heteroaromatics *via* hypervalent iodine-promoted decarboxylation. *Org. Lett.* 20: 3229–3232.
74 Zhang, X.-Y., Weng, W.-Z., Liang, H. et al. (2018). Visible-light-initiated, photocatalyst-free decarboxylative coupling of carboxylic acids with *N*-heterocycles. *Org. Lett.* 20: 4686–4690.
75 Dohi, T., Maruyama, A., Minamitsuji, Y. et al. (2007). First hypervalent iodine(III)-catalyzed C–N bond forming reaction: catalytic spirocyclization of amides to *N*-fused spirolactams. *Chem. Commun.* 43: 1224–1226.
76 Dohi, T., Minamitsuji, Y., Maruyama, A. et al. (2008). A new H_2O_2/acid anhydride system for the iodoarene-catalyzed C–C bond-forming reactions of phenols. *Org. Lett.* 10: 3559–3562.
77 Alla, S.K., Kumar, R.K., Sadhu, P., and Punniyamurthy, T. (2013). Iodobenzene catalyzed C–H amination of *N*-substituted amidines using *m*-chloroperbenzoic acid. *Org. Lett.* 15: 1334–1337.
78 Kashiwa, M., Sonoda, M., and Tanimori, S. (2014). Facile access to 1*H*-indazoles through iodobenzene-catalyzed C–H amination under mild, transition-metal-free conditions. *Eur. J. Org. Chem.* 2014: 4720–4723.
79 Alla, S.K., Sadhu, P., and Punniyamurthy, T. (2014). Organocatalytic syntheses of benzoxazoles and benzothiazoles using aryl iodide and oxone via C–H functionalization and C–O/S bond formation. *J. Org. Chem.* 79: 7502–7511.
80 Liang, D., Yu, W., Nguyen, N. et al. (2017). Iodobenzene-catalyzed synthesis of phenanthridinones via oxidative C–H amidation. *J. Org. Chem.* 82: 3589–3596.
81 Liang, D.D., Sersen, D., Yang, C. et al. (2017). One-pot sequential reaction to 2-substituted-phenanthridinones from *N*-methoxybenzamides. *Org. Biomol. Chem.* 15: 4390–4398.
82 Reddy, B.V.S., Reddy, C.R., Reddy, M.R. et al. (2015). Substrate directed C–H activation for the synthesis of benzo[*c*]cinnolines through a sequential C–C and C–N bond formation. *Org. Lett.* 17: 3730–3733.
83 He, Y., Huang, J., Liang, D. et al. (2013). C–H cycloamination of *N*-aryl-2-aminopyridines and *N*-arylamidines catalyzed by an *in situ* generated hypervalent iodine(III) reagent. *Chem. Commun.* 49: 7352–7354.

84 Fabry, D.C., Stodulski, M., Hoerner, S., and Gulder, T. (2012). Metal-free synthesis of 3,3-disubstituted oxoindoles by iodine(III)-catalyzed bromocarbocyclizations. *Chem. Eur. J.* 18: 10834–10838.

85 Rao, D.N., Rasheed, S., Vishwakarma, R.A., and Das, P. (2014). Hypervalent iodine(III) catalyzed oxidative C–N bond formation in water: synthesis of benzimidazole-fused heterocycles. *RSC Adv.* 4: 25600–25604.

86 Ito, M., Kubo, H., Itani, I. et al. (2013). Organocatalytic C–H/C–H′ cross-biaryl coupling: *C*-selective arylation of sulfonanilides with aromatic hydrocarbons. *J. Am. Chem. Soc.* 135: 14078–14081.

87 Morimoto, K., Sakamoto, K., Ohshika, T. et al. (2016). Organo-iodine(III)-catalyzed oxidative phenol–arene and phenol–phenol cross-coupling reaction. *Angew. Chem. Int. Ed.* 55: 3652–3656.

88 Maiti, S., Achar, T.K., and Mal, P. (2017). An organic intermolecular dehydrogenative annulation reaction. *Org. Lett.* 19: 2006–2009.

89 Maiti, S. and Mal, P. (2017). Dehydrogenative aromatic ring fusion for carbazole synthesis via C–C/C–N bond formation and alkyl migration. *Org. Lett.* 19: 2454–2457.

90 Bal, A., Maiti, S., and Mal, P. (2018). Iodine(III)-enabled distal C–H functionalization of biarylsulfonanilides. *J. Org. Chem.* 83: 11278–11287.

91 Manna, S. and Antonchick, A.P. (2014). Organocatalytic oxidative annulation of benzamide derivatives with alkynes. *Angew. Chem. Int. Ed.* 53: 7324–7327.

92 Pialat, A., Bergès, J., Sabourin, A. et al. (2015). Iodine(III)-mediated *para*-selective direct imidation of anilides. *Chem. Eur. J.* 21: 10014–10018.

93 Samanta, R., Bauer, J.O., Strohmann, C., and Antonchick, A.P. (2012). Organocatalytic, oxidative, intermolecular amination and hydrazination of simple arenes at ambient temperature. *Org. Lett.* 14: 5518–5521.

94 Manna, S., Serebrennikova, P.O., Utepova, I.A. et al. (2015). Hypervalent iodine(III) in direct oxidative amination of arenes with heteroaromatic amines. *Org. Lett.* 17: 4588–4591.

95 Shen, C., Yang, M., Xu, J. et al. (2017). Iodobenzene-catalyzed synthesis of aryl sulfonate esters from aminoquinolines *via* remote radical C–O cross-coupling. *RSC Adv.* 7: 49436–49439.

7

Design of Chiral Organoiodine(I/III) Catalysts for Asymmetric Oxidative Transformations

Andreas Boelke and Boris J. Nachtsheim

Institut für Organische und Analytische Chemie, Universität Bremen, Bremen, Germany

7.1 Introduction

The use of chiral aryl iodides as catalysts in asymmetric oxidative transformations is a young research area within the field of iodane chemistry. Chiral iodanes have been [1] described already in 1907, initiated by a report of Richard Pribam titled "Ueber das optische Drehungsvermögen des Jodoniumtartrates" (or "About the optical rotation of iodonium tartrate") [1]. Since then, their widespread application as *stoichiometric* chiral oxidants in enantioselective oxidative couplings could be well established in the following decades [2]. While the catalytic application of *non-chiral* aryl iodides as mediators in oxidative couplings is known since the 1990s [3, 4], the use of *chiral* aryl iodide catalysts in enantioselective oxidative transformations was initiated almost two decades later. In a theoretically straightforward, but until today, often hard to accomplish, catalytic cycle (see Scheme 7.1), a chiral aryl iodide in the presence of external ligands [**I-Ar + L$_n$**]* is oxidized by a terminal oxidant to generate a chiral hypervalent iodine compound **Int A*** *in situ*. The chiral iodane subsequently undergoes a ligand exchange process with a nucleophile (Nu^1) to generate an iodane-nucleophile complex **Int B*** as the reactive key intermediate. Due to the highly exergonic driving force of the iodine(III)–iodine(I) reduction, the iodane-bound nucleophile further couples with a second nucleophilic substrate Nu^2 in the final oxidative coupling, releasing the starting iodoarene catalyst. In this final coupling step, chirality is transferred from the chiral environment of the iodane onto the desired reaction product.

The key for a successful chirality transfer is the chemical nature of the *in situ* generated iodanes, in particular, once the first nucleophile is attached. A variety of parameters must be considered for a rational structural design of chiral iodoarene catalysts and, hence, the formation of these key intermediates (Scheme 7.2).

(1) *Oxidation State*

Organic hypervalent iodine reagents can either exist in a trivalent form as aryl-λ^3-iodanes or a pentavalent state as aryl-λ^5-iodanes. The trivalent state is characterized by a trivalent bipyramid with the arene and the two-electron lone

Iodine Catalysis in Organic Synthesis, First Edition. Edited by Kazuaki Ishihara and Kilian Muñiz.

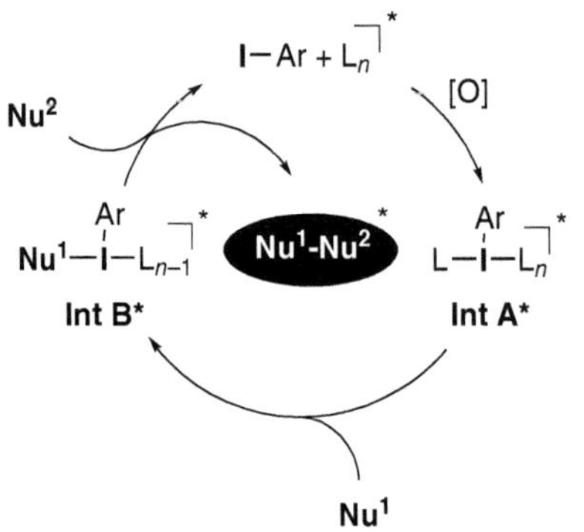

Scheme 7.1 General catalytic cycle for aryl iodide-catalyzed enantioselective oxidative couplings.

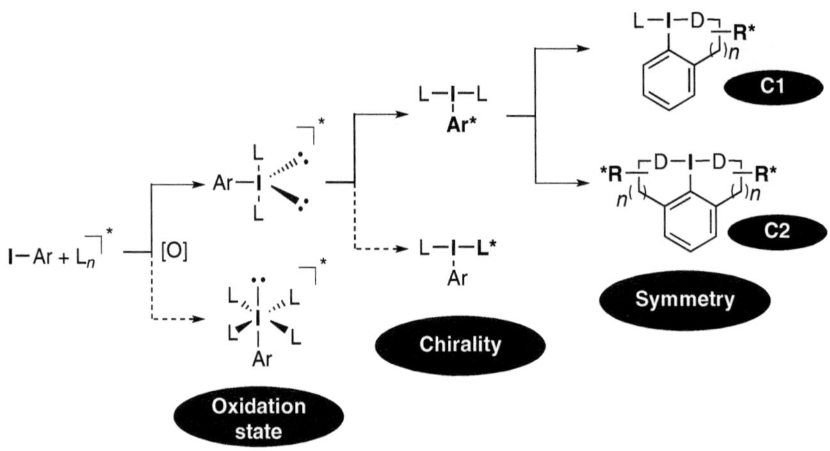

Scheme 7.2 Design concept of chiral iodoarene catalysts.

pairs in the equatorial position and the two Lewis basic ligands in the apical position spanning the 4-e-3-c bond together with the central iodine atom. In the pentavalent state, two 4-e-3-c bonds exist resulting in an overall tetragonal bipyramidal arrangement of the iodane. Here, the arene is situated in the apical position of the bipyramid.

Most oxidative couplings are mediated by trivalent iodanes, whereas pentavalent iodanes are convenient oxidants in dehydrogenative 2-electron oxidations. It is therefore not surprising that the majority of catalytic enantioselective reactions are based on the generation of trivalent iodanes, rendering the trigonal bipyramidal structure as the major design motif for the chiral iodoarene catalysts.

(2) *Origin of Chirality*

Chirality can be introduced into the key iodane intermediates by two different design approaches: (i) The combination of an achiral iodoarene, in the simplest case iodobenzene, and a chiral ligand L*. Due to the great abundance of potential chiral ligands (carboxylic acids, amino acids, amino alcohols, etc.), this seems to be a synthetically appealing and highly variable strategy. However, due to highly dynamic ligand exchange processes and resulting conformational switches among trivalent iodanes, the pure appliance of intermolecular attached chiral ligands is usually not successful in terms of chirality transfer. (ii) The other possibility to introduce chirality would be the modification of the arene with a chiral substituent **R***. This substituent usually contains an electron-donating functionality that coordinates to the iodane and forms a stabilized, and most important, rigidified (pseudo)cyclic chiral hypervalent iodine compound.

(3) *Symmetry*

Intramolecular stabilized chiral iodoarenes can be further divided into two major subclasses, according to the structure of the iodane: C_1- and C_2-symmetric derivatives. Both subclasses appeared nearly simultaneously. In 2007, Wirth and coworkers described a chiral C_1-symmetric iodoarene ether which catalyzed the enantioselective α-oxytosylation of ketones in the presence of *m*CPBA as stoichiometric co-oxidant [5]. One year later, the first efficient C_2-symmetric chiral aryl iodide was described and applied in the oxidative dearomatization of phenols by Kita and coworkers [6].

In this chapter, we will discuss the evolution of these chiral catalysts chronologically with a special focus on indeed rare structural information of the *in situ* generated iodane. A structural distinction is made by the symmetry of the parent aryl iodide (C_1- or C_2-symmetric). The performance of prominent catalysts in oxidative couplings will be analyzed in a comparative analysis.

7.2 C_1-Symmetric Iodoarenes

7.2.1 Structural Features

C_1-symmetric iodoarene catalysts can be divided into structures with distinct stereogenic chiral carbons (**Type I**), axial chiral (**Type II**), and planar chiral derivatives (**Type III**) (Figure 7.1). In **Type I**, double-substituted iodoarenes bearing a chiral functional group R* *ortho* to the central iodine atom and usually another non-chiral substituent R at the opposite *ortho*-position of the iodine are the core structural motifs (Figure 7.1a). For **Type II** binaphthyls, biphenyls and spirobiindanes have been utilized as atropisomeric chiral core structures (Figure 7.1b). In contrast to C_2-symmetric chiral iodoarenes, the development of C_1-symmetric iodoarenes

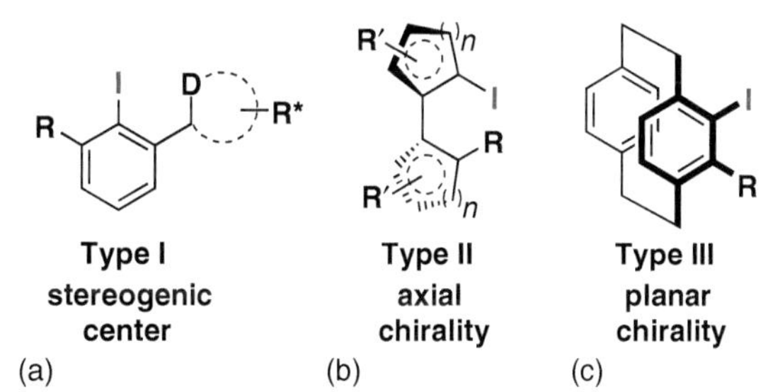

Figure 7.1 Design principles of C_1-symmetric iodoarene catalysts.

is focused on structures with "classical" chiral centers of **Type I**. Planar chiral iodoarenes of **Type III** based on paracyclophane are a comparable new development and, hence, have not been established yet [7].

A timeline demonstrating the evolution of the most important iodoarene catalysts divided by the different subtypes (**I**, **II**, and **III**) is shown in Figure 7.2. Chiral iodoarenes of **Type I** are not only the most common catalysts but also were the first literature-described derivatives thereof. While the application of stoichiometric amounts of chiral hypervalent iodine reagents in enantioselective oxidative couplings is known nearly as long as non-chiral hypervalent iodine reagents have been described, their *in situ* generation is a modern development. In 2007, iodoarenes **1** with a chiral benzyl ether were found to be an efficient catalyst for α-oxytosylations of ketones **14** [5]. Even though yields and enantioselectivities in this reaction using these early catalysts were only moderate, they were not only the blueprint for structurally related iodoarene catalysts but also initiated the whole field of iodoarene-catalyzed enantioselective couplings [8–11]. The general motif of having the chiral center in the benzylic position is still a major design principle among **Type I** catalysts. Very similar motifs can be found in iodoarenes, which are *ortho*-substituted with chiral benzyl esters **2**, pyridines **7a**, triazoles **7b** and, lately, in indanole-based catalysts **8**. Another strategy to include chiral information is the introduction of chiral amides, in particular, based on ephedrine derivatives **4** as chiral substituents [12].

Further cyclization of benzyl amides to oxazolines gives chiral catalysts **3**, which have been applied in great success in α-oxytosylations [13–15]. For chiral oxazolines **3**, the high importance of the other, non-chiral, *ortho*-substituent attached to the iodoarene has been rationalized for the first time. The structural rationale of this important ortho-effect, also called "hypervalent twist," is described in depth in Section 7.2.1.1 of this book chapter. Iodotetralones **5** have been rationally developed for *para*-phenol dearomatizations. Based on a density functional theory (DFT)-calculated structure of the central iodine(III) phenoxide, a rational design of this catalyst was possible [16]. In a later study, related hypervalent tetralone **22** could be synthesized and characterized *via* X-ray structure analysis (Figure 7.3) [17]. This study is of remarkable interest since it is a rare example in which the structure of **Type I** chiral iodoarenes could not only be verified through computational chemistry but crystallized and analyzed in its solid-state structure in the oxidized state. Interestingly, no interactions between the benzylic oxygen and the hypervalent iodine center were observed. The 3c-4e-bond is constructed by two coordinating acetate ligands and lies perpendicular to the arene plane (Fig. 7.3).

Another class of Type **I** C_1-symmetric catalysts are *o*-iodophenol-derived lactate ethers **6**. Although numerous applications of the corresponding iodanes as stoichiometric oxidants have been described, catalytic applications of the iodoarenes are still rare and restricted to hydroxylactonizations, in contrast to the widely applied C_2-symmetric 2-iodoresorcinols. The introduction of *N*-heterocycles allows the further stabilization and rigidification of the hypervalent iodine atom and was proven to be highly beneficial for the reactivity of chiral benzyl ethers [18]. As an example, pyridine-substituted benzyl ethers **7a** were successfully applied in enantioselective

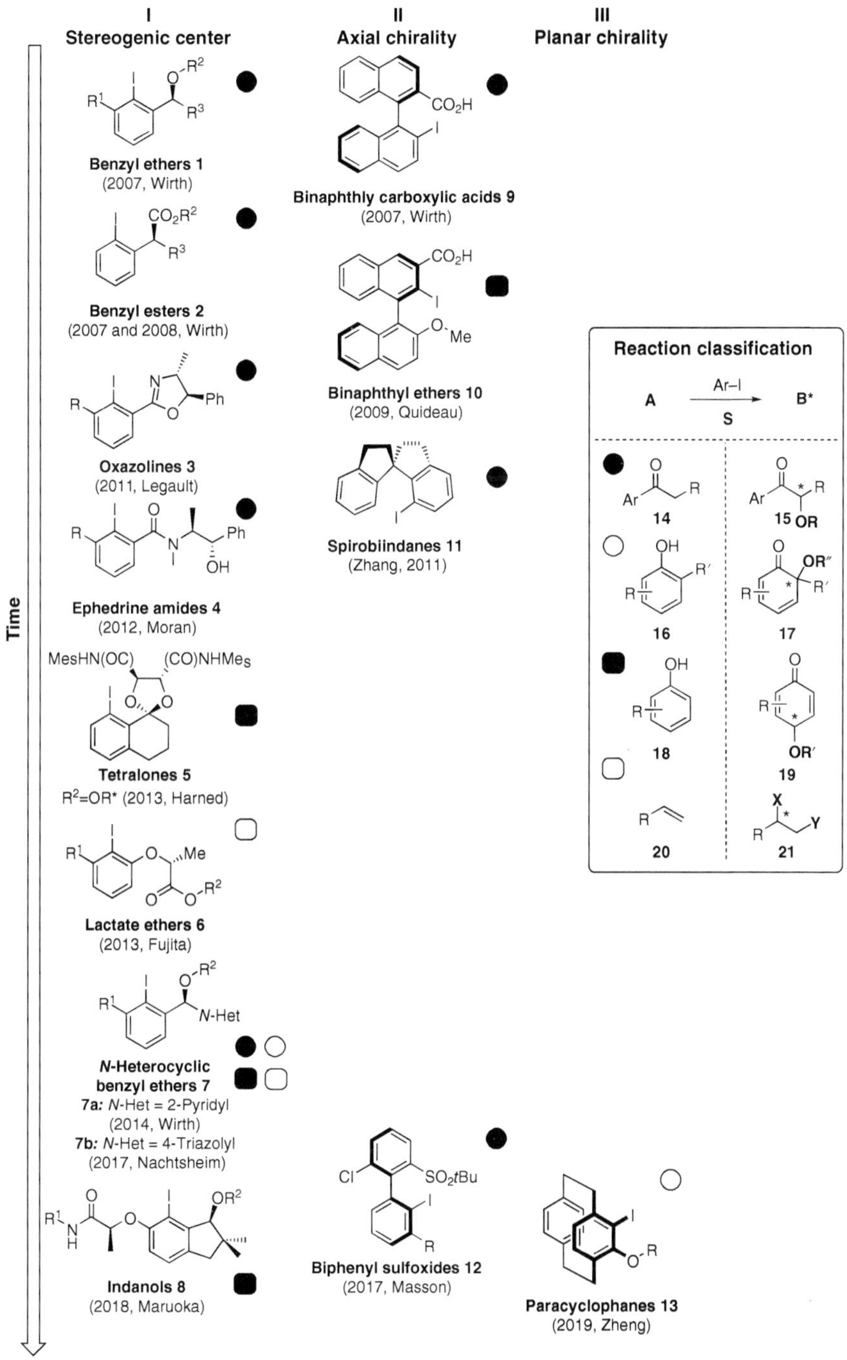

Figure 7.2 Catalyst evolution of chiral iodoarenes divided by the core chiral motif.

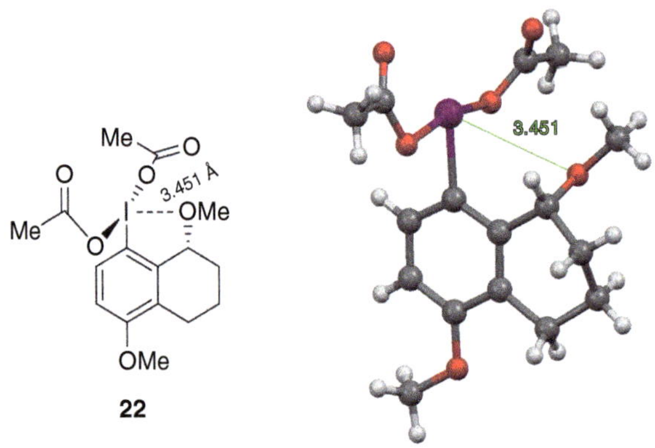

Figure 7.3 Solid state structure of a hypervalent iodotetralone **22**. *Source*: Structural data from CCDC-Database [17].

diaminations [19]. Triazole-substituted derivatives **7b** show good reactivities in a variety of transformations including phenol dearomatizations, α-oxytosylations, and lactonizations [20, 21]. DFT calculations clearly reveal that *N*-stabilization is strongly favored over *O*-coordination in the oxidized hypervalent state.

Axial chiral iodoarenes **9** of **Type II** based on binaphthyls have been described as early as the first benzyl ethers **1** in 2007 [5]. While these initially developed binaphthyl carboxylic acids in studies using *α*-oxytosylations as model reaction showed only poor results, the corresponding binaphthyl ethers **10** showed a promising selectivity in asymmetric hydroxylative phenol dearomatizations [22]. Short after, spirobiindanes **11** were developed and applied in *α*-oxytosylations [23]. The same reaction can be catalyzed in high enantioselectivities with unique sulfone-substituted iodobiphenyls **12** [24]. These catalysts have the significant advantage of a fast synthesis and thus render good synthetic accessibility, in particular, in direct comparison to their binaphthyl opponents. Planar chiral iodoarenes of **Type III** are so far underdeveloped. Although incidentally mentioned in the early reports from 2007, only one successful application of a 2,2-paracylophane-derived iodoarene **13** is known. Here it was applied in the dearomative lactonization of 1-naphthols [7].

7.2.2 The Influence of *ortho*-Substituents

Asymmetric induction in iodoarene-catalyzed oxidative couplings is directly correlated to the reactivity of the catalyst. Hence, strategies to improve the catalysts' reactivity is not only beneficial due to reduced reaction times, better yields, and fewer undesired side-reactions of the terminal oxidant but also gives the desired products in higher enantioselectivities. To rationally understand the reactivity of the chiral iodoarene, a general mechanistic description of oxidative couplings is drawn in Scheme 7.3. Due to the excess of the stoichiometric oxidant and external ligands **X**, the initial oxidation of the iodoarene into the active iodine(III)-species (**Int1**) is fast and therefore not shown. The next step is a ligand exchange of **X** with an incoming nucleophile (**Nu^1**) to give **Int2** (Scheme 7.3a). This process can be regarded as an S_N-reaction on the hypervalent iodine center, either through a dissociative or an associative mechanism. In any case, the major contribution of the energy barrier is a partial out-of-plane distortion of the 3c-4e-bond (X-I-D*) out of its ideal geometry of 180° and/or a distortion of the torsion angle between the hypervalent bond and the arene (ideally 0°). This can be rationalized by a π*-n-orbital interaction between the arenes' π-system and the lone pair of the hypervalent iodine atom. This interaction further stabilizes this bonding geometry and is significantly weakened by this distortion. To lower this barrier, a previous "bending" of the hypervalent bond is beneficial since this leads to a direct destabilization of the ground state for **Int1**. The same mode of destabilization also raises **Int2** in energy owing to the bending of the so-formed **Nu^1-I-D*** 3c-4e-bond and hence makes **Int2** more reactive for the final oxidative coupling with the second nucleophile **Nu^2** to regenerate the catalyst Ar-I* and the desired chiral coupling product **Nu^1-Nu^2***.

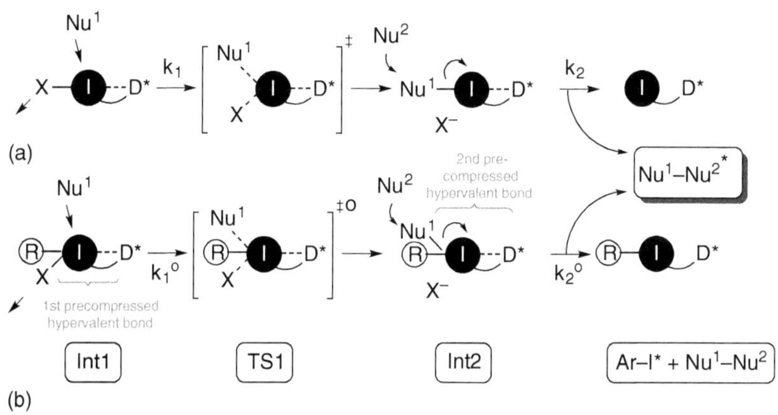

Scheme 7.3 Schematic representation of the *ortho*-effect in a comparison of iodanes with (a) and without (b) an *ortho*-substitutent R.

A "destabilization" through prebending of all potentially *in situ* formed hypervalent bonds can be achieved by design through the introduction of a substituent **R** *ortho* to the central iodine atom (Scheme 7.3b). This leads to a severe steric repulsion with **X** and results in the desired distortion or "bending" of the 180° angle of the 3c-4e-bond. Such *ortho*-substitution effects have been found empirically in the very earliest studies of iodoarene-catalyzed enantioselective reactions and were finally analyzed and, even more importantly, rationalized by Legault and coworkers by an obeserved "hypervalent twist" [15]. This distortion could be verified by X-ray structure analysis of achiral pseudocyclic amide-stabilized λ^3-iodanes **23a** (Figure 7.4a) and its *ortho*-substituted derivative **23b** (Figure 7.4b). The solid-state structure clearly reveals a significant larger dihedral angle between the arene plane (exemplified by the bond C6–C11) and the I–OH bond of 20.2° in the *ortho*-substituted amide **23b** compared to almost 0 in the unsubstituted molecule **23a**. Likewise, the 3c-4e-geometry is nearly ideal for **23a** (178.56°) and significantly distorted to 172.49° for **23b**. In addition, an increased C11–I1

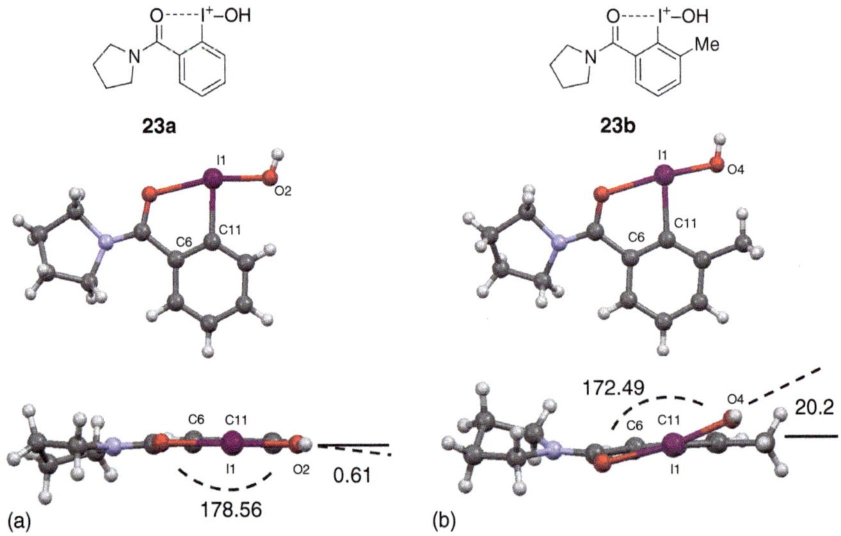

Figure 7.4 Verification of the "*ortho*-effect" through X-ray structure analysis of the achiral aryl(hydroxy)iodonium salts **23a-b** without (a) or with (b) an additional ortho-methyl substitutent. *Source*: Structural data from CCDC-Database [15].

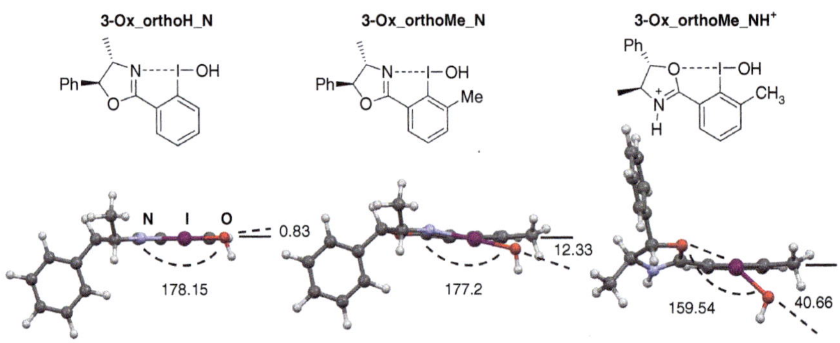

Figure 7.5 DFT-Optimized structures of chiral oxazoline-catalysts [13]. *Source*: Data from Guilbault et al. [13].

bond length is observed from 2.09 to 2.14 Å, implying an enhanced reactivity of **23b**.

Even though solid-state structures of comparable chiral λ^3-iodanes have not been described, due to their high reactivity, DFT-calculated structures reveal similar trends [15]. A DFT-calculated structure of the oxazoline-substituted iodoarenes **3-Ox_orthoH_N**, **-orthoMe_N** and **-orthoMe_NH$^+$** is shown in Figure 7.5 [13]. The α-oxytosylation of propiophenone was chosen as a model reaction. **3-Ox_orthoH_N**, with no additional *ortho*-substituent, shows a nearly ideal coplanarity between the arene and the hypervalent N–I–O bond. It is therefore the most stabilized derivative and consequently shows the worst performance among these substrates. The calculated structure of the *ortho*-methyl-substituted iodane **3-Ox_ortho_Me_N** is characterized by a significant torsion angle between the arene and the hypervalent bond of 12.33°. In the test reaction, the iodoarene precursor indeed outperforms the unsubstituted derivative. However, the most destabilized structure, with the largest torsion angle between the I–OH bond and the arene plane (40.66°) and the largest bending of the hypervalent bond itself (159.54°), is **3-Ox_orthoMe_NH$^+$**. Here, the oxazoline is *N*-protonated leading to a favored *O*-coordination of the heterocycle.

The O–I interaction is weaker than the corresponding N–I interaction, which further increases its reactivity through destabilization. DFT calculations of the initial step of the α-oxytosylation have also been performed that further could quantify the effect of *ortho*-substitution, at least for the first step of the α-oxytosylation of propiophenone. The significance of this *ortho*-effect is strongly affected by the underlying iodoarene structure. Empirically, the ring size of the *in situ*-formed pseudocycle is of particular importance. A strong influence can be observed for 5-membered pseudocyclic iodanes in which a strongly favored coplanarity of the 5-membered ring with iodoarene forces the second ligand X through the necessary 180° alignment of the hypervalent bond into the same coplanarity. As examples, strong *ortho*-substitution effects have been observed for chiral oxazoline (**3**) and triazole-substituted iodoarenes (**7c**). Due to the high reactivity of the *ortho*-substituted iodane, the solid-state structures of these derivatives are not available yet. However, DFT calculations strongly imply the same geometric effect as discussed for achiral derivatives in which solid structures were available. In larger rings, the donor-containing pseudocycle is intrinsically not coplanar with the iodoarene and thus the geometric restrictions of the hypervalent bond force the second ligand out of the arene plane, even in the absence of an *ortho*-substituent.

7.2.3 Catalyst Synthesis – Comparison Among C_1-Symmetric Derivatives

To adequately judge the overall applicability of a given catalyst, one not only has to consider its pure performance in an asymmetric reaction but also take its synthetic availability into account. In Table 7.1, an overview of the important synthetic parameters for the most important C_1-symmetric iodoarenes is given for a direct comparison. The table is ordered by the necessary synthetic steps based on commercially available starting materials. C_1-symmetric lactic acid-derived methyl esters **6a**, even though not performing very well catalytically, in particular, in direct comparison to their C_2-symmetric counterparts, have an outstanding synthetic availability since they can be generated in one step from *ortho*-iodophenol and lactic acid. Chiral iodoarenes **2a** with an α-chiral carboxylic acid function in benzylic position also have high synthetic availability with a good overall yield. However, they can only be obtained through a diastereoselective alkylation and hence the necessity of (*S*)-menthol as a chiral auxiliary. Iodoaryl amides **4a** can be synthesized efficiently as well through a direct peptide coupling between a chiral amine, such as ephedrine derivatives or α-amino acids and 2-iodo-3-methylbenzoic acid. The same starting

Table 7.1 General synthetic routes to C_1-symmetric chiral iodoarenes.

Commercial available starting material	Source of chirality or chiral building block	Steps	Overall yield	Final structure
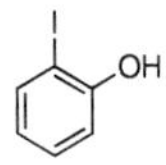	(–)-Lactic acid methylester	1	88%	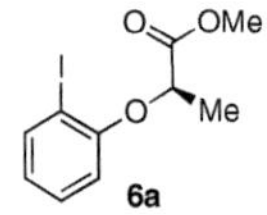 **6a**
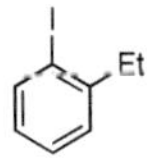	(S)-Menthol + diastereoselective alkylation	2	86%	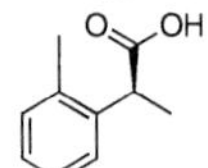 **2a**
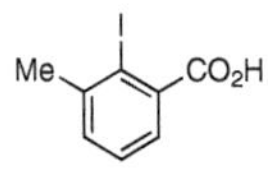	(1*S*,2*S*)-(+)-Pseudoephedrine	2	44%	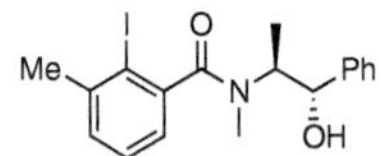 **4a**
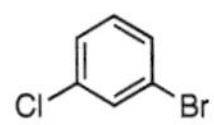 + 2-Bromoiodobenzene	(*S*)-Di-tert-butyl-thio-sulfinate	2	27%	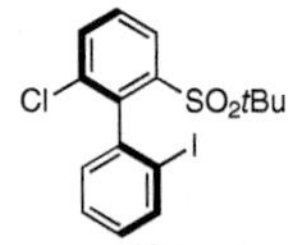 **12a**
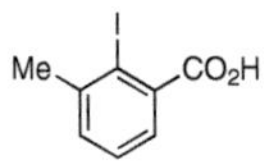	(1*S*,2*R*)-(+)-Norephedrine	3	58%	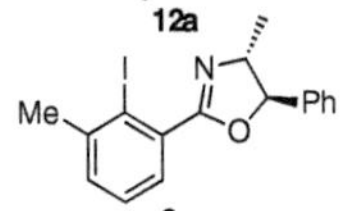 **3a**
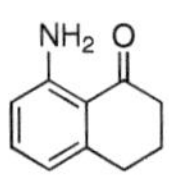	L -(+)-Tartaric acid	3	33%	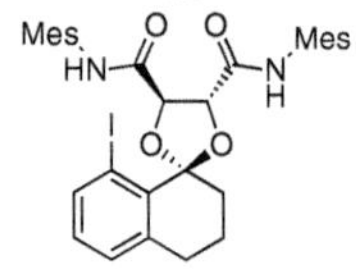 **5**
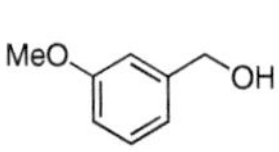	Kinetic resolution of propargylic alcohol with CALB	5	62%	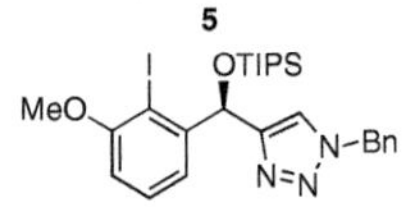 **7c**
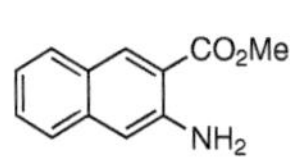 + 2-Naphthol	Chiral HPLC-separation	5	8%	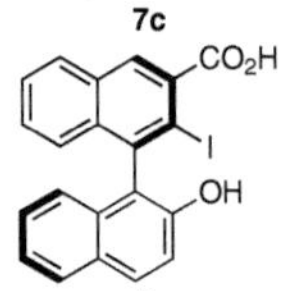 **9a**
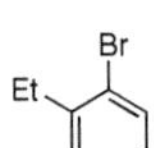	Borane reduction with (2)-(Ipc)-2-B-Cl	6	45%	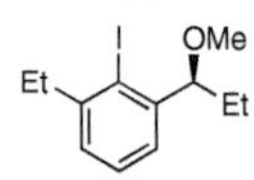 **1a**
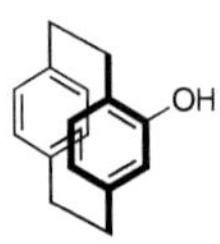	Chiral HPLC-seperation	7	28%	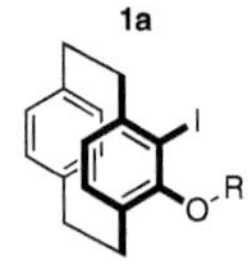 **13**
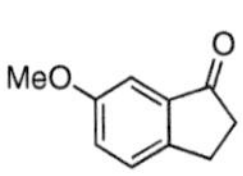	Diastereoselective esterification with *N*-Boc-L-Phe and (–)-Lactic Acid methylester	7	13%	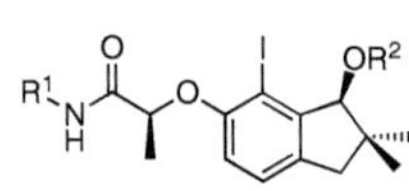 **8**
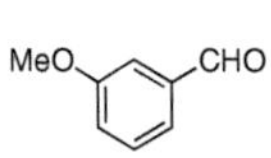	Diastereoselective esterification with *L*-menthyl chloroformate	14	7%	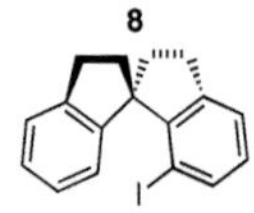 **11**

material can be used for chiral oxazoline-based catalysts **3a**. They are not only easy to synthesize but also show a good performance in enantioselective α-oxytosylations. Chiral sulfones **12** are also remarkable good catalysts for α-oxytoslyations. They can be generated in a fast 2-step sequence starting from 1-bromo-3-chlorobenzene *via* aryne chemistry and with (*S*)-di-*tert*-butyl-thio-sulfinate as the source of chirality. Iodotetralones **5** are accessible from an amino naphthalenone and can be directly treated with L-(+)-tartaric acid to give the desired chiral ketals. The synthesis of 2-(iodophenyl)propanoic acid **2a**, even though relying on a simple two-step sequence, has the small drawback of a chiral auxiliary-based alkylation which generates stoichiometric amounts of chiral waste. In this regard, enzymatic kinetic resolutions are highly beneficial for catalyst synthesis since such a key step makes both enantiomers available. This has, for example, been applied in the synthesis of chiral triazole-based iodoarenes **7b-c**. Even though a 5-step process is necessary for their synthesis, each step is highly robust with a good overall yield of 62%. In contrast, the most successful 2-iodo-binaphthyl-derived catalyst **9a**, even though can be achieved through a 5-step synthesis as well, is generated in 8% overall yield starting from methyl 3-amino-2-naphthoate and 2-naphthole. Here, a chiral HPLC separation is required to give the catalyst in its enantiopure form. Chiral benzyl ethers **1** require one more synthetic step but can be generated in a remarkably high overall yield of 45% starting from readily available 2-bromo ethylbenzene. Paracyclophane-derived catalysts **13** already require a 7-step synthesis with a moderate overall yield of 28% and again HPLC-separation of the enantiomers is required. Indalone **8**, one of the most efficient catalysts for phenol dearomatizations, also needs seven synthetic steps starting from the corresponding indanone and requires diastereoselective esterification. Synthesis of spirobiindanes **11** starts with 3-methoxy benzaldehyde as starting material. In summary, 14 consecutive steps are necessary for its preparation, including diastereoselective esterification to introduce the chiral information with an overall yield of 7% [23].

7.3 C_2-Symmetric Catalysts

7.3.1 Structural Features

As their C_1-symmetric opponents, C_2-symmetric pre-catalysts can be divided into structures with distinct stereogenic chiral carbons, here at both *ortho*-positions, or structures containing axial chiral or helical chiral information. The third type, which is unknown for C_1-symmetric catalysts, is defined by a unique helical environment around the I(III)-center, which is either intrinsic or based on two flexible *ortho*-substituents coordinating to the oxidized iodine or its ligands.

A timeline demonstrating the evolution of the most important C_2-symmetric iodoarene catalysts divided by the different subtypes is shown in Figure 7.6. One year after the initial finding of C_1-symmetric iodoarene catalysts, the first application of a C_2-symmetric pre-catalyst was reported [6]. An axial chiral spirobiindane **24a** was found to efficiently mediate phenol dearomatizations with good enantioselectivities.

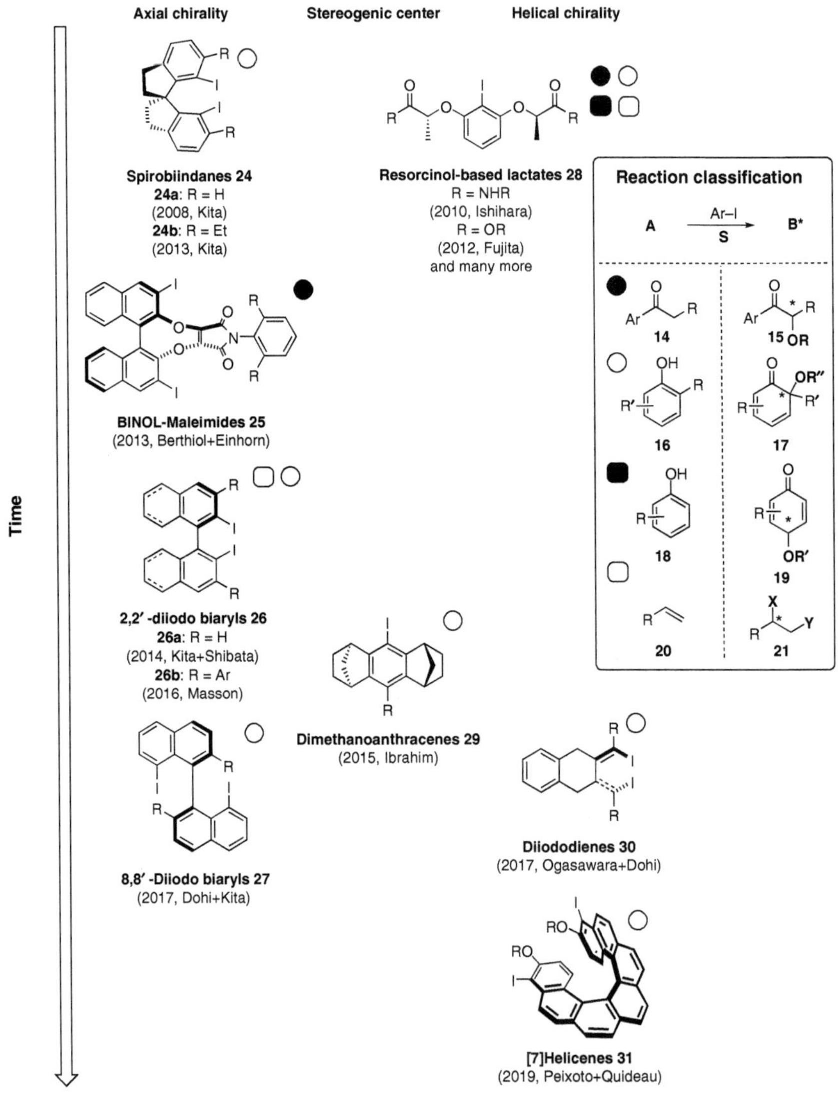

Figure 7.6 Catalyst evolution of C_2-symmetric chiral iodoarenes divided by the core chiral motif.

Five years later, an optimized catalyst structure **24b** bearing *ortho*-substituted ethyl groups was presented with drastically enhanced performance in the Kita-spirolactonization [25]. Following these reports, chiral biaryl frameworks received growing attention. In 2013, 3,3′-diiodo-BINOL-based maleimides **25** were developed and tested as catalysts in enantioselective α-oxytosylations of ketones **14** [26]. Stereoselective fluorinations were shortly after performed with 2,2′-binaphthalene **26a**, before sterically highly demanding 3,3′-aryl analogues **26b** emerged and were successfully applied in the Kita-spirolactonization [27, 28].

Binaphthyls **27** with a unique 8,8′-diiodo substitution pattern are known as well. These catalysts show a reactivity comparable to the initially developed spirobiindane catalyst [29].

Two years after the initial development of C_2-symmetric spirobiindanes, resorcinol-based aryl iodides **28** with chiral lactic acid-based side chains were introduced [30–33]. While lactate esters have initially been applied as stoichiometric chiral oxidants for oxylactonizations of alkenes, the corresponding lactamides were directly employed as chiral catalysts.

These resorcinol-based structures **28** are part of two C_2-symmetric subtypes as the stereogenic center of the side chains leads to an efficient helical environment around the I(III)-center. The underlying structure will be discussed in more detail in Section 7.2.2.1 of this book chapter. At this point, one can conclude that C_2-symmetric lactic acid-substituted resorcinols **28** are, in great contrast the C_1-symmetric opponents **6**, the most widely utilized iodoarene catalysts that have been developed so far. This is not only based on their excellent performance but also rests upon their straightforward synthesis.

Besides resorcinols, only one other successful C_2-symmetric catalyst structure with all-carbon stereogenic centers, the dimethanoanthracene-based aryl iodides **29**, have been described [34]. However, besides the *in situ* formed helical environment of the resorcinol-based catalysts, two examples with intrinsic helical chirality are known to date. Here, the non-biaryl atropisomeric diiododienes **30** and [7] helicene-based chiral aryl iodides **31** fall into this category. Both catalysts show a promising reactivity in phenol dearomatizations [35, 36].

7.3.2 X-Ray Structure Analysis and Computational Investigations

As with C_1-symmetric catalysts, deeper structural information of the active catalyst species or the underlying mechanisms, in particular, the stereoinduction, based on calculations and crystal structures are scarce and limited to spirobiindane and resorcinol-based catalysts. A stereochemical model for the performance of spirobiindanes in spirolactonizations was already proposed in their initial report (Scheme 7.4a). In this reaction, the unsubstituted spirobiindane **24a** gave the product **17a** with 65% ee. This value could later be improved up to 92% ee using a sterically more hindered derivative **24b** [25]. In the proposed reaction mechanism, the catalyst is first oxidized to the chiral λ^3-iodane **24b***, which then undergoes ligand exchange with the naphthol **16a**. After an intramolecular Re-face attack of the carboxylic acid due to the steric hindrance of the biindane-structure, the product **17a** is obtained with high stereoselectivity. Based on this proposal, a closer look at the stereo-induction process using quantum chemical calculations [37] revealed that the favored transition state is **TS-1** in which the naphthalene ring points away from the biindane backbone (Scheme 7.4b). In contrast, the disfavored transition state **TS-1′**, yielding the unfavored enantiomer, is destabilized by steric repulsion between the naphthyl moiety and the catalyst backbone. The *ortho* ethyl groups show no effect in this initial process, as the unsubstituted catalyst **24a** shows similar free energy differences.

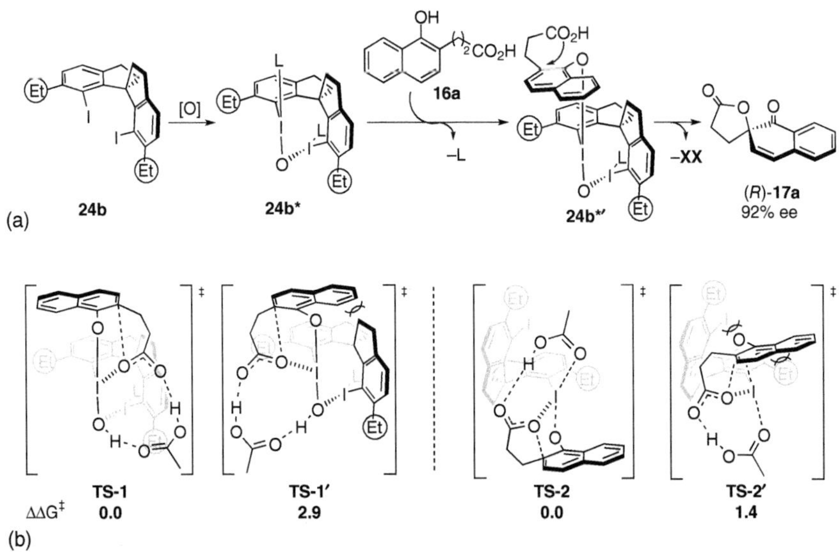

Scheme 7.4 Proposed mechanism (a) and revised transition states with relative calculated free energies (kcal·mol^{-1}) (b) for the Kita-spirolactonization.

The main difference in both catalysts lies in the previously neglected second spirolactonization process induced by the mono oxidized species. In the disfavored transition state **TS-2′** of this step, an additional steric repulsion *via* the ethyl group was observed beside the lone-pair repulsion between the iodine atom and the oxygen atom of naphthol **16a**. Due to this steric hindrance, an increased free energy gap of 1.4 kcal mol^{-1} between **TS-2** and **TS-2′** compared to 0.9 kcal mol^{-1} for the unsubstituted catalyst **24a** was calculated, explaining the better performance of catalyst **24b**. These calculations further demonstrate the low performance observed for C_1-symmetric spirobiindane **11** catalysts with only one iodine moiety (19% ee [25]). Here, the already lower stereoselectivity originating from the mono-oxidized species is even more decreased due to the missing lone-pair repulsion from the second iodine atom. The highly superior performance of the spirobiindanes **24** compared to 1,1′-binaphthyl scaffolds **26** is also based on the more rigid structure of the biindane backbone, which, in contrast to the more flexible biaryl system, is not able to avoid steric repulsion by adjusting its geometry [37].

Detailed structural information is also available for the hypervalent states of C_2-symmetric iodoresorcinol catalysts **28** [30, 32]. The effective creation of a chiral environment around the resorcinol-based catalysts lies within the ability of the employed side chains to efficiently form helical structures due to intramolecular coordination. Possible modes of interaction were first proposed by Ishihara and coworkers based on n–σ* interactions as shown in Scheme 7.5 (**A**) between the *in situ* formed I(III)-center and the carbonyl moiety as well as hydrogen-bonding interactions (**B**) between XH-groups of the side chains and the ligands at the I(III) center (Scheme 7.5 **B**).

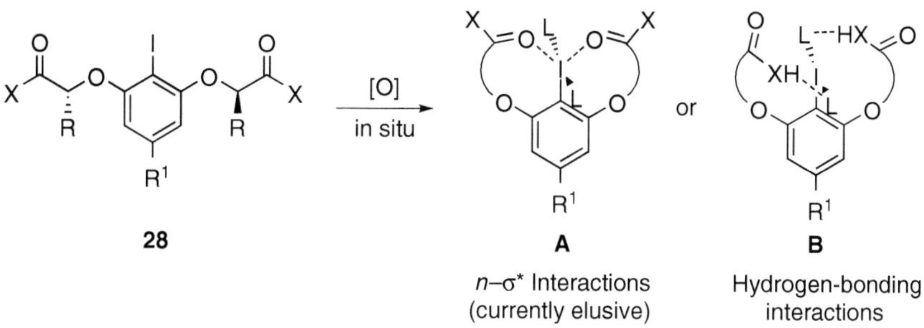

Scheme 7.5 Proposed self-interactions in resorcinol-based iodine(III) structures to create helical environments.

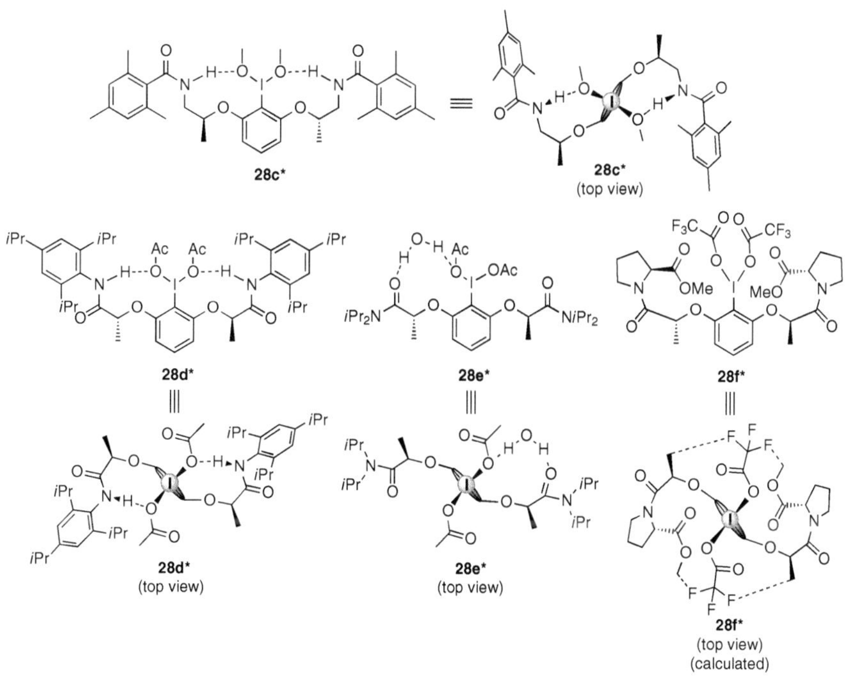

Figure 7.7 Elucidated structures of resorcinol-derived chiral iodine(III) species.

The X-ray structural analysis of **28c*** indeed shows strong H-bonding interactions between the NH-groups of the amide side chains with the oxygen atoms of the methoxy ligands creating helically chiral scaffolds around the hypervalent iodine atom (Figure 7.7) [38]. Similar interactions could be determined for the lactamide **28d*** with a sterically enlarged anilide group [39]. For the tertiary amide **28e***, hydrogen bonding via an incorporated water molecule was encountered [40]. However, the opposite relative helical chirality was observed. Hence, it can be concluded that the central chirality of the side chains is the decisive stereochemical control element and is transferred onto the supramolecular assembly *via* hydrogen bonding. This is an important finding for the further design of these catalysts.

Table 7.2 General synthetic routes to C_2-symmetric chiral iodoarenes.

Commercial available starting material	Source of chirality or chiral building block	Steps	Overall yield	Final structure
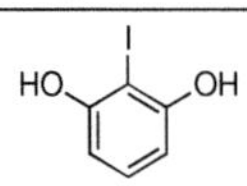	(−)-Lactic acid methylester	1	79%	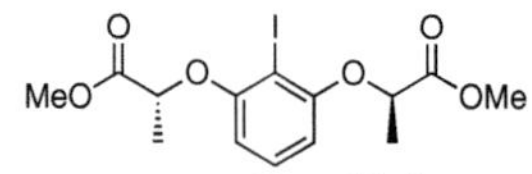 **28a** ⇣ 2 steps
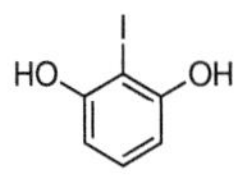	(−)-Lactic acid ethylester	3	70%	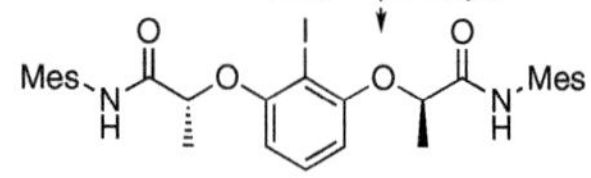 **28b**
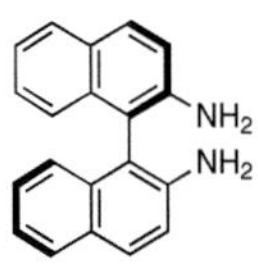	Commercially available (*R*)-BINAM	4	18%	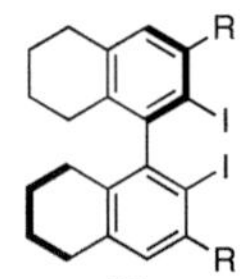 **26b**
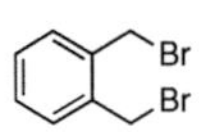	Chiral HPLC-separation	4	14%	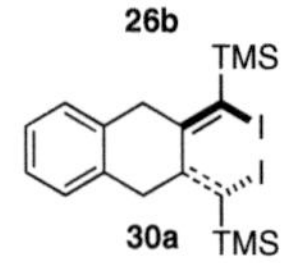 **30a**
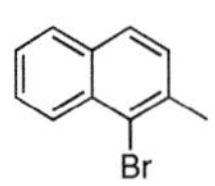	Chiral HPLC-separation	5	11%	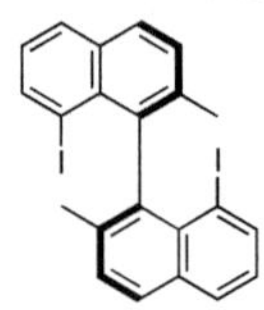 **27a**
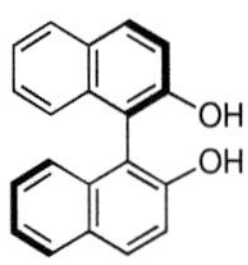	Commercially available (*R*)-BINOL	6	34%	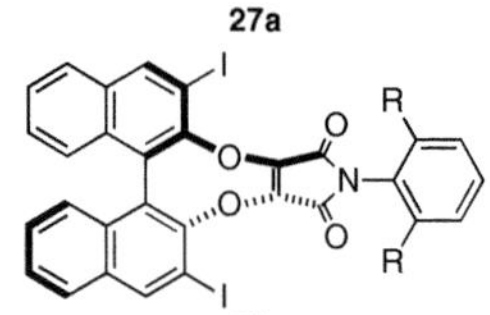 **25a**
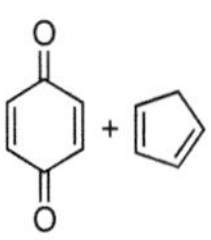	Separation of diastereomeres based on (*R*,*R*)-hydrobenzoin	9	11%	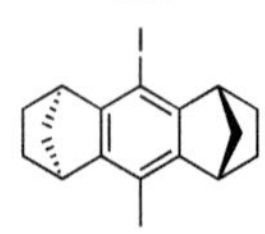 **29a**
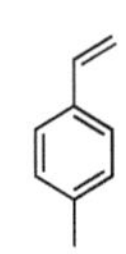	Separation of diastereomeres based on L-(+)-tartaric acid	12	2.4%	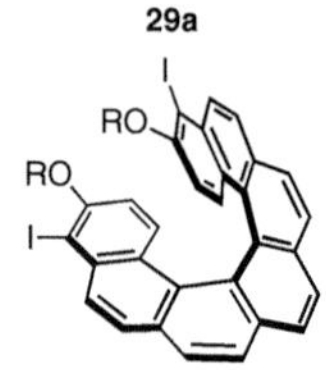 **31a**
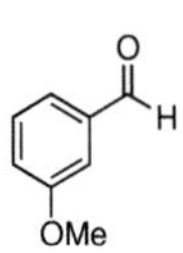	Resolution with *N*-benzylcinchonidinium chloride	13	3.8%	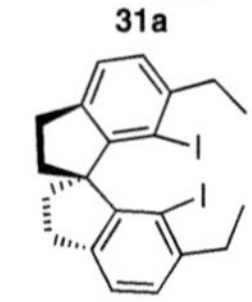 **24b**

The formation of defined supramolecular scaffolds may not be limited to these interactions. The prolinyl lactamide **28f*** was investigated using computational chemistry. Here, a network of attractive noncovalent interactions forming the same relative helical chirality as in **28d*** and **28e*** via CH-FC interactions was observed. This indicates a more general helical assembly of resorcinol-derived catalysts [41]. Further investigations should deliver additional insights and features of these promising catalysts, such as the proposed n–σ* interactions, which stayed elusive until now.

7.3.3 Catalyst Synthesis – Comparison Among C_2-Symmetric Derivatives

A comprehensive overview of the basic parameters for the synthesis of the most relevant C_2-symmetric iodoarenes is shown in Table 7.2.

Resorcinol-based lactic acid derivatives **28a** are easily accessible from commercially available 2-iodoresorcinol and methyl lactate in one step. Furthermore, lactamides **28b** can be obtained in three overall steps and high yield from the corresponding esters **28a** *via* saponification followed by amidation. The underlying concept is a highly modular approach, which allows an easily structural fine-tuning of the catalyst as a wide variety of chiral pool-derived auxiliaries such as lactic acid esters [33] as well as mono- and disubstituted amides [30, 42]. *N*-acylated amino alcohols [38, 43] and carbohydrates [44] can be efficiently implemented in one to three steps as well (Figure 7.8).

While unsubstituted (*R*)-2,2′-diiodonaphthalene [(*R*)-BINI] **26a** can be obtained from commercially available (*R*)-BINAM in one step, the synthesis of sterically

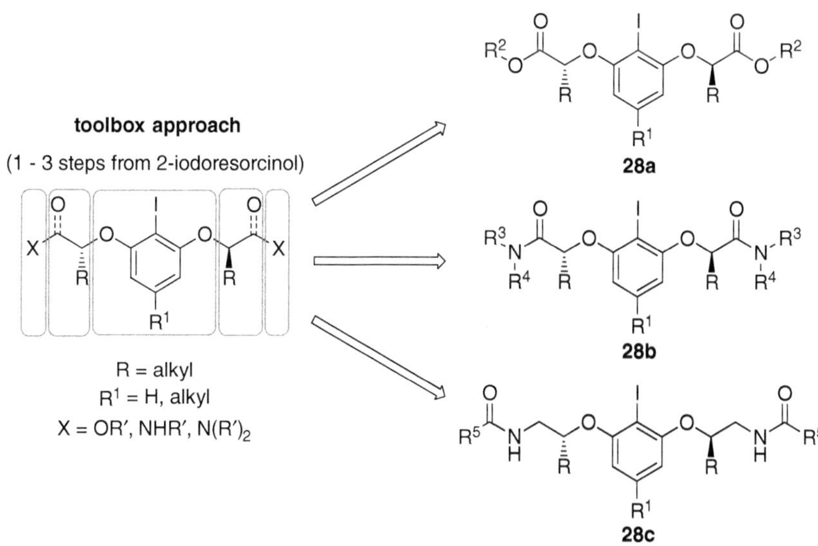

Figure 7.8 Toolbox approach based on 2-iodoresorcinol with chiral pool-derived auxiliaries.

demanding 3,3′-aryl-substituted (*R*)-[H_8]BINIs **26b** requires four steps with an overall yield of usually <20%. The same number of steps are needed for the synthesis of diiododiene **30a**. Here, chiral HPLC separation is required, limiting the overall yield to 14%. The binaphthyl frameworks of **27a** and **25a** demand five to six steps. While the 8,8′-iodo derivatives **27a** need to be separated by chiral HPLC, the synthesis of maleimides **25a** starts from (*R*)-BINOL following six high-yielding transformations. The synthesis of all-carbon dimethanoanthracene **29a** is based on benzoquinone and cyclopentadiene and requires nine linear steps including a separation of diastereomers which limits the overall yield to 11%. The syntheses of the [7]helicene **31a** and the spirobiindane **24b** require 12 and 13 linear steps resulting in overall yields below 5%.

7.4 Comparison of Catalyst Performance

In this chapter, we want to briefly give a comparative overview of the performance of the previously discussed structurally diverse C_1- and C_2-symmetric iodoarene catalysts in model oxidative enantioselective reactions. Among the excessive number of available oxidative enantioselective procedures, two reactions have gained special attention as benchmarks for hypervalent iodine catalysis. The first reaction class is the α-oxygenation of ketones **14**, which was the first catalytic transformation with induced enantioselectivities, and the second is the dearomatization of phenols and naphthols **16**, especially the Kita-spirolactonization. While the first was mainly tested among C_1-symmetric catalysts, the second one is the most popular benchmark system for C_2-symmetric iodoarenes.

Within the broad field of α-oxygenations, α-oxytosylation has gained particular interest. A comprehensive overview of the performance of different iodoarenes in this important transformation is given in Scheme 7.6. Here, the benzylic methyl ether **1a** shows a high reactivity, but only a moderate stereo induction with 27% ee [5]. Mono-iodinated spirobiindane **11** gave increased enantioselectivity with 53% ee, while the corresponding C_2-symmetric diiodo derivative **24a** performed worse (21% ee) [23]. Interestingly, the resorcinol-based lactamide **28b** shows only a low performance in the direct α-oxytosylation, despite their high utility in many other oxidative transformations [45, 46]. Oxazoline **3b** gave nearly the same enantiomeric excess as the C_1-symmetric spirobiindane **11**, but with an increased yield. As the best C_2-symmetric catalyst, BINOL-fused maleimide **25a** showed comparable stereo induction with 46% ee [26, 47]. A new highpoint was set by the application of the C_1-symmetric biphenyl catalyst **12a** giving **15a** in 67% ee [24] and afterward by triazole catalyst **7c** with enantioselectivity of 88% ee [21].

The performance of most C_2-symmetric catalysts has been tested in the dearomatization of phenol derivatives. In Scheme 7.7, the results for the most popular type of these reactions, the Kita-spirolactonization of 1-naphthols **16a**, are compared. Here, the optimized spirobiindane catalyst **24b** with *ortho* ethyl groups shows an enhanced stereoselectivity of 92% ee [25]. Similar results were achieved using the resorcinol-based mesitylamide **28b** (Scheme 7.7) [30, 31]. Later, *N*-acyl

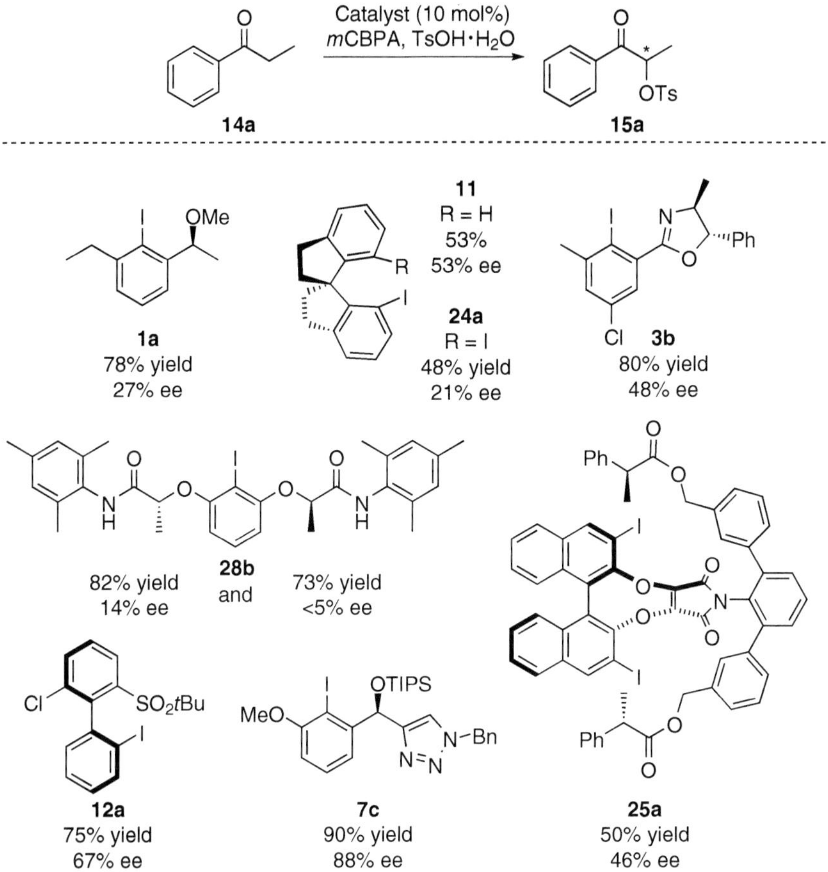

Scheme 7.6 α-Oxytosylation of propiophenone **14a**.

amine **28c** was found to be the most efficient C_2-selective catalysts with 98% ee [48]. In contrast, the dimethanoanthracene catalyst **29a** and the axial chiral biaryls **26b** and **27a** show significantly diminished enantioselectivities of 36–64% ee [28, 29]. The helical chiral tetralin motif **30a** gave an enantiomeric excess of 73% ee [35], while the helicene-based catalyst **31a** yielded **17a** in 56% ee [36]. As C_1-symmetric iodoarenes triazoles **7b-c** were investigated. Here, a remarkable *ortho*-effect was observed. While the *ortho*-unsubstituted derivative **7b** gave **17a** in 80% ee, the *ortho*-methoxy species **7c** showed a performance comparable to the successfully applied C_2-symmetric spirobiindane and resorcinol-based catalysts (99% ee) [20, 21].

Regarding versatility, it can be concluded that C_2-symmetric resorcinol-based catalysts **28** are the most widely applied catalysts. This is due to their easily available, chiral pool-derived side chains, their short and straightforward synthesis and meanwhile detailed structural information. These iodoresorcinols revolutionized the field of iodine(III)-promoted enantioselective transformations. Since their initial reports,

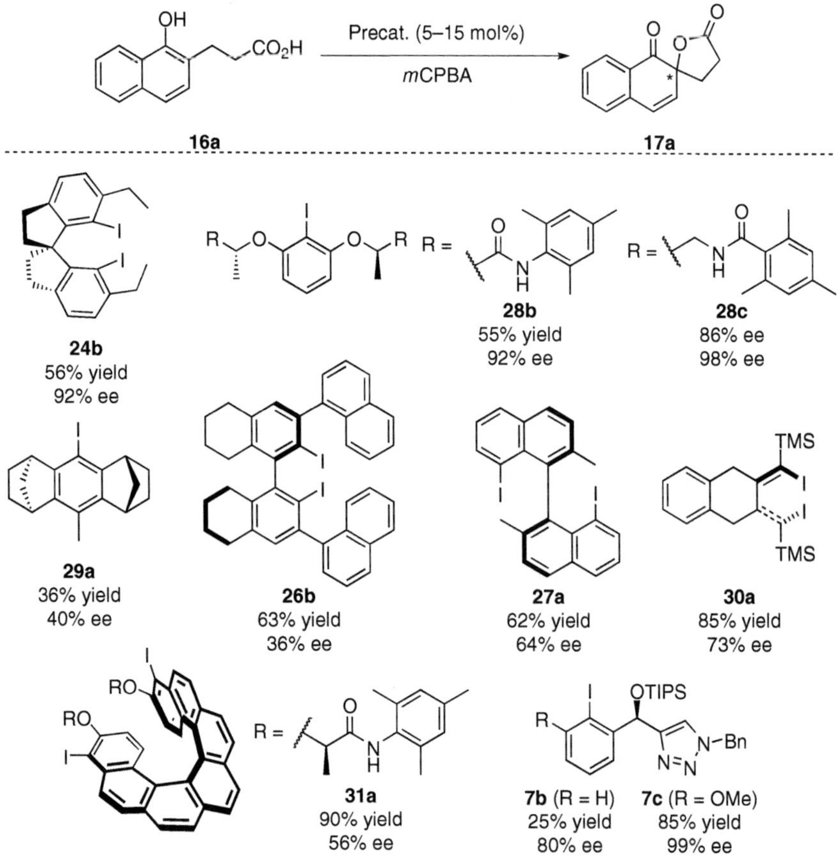

Scheme 7.7 Performance of iodoarene catalysts in the Kita-spirolactonization.

these catalysts were not only applied in the previously discussed lactonizations [33, 48–53], and spirocyclizations [41, 42, 54–56] but among others in highly efficient fluorinations [57–63], (di)acetoxylations and aminations (Figure 7.9) [39, 40, 64–66].

7.5 Conclusion

Within the last two decades of enantioselective iodoarene catalysis, nearly 20 chiral iodoarenes with highly diverse design features emerged. They can be divided into C_1- and C_2-symmetric catalysts, their underlying chirality as well as the further substitution patterns to stabilize the central iodine atom in its hypervalent state. Based on a well-balanced ratio between high performance and a straightforward synthesis, only a very few of them, in particular, resorcinol-derived lactic acids, have not only been benchmarked in α-oxygenations and phenol dearomatizations as typical test reactions but have also been applied in a variety of other enantioselective oxidative

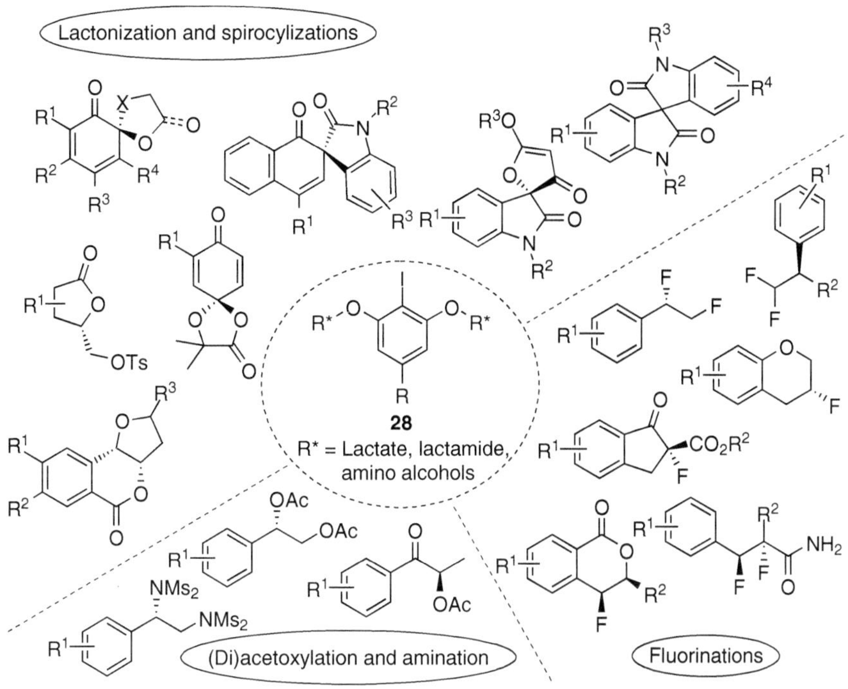

Figure 7.9 Examples for different enantioselective transformations with resorcinol-based catalysts.

couplings. A major factor is the existence of reliable structural information based on solid-state structures. This allows for a highly rationalized derivatization for specific reactivity needs and helps to better understand the underlying enantiodiscriminating steps. Since such information is missing for nearly all other chiral iodoarenes, this should be a major focus of future research in this area, together with high-quality DFT calculations. A plethora of synthetically useful enantioselective oxidative transformations is still waiting on the right catalyst. The design principles discussed in this chapter should help researchers in the near future to adapt existing catalysts or develop completely novel structures with excellent performance and good synthetic accessibility. Also, more immobilization techniques as well as the adaption of reaction parameters toward "large-scale" compatible easy-to-use oxidants should be a major focus of future research to open the door into industrial applications.

References

1 Pribram, R. (1907). Ueber das optische Drehungsvermögen des Jodoniumtartrates. *Justus Liebigs Ann. Chem.* 351: 481–485.

2 Parra, A. and Reboredo, S. (2013). Chiral hypervalent iodine reagents: synthesis and reactivity. *Chem. Eur. J.* 19: 17244–17260.

3 Ochiai, M. and Miyamoto, K. (2008). Catalytic version of and reuse in hypervalent Organo-lambda(3)- and -lambda(5)-iodane oxidation. *Eur. J. Org. Chem.* 2008: 4229–4239.
4 Dohi, T. and Kita, Y. (2009). Hypervalent iodine reagents as a new entrance to organocatalysts. *Chem. Comm.*: 2073–2085.
5 Richardson, R.D., Page, T.K., Altermann, S. et al. (2007). Enantioselective α-oxytosylation of ketones catalysed by iodoarenes. *Synlett*: 538–542.
6 Dohi, T., Maruyama, A., Takenaga, N. et al. (2008). A chiral hypervalent iodine(III) reagent for enantioselective dearomatization of phenols. *Angew. Chem. Int. Ed.* 47: 3787–3790.
7 Wang, Y., Zhao, C.-Y., Wang, Y.-P., and Zheng, W.-H. (2019). Enantioselective intramolecular dearomative lactonization of naphthols catalyzed by planar chiral iodoarene. *Synthesis* 51: 3675–3682.
8 Parra, A. (2019). Chiral hypervalent iodines: active players in asymmetric synthesis. *Chem. Rev.* 119: 12033–12088.
9 Flores, A., Cots, E., Bergès, J., and Muñiz, K. (2019). Enantioselective iodine(I/III) catalysis in organic synthesis. *Adv. Synth. Catal.* 361: 2–25.
10 Claraz, A. and Masson, G. (2018). Asymmetric iodine catalysis-mediated enantioselective oxidative transformations. *Org. Biomol. Chem.* 16: 5386–5402.
11 Ghosh, S., Pradhan, S., and Chatterjee, I. (2018). A survey of chiral hypervalent iodine reagents in asymmetric synthesis. *Beilstein J. Org. Chem.* 14: 1244–1262.
12 Rodriguez, A. and Moran, W.J. (2012). Chiral aryl iodide-catalyzed enantioselective alpha-oxidation of ketones. *Synthesis* 44: 1178–1182.
13 Guilbault, A.-A., Basdevant, B., Wanie, V., and Legault, C.Y. (2012). Catalytic enantioselective α-tosyloxylation of ketones using iodoaryloxazoline catalysts: insights on the stereoinduction process. *J. Org. Chem.* 77: 11283–11295.
14 Basdevant, B. and Legault, C.Y. (2015). Enantioselective iodine(III)-mediated synthesis of alpha-tosyloxy ketones: breaking the selectivity barrier. *Org. Lett.* 17: 4918–4921.
15 Guilbault, A.-A. and Legault, C.Y. (2012). Drastic enhancement of activity in iodane-based alpha-Tosyloxylation of ketones: iodine(III) does the hypervalent twist. *ACS Catal.* 2: 219–222.
16 Volp, K.A. and Harned, A.M. (2013). Chiral aryl iodide catalysts for the enantioselective synthesis of para-quinols. *Chem. Commun.* 49: 3001–3003.
17 Hokamp, T. and Wirth, T. (2019). Structurally defined α-tetralol-based chiral hypervalent iodine reagents. *J. Org. Chem.* 84: 8674–8682.
18 Vlasenko, Y.A., Yusubov, M.S., Shafir, A., and Postnikov, P.S. (2020). Hypervalent iodine in the structure of *N*-heterocycles: synthesis, structure, and application in organic synthesis. *Chem. Heterocycl. Compd.* 56: 854–866.
19 Mizar, P., Laverny, A., El-Sherbini, M. et al. (2014). Enantioselective diamination with novel chiral hypervalent iodine catalysts. *Chem. Eur. J.* 20: 9910–9913.
20 Hempel, C., Maichle-Moessmer, C., Pericas, M.A., and Nachtsheim, B.J. (2017). Modular synthesis of triazole-based chiral iodoarenes for enantioselective spirocyclizations. *Adv. Synth. Catal.* 359: 2931–2941.

21 Abazid, A.H. and Nachtsheim, B.J. (2020). A triazole-substituted aryl iodide with omnipotent reactivity in enantioselective oxidations. *Angew. Chem. Int. Ed.* 59: 1479–1484.

22 Quideau, S., Lyvinec, G., Marguerit, M. et al. (2009). Asymmetric hydroxylative phenol dearomatization through *in situ* generation of iodanes from chiral iodoarenes and m-CPBA. *Angew. Chem. Int. Ed.* 48: 4605–4609.

23 Yu, J., Cui, J., Hou, X.-S. et al. (2011). Enantioselective α-tosyloxylation of ketones catalyzed by spirobiindane scaffold-based chiral iodoarenes. *Tetrahedron: Asym.* 22: 2039–2055.

24 Levitre, G., Dumoulin, A., Retailleau, P. et al. (2017). Asymmetric α-sulfonyl- and α-phosphoryl-oxylation of ketones by a chiral hypervalent iodine(III). *J. Org. Chem.* 82: 11877–11883.

25 Dohi, T., Takenaga, N., Nakae, T. et al. (2013). Asymmetric dearomatizing spirolactonization of naphthols catalyzed by spirobiindane-based chiral hypervalent iodine species. *J. Am. Chem. Soc.* 135: 4558–4566.

26 Brenet, S., Berthiol, F., and Einhorn, J. (2013). 3,3′-Diiodo-BINOL-fused maleimides as chiral hypervalent iodine(III) organocatalysts. *Eur. J. Org. Chem.*: 8094–8096.

27 Suzuki, S., Kamo, T., Fukushi, K. et al. (2014). Iodoarene-catalyzed fluorination and aminofluorination by an Ar-I/HF·pyridine/*m*CPBA system. *Chem. Sci.* 5: 2754–2760.

28 Bekkaye, M. and Masson, G. (2016). Synthesis of new axially chiral iodoarenes. *Synthesis* 48: 302–312.

29 Dohi, T., Sasa, H., Miyazaki, K. et al. (2017). Chiral atropisomeric 8,8′-Diiodobinaphthalene for asymmetric dearomatizing spirolactonizations in hypervalent iodine oxidations. *J. Org. Chem.* 82: 11954–11960.

30 Uyanik, M., Yasui, T., and Ishihara, K. (2010). Enantioselective Kita oxidative spirolactonization catalyzed by *in situ* generated chiral hypervalent iodine(III) species. *Angew. Chem. Int. Ed.* 49: 2175–2177.

31 Uyanik, M., Yasui, T., and Ishihara, K. (2010). Chiral hypervalent iodine-catalyzed enantioselective oxidative Kita spirolactonization of 1-naphthol derivatives and one-pot diastereo-selective oxidation to epoxyspirolactones. *Tetrahedron* 66: 5841–5851.

32 Fujita, M., Yoshida, Y., Miyata, K. et al. (2010). Enantiodifferentiating endo-selective oxylactonization of ortho-alk-1-enylbenzoate with a lactate-derived aryl-λ_3-iodane. *Angew. Chem. Int. Ed.* 49: 7068–7071.

33 Fujita, M., Mori, K., Shimogaki, M., and Sugimura, T. (2012). Asymmetric synthesis of 4,8-dihydroxyisochroman-1-one polyketide metabolites using chiral hypervalent iodine(III). *Org. Lett.* 14: 1294–1297.

34 Murray, S.J. and Ibrahim, H. (2015). Asymmetric Kita spirolactonisation catalysed by anti-dimethanoanthracene-based iodoarenes. *Chem. Commun.* 51: 2376–2379.

35 Ogasawara, M., Sasa, H., Hu, H. et al. (2017). Atropisomeric chiral diiododienes (*Z*,*Z*)-2,3-Di(1-iodoalkylidene)tetralins: synthesis, enantiomeric resolution, and application in asymmetric catalysis. *Org. Lett.* 19: 4102–4105.

36 Antien, K., Pouységu, L., Deffieux, D. et al. (2019). Synthesis of [7]helicene enantiomers and exploratory study of their conversion into helically chiral iodoarenes and iodanes. *Chem. Eur. J.* 25: 2852–2858.

37 Zheng, H., Sang, Y., Houk, K.N. et al. (2019). Mechanism and origins of enantioselectivities in spirobiindane-based hypervalent iodine(III)-induced asymmetric dearomatizing spirolactonizations. *J. Am. Chem. Soc.* 141: 16046–16056.

38 Uyanik, M., Yasui, T., and Ishihara, K. (2013). Hydrogen bonding and alcohol effects in asymmetric hypervalent iodine catalysis: enantioselective oxidative dearomatization of phenols. *Angew. Chem. Int. Ed.* 52: 9215–9218.

39 Haubenreisser, S., Wöste, T.H., Martínez, C. et al. (2016). Structurally defined molecular hypervalent iodine catalysts for intermolecular enantioselective reactions. *Angew. Chem. Int. Ed.* 55: 413–417.

40 Muñiz, K., Barreiro, L., Romero, R.M., and Martínez, C. (2017). Catalytic asymmetric diamination of styrenes. *J. Am. Chem. Soc.* 139: 4354–4357.

41 Sreenithya, A., Patel, C., Hadad, C.M., and Sunoj, R.B. (2017). Hypercoordinate iodine catalysts in enantioselective transformation: the role of catalyst folding in stereoselectivity. *ACS Catal.* 7: 4189–4196.

42 Wu, H., He, Y.-P., Xu, L. et al. (2014). Asymmetric organocatalytic direct $C(sp^2)$-H/$C(sp^3)$-H oxidative cross-coupling by chiral iodine reagents. *Angew. Chem. Int. Ed.* 53: 3466–3469.

43 Jain, N., Xu, S., and Ciufolini, M.A. (2017). Asymmetric oxidative cycloetherification of naphtholic alcohols. *Chem. Eur. J.* 23: 4542–4546.

44 Imrich, M.R., Biehler, L.E., Maichle-Mössmer, C., and Ziegler, T. (2019). Carbohydrate-based chiral iodoarene catalysts: a survey through the development of an improved catalyst design. *Molecules* 24: 3883.

45 Rodríguez, A. and Moran, W. (2012). Chiral aryl iodide-catalyzed enantioselective α-oxidation of ketones. *Synthesis* 44: 1178–1182.

46 Basdevant, B. and Legault, C.Y. (2015). Enantioselective iodine(III)-mediated synthesis of α-Tosyloxy ketones: breaking the selectivity barrier. *Org. Lett.* 17: 4918–4921.

47 Brenet, S., Minozzi, C., Clarens, B. et al. (2015). 3,3′-Diiodobinaphthol and 3,3′-Diiodobiphenol derivatives as hypervalent iodine organocatalysts for the α-Oxytosylation of ketones. *Synthesis* 47: 3859–3873.

48 Uyanik, M., Yasui, T., and Ishihara, K. (2017). Chiral hypervalent organoiodine-catalyzed enantioselective oxidative spirolactonization of naphthol derivatives. *J. Org. Chem.* 82: 11946–11953.

49 Fujita, M., Mori, K., Shimogaki, M., and Sugimura, T. (2013). Total synthesis of (12R)- and (12S)-12-hydroxymonocerins: stereoselective oxylactonization using a chiral hypervalent iodine(III) species. *RSC Adv.* 3: 17717-17725.

50 Shimogaki, M., Fujita, M., and Sugimura, T. (2013). Enantioselective oxidation of alkenylbenzoates catalyzed by chiral hypervalent iodine(III) to yield 4-Hydroxyisochroman-1-ones. *Eur. J. Org. Chem.* 2013: 7128–7138.

51 Woerly, E.M., Banik, S.M., and Jacobsen, E.N. (2016). Enantioselective, catalytic fluorolactonization reactions with a nucleophilic fluoride source. *J. Am. Chem. Soc.* 138: 13858–13861.

52 Deng, Q., Xia, W., Hussain, M.I. et al. (2020). Synthesis of polycyclic cyclohexadienone through alkoxy-oxylactonization and dearomatization of 3′-Hydroxy-1,1′-biphenyl-2-carboxylic acids promoted by hypervalent iodine. *J. Org. Chem.* 85: 3125–3133.

53 Gelis, C., Dumoulin, A., Bekkaye, M. et al. (2017). Chiral hypervalent iodine(III) catalyst promotes highly enantioselective sulfonyl- and phosphoryl-oxylactonizations. *Org. Lett.* 19: 278–281.

54 Zhang, D.-Y., Xu, L., Wu, H., and Gong, L.-Z. (2015). Chiral iodine-catalyzed dearomatizative spirocyclization for the enantioselective construction of an all-carbon stereogenic center. *Chem. Eur. J.* 21: 10314–10317.

55 Cao, Y., Zhang, X., Lin, G. et al. (2016). Chiral aryliodine-mediated enantioselective organocatalytic spirocyclization: synthesis of spirofurooxindoles via cascade oxidative C–O and C–C bond formation. *Org. Lett.* 18: 5580–5583.

56 Uyanik, M., Sasakura, N., Mizuno, M., and Ishihara, K. (2017). Enantioselective synthesis of masked benzoquinones using designer chiral hypervalent organoiodine(III) catalysis. *ACS Catal.* 7: 872–876.

57 Molnár, I.G. and Gilmour, R. (2016). Catalytic difluorination of olefins. *J. Am. Chem. Soc.* 138: 5004–5007.

58 Banik, S.M., Medley, J.W., and Jacobsen, E.N. (2016). Catalytic, diastereoselective 1,2-Difluorination of alkenes. *J. Am. Chem. Soc.* 138: 5000–5003.

59 Banik, S.M., Medley, J.W., and Jacobsen, E.N. (2016). Catalytic, asymmetric difluorination of alkenes to generate difluoromethylated stereocenters. *Science* 353: 51–54.

60 Mennie, K.M., Banik, S.M., Reichert, E.C., and Jacobsen, E.N. (2018). Catalytic diastereo- and enantioselective fluoroamination of alkenes. *J. Am. Chem. Soc.* 140: 4797–4802.

61 Zhou, B., Haj, M.K., Jacobsen, E.N. et al. (2018). Mechanism and origins of chemo- and stereoselectivities of aryl iodide-catalyzed asymmetric difluorinations of β-substituted styrenes. *J. Am. Chem. Soc.* 140: 15206–15218.

62 Scheidt, F., Schäfer, M., Sarie, J.C. et al. (2018). Enantioselective, catalytic vicinal difluorination of alkenes. *Angew. Chem. Int.l Ed.* 57: 16431–16435.

63 Sarie, J.C., Thiehoff, C., Neufeld, J. et al. (2020). Enantioselective synthesis of 3-fluorochromanes via iodine(I)/iodine(III) catalysis. *Angew. Chem. Int. Ed.* 59: 15069–15075.

64 Wöste, T. and Muñiz, K. (2016). Enantioselective vicinal diacetoxylation of alkenes under chiral iodine(III) catalysis. *Synthesis* 48: 816–827.

65 Hokamp, T. and Wirth, T. (2020). Hypervalent iodine(III)-catalysed enantioselective α-acetoxylation of ketones. *Chem. Eur. J.* 26: 10417–10421.

66 Muñiz, K. and Fra, L. (2017). Enantioselective 4-hydroxylation of phenols under chiral organoiodine(I/III) catalysis. *Synthesis* 49: 2901–2906.

8

Catalytic and Asymmetric Dearomatization Reactions Employing Hypervalent Iodine Reagents

Toshifumi Dohi[1,2] *and Yasuyuki Kita*[2]

[1] *College of Pharmaceutical Sciences, Ritsumeikan University, Kusatsu, Shiga, Japan*
[2] *Research Organization of Science and Technology, Ritsumeikan University, Kusatsu, Shiga, Japan*

8.1 Introduction: Phenol Dearomatization

Hypervalent iodine reagents have recently attracted enormous attention as a mild, safe, and economical alternative to heavy-metal reagents in organic syntheses. The pioneering study on hypervalent iodine-induced reactions for synthesizing natural products was reported by Kishi and coworkers who employed phenyliodine(III) diacetate (PIDA) to synthesize sporidesmin A in 1973 [1]. In the early 1980s, Szantáy's and White's groups separately accomplished the total syntheses of bioactive alkaloids, salutaridine [2a], 6a-epipretazettine [2b], and (−)-codeine [2c] employing the oxidative intramolecular coupling of phenol with a hypervalent iodine reagent as the key step. Although the syntheses involved an interesting iodine(III)-induced coupling reaction, they did not attract much attention because of their low yields (13–30%).

However, they began to attract attention when an efficient phenolic oxidation reaction to produce the phenol-dearomatized coupling products (*p*-quinone monoacetals and spirocyclohexadienones) was developed by Kita and coworkers in the late 1980s [3a, b]. The yields have also been remarkably improved by changing the solvent into polar and weakly nucleophilic ones, such as 2,2,2-trifluoroethanol (TFE) and 1,1,1,3,3,3-hexafluoroisopropanol (HFIP) (Scheme 8.1) [3c, d, 4]. With these discoveries affording a turning point, hypervalent iodine-induced reactions (phenolic oxidations, as well as other different fundamental reactions) have been utilized widely in natural product syntheses [5–11].

To evaluate the high functional-group availability and low toxicity of the reagents, this method was widely employed in the total syntheses of natural products exhibiting important biological properties. Kita and coworkers further reported the total syntheses of discorhabdin C [5a], as well as the more complex sulfur-linked discorhabdin A [5b, c], based on the diastereoselective constructions of the spirocycle cores via phenolic dearomative coupling employing PIFA in the fluoroalcohol. The effective spirocyclization of *para*-substituted phenols exhibiting a cyclic enamide tether would efficiently generate a spirocyclic C–C bond (Scheme 8.2, Eq. 1).

Iodine Catalysis in Organic Synthesis, First Edition. Edited by Kazuaki Ishihara and Kilian Muñiz.

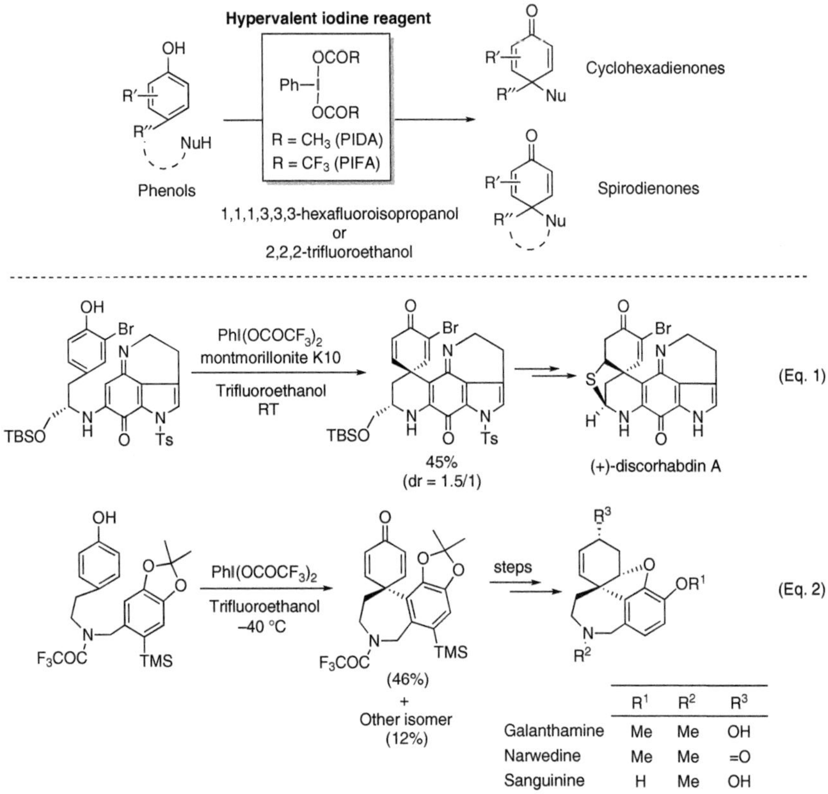

Scheme 8.1 1 PIDA- and PIFA-induced dearomatization of phenols in fluoroalcohols. *Source*: Kita et al. [3c], Kita et al. [3d], Kita et al. [4a], Eberson et al. [4b], Bégué et al. [4c]; Dohi et al. [4d].

A similar treatment of norbelladine derivatives with PIFA affords a mild and efficient method for preparing *Amaryllidaceae* alkaloids, galanthamine (for treating Alzheimer's disease as a selective acetylcholinesterase inhibitor), and other related alkaloids, such as narwedine, norgalanthamine, sanguinine, lycoramine, and maritidine (Scheme 8.2, Eq. 2) [6].

Based on the exhaustive contributions of Kita and coworkers, PIDA and PIFA have assumed crucial roles in the reproduction of biomimetic phenolic oxidation processes under mild conditions for the total syntheses of biologically beneficial natural products and their valuable intermediates [7]. Based on recent desires to develop greener synthetic processes, hypervalent iodine reagents are duly attracting increased attention in organic syntheses, as detailly documented in several recent Wiley books [8–11]. Many trivalent and pentavalent iodine reagents, *e.g.* PIDA, PIFA, iodosobenzene (PhIO), [hydroxy(tosyloxy)iodo]benzene (PhI(OH)OTs, Koser reagent), Dess–Martin periodinane (DMP), and 2-iodoxybenzoic acid (IBX), are now commercially available; they are a valuable synthetic tool, which exhibits

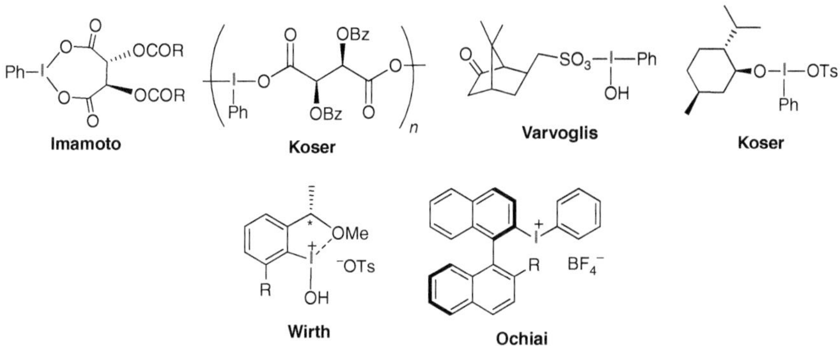

Scheme 8.2 Optically active hypervalent iodine(III) compounds reported by previous studies. *Source*: Imamoto and Koto [13], Hatzigrigoriou et al. [14], Ray and Koser [15a], Ray and Koser [15b], Ray and Koser [15c], Wirth and Hirt [16a], Hirt et al. [16b], Hirt et al. [16c].

similar reactivities to those of a series of oxidants that are based on heavy metals (lead(IV), mercury(II), cadmium(IV), and thallium(III)).

In this chapter, the history and recent investigations of the aforementioned hypervalent iodine reagents in achieving the asymmetric dearomative couplings of phenols were introduced. Their recent successes in efficient catalytic applications could particularly inspire the design of chiral hypervalent iodine compounds as a new series of novel organocatalysts, which would exhibit new applications and further develop oxidative coupling reactions.

8.2 Asymmetric Dearomative Coupling: A Turning Point

Despite the significant advances in organic syntheses, asymmetric induction controlled by hypervalent iodine reagents is still challenging [12]. Previously designed chiral hypervalent iodine compounds for asymmetric reactions are shown in Scheme 8.2 [13–17]. Thus, many efforts have been devoted to the design of new chiral reagents for asymmetric oxidations [13–21], although no effective chiral hypervalent iodine compound has been reported in the twentieth century.

Regarding the phenol dearomatization employing Imamoto chiral λ^3-iodane, which was prepared from PIDA and (+)-dibenzoyl tartaric acid, Pelter's group reported the first attempt at performing an enantioselective oxidative dearomatization reaction (Scheme 8.3) [22]. Although this reagent and its antipode successfully formed the desired *ortho*-quinone monoacetals, the product was isolated as a perfect racemic mixture. Pelter explained that the formation of the racemic product can be interpreted as evidence of the formation of the phenoxenium ions by releasing the iodine molecule, as well as the chiral tartaric acid, during the reactions.

A series of optically active iodine(III) carboxylates and sulfonates (Scheme 8.3) was examined, although these reagents, which exhibited a liberating chiral ligand, obtained undesired enantiomeric excess (*ee*) values for the same reactions (almost a

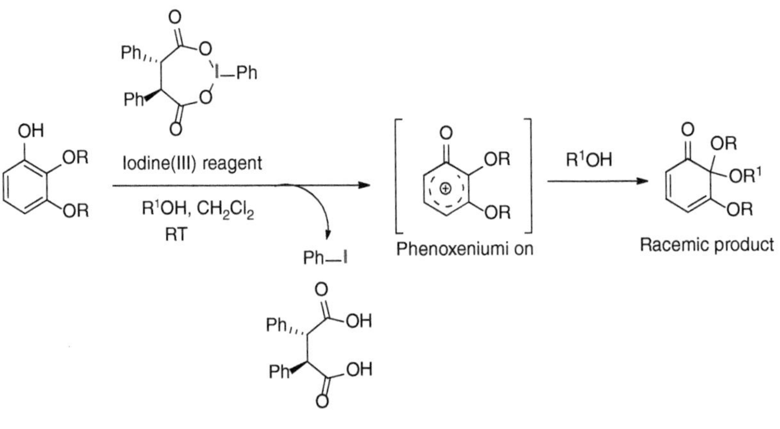

Scheme 8.3 Pelter's attempted asymmetric phenol dearomatization. *Source*: Kürti et al. [22a], Pelter and Ward [22b].

racemic mixture). Further, Wirth's and Ochiai's pseudocyclic chiral iodine(III) and bi-naphthyl compounds, respectively, exhibited very low *ee* values.

The breakthrough in the asymmetric controls employing a hypervalent iodine reagent was achieved by Kita and coworkers, who first reported asymmetric dearomative *ortho*-spirocyclization to achieve high enantioselectivities [23]. The reactions of naphthols **1** bearing an *ortho*-carboxylic acid moiety as the intramolecular nucleophilic tether proceeded at unprecedented excellent levels of asymmetric inductions (stoichiometric conditions: up to 86% ee value at −50 °C) employing a novel chiral hypervalent iodine(III) compound **3a** based on the spirobiindane backbone (Scheme 8.4). Along with recent progress in the catalytic utilization of hypervalent iodine reagents, further significant advances have been recently recorded in the asymmetric dearomative transformations of phenols with the proposal of new chiral hypervalent iodine reagents, as well as the suitable rationalization of the reaction systems [24, 25]. Interestingly, the μ-oxo hypervalent iodine(III) structure exhibited higher reactivity in the phenolic oxidation and dearomatization reactions compared with the non-μ-oxo-type reagents, such as PIDA and PIFA [26].

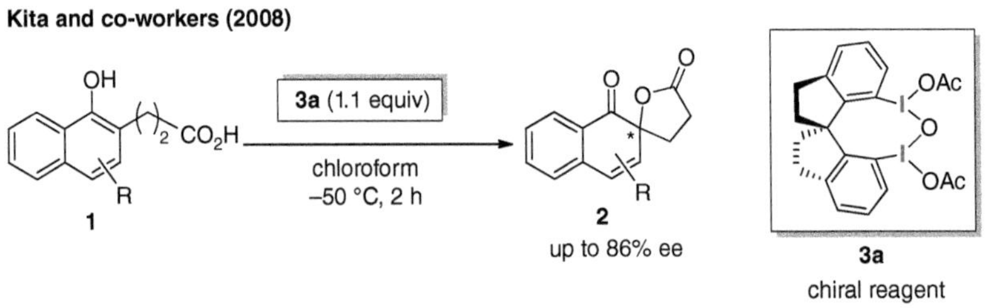

Scheme 8.4 Enantioselective dearomative coupling of naphthol 2-carboxylic acids employing a chiral spirobiindane iodine(III) reagent to obtain spirodienone lactones.

8.3 Catalytic Asymmetric Dearomative Couplings

After the discovery of the first hypervalent iodine compound (Willgerodt's reagent) more than 100 years ago [27], stoichiometric amounts have been generally employed when utilizing them in reactions. Fortunately, hypervalent iodine reagents have recently satisfied new criteria as new organocatalysts in the twenty-first century, following their adherence to reliable catalytic conditions [28–30]. The catalytic strategy is now available for performing many representative iodine(III)- and iodine(V)-induced oxidations employing the suitable terminal oxidants and reoxidation conditions (Scheme 8.5) [31].

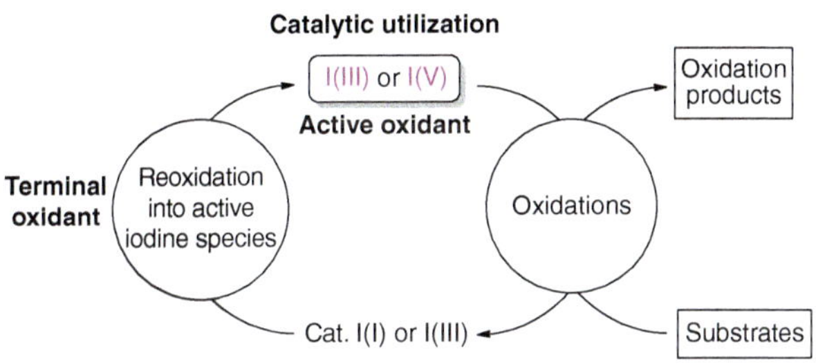

Scheme 8.5 Catalytic utilizations of hypervalent iodine reagents. *Source*: Richardson and Wirth [31a], Ochiai and Miyamoto [31b], Dohi and Kita [31c], Uyanik and Ishihara [31d], Yusubov and Zhdankin [31e].

Kita and coworkers found that their method employing *m*-chloroperbenzoic acid (*m*CPBA) [32] could act as optimum regenerating conditions for producing iodine(III), which could be applied to catalyze the phenol dearomative spirocyclizations (Scheme 8.6) [28]. The main active catalytic species for the reactions, 4-Tol-I$(OCOCF_3)_2$, was effectively generated *in situ* from 4-iodotoluene and trifluoroacetic acid via the action of *m*CPBA. The reaction achieved a high yield of the product even with 1 mol% 4-iodotoluene when trifluoroacetic acid was utilized as the solvent scale; the maximum turnover number of the catalyst reached 71 in this case. Notably, the yields of the spirocyclyzed products were comparable to or even better than the results obtained with stoichiometric amounts of PIFA [3a].

Furthermore, the *ortho* versions of the catalytic phenol and naphthol dearomative spirocyclizations [23a] were also successfully developed. A 0.55 equivalent of the

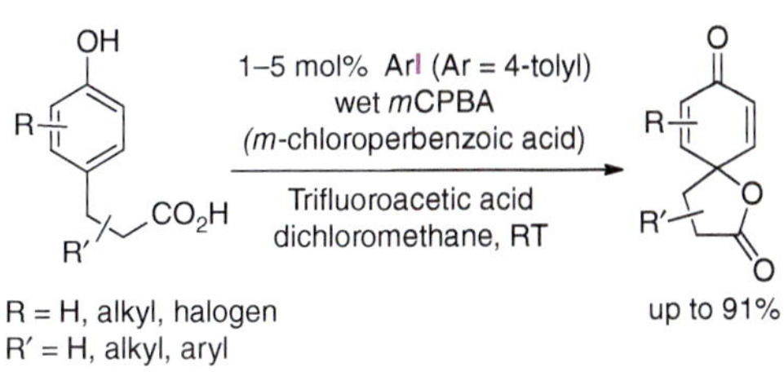

Scheme 8.6 Catalytic hypervalent iodine-induced dearomative spirocyclization of phenols. *Source*: Based on Dohi et al. [28].

new spirobiindane-based chiral iodine(III) reagent **3a** (Scheme 8.4) was generally utilized for the asymmetric reactions. The asymmetric dearomative cyclization reaction was developed employing the catalytic spirobiindane chiral iodoarene **pre-3a** as the precatalyst in the presence of *m*CPBA and acetic acid at 0 °C (Scheme 8.7). The spirolactones, **2a** (R = H) and **2b** (R = 4-Br), were obtained in 65% ee and 69% ee, respectively [23a], and this was a landmark for chiral organoiodine compounds as a new candidate for the asymmetric organocatalysis of phenolic oxidation reactions.

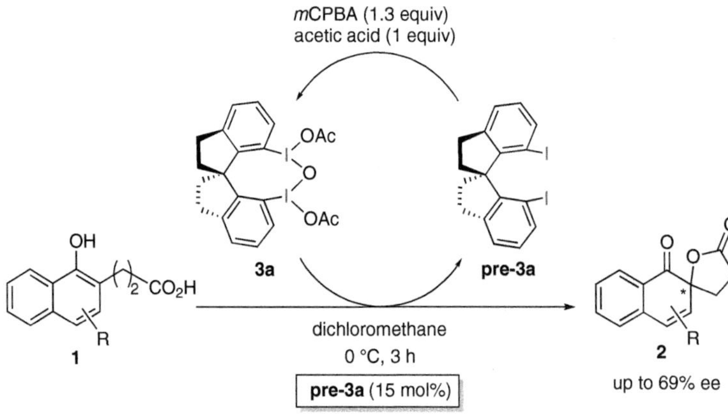

Scheme 8.7 Asymmetric dearomative coupling employing the catalytic chiral spirobiindane iodine(III) precursor.

The same research group has thoroughly investigated the effect of the structure of the spirobiindane substituents on the improvement of enantioselectivity and proposed modified precatalysts containing *ortho*-substituents on the two iodine groups [23b]. The corresponding diamines were the synthetic precursor, and these *ortho*-substituted spirobiindanes could be obtained via the Sandmeyer-type iodination. Employing *m*CPBA as a terminal oxidant, all the synthesized iodoarene compounds (>99% ee) were examined as precatalysts in the model reaction of naphthol (**1a**) to spirolactone **2a** under the same conditions. The installation of the ethyl substituents obtained the best result regarding the yields and *ee*, and a series of naphthol derivatives (**1a–m**) were utilized with the best precatalyst **pre-3b** (Scheme 8.8). The spirolactone structure is the key motif of biologically active natural products, such as arnottin II and lactonamycin [33]. The absolute configuration of **2a**, which was produced with the (*R*)-catalyst **3b** was determined as (*R*)-isomer via single X-ray crystal analysis.

In most cases, the spirolactone products **2** were obtained in good yields with better *ee* compared with the cases involving the original non-substituted catalysts **3a** (Scheme 8.7). A series of the substituted naphthol carboxylic acids bearing bromo, benzyl, aryl, azide, and benzyloxy groups was applied to the effective asymmetric transformations, although the outcomes of the reaction were certainly influenced by the substituent in the naphthyl ring of the substrates. The desired spirolactonization barely proceeded when a methoxy group was installed at the C-4 position of the naphthol ring in substrate **1c**, and a poor yield of the racemic mixture of **2c** was obtained with the enantiopure iodoarene catalyst. Further, **2m** was obtained in a

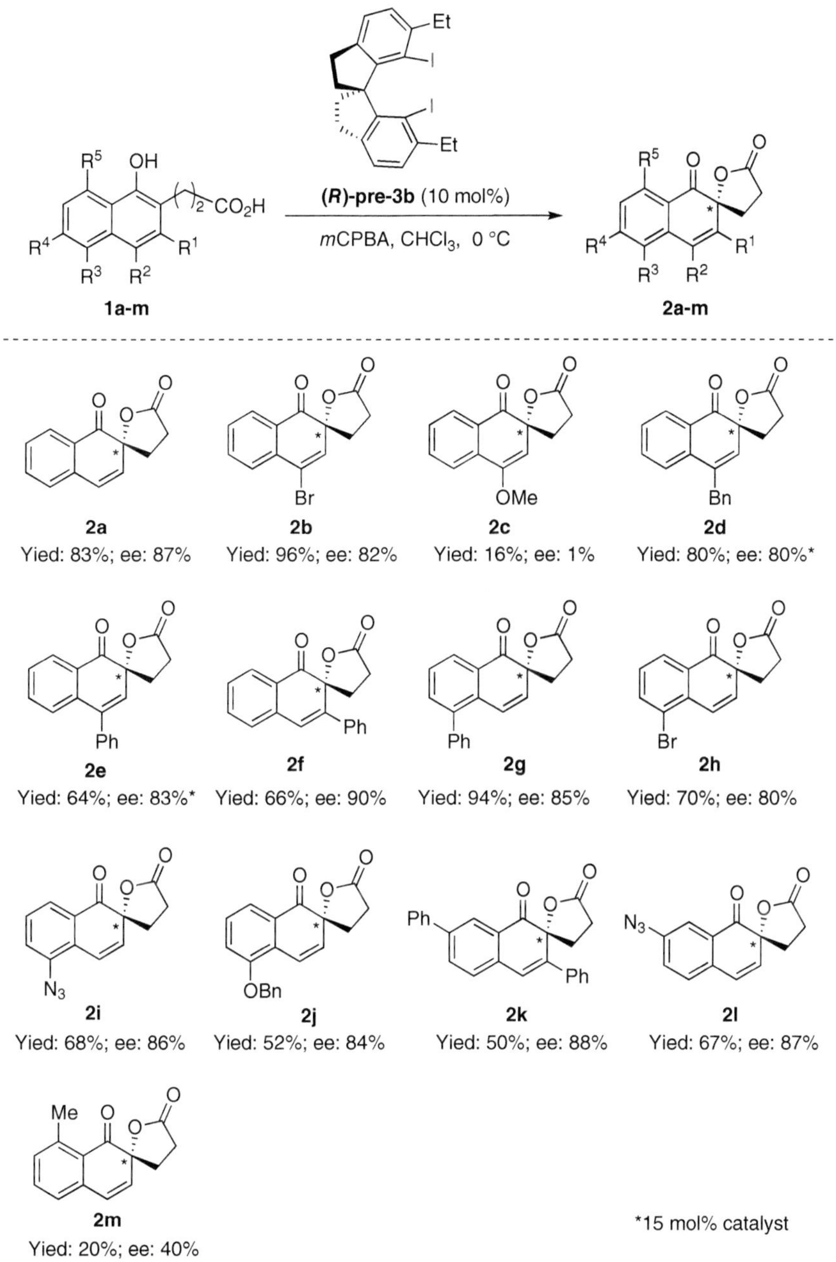

Scheme 8.8 Enantioselective dearomative *ortho*-spirocyclization employing the catalytic Et-substituted spirobiindane iodine compound (**pre-3b**).

20% yield and 40% ee when a methyl group was present at the C-8 position. In other successful cases, the spirolactone derivatives were obtained in enantiomerically pure form after single recrystallization.

It was also observed that the yield and selectivity were affected by the catalytic loading, as well as the solvent combination. A significant improvement was observed in the *ee* when the catalytic loading was increased to 15 mol%. However, the yield was notably reduced. Conversely, only a slight decrease was observed in the yield with a similar *ee* at a catalytic loading of 5 mol%. Additionally, the reaction was performed in a solvent comprising chloroform and hexane (1:1), and a notable increase and slight decrease were observed in the product yield and enantioselectivity, respectively.

Bekkaya and Masson reported a series of axially chiral iodoarenes (Scheme 8.9) [34]. These precatalysts significantly catalyzed the Kita spirolactonization reaction, but only a maximum *ee* of 36% was obtained for the conversion of naphthol carboxylic acid **1a** to spirolactone **2a**. Notably, 1,1′-diiodonaphthalene (**pre-5a**) exhibited only 2% *ee*, as also reported by Kita and coworkers [23a].

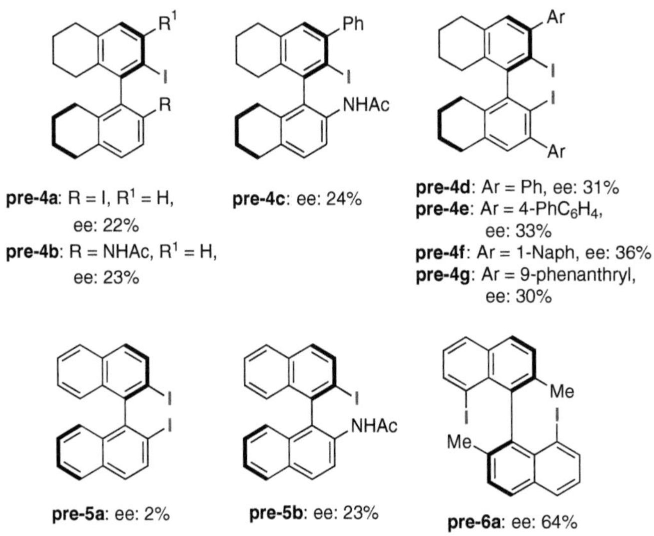

Scheme 8.9 Series of atropisomeric chiral binaphthyl iodides. *Source*: Based on Bekkaye and Masson [34].

Considering the structure of the spirobiindane catalyst (**3**, Scheme 8.10), Dohi's and Kita's groups alternatively developed atropisomeric 8,8′-diiodobinaphthyl compounds [35]. Thereafter, these new types of 8,8′-difunctionalized chiral compounds, as the optically pure forms, were examined for the spirocyclization of the model substrate (**1a**), and it was observed that diiodide (**pre-6a**) exhibited a good *ee* of 64% (Scheme 8.9). This enantioselectivity correlated with that, which was obtained by the spirobiindane catalyst (**pre-3a**), although the scope of the substrate for an effective asymmetric process was quite limited regarding the product yield and enantioselectivity probably because of the conformational flexibilities of the

1,1′-linked binaphthyl molecules. Hence, the rigid structure of the spirobiindane backbone that is required to maintain the chiral environment around iodine(III) during the dearomatization process generally favors asymmetric induction and the effective formation of the product.

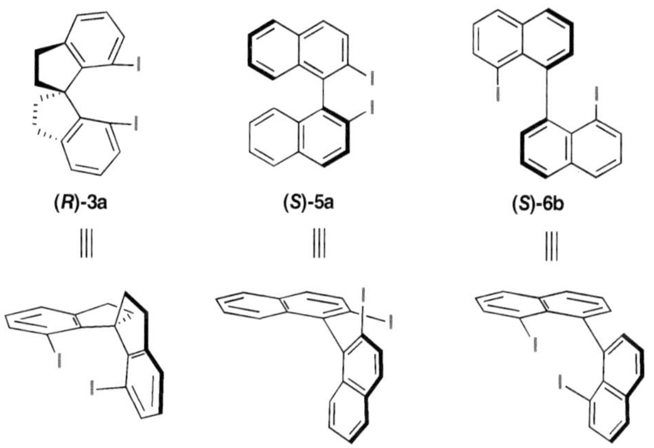

Scheme 8.10 Structural comparison of **pre-3a**, 2,2′-diiodo-1,1′-binaphthalene (**pre-5a**), and 8,8′-diiodo-1,1′-binaphthalene **pre-6b**.

The mechanism and origins of the enantioselectivities of these spirobiindane- and binaphthalene-based hypervalent iodine(III) reagents in Kita's spirolactonizations, as well as the high reactivities of the μ-oxo structures, have been suitably rationalized via recent theoretical calculations by Houk and Xue [36, 37a]. The computational studies of the mechanisms and stereocontrolling models of the spirobiindane catalysts (**3a** and **3b**) during the asymmetric oxidative spirolactonization indicates that the direct attack of the *ipso*-position of the naphthol ring by the pendant carboxylic acid to yield the spirolactonization product (**2a**) exhibited an activation-free energy barrier that is, as high as 48 kcal/mol (Scheme 8.11, **Int-1**). Therefore, the transfer of protons to the μ-oxygen group or the ligand of the iodine(III) species to enhance the nucleophilicity of the carboxylic oxygen and the nucleofugality of the iodoarene could be crucial to the dearomative spirolactonization (**TS-1**). The halogen bonding between the resulting carboxylate and electron-deficient iodine(III) center could further stabilize the transition states of the dearomatization (**TS-2**). Consequently, the overall activation barrier can be reduced to 21.0 kcal/mol by introducing an acetic acid molecule as a hydrogen-bonding bridge between the naphthol carboxylic acid group and the bridging oxygen of **3a**. This efficient stereochemical model could be applied to other naphthol carboxylic acid substrates (**1**), as well as other axially chiral biaryl catalysts, such as **pre-6a**.

8.4 Further Breakthrough and Recent Advances

A further breakthrough in the expansion of the application of asymmetric hypervalent iodine catalysis includes the development of Ishihara's C_2-symmetric

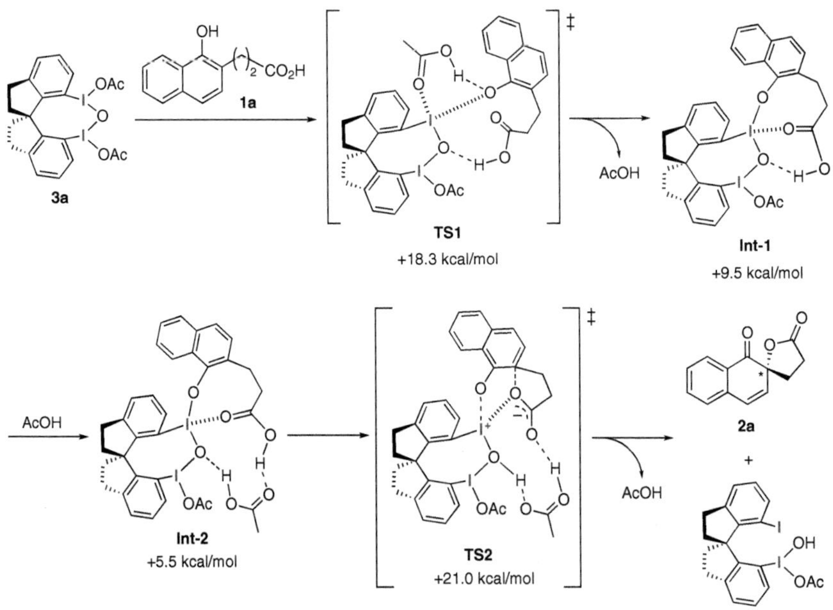

Scheme 8.11 Computational mechanistic study of the asymmetric dearomative spirolactonization of naphthol carboxylic acid (**1a**) catalyzed by **3a.**

hypervalent iodine catalysts. In their evaluation of the asymmetric Kita dearomative spirolactonization of naphthol, Ishihara's group discovered an elegant hypervalent iodine catalyst *bearing the two chiral lactate groups* in 2010 [38]. Notably, Ishihara's catalyst afforded the same spirolactone product with up to 92% ee with a catalytic loading of 10–15 mol%, and that was the best asymmetric score for this catalytic dearomatization process (Scheme 8.12).

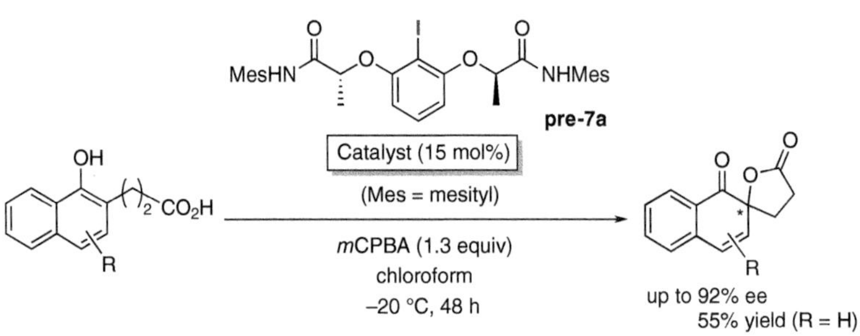

Scheme 8.12 Ishihara C_2-symmetric hypervalent iodine catalyst for the asymmetric Kita spirolactonization of naphthol.

The *ortho* chiral tether-type aryliodine compounds were first observed by Wirth's group during their study of the chemistry of hypervalent iodine as a new class of pseudocyclic reagents (Scheme 8.2) for the enantioselective α-oxygenation of carbonyl compounds, as well as the oxidative difunctionalization of alkenes [16]. Later, Fujita's group proposed several chiral *ortho* lactate-type hypervalent iodine(III) compounds for the intramolecular oxycyclization of alkenes exhibiting functional groups [21]. Ishihara and coworkers have successfully designed C_2-symmetric chiral iodoarenes exhibiting *ortho* chiral moieties (Scheme 8.13). To obtain molecular diversity, the iodoarene was spliced into three basic units (the iodoaryl moiety (**A**), chiral linker (**B**), and subfunctional group (**C**)). Moreover, iodoarene could be readily oxidized *in situ* into active hypervalent iodine(III) species by *m*CPBA. It is expected that this hypervalent iodine species would exhibit intramolecular n–σ* interactions between the electrophilic iodine(III) center and carbonyl oxygen in **B** (**7a-I**). The intramolecular hydrogen bonding between the acidic hydrogen of **C** and the ligand that was attached to the iodine atom is also assumed as another possibility (**7a-II**). These highly organized chiral environments around iodine(III) could effectively facilitate chiral induction during the asymmetric phenol dearomative cyclization processes.

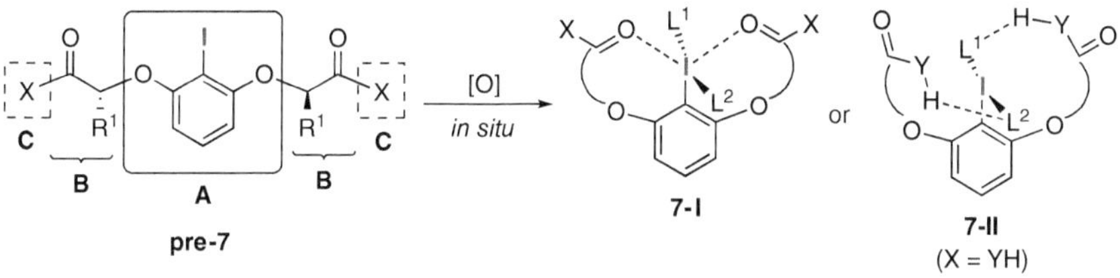

Scheme 8.13 Structure of the Ishihara C_2-symmetric chiral hypervalent iodine catalysts.

The preparation of the Ishihara catalyst generally includes three chemical steps starting first with the commercially available 2-iodoresorcinol. The structure of the catalyst has been further optimized to facilitate an efficient dearomatization into the spirocyclic compound (**2**) by screening many of the obtained C_2-symmetric iodoarene derivatives. Iodoarene (**pre-7b**) consisting of chiral aminoalcohol is known as Ishihara's 2nd-generation catalysts (Scheme 8.14) [39]. The dearomative spirocyclizations proceeded very smoothly with this catalyst, and the spirolactones (**2**) were obtained in high yields with an enhanced *ee* of up to 98% (Scheme 8.14, Eq. 1). As previously reported [40], ethanol was utilized as an additive, which was vital to obtaining high yields and *ee*. The asymmetric induction was significantly influenced by the position of the substituents of the naphthol substrates and their electronic nature, although the yields of the electron-rich substituents were reduced. Moreover, the 2-naphthol-1-carboxylic acids (**1′**) were also applicable substrates, and the pericyclic spirolactones (**2′**) were obtained with excellent enantiomeric control (Scheme 8.14, Eq. 2). The product (**2′**) was obtained in good yields and up to 95% ee under the standard catalytic conditions, although the more acidic hexafluoro-2-propanol (HFIP) was selected as a fascinating additive. Similarly, the enantioselectivity was not greatly influenced by the naphthol substituents, but **2′** were generally formed with the electron-rich substituents in lowered yield.

The excellent feature of the Ishihara-type catalysts is that the effective asymmetric induction could proceed even in the phenol dearomative spirocyclization process.

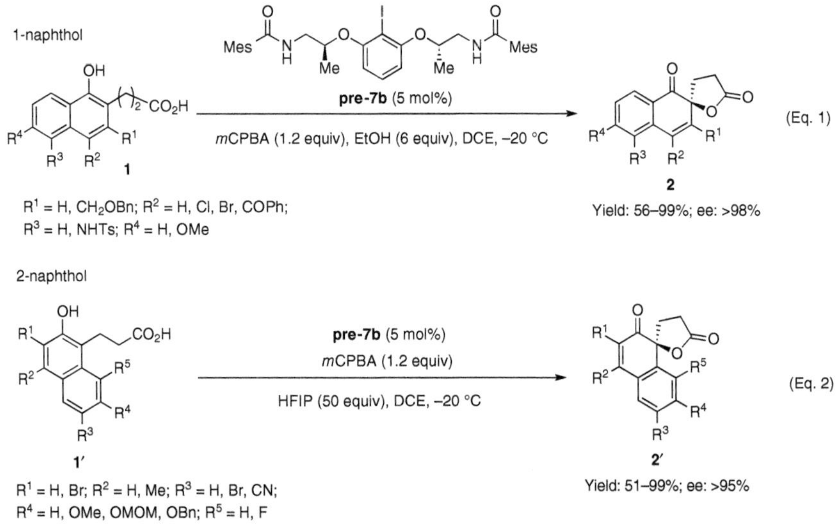

Scheme 8.14 Second-generation Ishihara catalyst for the asymmetric dearomative spirocylization of naphthols. *Source*: Based on Uyanik et al. [39].

In 2013, the corresponding enantioselective dearomative cyclizations of phenols (**8**) were accomplished utilizing the 2nd-generation catalyst (**pre-7b**) to afford *ortho*-spirocyclohexadienones (**9**) in good yields and ~98% ee (Scheme 8.15, Eq. 1) [40]. The reactions were performed under two reaction conditions (**A** and **B**). Methanol and HFIP were utilized as the additives under **A** and **B**. Surprisingly, the catalytic loading was reduced to 1 mol% without reducing the product yield and enantioselectivity of the phenol substrate (**8g**). Thus, the intramolecular hydrogen bonding, as well as the addition of the alcohol mainly contributed to the effective chiral environment around the electrophilic iodine(III) center. The synthesized cyclohexadienones (**9a–g**) can be utilized as chiral dienes for the Diels–Alder reaction employing different dienophiles. The 2nd-generation catalyst was also effective in the spirolactonization of the phenol ether compounds (**10**; Scheme 8.15 and Eq. 2) [41]. The spiroketal compounds (**11**) were produced with high enantioselectivity (~96% ee) after the addition of methanol. However, prolonged reaction time was required since methanol suppressed the regeneration of the iodine(III) species from **pre-7b**. Conversely, the dearomatization of the naphthol ether (**12**) proceeded well without methanol, and the spironolactone ketal (**13**) was obtained in an 88% yield with 86% ee (Scheme 8.15 and Eq. 3) [41].

In 2018, Suzuki's and Tanino's groups accomplished the total synthesis of (–)-maldoxin via the asymmetric oxidative dearomatization of pestheic acid (**14**) employing the Ishihara catalyst (**pre-7b**) with *m*CPBA as the stoichiometric terminal oxidant (Scheme 8.16) [42]. The desired product, (–)-maldoxin, was obtained in a 93% yield with 99% ee. The pestheic acid, **14**, was prepared in six synthetic steps starting with 3,5-difluorotoluene; the overall yield of (–)-maldoxin was 23% yield in this sub-gram-scale synthesis.

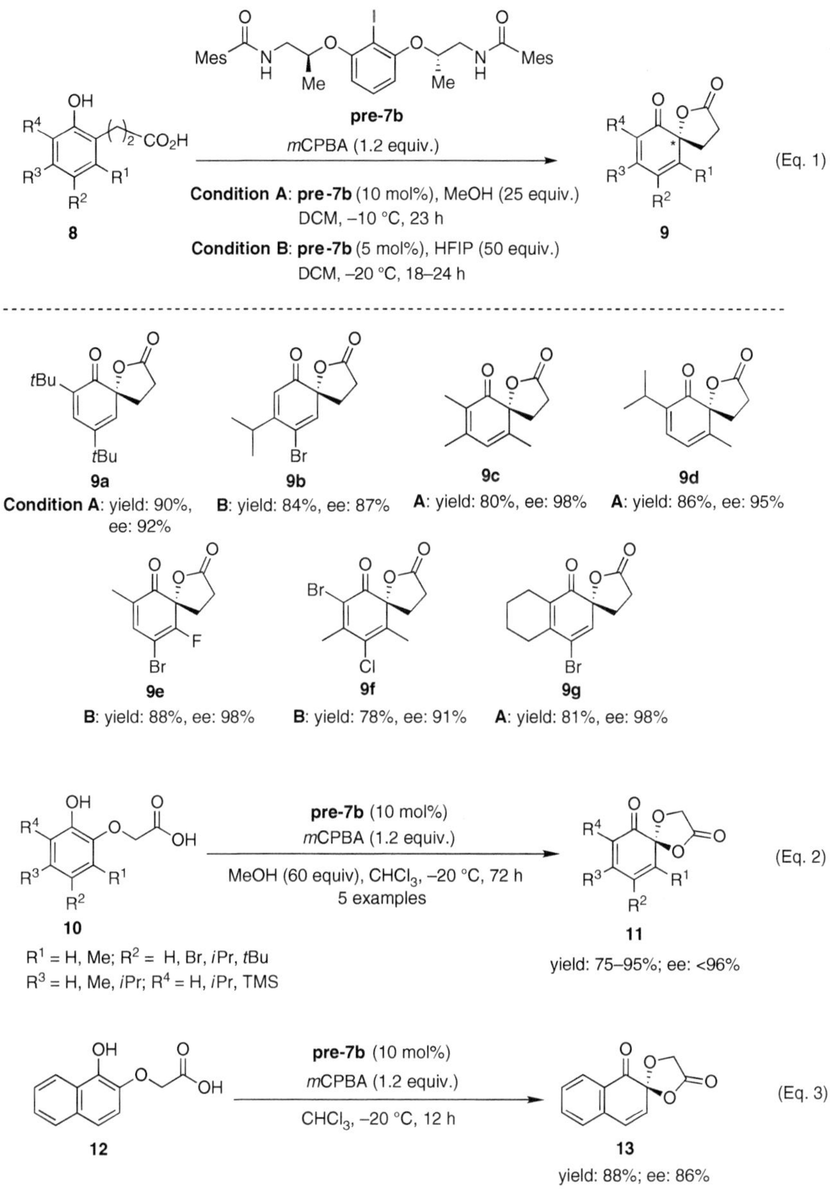

Scheme 8.15 Asymmetric dearomative *ortho*-spirocyclization of phenols employing the Ishihara catalyst. *Source*: Uyanik et al. [40], Uyanik et al. [41].

It was also revealed that the dearomative spirocyclization strategy is effective even with intramolecular alcohols for synthesizing spirocyclic ketals. Thus, the spiroketals (**15**) were obtained in good yields and up to 93% ee via dearomative cyclization (Scheme 8.17) [41]. For the pretreatment, a toluene–water (5:1) mixture was sonicated for 1 h to homogenate the solvent particles before the reaction started. The influence of the ring substituents in substrates **15** on the enantioselectivity was

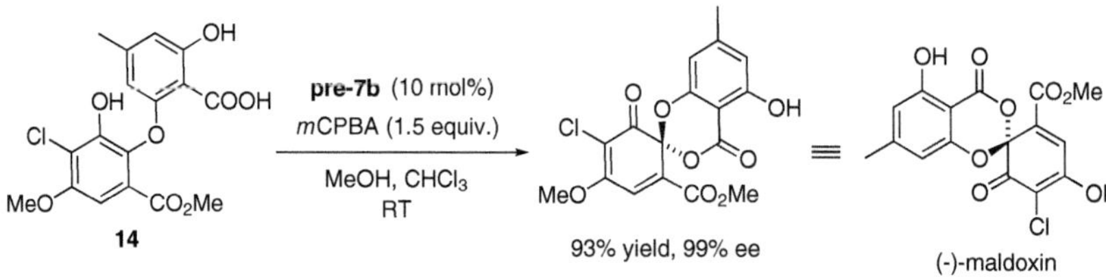

Scheme 8.16 Application of **pre-7b** to the total synthesis of (−)-maldoxin. *Source*: Suzuki et al. [42].

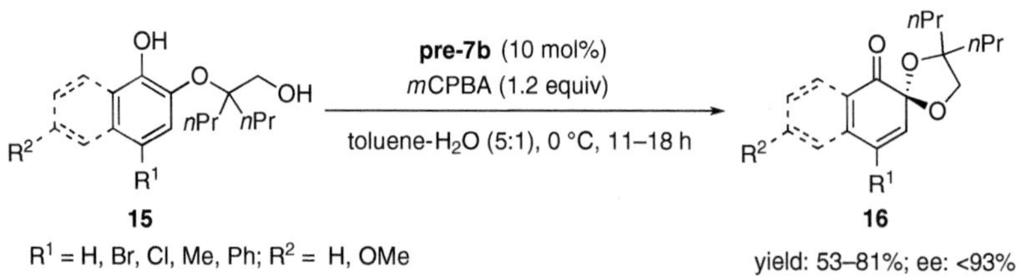

Scheme 8.17 Asymmetric dearomative *ortho*-spirocyclic cycloetherification of phenol alcohols (**15**). *Source*: Based on Uyanik et al. [41].

similar to that, which was observed in the aforementioned spirolactonization of 1-naphthol-2-carboxylic acids (**1**). Furthermore, the asymmetric induction that was catalyzed by **pre-7b** was also suitable for the formation of the corresponding phenol substrates.

Ciufolini and coworkers reported the enantioselective spiroetherification of naphtholic alcohols (**17**) in the presence of a C_2-symmetric chiral iodoarene derivative (**pre-7c**) as the precatalyst (Scheme 8.18) [43]. Spirocyclic ethers (**18**) were obtained in moderate-to-good yields with up to 98% ee. In most cases, the products were obtained with excellent *ee* except for the substrate exhibiting an anisole moiety at the C-4 position of the naphthol ring. The utilized precatalyst (**pre-7c**) was recovered in a yield of >90%, after which it was reused without losing its catalytic efficiency. The major isomer of the product (**18**) was assigned as an *R*-enantiomer for this transformation.

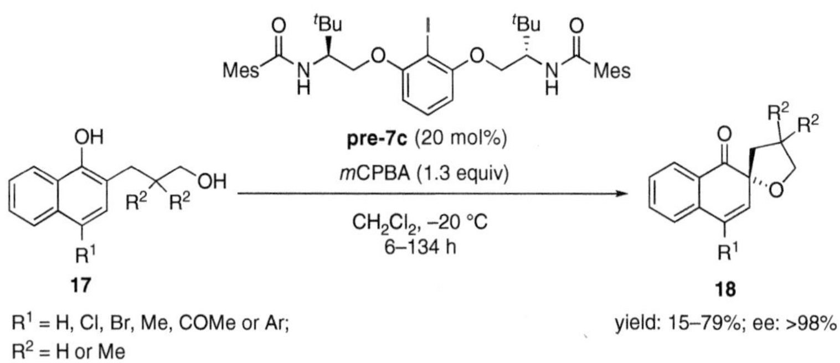

Scheme 8.18 Asymmetric spiroetherification of **17** employing the modified catalyst (**pre-7c**). *Source*: Based on Jain et al. [43].

Concurrently, this research group briefly evaluated the same iodine catalyst for the enantioselective oxidative spirocyclization of naphthol sulfonamide (**19a**, R^1 = H, R^2 = Me), which yielded a spirocyclic amide (**20a**) with 67% ee but in a 20% yield after 15 h under the following conditions: 20 mol% **pre-7c** with *m*CPBA (1.3 equiv.) in dichloromethane at −20 °C (Scheme 8.19) [43]. Subsequently, the reaction was detailly studied by changing the substrates with different electron-donating and -withdrawing groups at the C-4 position of the naphthalene ring [44]. The presence of the electron-donating and -withdrawing substituents at the C-4 position significantly decreased the enantioselectivity.

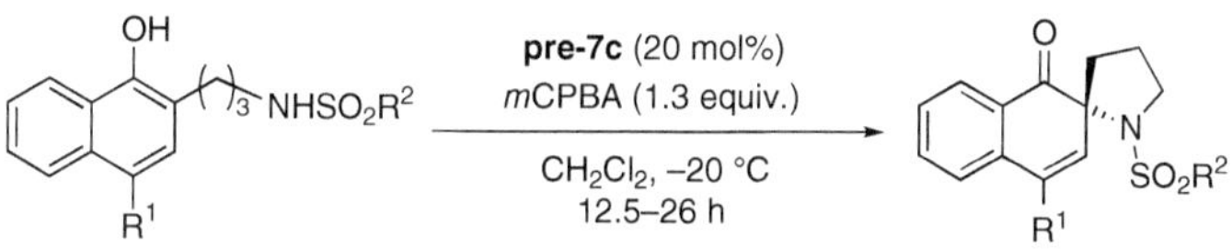

19

R^1 = H, Br, Cl, Ph, Ar; R^2 = Me, Ar, Bn

20

yield: 13–45%; ee: >67%

Scheme 8.19 Asymmetric spirocyclic amination of the naphtholic sulfonamide amine (**19**). *Source*: Based on Jain et al. [43].

Employing the C_2-symmetric chiral lactate (**pre-7a**), the spirooxindole derivative (**22**) was obtained with high-to-excellent enantioselectivities by stereoselectively generating the all-carbon spirostereogenic center during the dearomative spirocyclization of the 1-hydroxy-*N*-aryl-2-naphthamides (**21**), which was reported by Gong and coworker (Scheme 8.20) [45]. This is the first successful catalytic asymmetric dearomative spirocyclization of phenol accompanying the formation of a C–C bond.

The Kita dearomatizing reactions and catalytic systems have been applied to further catalyst screening by other researchers, and the selected examples are shown in Scheme 8.21. Murray and Ibrahim developed new *anti*-dimethanoanthracene-based iodoarenes (**pre-23**) [46]. The precatalyst (**pre-23a**) exhibited slightly higher enantioselectivity compared with the methoxy-substituted derivative (**pre-23b**). The spirolactones (**2**) were obtained in moderate yields with up to 67% ee employing a catalyst loading of 10 mol%. Imrich and Ziegler reported the one-step syntheses of carbohydrate-based chiral iodoarenes (**pre-24**) from known compounds [47]. The catalyst (15 mol%) (**pre-24** (R^2 = (1-naphthyl)CH_2) afforded spirolactone (**2a**)

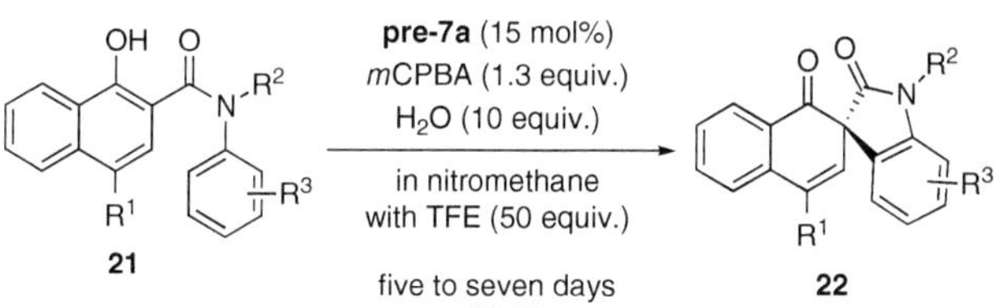

21

22

up to 80% yield, 92% ee

Scheme 8.20 Asymmetric spirocyclic arylation of **21** to **22**. *Source*: Zhang et al. [45].

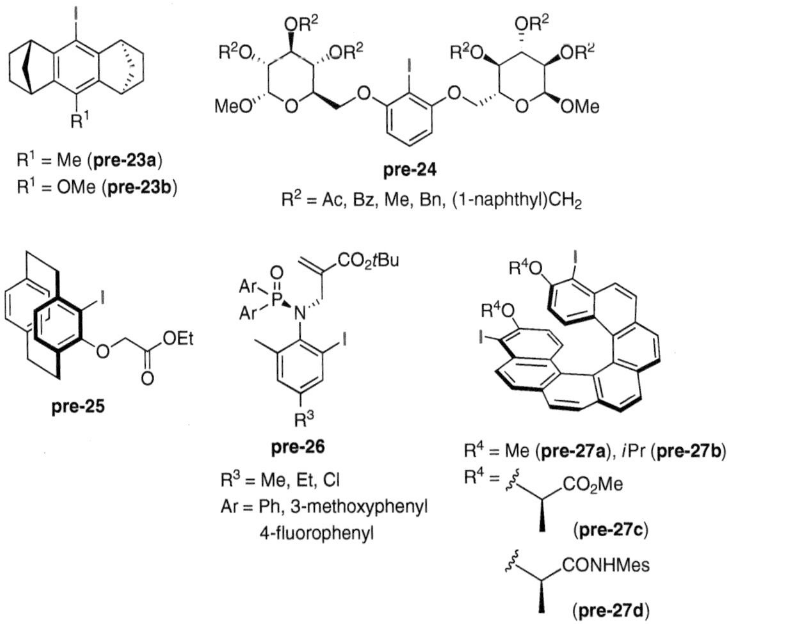

Scheme 8.21 New chiral hypervalent iodine(III) catalysts for the asymmetric Kita spirolactonization.

in a 75% yield and enantiomeric ratio (*er*) of up to 88:12. Zheng's group developed [2.2]-paracyclophane-type planar chiral iodoarenes (**pre-25**) [48]. The spirolactones (**2**) were obtained in moderate yields with up to 72% ee at 0 °C employing a 15 mol% amount of the catalyst. Li and coworkers reported the utilization of their newly synthesized axially chiral *ortho*-iodine-substituted phosphamides (**pre-26**) as an efficient catalyst for Kita's spirolactonization (up to 75% ee employing 20 mol% **pre-26**) [49]. Peixoto and Quideau also reported [7]-helicenic chiral iodoarene compounds (**pre-27**); they were evaluated as a catalyst (15 mol%) for the Kita spirolactonization under the standard conditions [50]. The spirolactone (**2a**) was obtained (almost as a racemic mixture) in the presence of catalysts **pre-27a** and **pre-27b**, while the utilization of (*P*)-**pre-27d** with the pendent chiral groups improved the yield and *ee* up to 90% and 66%, respectively.

Additionally, interesting nonaromatic chiral iodides (**pre-28**) were synthesized and employed as pre-catalysts for the Kita spirolactonization (Scheme 8.22) [51]. The unique C_2-symmetric tetralin-fused atropisomeric diiodides, (*M*)-**pre-28** (10 mol%), adequately catalyzed the dearomative spirolactonization of the naphthols (**1**) in which the silyl-substituted **pre-28c** exhibited the best enantioselectivity (73% ee).

The triazole-based chiral iodoarene (**pre-29**) was recently synthesized and employed in the Kita spirolactonization reaction (Scheme 8.23) [52]. The active catalytic hypervalent species was generated *in situ* via the oxidation of chiral iodoarenes employing *m*CPBA. Thus, the catalytic reaction (15 mol%) was facilitated by applying the standard conditions, although the yields and enantioselectivity

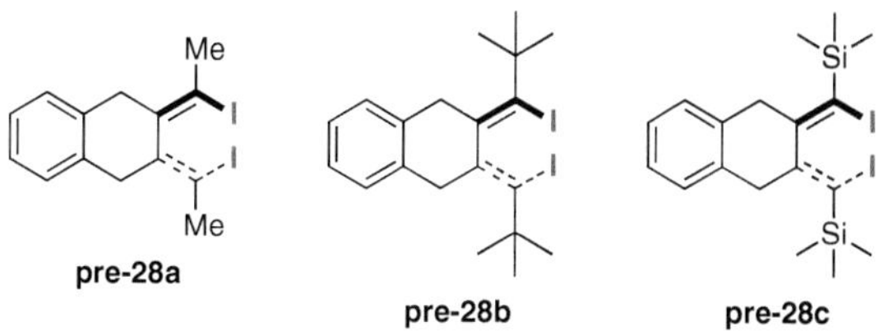

Scheme 8.22 Unique nonaromatic chiral diiodides (*M*)–**pre-28a–c**. *Source*: Based on Ogasawara et al. [51].

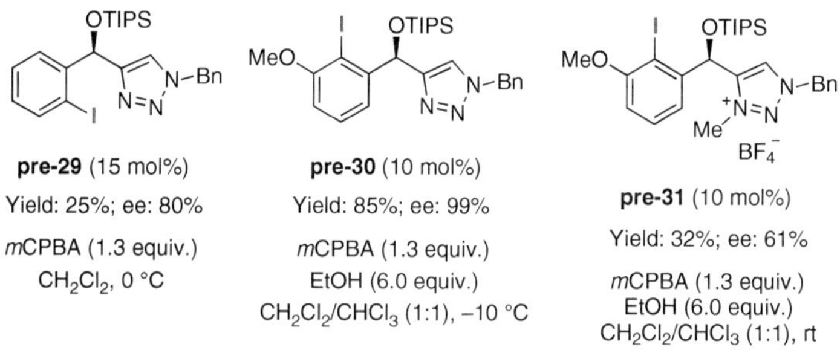

Scheme 8.23 New triazole-based chiral aryl iodides synthesized by Nachtsheim's group. *Source*: Boelke et al. [52a], Abazid and Nachtsheim [52b].

were decreased compared with those of the stoichiometric version. More recently, Abazid and Nachtsheim improved the structure of the catalyst, after which **2a** was obtained with >99% ee employing the 2nd-generation catalyst (**pre-30**) [52b]. The addition of ethanol (6 equiv.) could promote the catalytic reactions probably because the covalently attached triazole functionality further stabilizes the *in situ*-generated cationic iodine(III) species. Moreover, the benefit of the N–I interaction for the performance of this catalyst was confirmed by synthesizing *N*-methyl triazolium tetrafluoroborate (**pre-31**). Indeed, this salt exhibited diminished reactivity, affording **2** in a yield of only 32% with an *ee* of 61%. Notably, this new triazole-based chiral iodoarene catalyst exhibited extensive application in the formation of dearomative quinol (vide infra), as well as other types of hypervalent iodine(III) transformations.

The unique design of the catalyst for the asymmetric *para*-dearomatization of phenol was proposed by Harned and Volp employing a newly synthesized tetrahydronaphthalene-based tricyclic iodoarene (**pre-32**) to restrict the conformational freedom of the chiral moiety [53]. Additionally, the amide group was employed to facilitate the possible formation of a hydrogen bond between the carbonyl group and the ligand that is attached to iodine after the *in situ* generation of the catalytic hypervalent iodine species. The chiral iodoarene (**pre-32**) was synthesized in two chemical steps via the reaction of *L*-(+)-dimethyltartrate with 8-iodotetralone, followed by mesityl amine. Although the enantioselectivity was moderate, Harned and Volp achieved the catalytic asymmetric *para*-spirocyclizations with this catalyst for the first time (Scheme 8.24). The catalytic reaction proceeded well, and a spirocyclic ether (**34**) was obtained in a 50% yield with 40% ee via the reaction of

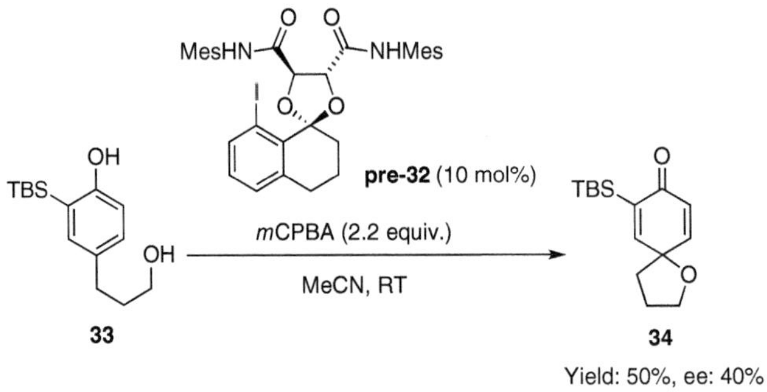

Scheme 8.24 Dearomative *para*-spirocyclization of phenols employing the chiral acetal catalyst (**pre-32**).

phenol *para*-carboxylic acid (**33**). This type of asymmetric *para*-spirocyclizations is also possible in the presence of the Ishihara catalyst (**pre-7b**) [39].

Cai and coworkers reported a new desymmetrization strategy for performing asymmetric *para*-spirolactamizations employing a chiral hypervalent iodine reagent (Scheme 8.25) [54]. A Kita-type chiral spirobiindane catalyst (**pre-3c**) was employed for the prochiral substrates (**35**). A series of spirolactams (**36**) was obtained in good yields with moderate-to-high *ee* via this desymmetric dearomatization process. It was explained that the formation of the desymmetric C–N bond would occur via the attack of the anisole group to the chiral nitrenium species, which was formed by reacting the amide group with the hypervalent iodine(III) species. The R^2 group was replaced with water during the dearomatization process, thus affording these *para*-spirodienone lactam compounds (**36**).

The catalytic asymmetric dearomatization of phenol employing an intermolecular nucleophile is more challenging than that employing an intramolecular one. In 2008, Birman and coworkers synthesized an iodoxybenzene derivative (**37**) with

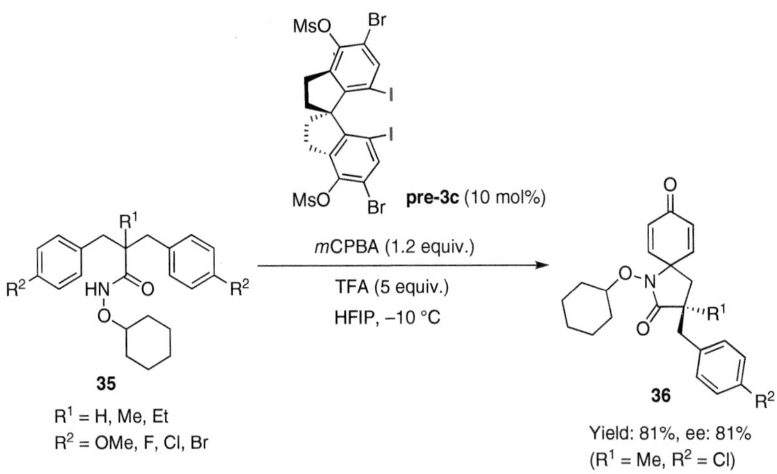

Scheme 8.25 Desymmetrization strategy for performing asymmetric dearomative *para*-spirolactamization. *Source*: Based on Ding et al. [54].

chiral isoxazoline groups at the *ortho* position of the iodine group. Afterward, it was applied to the oxidative formation of *o*-quinol (Scheme 8.26) [55]. The transformations of *o*-alkylphenols (**38**) into *ortho*-quinol Diels–Alder dimer (**39**) were observed. The stoichiometric versions of a few substrates afforded the Diels–Alder products with significant levels of asymmetric inductions of up to 77% ee.

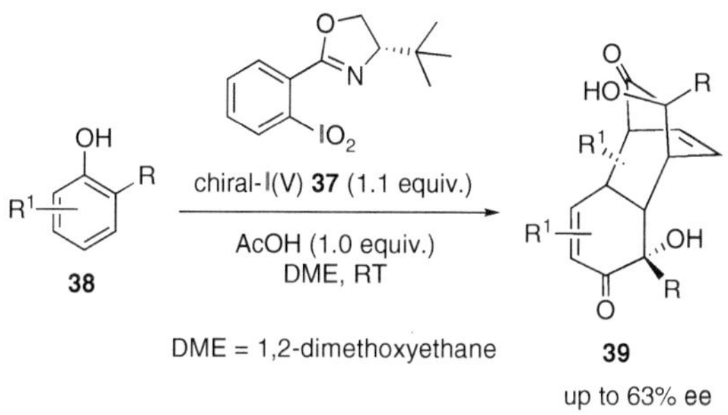

Scheme 8.26 Birman's chiral isoxazoline catalyst for the asymmetric dearomative hydroxylation of phenols (a previous stoichiometric case). *Source*: Modified from Boppisetti and Birman [55].

Following their previous stoichiometric success, Quideau and coworkers reported the catalytic asymmetric oxidations of *ortho*-substituted-1-naphthol (**40**) in 2009. The hydroxylative dearomatization of phenol was developed asymmetrically employing a chiral iodobinaphtyl catalyst (**pre-4h**, 10 mol%) (Scheme 8.27). The subsequent regio- and diastereoselective epoxidations of the *ortho*-quinols in the presence of *m*CPBA afforded the optically active products with 29% *ee*; a maximum *ee* of 50% was obtained under the stoichiometric conditions [56].

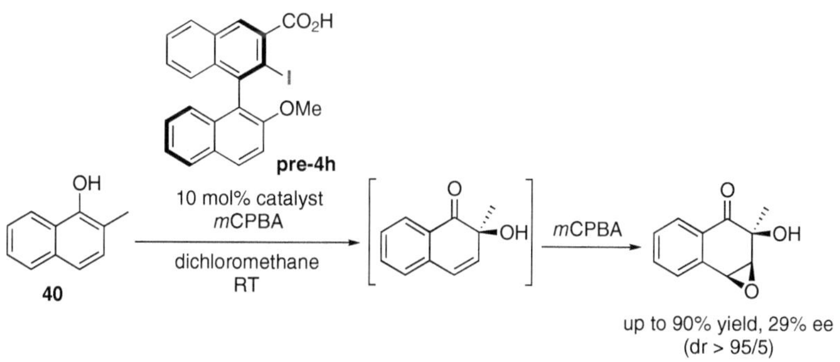

Scheme 8.27 First catalytic case for the formation of enantioselective *o*-quinol by Quideau and coworkers.

In 2013, Harned and coworkers reported the investigation of a new chiral λ^3-iodane (**pre-32**) for the formation of an asymmetric *p*-quinol [53]. The utilization of a 10 mol% spirocyclic acetal catalyst (**pre-32**) with 2.2 equivalents of *m*CPBA could generate the asymmetric hydroxylated center of the *p*-quinols (**42**) despite the low-to-moderate *ee* values (Scheme 8.28 and Eq. 1). Among the tested catalysts,

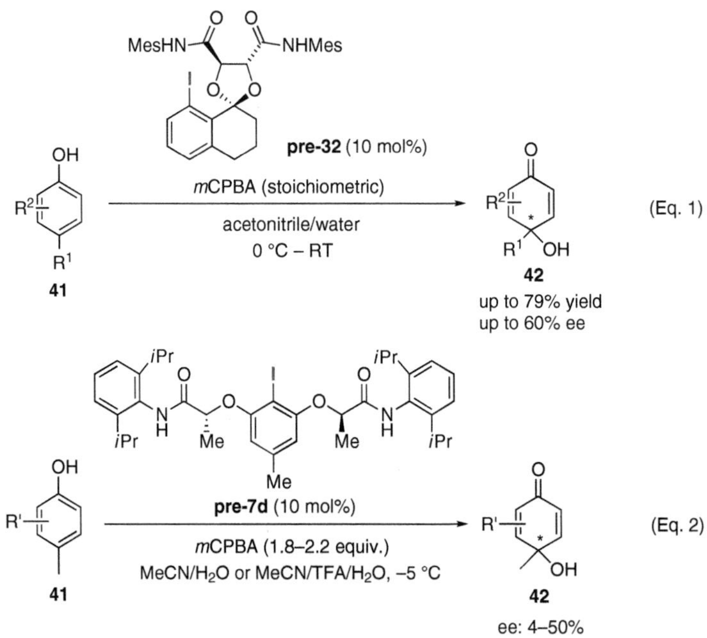

Scheme 8.28 Enantioselective formations of *p*-quinol by Harned and Muñiz. *Source*: Muniz and Fra [57].

(**pre-32**), which was rooted in the tetrahydronaphthalene moiety, afforded **42** with good enantioselectivity (up to 60% ee). Muñiz *et al.* also reported the formation of an intermolecular enantioselective dearomative *p*-quinol employing water as an external nucleophile in the presence of the Ishihara-type C_2-symmetric chiral iodide (**pre-7d**, Scheme 8.28 and Eq. 2) [57]. The hydroxylative dearomatizations of different phenols (**41**) yielded the corresponding *p*-quinol products with up to 50% ee. The scope of this reaction was also extended for *N*-tosylated aniline derivatives, and the corresponding *p*-iminoquinols were formed with up to 34% ee.

In 2018, Marouka's and Hashimoto's group developed a suitably architectured chiral iodine catalyst (**pre-43**) exhibiting an indanol scaffold (Scheme 8.29) [58]. The enantiopure catalyst exhibiting this suitable design was employed for the intermolecular dearomatization of phenols (**41**) to yield **42** with good enantioselectivity (*ee* of up to 84%). Following the chromatography-free, scalable synthesis of enantiopure indanol, the authors designed the synthesis of an indanol-based organoiodine catalyst library.

Nachtsheim's triazole-based chiral iodoarene (**30**), which was reported for Kita's spirolactonization, can also facilitate the asymmetric dearomative *para*-hydroxylation of phenols effectively [52b]. Although only one substrate (**41b**) was examined at the present stage, the reaction in the presence of 10 mol% **pre-30** proceeded efficiently to afford a corresponding *para*-quinol (**42b**) in a 92% yield and 93% ee (Scheme 8.30).

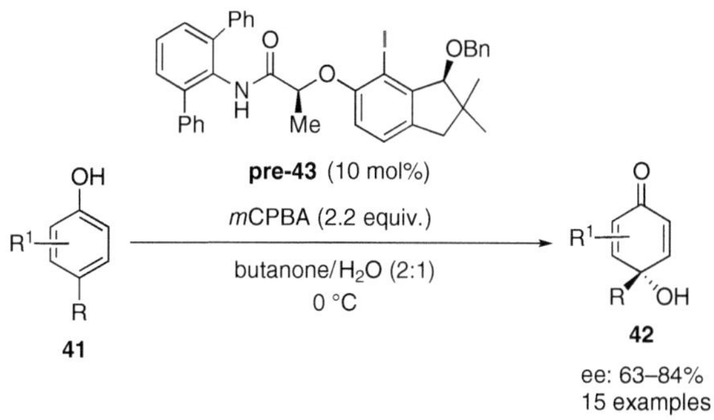

Scheme 8.29 Indane-based chiral hypervalent iodine catalyst for the dearomative *para*-hydroxylation of phenol. *Source*: Based on Hashimoto et al. [58].

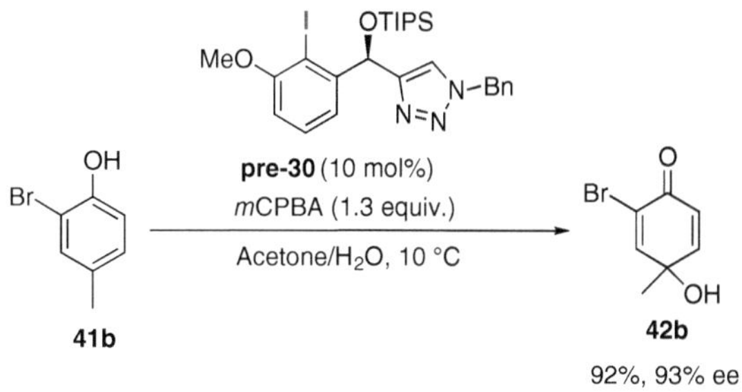

Scheme 8.30 Indane-based chiral hypervalent iodine catalyst for the dearomative *para*-hydroxylation of phenol.

Xiong and coworkers recently reported the tandem intramolecular *p*-lactonization and asymmetric dearomative *para*-hydroxylation of phenol employing the Ishihara-type catalyst (**pre-7d**, Scheme 8.32) [59]. First, it was observed that the reaction of phenolic lactone (**43a**) in the presence of 20 mol% **pre-7d** and *m*CPBA (1.2 equiv.) induced efficient dearomative *para*-hydroxylation to afford a cyclohexadiene lactone (**44a**) in a good yield with 56% ee (Scheme 8.31 and Eq. 1). A noncyclic phenolic carboxyclic acid (**45a**) could also produce **44a** under the same catalytic conditions employing **43a** as an intermediate (Scheme 8.31 and Eq. 2). In this reaction, the incorporation of different alcohols afforded the corresponding products (**44a–d**) with 40–52% ee.

8.5 Essential Mechanistic Guide

Regarding the oxidations of phenols employing the hypervalent iodine reagents, PIDA and PIFA, the reactions are typically explained by the two-electron-transfer processes involving the initial ligand exchange at the iodine(III) centers. Thus, PIDA and PIFA reacted with the phenolic oxygens to yield the phenoxyiodine(III) intermediates (**A**, Scheme 8.32). The introductions of the inter- and intramolecular nucleophiles (alcohols, amides, carboxylic acids, water, oximes, alkenes and

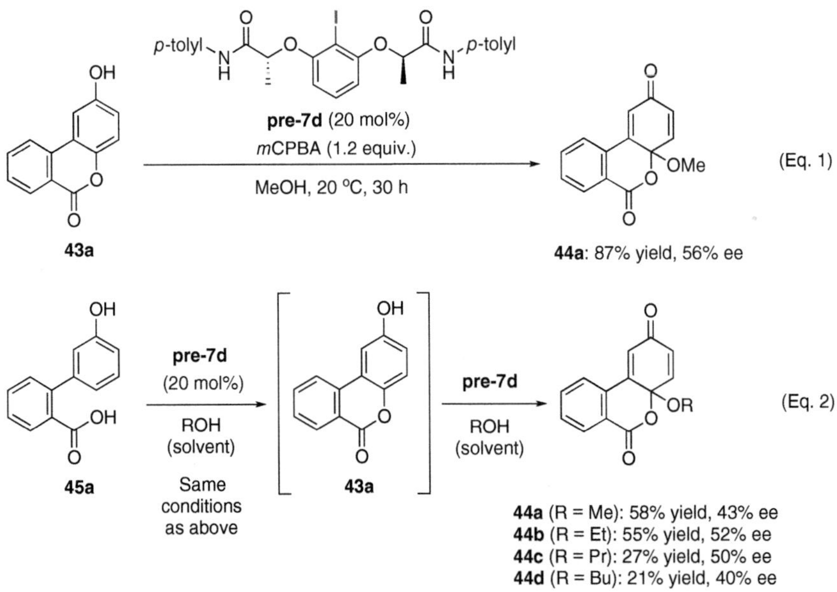

Scheme 8.31 Tandem intramolecular *p*-lactonization and asymmetric dearomative *para*-hydroxylation of phenols. *Source*: Based on Deng et al. [59].

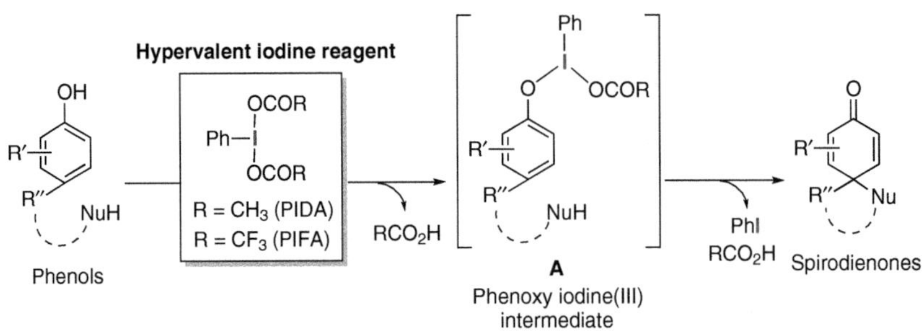

Scheme 8.32 General pathways and mechanism of the dearomatization of phenol.

alkynes or enamides, electron-rich aromatic rings, fluoride ions, etc.) proceeded because of the rapid elimination of iodobenzene (PhI) to yield the dearomatized cyclohexadienones-bearing nucleophiles and spiroannulated structures at the *para*- or *ortho*-positions of the original phenol group [7]. The excellent leaving ability of the high-valent iodine atoms to produce very stable iodobenzene can smoothly generate phenoxenium ions in polar solvents; these ions are subsequently trapped by different concomitant nucleophiles to complete the dearomatization reactions. Thus, the highly polar and non-nucleophilic fluoroalcohols (HFIP and trifluoroethanol (TFE)) can significantly accelerate these oxidation processes [4].

Conversely, asymmetric induction during the dearomatization of phenol must be rationalized, proceeding with the associative attack of a nucleophile on intermediate **A** in suitable chiral environments rather than the formation of the discrete

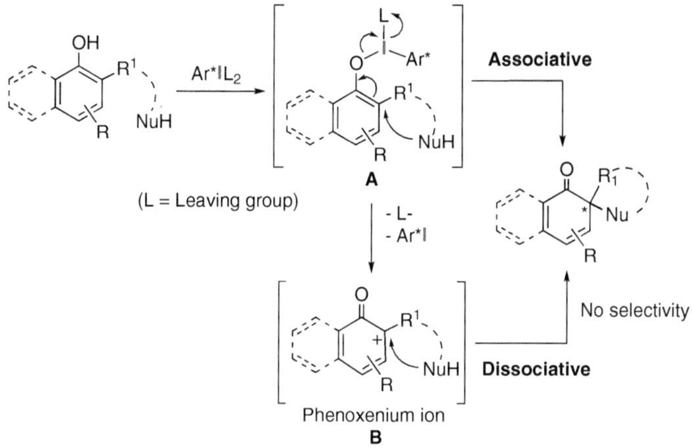

Scheme 8.33 Asymmetric dearomatization of phenol via the associative pathway.

phenoxenium ion via the dissociation of the chiral iodine(III) (Scheme 8.33). Alternatively, the oxidation could include the discrete phenoxenium ions (**B**), which are formed after the dissociation of iodobenzene from the **A**, and subsequent attack by the nucleophile to afford the product. The elucidation and control of the mechanisms of the dearomatization reaction of phenol is a basis for developing a highly efficient asymmetric reaction system involving hypervalent iodine catalysis. Despite the modest experimental evidence for either of the two mechanisms [22, 23], researchers have adopted these pathways to rationalize the observed reactivities and enantioselectivities. Moreover, this mechanistic course has been rationalized in many recent theoretical studies [36, 37].

8.6 Summary

A variety of biologically active natural products were synthesized via the hypervalent iodine-induced asymmetric dearomatization of phenols, which has become increasingly valuable and beneficial to the practical productions of chemicals, pharmaceutical drugs, and agrochemicals owing to their versatility, mild reactivity, availability, high yields, and safety. The history and recent investigations of these hypervalent iodine reagents in asymmetric phenol dearomative couplings, especially for the catalytic cases involving specific chiral hypervalent iodine(III) species, were introduced in this chapter. The metal catalyst-free coupling studies employing the organocatalysis of hypervalent iodine species combined with stoichiometric amounts of practical terminal oxidants are now significant in this field. The first example of asymmetric dearomative spirolactonization was accomplished with a chiral hypervalent iodine catalyst in 2008. Afterward, the scope of catalytic asymmetric spirocyclization through the formation of dearomative bonds was extended to transition-metal catalysis and other organocatalysis employing many other substrates. Additionally, the alternative role of hypervalent iodine reagents in asymmetric transformations involves their cooperative applications with asymmetric catalysts. Inspired by hypervalent-iodine catalysis, many new conceptual chiral organoiodine catalysts, such as chiral ammonium iodides and halogen-bonding iodine catalysts, have also emerged recently.

The prospective role of hypervalent iodine reagents in the twentieth century involves the simple replacement of heavy-metal oxidants, which transcends being mere alternatives to the metal catalysts. Their unique reactivities and selectivities have been pioneered. Based on such contributions, the chemistry of hypervalent iodine will continue to exert significant impacts in future studies.

References

1 Kishi, Y., Nakatsuka, S., Fukuyama, T., and Havel, M. (1973). Total synthesis of sporidesmin A. *J. Am. Chem. Soc.* 95: 6493–6495.

2 (a) Szantáy, C., Blaskó, G., and Bárczai-Beke, M. (1980). Studies aiming at the synthesis of morphine II: studies on phenolic coupling of *N*-norreticuline derivatives. *Tetrahedron Lett.* 21: 3509–3512. (b) White, J.D., Chong, W.K.M., and Thirring, K. (1983). Phenolic oxidative coupling with hypervalent iodine. A synthesis of 6a-epipretazettine. *J. Org. Chem.* 48: 2300–2302. (c) White, J.D., Caravatti, G., Kline, T.B. et al. (1983). Biomimetic total synthesis of (−)-codeine. *Tetrahedron* 39: 2393–2397.

3 (a) Tamura, Y., Yakura, T., Haruta, J., and Kita, Y. (1987). Hypervalent iodine oxidation of *p*-alkoxyphenols and related compounds: a general route to *p*-benzoquinone monoacetals and spiro lactones. *J. Org. Chem.* 52: 3927–3930. (b) Kita, Y., Yakura, T., Tohma, H. et al. (1989). A synthetic approach to discorhabdin alkaloids: hypervalent iodine oxidation of *p*-substituted phenol derivatives to azacarbocyclic spirodienones. *Tetrahedron Lett.* 30: 1119–1120. (c) Kita, Y., Tohma, H., Kikuchi, K. et al. (1991). Hypervalent iodine oxidation of *N*-acyltyramines: synthesis of quinol ethers, spirohexadienones, and hexahydroindol-6-ones. *J. Org. Chem.* 56: 435–438. (d) Kita, Y., Takada, T., Ibaraki, M. et al. (1996). An intramolecular cyclization of phenol derivatives bearing aminoquinones using a hypervalent iodine reagent. *J. Org. Chem.* 61: 223–227.

4 For early accounts on utility of fluoroalcohols in organic synthesis, see: (a) Kita, Y., Takada, T., Tohma, H. (1996). Hypervalent iodine reagents in organic synthesis: nucleophilic substitution of *p*-substituted phenol ethers. *Pure Appl. Chem.* 68: 627-630; (b) Eberson, L., Hartshorn, M.P., Persson, O., and Radner, F. (1996). Making Radical Cations Live Longer. *Chem. Commun.*: 2105–2112. (c) Bégué, J., Bonnetdelpon, D., and Crousse, B. (2004). Fluorinated alcohols: a new medium for selective and clean reaction. *Synlett* 2004: 18–29. (d) Dohi, T., Yamaoka, N., and Kita, Y. (2010). Fluoroalcohols: versatile solvents in hypervalent iodine chemistry and syntheses of diaryliodonium(III) salts. *Tetrahedron* 66: 5775–5785.

5 (a) Kita, Y., Tohma, H., Inagaki, M. et al. (1992). Total synthesis of discorhabdin C: a general aza spiro dienone formation from *O*-silylated phenol derivatives using a hypervalent iodine reagent. *J. Am. Chem. Soc.* 114: 2175–2180. (b) Tohma, H., Harayama, Y., Hashizume, M. et al. (2002). Synthetic studies on the sulfur-cross-linked core of antitumor marine alkaloid, discorhabdins: total synthesis of discorhabdin A. *Angew. Chem. Int. Ed.* 41: 348–350. (c) Tohma, H., Harayama, Y., Hashizume, M. et al. (2003). The first total synthesis of discorhabdin A. *J. Am. Chem. Soc.* 125: 11235–11240. (d) Wada, Y., Otani, K., Endo, N. et al. (2009). Synthesis of antitumor marine alkaloid discorhabdin A oxa analogues. *Org. Lett.* 11: 4048–4050.

6 (a) Kita, Y., Takada, T., Gyoten, M. et al. (1996). An oxidative intramolecular phenolic coupling reaction for the synthesis of amaryllidaceae alkaloids using a hypervalent iodine(III) reagent. *J. Org. Chem.* 61: 5857–5864. (b) Kita, Y., Arisawa, M., Gyoten, M. et al. (1998). Oxidative intramolecular phenolic coupling reaction induced by a hypervalent iodine(III) reagent: leading to galanthamine-type amaryllidaceae alkaloids. *J. Org. Chem.* 63: 6625–6633. (c) Account: Harayama, Y., Kita, Y. (2005). Pyrroloiminoquinone alkaloids. Discorhabdins and makakuvamines. *Curr. Org. Chem.* 9: 1567–1588.

7 Accounts and reviews: (a) Kita, Y., Tohma, H., Yakura, T. (1992). Hypervalent iodine reagents in organic synthesis: development of novel reactions and their application to biologically active natural products. *Trends Org. Chem.* 3: 113-128; (b) Tohma, H. and Kita, Y. (2003). Synthetic applications (total synthesis and natural product synthesis). *Top. Curr. Chem.* 224: 209–248. (c) Dohi, T., Ito, M., Yamaoka, N. et al. (2009). Hypervalent iodine(III): selective and efficient single-electron-transfer (SET) oxidizing agent. *Tetrahedron* 65: 10797–10815. (d) Dohi, T. and Kita, Y. (2016). Hypervalent iodine-induced oxidative couplings (new metal-free coupling advances and their applications in natural product syntheses). *Top. Curr. Chem.* 373: 1–24. (e) Dohi, T. and Kita, Y. (2018). Oxidative C–C bond formations (couplings, cyclizations, cyclopropanaton, etc.). In: *Hypervalent Halogen Compounds, Patai's Chemistry and Functional Groups, Chichester* (eds. B. Olofsson, I. Marek and Z. Rappoport). UK: John Wiley & Sons Ltd.

8 Zhdankin, V.V. (2014). *Hypervalent Iodine Chemistry: Preparation, Structure, and Synthetic Applications of Polyvalent Iodine Compounds.* John Wiley & Sons Ltd: Chichester, UK.

9 Kaiho, T. (ed.) (2015). *Iodine Chemistry and Applications.* Hoboken, New Jersey: John Wiley & Sons Ltd.

10 Olofsson, B., Marek, I., Rappoport, Z., editors. *The Chemistry of Hypervalent Halogen Compounds*, Patai's Chemistry and Functional Groups, Chichester, UK: John Wiley & Sons Ltd, 2018.

11 Wirth, T. (ed.) (2016). *Hypervalent Iodine Chemsitry.* Nettetal, Nordrhein-Westfalen, Germany: Springer GmbH.

12 For early commentaries, see: (a) Ngatimin, M., Lupton, D. W. (2010). The discovery of catalytic enantioselective polyvalent iodine mediated reactions. *Aust. J. Chem.* 63: 653-658; (b) Liang, H. and Ciufolini, M.A. (2011). Chiral hypervalent iodine reagents in asymmetric reactions. *Angew. Chem., Int. Ed.* 50:

11849–11851. (c) Parra, A. and Reboredo, S. (2013). Chiral hypervalent iodine reagents: synthesis and reactivity. *Chem. Eur. J.* 19: 17244–17260.

13 Imamoto, T. and Koto, H. (1986). Asymmetric oxidation of sulfides to sulfoxides with trivalent iodine reagents. *Chem. Lett.* 15 (6): 967–968.

14 Hatzigrigoriou, E., Varvoglis, A., and Bakola-Christianopoulou, M. (1990). Preparation of [Hydroxy(((+)-10-camphorsulfonyl)oxy)iodo]benzene and its reactivity toward carbonyl compounds. *J. Org. Chem.* 55: 315–318.

15 (a) Ray, D.G. and Koser, G.F. (1990). Iodinanes with iodine(III)-bound homochiral alkoxy ligands: preparation and utility for the synthesis of alkoxysulfonium salts and chiral sulfoxides. *J. Am. Chem. Soc.* 112: 5672–5673. (b) Ray, D.G. and Koser, G.F. (1992). Iodinanes with chiral ligands. Synthesis and structure of iodine(III) dibenzoyl tartrates. *J. Org. Chem.* 57: 1607–1610. (c) Ray, D.G. and Koser, G.F. (1996). Facile synthetic entry into the 1,3-dihydro-3-methyl-3-phenyl-1,2-benziodoxole family of λ^3-iodanes. *Tetrahedron Lett.* 37: 6453–6456.

16 (a) Wirth, T. and Hirt, U.H. (1997). Chiral hypervalent iodine compounds. *Tetrahedron: Asymmetry* 8: 23–26. (b) Hirt, U.H., Spingler, B., and Wirth, T. (1998). New chiral hypervalent iodine compounds in asymmetric synthesis. *J. Org. Chem.* 63: 7674–7679. (c) Hirt, U.H., Schuster, M.F.H., French, A.N. et al. (2001). Chiral hypervalent organo-iodine(III) compounds. *Eur. J. Org. Chem.* 2001: 1569–1579.

17 Ochiai, M., Kitagawa, Y., Takayama, N. et al. (1999). Synthesis of chiral diaryliodonium salts, 1,1′-binaphthyl-2-yl(phenyl)iodonium tetrafluoroborates: asymmetric α-phenylation of β-keto ester enolates. *J. Am. Chem. Soc.* 121: 9233–9234.

18 Ladziata, U., Carlson, J., and Zhdankin, V.V. (2006). Synthesis and oxidative reactivity of new chiral hypervalent iodine(V) reagents based on (*S*)-proline. *Tetrahedron Lett.* 47: 6301–6304.

19 (a) Tohma, H., Takizawa, S., Watanabe, H., and Kita, Y. (1998). Hypervalent iodine(III) oxidation catalyzed by quaternary ammonium salt in micellar systems. *Tetrahedron Lett.* 39: 4547–4550. (b) Tohma, H., Takizawa, S., Watanabe, H. et al. (1999). Hypervalent iodine(V)-induced asymmetric oxidation of sulfides to sulfoxides mediated by reversed micelles: novel nonmetallic catalytic system. *J. Org. Chem.* 64: 3519–3523. (c) Tohma, H., Takizawa, S., Morioka, H. et al. (2000). Novel catalytic asymmetric sulfoxidation in water using the hypervalent iodine reagent iodoxybenzene. *Chem. Pharm. Bull.* 48: 445–446.

20 (a) Richardson, R.D., Page, T.K., Altermann, S. et al. (2007). Enantioselective α-oxytosylation of ketones catalysed by iodoarenes. *Synlett* 2007: 538–542. (b) For a recent work on this catalytic reaction, see: Guilbault, A. A., Basdevant, B., Wanie, V., Legault, C. Y. (2012). Catalytic enantioselective α-tosyloxylation of ketones using iodoaryloxazoline catalysts: insights on the stereoinduction process. *J. Org. Chem.* 77: 11283-11295.

21 (a) Fujita, M., Okuno, S., Lee, H.J. et al. (2007). Enantiodifferentiating tetrahydrofuranylation of but-3-enyl carboxylates using optically active hypervalent iodine(III) reagents *via* a 1,3-dioxan-2-yl cation intermediate. *Tetrahedron Lett.*

48: 8691–8694. (b) Fujita, M., Yoshida, Y., Miyata, K. et al. (2010). Enantiodifferentiating *endo*-selective oxylactonization of *ortho*-alk-1-enylbenzoate with a lactate-derived aryl-λ^3-iodane. *Angew. Chem., Int. Ed.* 49: 7068–7071.

22 (a) Kürti, L., Herczegh, P., Visy, J. et al. (1999). New insights into the mechanism of phenolic oxidation with phenyliodonium(III) reagents. *J. Chem. Soc., Perkin Trans.* 1: 379–380. (b) Pelter, A. and Ward, R.S. (2001). Two-electron phenolic oxidations using phenyliodonium dicarboxylates. *Tetrahedron* 57: 273–282.

23 (a) Dohi, T., Maruyama, A., Takenaga, N. et al. (2008). A chiral hypervalent iodine(III) reagent for enantioselective dearomatization of phenols. *Angew. Chem. Int. Ed.* 47: 3787–3790. (b) Dohi, T., Takenaga, N., Nakae, T. et al. (2013). Asymmetric dearomatizing spirolactonization of naphthols catalyzed by spirobiindane-based chiral hypervalent iodine species. *J. Am. Chem. Soc.* 135: 4558–4566.

24 For recent reviews and accounts on asymmetric hypervalent iodine catalysis, see: (a) Parra, A. (2019). Chiral hypervalent iodines: active players in asymmetric synthesis. *Chem. Rev.* 119: 12033-12088; (b) Flores, A., Cots, E., Berges, J., and Muñiz, K. (2019). Enantioselective iodine(I/III) catalysis in organic synthesis. *Adv. Synth. Catal.* 361: 2–25. (c) Claraz, A. and Masson, G. (2018). Asymmetric iodine catalysis-mediated enantioselective oxidative transformations. *Org. Biomol. Chem.* 16: 5386–5402. (d) Ghosh, S., Pradhan, S., and Chatterjee, I. (2018). A survey of chiral hypervalent iodine reagents in asymmetric synthesis. *Beilstein J. Org. Chem.* 14: 1244–1262. (d) Gao, G.L. and Chen, Q. (2017). Recent advances in utilities of active iodine reagents as organo-catalysts in organic synthesis. *Curr. Organocat.* 4: 33–47. (e) Quideau, S., Pouységu, L., Peixoto, P.A., and Deffieux, D. (2016). Phenol dearomatization with hypervalent iodine reagents. *Topics in Current Chemistry* 373: 25–74. (f) Kumar, R. and Wirth, T. (2016). Asymmetric synthesis with hypervalent iodine reagents. *Top. Curr. Chem.* 373: 243–261.

25 For recent reviews dealing with asymmetric daromatizations, see: (a) An, J., Bandini, M. (2020). Recent advances in the catalytic dearomatization of naphthols. *Eur. J. Org. Chem.* 2020 4087-4097; (b) Kikushima, K., China, H., and Dohi, T. (2019). Asymmetric construction of heterocycles *via* dearomative coupling and addition reactions of phenol and aniline derivatives. *Heterocycles* 98: 1489–1511. (c) Shi, L., Zhang, W., Chen, S. et al. (2018). The role of *ortho*-dearomatization reaction in constructing spirocyclic scaffolds with an all-carbon ring junction. *Curr. Org. Synth.* 15: 904–923. (d) Sun, W., Li, G., Hong, L., and Wang, R. (2016). Asymmetric dearomatization of phenols. *Org. Biomol. Chem.* 14: 2164–2176. (e) Pantaine, L., Moreau, X., Coeffard, V., and Greck, C. (2016). The impact of asymmetric organocatalysis in dearomatization and aromatization of carbocycles: increasing molecular complexity and diversity. *Tetrahedron Lett.* 57: 2567–2574. (f) Wu, W.T., Zhang, L., and You, S.L. (2016). Catalytic asymmetric dearomatization (CADA) reactions of phenol and aniline derivatives. *Chem. Soc. Rev.* 45: 1570–1580. (g) Roche, S.P. and Porco, J.A. Jr., (2011). Dearomatization strategies in the synthesis of complex natural products. *Angew. Chem. Int. Ed.* 50: 4068–4093.

26 (a) Takenaga, N., Uchiyama, T., Kato, D. et al. (2011). Efficient phenolic oxidations to construct *ortho*-spirolactone structures using oxo-bridged hypervalent iodine(III) compound. *Heterocycles* 82: 1327–1336. (b) Dohi, T., Uchiyama, T., Yamashita, D. et al. (2011). Efficient phenolic oxidations using μ-oxo-bridged phenyliodine trifluoroacetate. *Tetrahedron Lett.* 52: 2212–2215. (c) Dohi, T., Takenaga, N., Fukushima, K. et al. (2010). Designer μ-oxo-bridged hypervalent iodine(III) organocatalysts for greener oxidations. *Chem. Commun.* 46: 7697–7699. (d) Dohi, T., Kato, D., Hyodo, R. et al. (2011). Discovery of stabilized bisiodonium salts as intermediates in the carbon-carbon bond formation of alkynes. *Angew. Chem. Int. Ed.* 50: 3784–3787.

27 Willgerodt, C. (1886). Űber einige aromatische jodidchloride. *J. Prakt. Chem.* 33: 154–160.

28 Dohi, T., Maruyama, A., Yoshimura, M. et al. (2005). Versatile hypervalent-iodine(III)-catalyzed oxidations with *m*-chloroperbenzoic acid as a cooxidant. *Angew. Chem. Int. Ed.* 44: 6193–6196.

29 Ochiai, M., Takeuchi, Y., Katayama, T. et al. (2005). Iodobenzene-catalyzed α-acetoxylation of ketones. *In Situ* generation of hypervalent (diacyloxyiodo)benzenes using *m*-chloroperbenzoic acid. *J. Am. Chem. Soc.* 127: 12244–12245.

30 Thottumkara, A.P., Bowsher, M.S., and Vinod, T.K. (2005). *In Situ* generation of *o*-iodoxybenzoic acid (IBX) and the catalytic use of it in oxidation reactions in the presence of oxone as a co-oxidant. *Org. Lett.* 7: 2933–2936.

31 For reviews and accounts, see: (a) Richardson, R.D., Wirth, T. (2006). Hypervalent iodine goes catalytic. *Angew. Chem. Int. Ed.* 45: 4402-4404; (b) Ochiai, M. and Miyamoto, K. (2008). Catalytic version of and reuse in hypervalent organo-λ^3- and -λ^5-iodane oxidation. *Eur. J. Org. Chem.*: 4429–4239. (c) Dohi, T. and Kita, Y. (2009). Hypervalent iodine reagents as a new entrance to organocatalysts. *Chem. Commun.* 2009: 2073–2085. (d) Uyanik, M. and Ishihara, K. (2009). Hypervalent iodine-mediated oxidation of alcohols. *Chem. Commun.*: 2086–2099. (e) Yusubov, M.S. and Zhdankin, V.V. (2010). Development of new recyclable reagents and catalytic systems based on hypervalent iodine compounds. *Mendeleev Commun.* 20: 185–191.

32 (a) Tohma, H., Maruyama, A., Maeda, A. et al. (2004). Preparation and reactivity of 1,3,5,7-tetrakis[4- (diacetoxyiodo)phenyl]adamantane, a recyclable hypervalent iodine(III) reagent. *Angew. Chem., Int. Ed.* 43: 3595–3598. (b) Dohi, T., Fukushima, K., Kamitanaka, T. et al. (2012). An excellent dual recycling strategy for the hypervalent iodine/nitroxyl radical mediated selective oxidation of alcohols to aldehydes and ketones. *Green Chem.* 14: 1493–1501.

33 (a) Ding, A., Meazza, M., Guo, H. et al. (2018). New development in the enantioselective synthesis of spiro compounds. *Chem. Soc. Rev.* 47: 5946–5996. (b) Quintavalla, A. (2018). Spirolactones: recent advances in natural products, bioactive compounds and synthetic strategies. *Curr. Med. Chem.* 25: 917–962. (c) Reddy, C.R., Prajapti, S.K., Warudikar, K. et al. (2017). *ipso*-Cyclization: an emerging tool for multifunctional spirocyclohexadienones. *Org. Biomol. Chem.*

15: 3130–3151. (d) Moschitto, M.J., Anthony, D.R., and Lewis, C.A. (2015). Syntheses of arnottin I and arnottin II. *J. Org. Chem.* 80: 3339–3342. (e) Rios, R. (2012). Enantioselective methodologies for the synthesis of spiro compounds. *Chem. Soc. Rev.* 41: 1060–1074.

34 Bekkaye, M. and Masson, G. (2016). Synthesis of new axially chiral iodoarenes. *Synthesis* 48: 302–312.

35 Dohi, T., Sasa, H., Miyazaki, K. et al. (2017). Chiral atropisomeric 8, 8′-diiodobinaphthalene for asymmetric dearomatizing spirolactonizations in hypervalent iodine oxidations. *J. Org. Chem.* 82: 11954–11960.

36 Zheng, H., Sang, Y., Houk, K.N. et al. (2019). Mechanism and origins of enantioselectivities in spirobiindane-based hypervalent iodine(III)-induced asymmetric dearomatizing spirolactonizations. *J. Am. Chem. Soc.* 141: 16046–16056.

37 (a) Zheng, H. and Xue, X.S. (2020). Recent computational studies on mechanisms of hypervalent iodine(III)-promoted dearomatization of phenols. *Curr. Org. Chem.* 24: 2106–2117. (b) Ganji, B. and Ariafard, A. (2019). DFT mechanistic investigation into phenol dearomatization mediated by an iodine(III) reagent. *Org. Biomol. Chem.* 17: 3521–3528. (c) Tang, T. and Harned, A.M. (2018). Experimental evidence for the formation of cationic intermediates during iodine(III)-mediated oxidative dearomatization of phenols. *Org. Biomol. Chem.* 16: 6871–6874. (d) Harned, A.M. (2018). Concerning the mechanism of iodine(III)-mediated oxidative dearomatization of phenols. *Org. Biomol. Chem.* 16: 2324–2329. (e) Harned, A.M. (2014). Asymmetric oxidative dearomatizations promoted by hypervalent iodine(III) reagents: an opportunity for rational catalyst design? *Tetrahedron Lett.* 55: 4681–4689.

38 (a) Uyanik, M., Yasui, T., and Ishihara, K. (2010). Enantioselective Kita oxidative spirolactonization catalyzed by *in situ* generated chiral hypervalent iodine(III) species. *Angew. Chem. Int. Ed.* 49: 2175–2177. (b) Uyanik, M., Yasui, T., and Ishihara, K. (2010). Chiral hypervalent iodine-catalyzed enantioselective oxidative Kita spirolactonization of 1-naphthol derivatives and one-pot diastereoselective oxidation to epoxyspirolactones. *Tetrahedron* 66: 5841–5851.

39 Uyanik, M., Yasui, T., and Ishihara, K. (2017). Chiral hypervalent organoiodine-catalyzed enantioselective oxidative spirolactonization of naphthol derivatives. *J. Org. Chem.* 82: 11946–11953.

40 Uyanik, M., Yasui, T., and Ishihara, K. (2013). Hydrogen bonding and alcohol effects in asymmetric hypervalent iodine catalysis: enantioselective oxidative dearomatization of phenols. *Angew. Chem. Int. Ed.* 52: 9215–9218.

41 Uyanik, M., Sasakura, N., Mizuno, M., and Ishihara, K. (2017). Enantioselective synthesis of masked benzoquinones using designer chiral hypervalent organoiodine(III) catalysis. *ACS Catal.* 7: 872–876.

42 Suzuki, T., Watanabe, S., Uyanik, M. et al. (2018). Asymmetric total synthesis of (−)-maldoxin, a common biosynthetic ancestor of the chloropupukeananin family. *Org. Lett.* 20: 3919–3922.

43 Jain, N., Xu, S., and Ciufolini, M.A. (2017). Asymmetric oxidative cycloetherification of naphtholic alcohols. *Chem. Eur. J.* 23: 4542–4546.

44 Jain, N., Hein, J.E., and Ciufolini, M.A. (2019). Oxidative cyclization of naphtholic sulfonamides mediated by a chiral hypervalent iodine reagent: asymmetric synthesis versus resolution. *Synlett* 30: 1222–1227.

45 Zhang, D.Y., Xu, L., Wu, H., and Gong, L.Z. (2015). Chiral iodine-catalyzed dearomatizative spirocyclization for the enantioselective construction of an all-carbon stereogenic center. *Chem. Eur. J.* 21: 10314–10317.

46 (a) Murray, S.J., Mueller-Bunz, H., and Ibrahim, H. (2012). Congested C_2-symmetric aryliodanes based on an *anti*-dimethanoanthracene backbone. *Chem. Commun.* 48: 6268–6270. (b) Murray, S.J. and Ibrahim, H. (2015). Asymmetric kita spirolactonisation catalysed by *anti*-dimethanoanthracene-based iodoarenes. *Chem. Commun.* 51: 2376–2379.

47 Imrich, M.R., Biehler, L.E., Maichle-Moessmer, C., and Ziegler, T. (2019). Carbohydrate-based chiral iodoarene catalysts: a survey through the development of an improved catalyst design. *Molecules* 24: 3883–3901.

48 Wang, Y., Zhao, C.Y., Wang, Y.P., and Zheng, W.H. (2019). Enantioselective intramolecular dearomative lactonization of naphthols catalyzed by planar chiral iodoarene. *Synthesis* 51: 3675–3682.

49 Yang, G.H., Zheng, H., Li, X., and Cheng, J.P. (2020). Asymmetric synthesis of axially chiral phosphamides via atroposelective *N*-allylic alkylation. *ACS Catal.* 10: 2324–2333.

50 Antien, K., Pouységu, L., Deffieux, D. et al. (2019). Synthesis of [7]helicene enantiomers and exploratory study of their conversion into helically chiral iodoarenes and iodanes. *Chem. Eur. J.* 25: 2852–2858.

51 Ogasawara, M., Sasa, H., Hu, H. et al. (2017). Atropisomeric chiral diiododienes (*Z,Z*)-2,3-di(1-iodoalkylidene)tetralins: synthesis, enantiomeric resolution, and application in asymmetric catalysis. *Org. Lett.* 19: 4102–4105.

52 (a) Hempel, C., Maichle-Mössmer, C., Pericàs, M.A., and Nachtsheim, B.J. (2017). Modular synthesis of triazole-based chiral iodoarenes for enantioselective spirocyclizations. *Adv. Synth. Catal.* 359: 2931–2941. (b) Abazid, A.H. and Nachtsheim, B.J. (2020). A triazole-substituted aryl iodide with omnipotent reactivity in enantioselective oxidations. *Angew. Chem. Int. Ed.* 59: 1479–1484.

53 Volp, K.A. and Harned, A.M. (2013). Chiral aryl iodide catalysts for the enantioselective synthesis of *para*-quinols. *Chem. Commun.* 49: 3001–3003.

54 Ding, Q., He, H., and Cai, Q. (2018). Chiral aryliodine-catalyzed asymmetric oxidative C–N bond formation *via* desymmetrization strategy. *Org. Lett.* 20: 4554–4557.

55 Boppisetti, J.K. and Birman, V.B. (2009). Asymmetric oxidation of *o*-alkylphenols with chiral 2-(*o*-iodoxyphenyl)-oxazolines. *Org. Lett.* 11: 1221–1223.

56 (a) Quideau, S., Lyvinec, G., Marguerit, M. et al. (2009). Asymmetric hydroxylative phenol dearomatization through *in situ* generation of iodanes from chiral iodoarenes and *m*-CPBA. *Angew. Chem. Int. Ed.* 48: 4605–4609. (b) Bosset, C., Coffinier, R., Peixoto, P.A. et al. (2014). Asymmetric hydroxylative phenol dearomatization promoted by chiral binaphthylic and biphenylic iodanes. *Angew. Chem. Int. Ed.* 53: 9860–9864.

57 Muñiz, K. and Fra, L. (2017). Enantioselective 4-hydroxylation of phenols under chiral organoiodine(I/III) catalysis. *Synthesis* 49: 2901–2906.
58 Hashimoto, T., Shimazaki, Y., Omatsu, Y., and Maruoka, K. (2018). Indanol-based chiral organoiodine catalysts for enantioselective hydrative dearomatization. *Angew. Chem. Int. Ed.* 57: 7200–7204.
59 Deng, Q., Xia, W., Hussain, M.I. et al. (2020). Synthesis of polycyclic cyclohexadienone through alkoxy-oxylactonization and dearomatization of 3′-Hydroxy-[1,1′-biphenyl]-2-carboxylic acids promoted by hypervalent iodine. *J. Org. Chem.* 85: 3125–3133.

9

Catalytic Alkene Difunctionalization Reactions

John M. Ovian and Eric N. Jacobsen

Department of Chemistry and Chemical Biology, Harvard University, Cambridge, Massachusetts, United States

9.1 Introduction

The discovery of selective transformations of alkenes has played a central role in the advancement of organic synthesis. A continuously growing list of reagents and reactions has been developed to leverage the reactivity of the C_{sp2}–C_{sp2} π-bond in valuable bond constructions. Further, the adjacent prochiral centers of the π-system provide a platform for the efficient generation of stereochemical complexity through stereospecific and stereoselective pathways, and identification of alkene addition reactions that proceed with a high degree of relative and/or absolute stereocontrol has stood as an important objective throughout the history of the field [1–5].

Alkene difunctionalization reactions involve the installation of two functional groups either vicinally, or geminally with rearrangement, in a single transformation and are particularly potent from the standpoint of complexity generation. Traditionally, transition-metal catalysis has played a dominant role in alkene difunctionalization chemistry, especially with respect to enantioselective processes [6]. However, over the past several years, advances in asymmetric hypervalent iodine chemistry have introduced several attractive complements to transition-metal catalysis. Aryl iodide organocatalysts are generally readily accessible and tunable and have been applied successfully in a diverse array of enantioselective alkene difunctionalization reactions [7–10].

The focus of this chapter will be on enantioselective alkene difunctionalization reactions catalyzed by organoiodides. While certain important precedents of transformations promoted by stoichiometric hypervalent iodine reagents will also be discussed, the reader is referred to the many excellent books, chapters, and reviews that have been published for more comprehensive treatments of that topic [11–23].

9.2 General Mechanistic Considerations

Alkene difunctionalization reactions mediated by hypervalent iodide reagents are generally proposed to proceed via dissociative substitution to generate an

Iodine Catalysis in Organic Synthesis, First Edition. Edited by Kazuaki Ishihara and Kilian Muñiz.

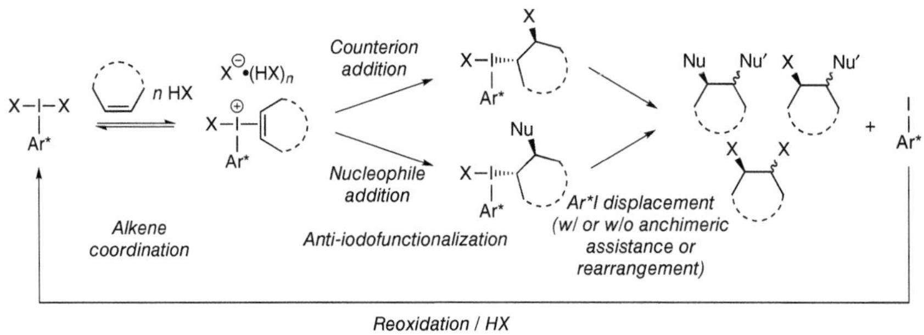

Figure 9.1 General mechanistic considerations in iodine(III)-catalyzed alkene difunctionalization.

electrophilic π-complex [24–27], with the possibility of rate acceleration by Brønsted or Lewis acids (Figure 9.1) [26–30]. *Anti*-addition of either the dissociated counterion or an external nucleophile to the complexed alkene affords an alkyl iodane intermediate in what has been proposed to represent the first irreversible, and, therefore, enantiodetermining step [31, 32]. The diverse array of transformations that will be described in this chapter leverages the extraordinary leaving group ability of the I(III) center in the alkyl iodane, which is on the order of 10^6 times that of a triflate [33, 34], to enable a wide assortment of substitution or rearrangement pathways leading ultimately to the difunctionalized products. The synthetic utility of these processes is heightened by several practical and fundamental features, including the potential stereospecificity of each of the bond-forming steps; the high electrophilicity of both the π-complex and alkyl iodane intermediates, allowing engagement of weakly nucleophilic addition partners and possible intervention of interesting anchimeric assistance or carbon-framework rearrangement pathways; and the possibility of reoxidizing the reduced aryl iodide by-product under the reaction conditions to regenerate the reactive ArI(III) reagent, thereby permitting the use of the aryl iodide as a catalyst. The rich chemistry of the electrophilic intermediates also gives rise to multiple and sometimes competing reactivity channels, and with few exceptions, most mechanistic analyses of aryl-iodide-catalyzed alkene difunctionalization reactions are primarily rationalizations of observed product and stereochemical outcomes [23]. These analyses may be further complicated by the fact that the terminal oxidant often can oxidize the substrate directly, leading to either the desired product in a racemic form or competing side-products. The state of mechanistic insight into the various transformations described in this chapter will be presented as appropriate.

9.3 Oxyfunctionalization

Vicinal dioxygenation and oxyfunctionalization reactions constitute the most well-established and thoroughly explored enantioselective hypervalent-iodine-catalyzed alkene difunctionalization reactions. In a seminal 1997 report, Wirth

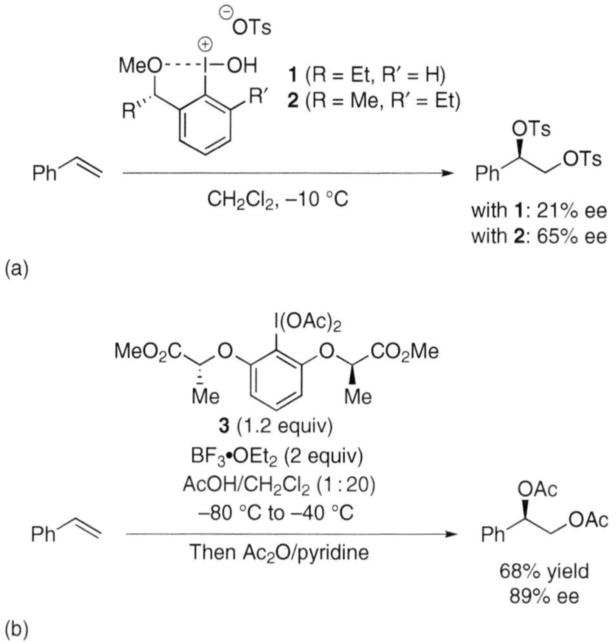

Figure 9.2 Key precedents in enantioselective intermolecular vicinal dioxygenations of alkenes promoted by stoichiometric iodine(III) reagents. (a) Styrene ditosylation developed by Wirth. (b) Styrene diacetoxylation reaction developed by Fujita.

and coworkers reported that the chiral iodine(III) reagent **1** induced the ditosylation of styrene in up to 21% enantiomeric excess (ee) (Figure 9.2a) [35]. Subsequently, this framework was optimized through variation of the 6-position on the iodoarene to reagent **2** to afford up to 65% ee in the same reaction [36]. Fujita subsequently achieved the enantioselective diacetoxylation of styrene [37] using the resorcinol-derived reagent **3** (Figure 9.2b), a representative of a class of C_2-symmetric aryl iodide derivatives first developed by Ishihara and now commonly known as Ishihara-type catalysts [38].

Drawing on the Wirth and Fujita precedents, Muñiz and coworkers achieved the catalytic vicinal diacetoxylation of styrenes using peracetic acid as the stoichiometric oxidant (Figure 9.3) [39]. The amide-substituted Ishihara-type catalyst **4** was found to afford the highest enantioselectivities. The secondary amide protons (N–Hs) were seen to engage in hydrogen-bonding interactions with the acetoxy ligands in the crystal structure of the active $ArI(OAc)_2$ intermediate, resulting in a well-defined helical environment at the I(III) reaction site. Substrates bearing slightly electron-donating to strongly electron-withdrawing substituents at the *ortho*, *meta*, and *para* positions participated in the reaction and provided the difunctionalized products in up to 94% ee and 94% yield. Co-catalytic triflic acid was required to activate the iodine(III) intermediate and minimize the formation of the racemic product arising from initial direct substrate epoxidation. The choice of peracetic acid as the stoichiometric oxidant was important, as *m*-CPBA produced greater amounts of direct substrate

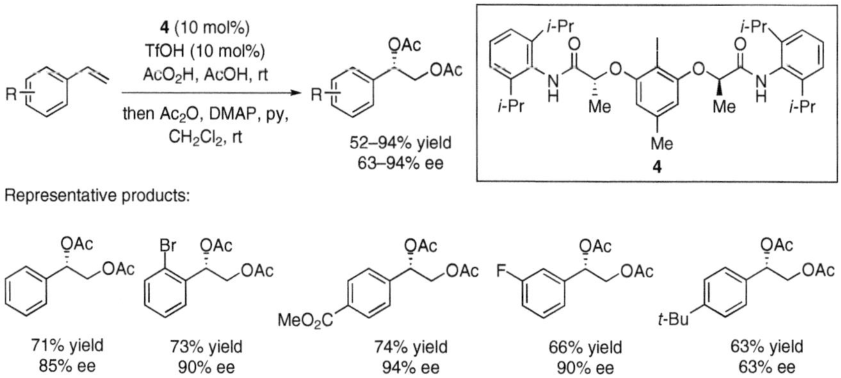

Figure 9.3 Catalytic, enantioselective diacetoxylation of unsubstituted styrenes reported by Muñiz.

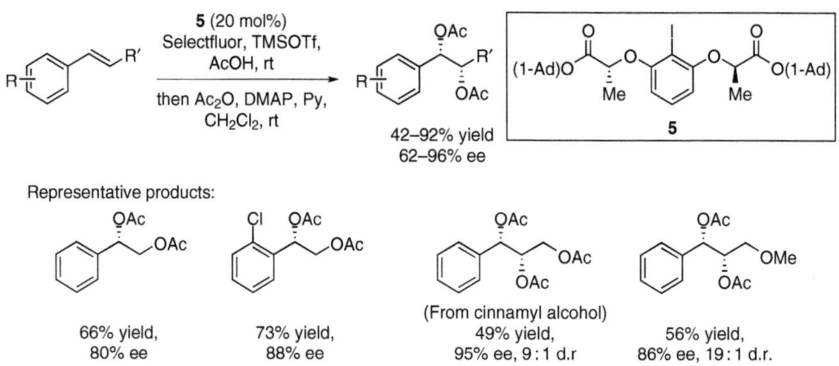

Figure 9.4 Enantioselective catalytic diacetoxylations of styrene derivatives using Selectfluor as the stoichiometric oxidant. 1-Ad = 1-adamantyl.

oxidation products. Two regioisomeric acetoxy alcohols are generated under the catalytic conditions, and treatment of the crude reaction mixtures with acetic anhydride afforded the desired diacetoxy products.

Selectfluor is also commonly used as a stoichiometric oxidant in iodoarene catalysis, as illustrated by the Muñiz laboratory in vicinal diacetoxylations of styrenes using the lactic-acid-derived Ishihara-type catalyst **5** (Figure 9.4) [40]. Additionally, the authors were able to extend the scope to cinnamyl alcohol and cinnamyl ether substrates with high ee albeit in modest yield. As in the initial report, a Brønsted or Lewis acid activator was required to achieve good reactivity, with trimethylsilyl trifluoromethanesulfonate (TMSOTf) proving optimal in this case. A superstoichiometric amount of Lewis acid was required to offset neutralization by the quinuclidine by-product formed from Selectfluor. The adamantyl ester-bearing catalyst **5** was compatible with TMSOTf, whereas the corresponding *tert*-butyl ester derivative was unstable and afforded diminished yields of the desired product.

Both diacetoxylated and hydroxyacetoxylated products are produced as the primary products of the reaction prior to treatment with acetic anhydride, both

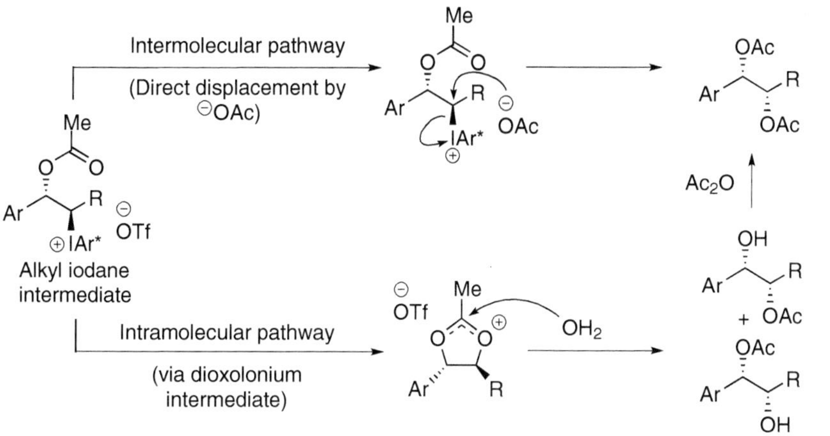

Figure 9.5 Competing pathways in the enantioselective, catalytic diacetoxylation of styrenes.

with the same sense and nearly identical degrees of enantioinduction. To account for these results, the authors proposed participation of two competing pathways proceeding through a common alkyl iodane intermediate, which is formed via enantiodetermining *anti*-acetoxyiodination of the alkene substrate (Figure 9.5). Stereospecific, intermolecular displacement of the iodane by acetate affords the observed syn-diacetoxylated product directly. Alternatively, intramolecular displacement of the iodane by neighboring acetate affords the dioxolonium intermediate, which is opened by water to give the regioisomeric hydroxyacetoxylated products. Subsequent acetylation affords the same *syn*-diacetoxylated product.

Fujita and coworkers developed an extensive array of stoichiometric enantioselective oxylactonization reactions based on chiral hypervalent iodine reagents [41–45], and subsequently reported the first examples of catalytic variants in 2012 [46]. Remarkable transformations involving intramolecular delivery of two different nucleophiles to furnish tetrahydrofuran-fused isochromanones rely on Ishihara-type catalyst **6**, *m*-CPBA as the terminal oxidant, and superstoichiometric trifluoroacetic acid as an activator (Figure 9.6). A catalyst-controlled diastereoselective variant of this method was applied successfully to the synthesis of the polyketide metabolite monocerin.

The same group devised a similar catalytic protocol for the diastereo- and enantioselective synthesis of 4-hydroxyisochromanones, a core motif in a wide variety of polyketide natural products. Drawing on their earlier discovery of stoichiometric endo-selective tosyloxy- and acetoxy-lactonizations with chiral iodoarene diacetate reagents [45], Fujita and coworkers demonstrated that *ortho*-alkenyl benzoate esters undergo net hydroxylactonization under the conditions depicted in Figure 9.7 to afford 4-hydroxyisochromanone products in up to 98% ee and with generally high *endo*/*exo* regioselectivity [47]. Direct oxidation by the stoichiometric *m*-CPBA oxidant was shown to give rise to racemic mixtures of *endo* and *exo* lactonization products possessing *anti*-stereochemistry, presumably via an intermediate epoxide.

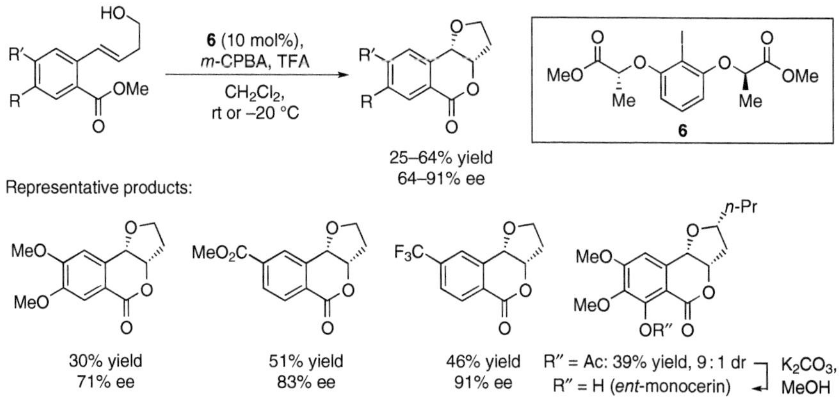

Figure 9.6 Catalytic, enantioselective alkoxylactonization of internal alkenes reported by Fujita.

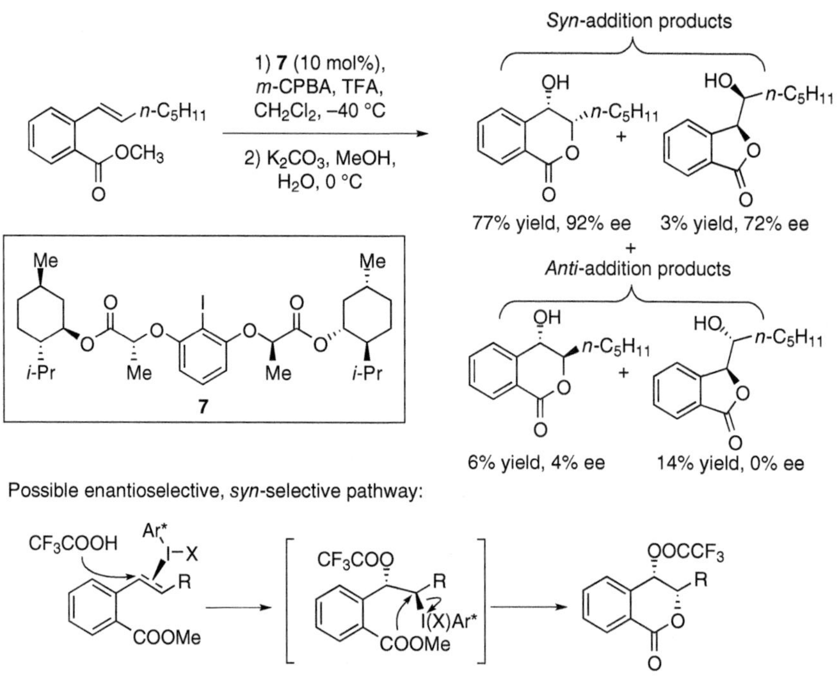

Figure 9.7 Catalytic, enantioselective hydroxylactonization of alkenyl benzoates.

Under the catalytic conditions, the *anti*-diastereomers were also generated in nearly racemic form, in a pathway thus ascribable to uncatalyzed reaction with *m*-CPBA. In contrast, the *syn* diastereomers of the major *endo* and minor *exo* regioisomers were generated enantioselectively via the Ar*I-catalyzed pathway.

Interest in the incorporation of fluorine atoms to modulate pharmacological properties of molecules motivated the exploration of synthetic methods to access 4-fluoro analogs of the 4-hydroxyisochromanone class of biologically active natural

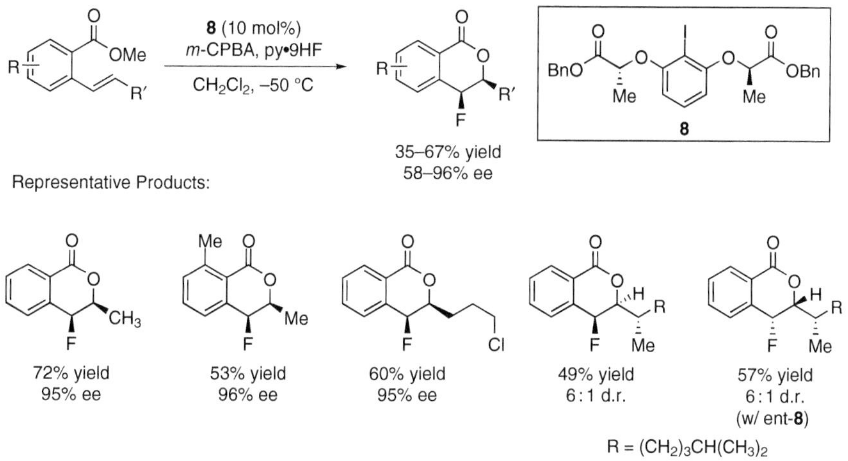

Figure 9.8 Catalytic, enantioselective fluorolactonizations reported by Jacobsen.

products. Fluorolactonization of alkenyl benzoates was achieved by Jacobsen and coworkers using hydrogen fluoride-pyridine (HF-pyridine) as a nucleophilic fluoride source and Ishihara-type catalyst **8** with generally excellent enantioselectivity and high syn diastereocontrol (Figure 9.8) [48]. The protocol using hypervalent iodine catalysis afforded the *endo* cyclization products exclusively, and thus provide a complementary reactivity profile to fluorolactonizations with electrophilic fluorinating reagents such as Selectfluor, which afford γ-butyrolactone products via *exo*-selective cyclization.

A similar protocol was applied using catalyst **9** to highly enantio- and *syn*-diastereoselective 1,2-oxyfluorinations of cinnamyl alcohol and amine derivatives (Figure 9.9) [49]. In each case, neighboring group participation pathways were invoked to account for the observed product and stereochemical outcomes. Thus, benzyl cinnamyl ether **10** was transformed to the 1,3-difluoride **11** through the proposed intermediacy of an oxiranium ion (Figure 9.9a). Allylic carbamate **12** (Figure 9.9b) and acetate **14** (Figure 9.9c) underwent fluorofunctionalization with intramolecular carbonyl trapping to afford the cyclic carbonate **13** and the hydroxy acetate (isolated as bis-acetoxylated product **15**), respectively. In each case, only the *syn*-difunctionalized products were observed. Catalyst **9** has also been applied successfully in a variety of other enantioselective fluorinative reactions of alkenes (discussed in Sections 9.2.1.1–9.2.1.3, Figures 9.16, 9.22, 9.23 and 9.29 [31, 49, 75, 82]). The presence of the para carbomethoxy substituent in **9** was found to increase the robustness of the catalyst to the oxidative conditions at the temperatures required for those reactions (≥ −30 °C), leading to improved enantioselectivities and yields relative to the unsubstituted analog (representative data provided in Table S1 of ref. [82]).

Masson and coworkers extended hypervalent-iodine-catalyzed enantioselective cyclizations to *exo*-selective sulfonyl- and phosphoryloxylactonizations of 4-pentenoic acids to generate γ-butyrolactone products (Figure 9.10) [50]. In the presence of 20 mol% of the *N*-mesityl lactamide Ishihara-type catalyst **16**, aliphatic

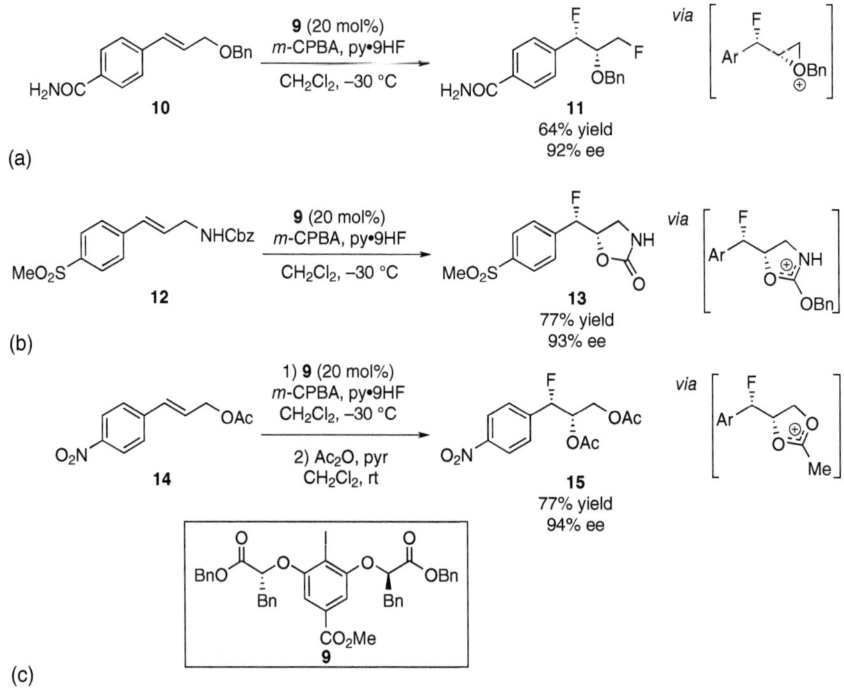

Figure 9.9 Catalytic, enantioselective oxyfluorination reactions. (a) 1,3-difluorination of a benzyl cinnamyl ether with benzyloxy migration. (a) Fluorination of carboxybenzoyl-protected cinnamyl amine to afford the cyclic carbamate product. (c) Fluorination of acetate-protected cinnamyl alcohol yields bis-acetoxylated product after acylation.

acids bearing α,α- or β,β-disubstitution underwent enantioselective oxylactonization in up to 75% ee for sulfonic acid nucleophiles (Figure 9.10a) and 80% ee for phosphoric acid nucleophiles (Figure 9.10b). Higher enantioselectivities were obtained at −45 °C (up to 93% ee); however, the oxidation of the iodoarene was prohibitively slow under those conditions, and a stoichiometric amount of the iodoarene and extended reaction times were required to achieve acceptable yields.

Moran and coworkers applied a similar strategy to forge valuable heterocycles such as oxazolines, dihydrooxazines, and tetrahydrooxazepines in alkene hydroxyfunctionalization reactions [51]. A racemic, catalytic protocol engaging 2-iodoanisole as the catalyst and Selectfluor as the stoichiometric oxidant was identified for the hydroxycyclization of terminal alkenyl benzamides (Figure 9.11a). While 5- and 6-membered-ring products were formed in good yields, a 7-membered ring analog was produced in only 30% yield and more distal alkenes did not undergo cyclization. Moderate enantioselectivity in the cyclization to dihydrooxazines was achieved using Ishihara-type catalyst **17** bearing *N*,*N*-dimethyl lactamide groups (Figure 9.11b). The diisopropyl lactamide analog **18** catalyzed the formation of the corresponding isooxazoline in 69% ee, but only 11% yield (Figure 9.11c).

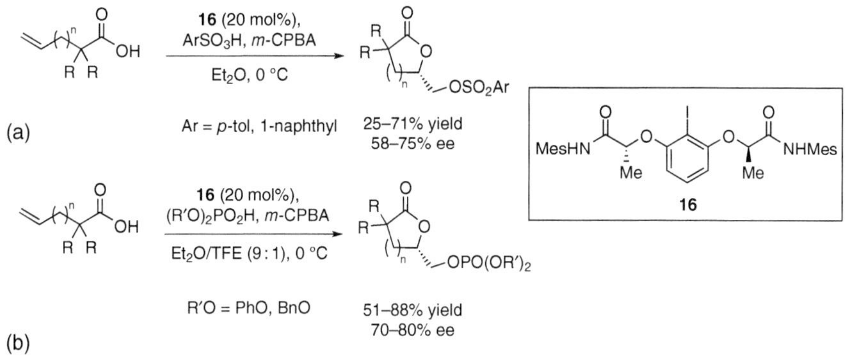

Figure 9.10 (a) Catalytic, enantioselective sulfonyloxylactonization of 4-pentenoic acids. (b) Catalytic, enantioselective phosphoryloxylactonization.

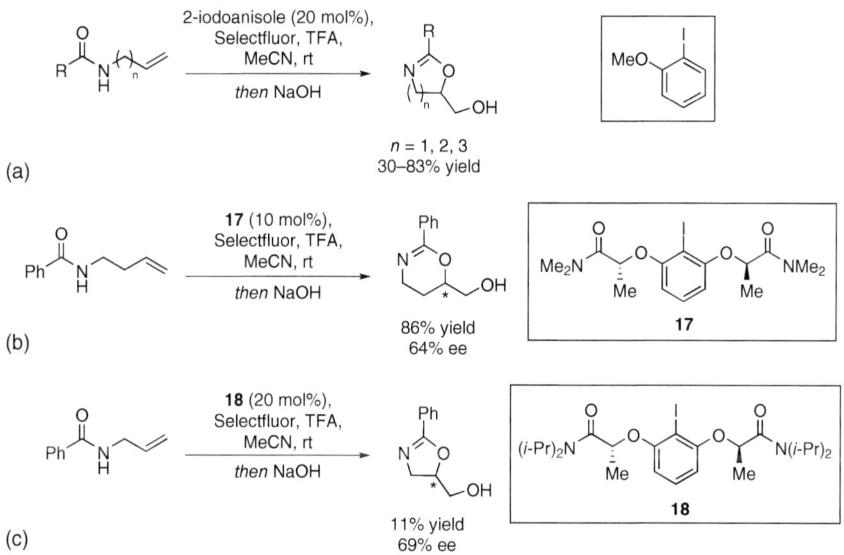

Figure 9.11 (a) Catalytic, racemic cyclization of *N*-alkenyl benzamides. (b) Enantioselective synthesis of dihydrooxazines using a tertiary-amide iodoarene catalyst. (c) Catalytic, enantioselective isooxazoline synthesis from *N*-allyl benzamide.

Following from precedents developed by Wirth involving iodine(III)-promoted asymmetric oxidative rearrangements of chalcones and 1,1-disubstituted alkenes [52–56], Gong and coworkers reported the aryl iodide-catalyzed enantioselective oxidative rearrangement of tertiary allylic alcohols to afford α-carboalkoxy ketones (Figure 9.12a) [57]. Ishihara-type catalysts such as **16** (Figure 9.10) bearing secondary amides did not afford any product; however, analogous tertiary amide derivatives were effective, with the proline-derived **19** providing up to 94% ee and 57% yield in the model reaction. Sterically and electronically diverse alcohol nucleophiles, including trifluoroethanol, *tert*-butanol, and various benzyl alcohols, participated effectively in the tandem rearrangement/oxidation sequence, affording

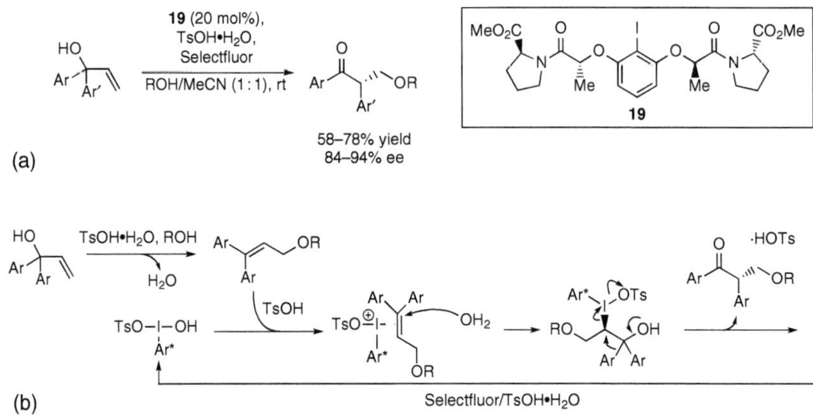

Figure 9.12 (a) Tandem alkoxylation/oxidative rearrangement of 1,1-diaryl allyl alcohols. (b) Proposed mechanism.

the desired products in good yield and high enantioselectivity. Variation of the electronic properties of the aryl rings of the allylic alcohol was demonstrated with little effect on enantioselectivity and yield. With substrates bearing two different arenes at the allylic position, the reactions afforded a mixture of regioisomers with modest selectivity, with both products formed with high enantioselectivity and with the major isomer arising from migration of the more electron-rich arene. Control experiments performed by omitting the iodoarene revealed that the tertiary allylic alcohol substrate undergoes rapid Brønsted acid-catalyzed conversion to the primary allylic ether **20**. Strong evidence was obtained that the alkoxylation step is irreversible, that **20** is the actual substrate for the oxidative rearrangement, and that the intermolecular addition of water provides the source of the carbonyl oxygen. A mechanism proposed by the authors and consistent with the experimental results is depicted in Figure 9.12b.

9.4 Aminofunctionalization

Chiral amines feature prominently among bioactive compounds and FDA-approved pharmaceuticals, and enantioselective diamination and aminofunctionalization of alkenes can provide highly appealing approaches to those targets. In that context, important progress has been achieved in the direct introduction of nitrogen-containing groups via hypervalent-iodine-mediated alkene difunctionalization catalysis. In 2014, Wirth and coworkers reported enantioselective, intramolecular hypervalent-iodine-mediated formal diaminations of alkenes using diaminosulfones and guanidines as nucleophilic *N*-equivalents (Figure 9.13a) [58]. A strongly chelating ligand on the aryliodide reagent was shown to be required for achieving high enantioselectivities, and hypervalent iodine reagent **20** in combination with a 1:1 mixture of boron trifluoride etherate and trimethylsilyl triflate was found to afford access to the cyclized 1,2-diamine precursors in up to 94% ee.

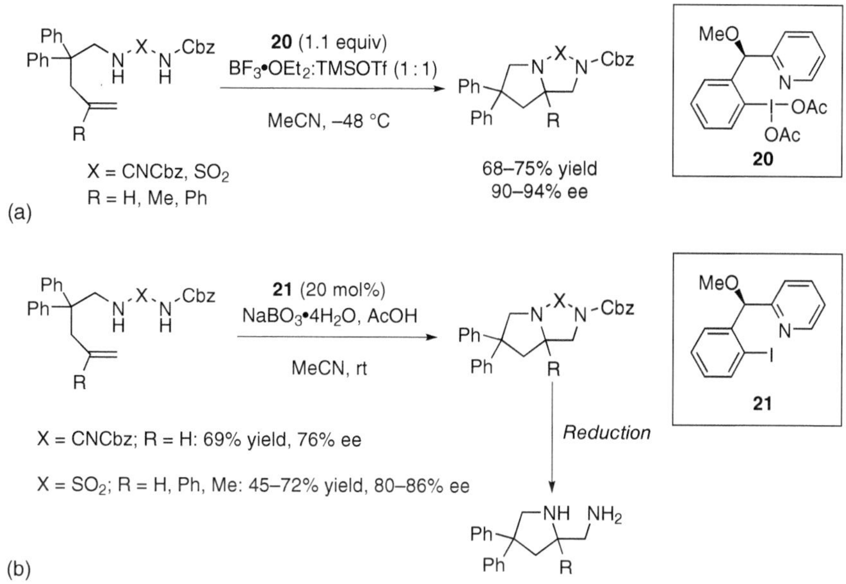

Figure 9.13 (a) Enantioselective, intramolecular diamination reactions promoted by stoichiometric iodine(III) reagent. (b) Catalytic, enantioselective diamination of guanidines and diaminosulfones.

The use of 2,2-disubstituted alkenyl substrates afforded products bearing α-tertiary amine stereocenters in 90–94% ee. However, the substrate scope was generally limited to alkenyl amine derivatives bearing *gem*-diphenyl substitution on the alkyl linker, with unsubstituted and *gem*-dimethyl- and *gem*-dicarbomethoxy-substituted substrates giving only very low yields of the desired products. A catalytic variant using aryl iodide **21**, sodium perborate as the stoichiometric oxidant, and acetic acid as the acid activator was also devised (Figure 9.13b). Enantioselectivities achieved in the catalytic reaction were lower than those obtained in the stoichiometric reaction, a fact that may be ascribed at least in part to the higher temperatures and different acid activators needed to achieve catalytic efficiency. The bicyclic products were readily converted to the corresponding free 1,2-diamines using standard reduction protocols.

The first examples of enantioselective intermolecular C–N functionalizations of alkenes using a chiral hypervalent iodine reagent were reported by Muñiz and coworkers in 2011 [59]. Using bismesylimide as the amine equivalent source and stoichiometric hypervalent iodine reagent **22**, the formal enantioselective diamination of styrenes was achieved under mild conditions (Figure 9.14a). A wide array of electronically and sterically differentiated styrenes were engaged, with the 1,2-disulfonimides obtained in good yield and with high enantioselectivity. In contrast, non-styrenyl substrates were oxidized with very low enantioselectivities, with 1-octene affording the diaminated product in only 5% ee. The enantiomeric excess of many of the styrenyl-derived products could be upgraded by a single crystallization to afford enantiopure 1,2-diamines after removal of the methanesulfonyl groups

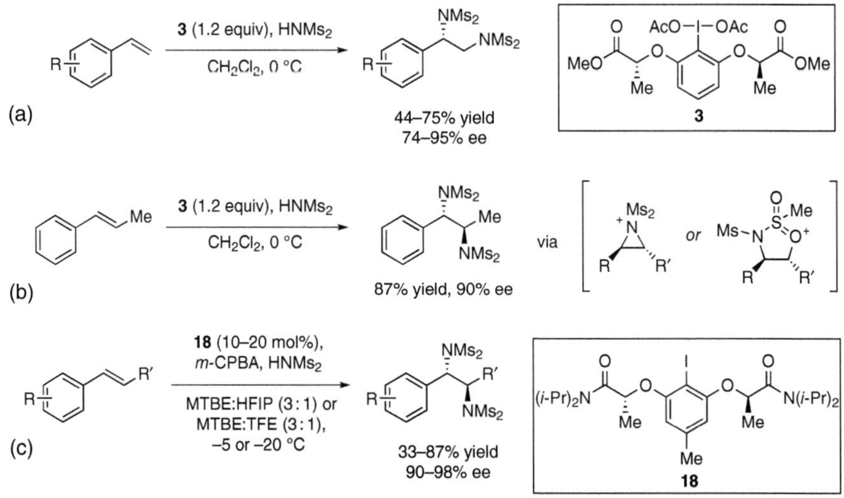

Figure 9.14 (a) Enantioselective, intermolecular diamination of styrenes promoted by a stoichiometric iodine(III) reagent. (b) The enantioselective diamination of β-methyl styrene yields the *anti*-diastereomer, likely through anchimeric assistance by the N or O of the sulfonamide. (c) Catalytic, enantioselective diaminations of β-substituted styrenes.

by a reductive sequence. 1,2-disubstituted alkenes such as *trans*-β-methyl styrene underwent diamination to afford the *anti*-diastereomer as a single observable stereoisomer in 90% ee (Figure 9.14b). The stereochemical outcome may be ascribed to the net retentive displacement of the alkyl iodane intermediate via a doubly invertive substitution mechanism, with anchimeric assistance by the NMs_2 to give a 3-membered ring aziridinium or a 5-membered ring dioxooxathiazolidinium intermediate.

Following this seminal report, the same group described a catalytic variant of the diamination reaction in 2017 [60]. In the presence of *m*-CPBA as the stoichiometric oxidant and bis-amide iodoarene catalyst **18**, a broad scope of terminal and β-substituted styrenes underwent diamination in up to 98% ee (Figure 9.14c). As in the stoichiometric reactions, (*E*)-styrenes afforded the *anti*-diaminated product through presumed anchimeric assistance by the sulfonimide. Background epoxidation by *m*-CPBA was minimized through temperature control and the use of a *tert*-butyl methyl ether/hexafluoroisopropanol solvent blend.

Fluoroaminations of alkenes have traditionally relied on the use of electrophilic fluoride sources with nitrogen-based nucleophiles. In 2013, Nevado and coworkers reported an alternative strategy based on hypervalent iodine chemistry in the enantioselective aminofluorination of pent-4-enamines in up to 88% ee (>99% after recrystallization) using the iodoarene difluoride reagent **22** in toluene at ambient temperature (Figure 9.15a) [61]. Kita and Shibata subsequently reported a catalytic version of this transformation using the binaphthyl diiodide catalyst **23** (Figure 9.15b) [62]. Fluoropiperidines were thus synthesized in up to 70% ee using aqueous hydrogen fluoride as a nucleophilic fluoride source and *m*-CPBA as the stoichiometric oxidant.

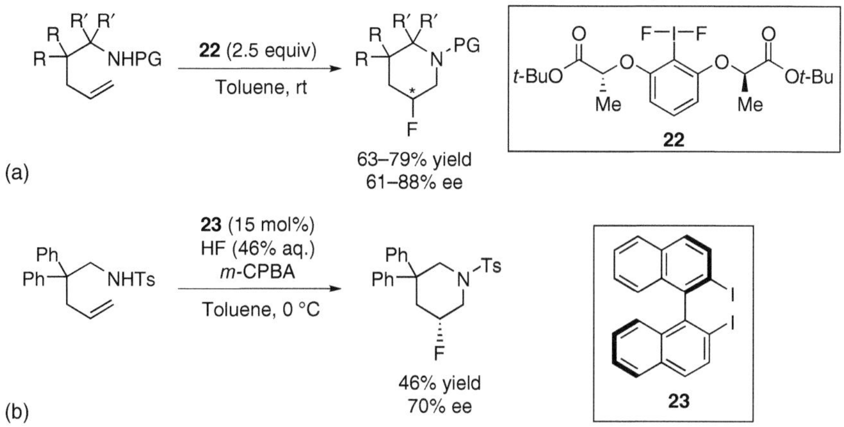

Figure 9.15 (a) Enantioselective fluoroamination promoted by a stoichiometric iodoarene difluoride reagent **22** reported by Nevado. (b) Kita and Shibata's catalytic, enantioselective fluoroamination catalyzed by the binaphthyl diiodide **23**.

The scope of intramolecular fluoroaminations was expanded by Jacobsen and coworkers in the highly enantioselective synthesis of β-fluorinated aziridines from *N*-sulfonyl cinnamyl amines [49]. In the presence of the Ishihara-type catalyst **9**, the asymmetric fluoroamination was achieved with complete syn diastereoselectivity using HF-pyridine as the fluoride source and *m*-CPBA as the stoichiometric oxidant (Figure 9.16a). The scope of cinnamyl sulfonamides was limited to ones bearing electron-withdrawing arene substituents, as electron-rich arenes underwent preferential aryl migration to the 1,1-difluoride products (discussed in Section 9.2.1.2). Indeed, a direct correlation between the product ratio and the σ^+ value of the arene substituent was observed. However, cross-coupling handles such as aryl bromide and triflate were tolerated in the fluoroaziridination reaction, with the latter providing an indirect route to phenolic derivatives that are commonly present in biologically relevant arylethylamines. Variation of the *N*-sulfonyl-protecting group had no significant effect on enantioselectivity but resulted in diminished yields. Under the catalytic conditions, substrates bearing a stereocenter at the allylic position, combined with the use of larger excesses of HF-pyridine, led to sequential fluoroaziridination and ring-opening to afford 1,3-difluorides, with the chiral catalyst allowing the highly selective generation of either *anti*- or *syn* diastereomers possessing distinct conformational properties (Figure 9.16b).

Intramolecular aminofluorination was also achieved using alkenyl substrates bearing more distal sulfonamides, with a β-fluoropyrrolidine product obtained in 86% ee and high diastereoselectivity (Figure 9.16c). In contrast to the completely *syn*-selective fluoroaziridination reaction, the *anti*-fluoroamination product was obtained as the exclusive product. This difference in outcomes, albeit with essentially complete diastereoselectivity in each case, indicates that the two reactions proceed by distinct mechanisms. A rationalization is provided in Figure 9.17. Both pathways are expected to proceed via initial dissociative exchange to afford the alkene complex common to all alkene difunctionalization reactions promoted

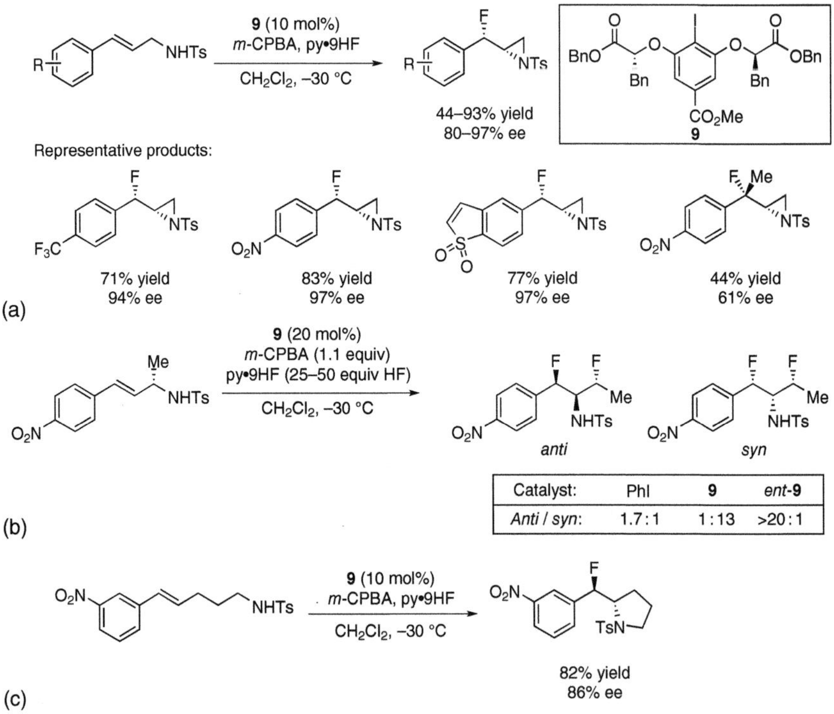

Figure 9.16 (a) Catalytic, enantioselective fluoroamination of cinnamyl tosylamides to yield β-fluoro aziridines. (b) Catalyst-controlled diastereoselectivity in the fluoroamination of a chiral, branched tosylamide. (c) Cyclization of a 4-pentenyl-1-sulfonamide yields a β-fluoropyrrolidine with opposite diastereoselectivity to the fluoroaziridination reaction.

by hypervalent iodine reagents (Figure 9.1). Nucleophilic addition to the alkene complex can occur either by fluoride addition or by intramolecular addition of the sulfonamide. Initial antifluoroiodination, followed by intramolecular, stereoinvertive substitution of the alkyl iodane by the tosylamide would give rise to the observed *syn*-fluoroaziridination product (Figure 9.17, top). In contrast, the fluoropyrrolidine product likely arises through initial aminoiodination (Figure 9.17, bottom). Anchimeric assistance from the sulfonamide and subsequent opening of the resulting dioxooxathiazolidinium intermediate by fluoride would furnish the observed *anti*-stereochemistry in the fluoropyrrolidine product. A more detailed mechanistic analysis of such competing pathways in alkene difunctionalization reactions was performed in the context of difluorinative Wagner–Merwein rearrangements and will be discussed in Section 9.2.1.3.

Szabó and coworkers reported analogous enantioselective fluorocyclizations of 2,2-disubstituted alkenes to furnish products bearing tertiary fluoride stereocenters (Figure 9.18) [63]. In the presence of iodoarene catalyst **24** and HF-pyridine, electron-deficient styrenes bearing pendant nosylamides and alcohols underwent amino- and oxyfluorination to afford pyrrolidine and tetrahydrofuran products, respectively, in good yield and high enantioselectivity. In contrast, the corresponding

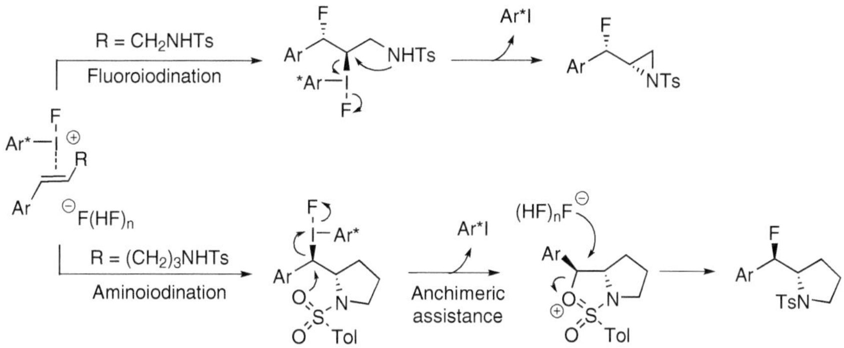

Figure 9.17 Divergent mechanisms in fluoroaziridine and fluoropyrrolidine products leading to stereochemically distinct outcomes.

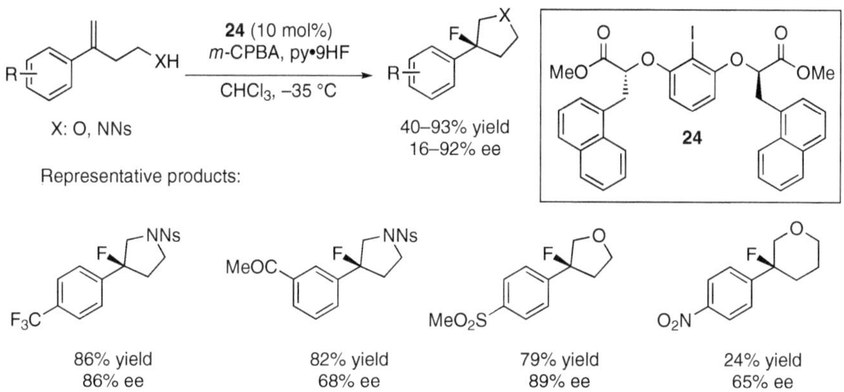

Figure 9.18 Catalytic enantioselective amino- and oxyfunctionalization of α-substituted styrenes reported by Szabó.

oxyfluorination to afford six-membered fluorinated tetrahydropyrans proceeded in low yield and ee.

9.5 Dihalogenation

Although dichloroiodobenzene was the first reported hypervalent iodine compound [64] and alkene dichlorination was among the first identified iodine(III)-promoted reactions [65, 66], most of the key advances in aryl iodide-catalyzed dihalogenations have, in fact, occurred in the arena of alkene difluorination. The discovery of methods for the direct synthesis of 1,2-difluorides has drawn considerable research interest motivated to a significant extent by the interesting conformational properties of those compounds (the "gauche effect") and the practical and fundamental challenges associated with the use of electrophilic fluorinating agents.

The seminal advance in alkene difluorination with iodine-based reagents was made by Hara and Yoneda with the 1998 discovery that the vicinal difluorination

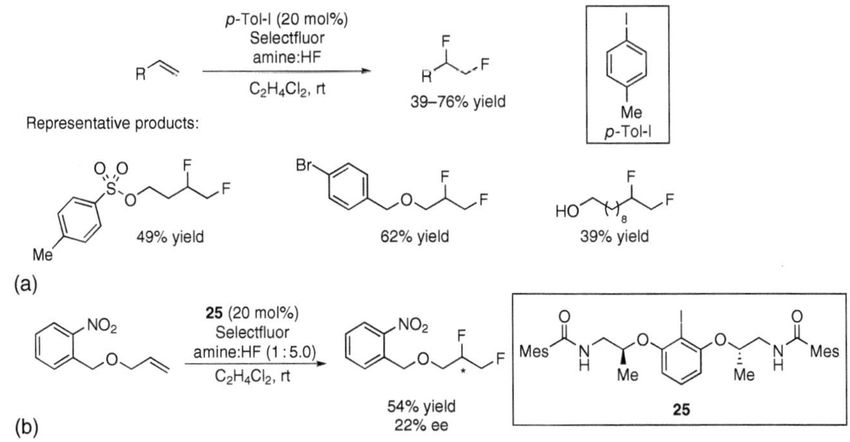

Figure 9.19 (a) Catalytic, vicinal difluorination of terminal alkenes reported by Gilmour. (b) Initial enantioselective, catalytic difluorination result.

of simple terminal alkenes is effected with iodotoluene difluoride in the presence of excess Et_3N-5HF [67]. Nearly two decades later, Gilmour [68] and Jacobsen [69] independently reported the first catalytic variants of the vicinal difluorination reaction. Using triethylamine-HF as a fluoride source and Selectfluor as the stoichiometric oxidant, Gilmour and coworkers achieved the difluorination of terminal olefins with *p*-iodotoluene as the catalyst (Figure 9.19a). Optimal yields with most substrates were obtained using ratios of HF to amine of ca. 5:1. Both terminal and alkyl- and aryl-substituted alkenes were shown to undergo 1,2-difluorination, with tolerance displayed toward functional groups such as α,β-unsaturated esters, tosylates, phthalimides, and free alcohols. Allyl ethers also underwent difluorination in high yields, and a preliminary result in enantioselective difluorination of that substrate class was provided using the bis-amide catalyst **25** that was developed originally for enantioselective dearomatizations by Ishihara (Figure 9.19b) [70–72].

The racemic, catalytic 1,2-difluorination protocol developed by Jacobsen and coworkers incorporated excess of pyridine•9HF (Olah's Reagent) as the fluoride source [73], *m*-CPBA as the stoichiometric oxidant, and either iodotoluene or an achiral variant of an Ishihara-type iodoarene (**26**) as the catalyst (Figure 9.20a). The incorporation of the para-carbomethoxy substituent on the catalyst resulted in improved catalyst stability and significantly increased yields in the difluorination reaction (70 vs. 32% yield in the model reaction). A wide variety of simple and highly functionalized terminal alkenes underwent difluorination, with some substrates requiring added pyridine to achieve useful reactivity. Reactions proceeded to full conversion of alkene, with minor by-products ascribable to competitive epoxidation/ring opening or to *m*-chlorobenzoic acid addition to the putative alkyl iodonium intermediates. Tolerated functional groups included ethers, esters, sulfonamides, basic-nitrogen-containing heteroaromatic rings, and tertiary amines, the latter two likely undergoing *in situ* protection via protonation under the acidic reaction conditions. Electron-rich aromatic rings were found to be incompatible with the highly oxidizing conditions. Whereas simple disubstituted olefins did not undergo the desired reaction, a variety of trisubstituted alkenes provided tertiary fluoride products in good yields. Electron-deficient β-substituted styrenes and a wide variety of substituted acrylamide derivatives underwent highly diastereoselective difluorination. In the case of substrates lacking neighboring Lewis basic functionality, the *syn*-1,2-difluorides were obtained as expected according to the generally accepted alkene difunctionalization mechanism (Figure 9.1). However, substrates bearing groups capable of neighboring group participation afforded the anti-1,2-difluorides with excellent stereocontrol via a proposed doubly invertive, anchimeric assistance pathway. A preliminary enantioselective variant was also disclosed, with a β-isopropyl primary cinnamamide substrate affording the *anti*-1,2-difluoride in 93% ee in the presence of Ishihara-type catalyst **27** (Figure 9.20b).

Gilmour and coworkers followed up their initial catalytic difluorination disclosure with the report in 2018 of catalytic, enantioselective vicinal difluorinations promoted by bis-amide catalyst **28** (Figure 9.21) [74]. Electron-deficient styrenes were demonstrated to undergo difluorination in up to 88% ee. The HF-to-amine ratio was found to have a dramatic effect on selectivity for the 1,2-difluoride products versus the geminal difluoride products generated via aryl migration. Thus, the use of the HF:amine ratio of 4.5:1 provided the vicinal difluoride in optimal ee's and high (>20:1) chemoselectivity, whereas increasing the HF:amine ratio to 9.2:1 led to a complete reversal to favor the exclusive formation of the 1,1-product. The authors further demonstrated

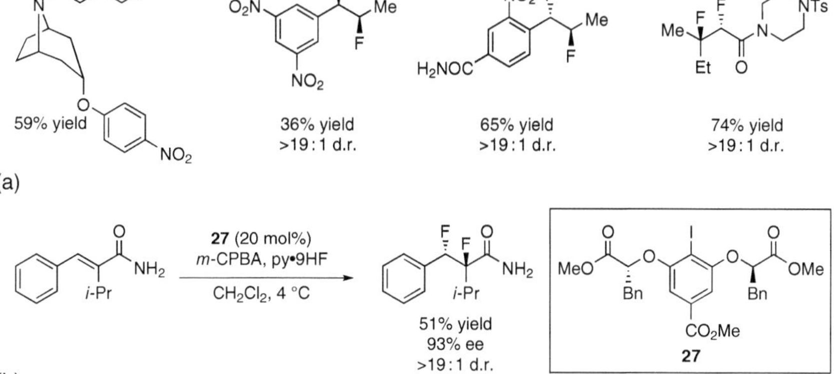

Figure 9.20 (a) Catalytic, diastereoselective difluorination of internal and terminal alkenes reported by Jacobsen. (b) Initial catalytic, enantioselective difluorination of cinnamamides.

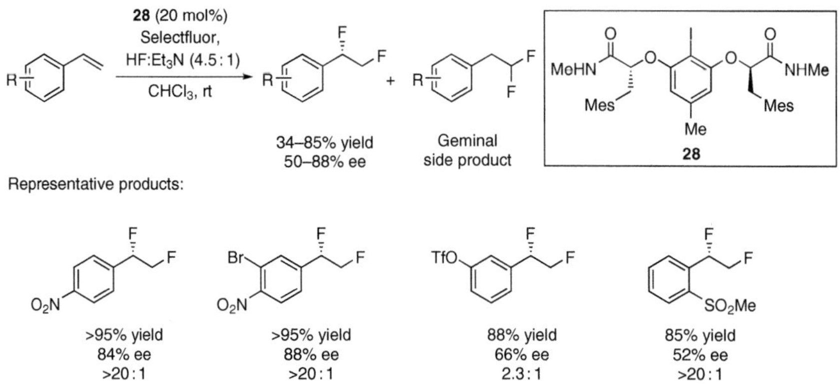

Figure 9.21 Catalytic, enantioselective difluorination of unsubstituted styrenes reported by Gilmour.

the possibility of upgrading the enantiomeric purity of the 1,2-difluorides through recrystallization.

The Jacobsen group followed up its preliminary enantioselective result with a full study of asymmetric catalytic vicinal difluorination of cinnamamide derivatives (Figure 9.22a) [75]. Several β-alkylcinnamamides underwent vicinal difluorination in up to >100:1 product selectivity and 98% ee. As in the Gilmour system discussed above, selectivity for vicinal difluorination versus geminal difluorination with aryl migration proved to be a critical consideration. The size of the β-alkyl substituent, the nucleophilicity of the arene, and both the steric and electronic properties of the amide were all found to correlate directly with selectivity between the two pathways. For example, a strong correlation between the Charton value of the secondary-amide substituent and the selectivity for the 1,2-product was observed. The IR-stretching frequency of the amide carbonyl in cinnamamides bearing differently fluorinated *N*-ethyl substituents was also found to correlate with product selectivity. On the basis of these and other experimental observations, chemoselectivity between the 1,2- and 1,1-difluorination pathways was concluded to hinge on the competition between phenonium ion formation versus anchimeric assistance by the amide in the breakdown of the key alkyl iodane intermediate (Figure 9.22b).

The propensity for β-aryl alkyl iodanes to undergo rearrangement via phenonium ion formation [76–81] was harnessed by Jacobsen and coworkers in the development of an enantioselective catalytic protocol for the synthesis of geminally difluorinated compounds (Figure 9.23) [82]. Difluoromethyl groups can serve as oxidatively stable bioisosteres of sensitive protic functionalities such as alcohols and thiols, and the stereoselective construction of compounds bearing such motifs is therefore of considerable interest. In the presence of HF-pyridine and *m*-CPBA and iodoarene catalyst **9**, disubstituted cinnamamides and di- and tri-substituted cinnamate esters were found to undergo difluorination with aryl migration to afford the rearrangement products in high yields and enantioselectivities. The incorporation of a para methoxycarbonyl group on the iodoresorcinol core had a marked impact on ee but minimal influence on yield. Cinnamate esters bearing β-alkyl substitution provided

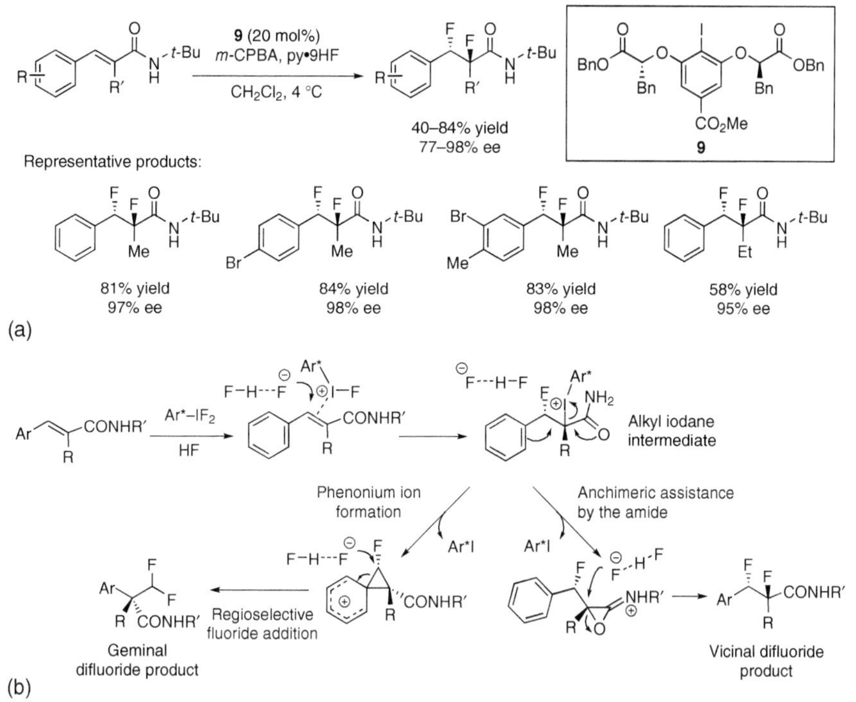

Figure 9.22 (a) Catalytic, enantioselective difluorination of *N-tert*-butylcinnamamides reported by Jacobsen. (b) Proposed basis for portioning between 1,2- and 1,1-difluorination pathways.

geminally difluorinated products bearing adjacent quaternary stereocenters in up to 97% ee. The difluorinated products were shown to be readily diversifiable through hydrolysis, reduction, oxidation, and transamidation sequences.

The presence of electron-rich benzyl groups at the stereogenic centers in iodoarene catalyst **9** was found to be essential for high enantioselectivity. An Eyring analysis led the authors to the conclusion that enantioselectivity was primarily due to enthalpic factors and to the proposal that stabilizing cation-π interactions may play a direct role in enantiocontrol. An extensive computational analysis by Jacobsen, Houk, and Xue identified the alkyl iodane formation common to both 1,2- and 1,1-difluorination pathways (Figure 9.22b) as enantiodetermining and supported the hypothesis that multiple attractive non-covalent interactions such as C–H--π, C–H--O, and slipped π–π stacking underlie enantiodiscrimination [32].

An alternative difluorinative rearrangement pathway was engaged by Jacobsen, Sigman, and coworkers in the geminal difluorination of α-bromostyrenes (Figure 9.24) [83]. These substrates underwent rearrangement via bromonium rather than phenonium ion formation to provide β,β-difluorinated alkyl bromide-building blocks. A catalyst decomposition pathway was identified involving alkylation of the ester groups by the bromonium ion intermediate. The incorporation of electron-deficient benzyl esters led to suppression of that

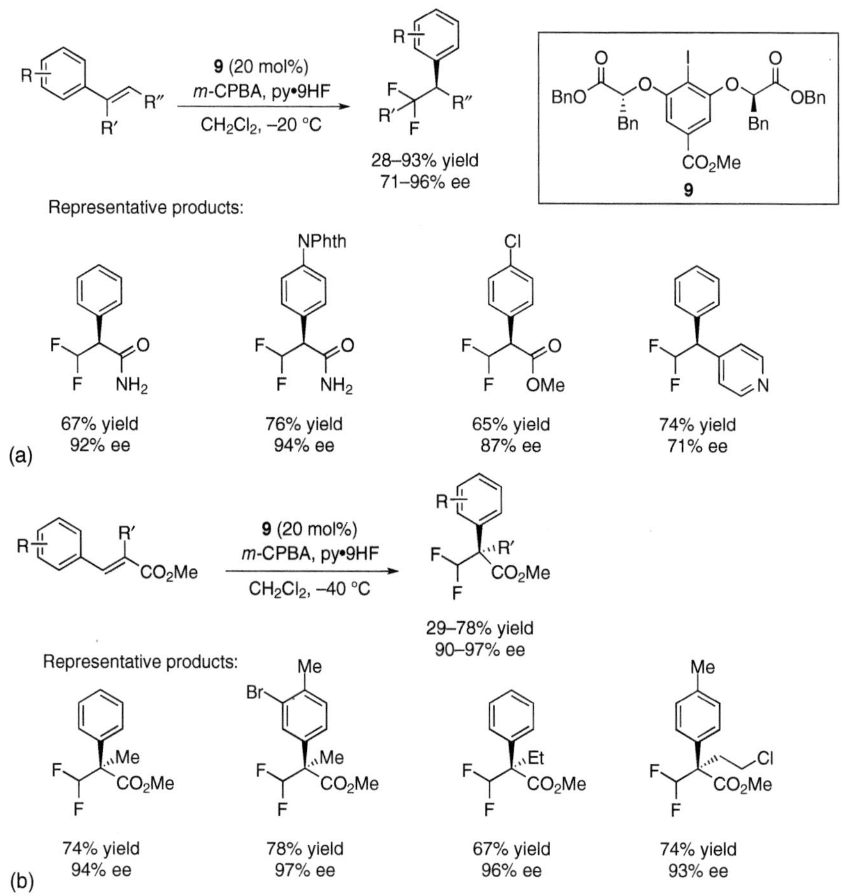

Figure 9.23 (a) Catalytic, enantioselective difluorinative rearrangement of styrene derivatives. (b) Catalytic, enantioselective 1,1-difluorination of trisubstituted cinnamate esters.

undesired pathway and allowed the use of reduced catalyst loadings. Reaction enantioselectivity proved highly responsive to substitution at the stereogenic benzylic positions, with electron-rich, sterically demanding groups such as *t*-butyl or 1-adamantyl affording highest ees. In this manner, the optimal catalyst **29** was identified and shown to promote geminal difluorination in up to 93% ee and 98% yield.

A set of analogs of **29** bearing varying substitution at the benzyl-group-bearing stereocenters was evaluated in the geminal difluorination reaction. A linear free-energy correlation was identified between experimentally determined enantioselectivities and computed C–H--π interaction energies between the catalyst arene and an orthogonally disposed benzene probe. A correlation between the LUMO energy of the substrate and reaction enantioselectivity was observed. LUMO energies were, in turn, found to correlate with π–π interaction energies. The evidence that both C–H--π and π–π interactions may play a critical role in

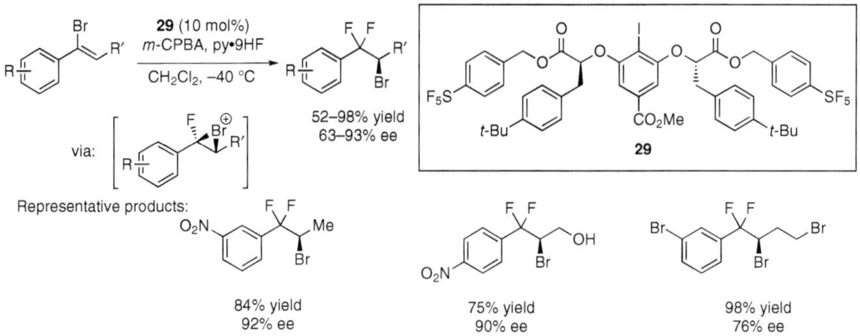

Figure 9.24 Catalytic, enantioselective 1,1-difluorination of α-bromostyrenes reported by Jacobsen and Sigman.

enantioinduction is consistent with the computational study by Jacobsen, Houk, and Xue noted above [32] and supports the hypothesis that a complex network of non-covalent interactions provides selective stabilization of the transition state leading to the major enantiomer.

In contrast to the advances outlined above in catalytic enantioselective difluorinations promoted by aryl iodide catalysts, only limited progress has been made thus far in analogous dichlorination reactions. Indeed, despite the fact that dichloroiodobenzene was the first reported hypervalent iodine compound [64] and alkene dichlorination was among the first iodine(III)-promoted reactions discovered [65, 66], it was not until 2019 that Gilmour and coworkers reported the catalytic vicinal dichlorination of diverse alkenes (Figure 9.25a) [84]. In the presence of cesium chloride and Selectfluor, the racemic dichlorination of the terminal and 1,2-disubstituted olefins was achieved in good yields and with no observable reactivity in the absence of the iodoarene. Hexafluoroisopropanol (HFIP) was identified as an essential additive, presumably facilitating chloride dissociation from the $ArICl_2$ intermediate through hydrogen bonding. Internal alkenes underwent dichlorination stereospecifically, with (*E*)-alkenes giving anti-1,2-dichlorides and (*Z*)-alkenes, syn dichlorides (Figure 9.26b). A preliminary example of an enantioselective variant was identified using Ishihara-type iodoarene catalyst **5**, with 4-chlorostyrene undergoing dichlorination in 28% ee (Figure 9.26c).

9.6 Carbofunctionalization

Alkene carbofunctionalization reactions are relatively underexplored in hypervalent iodine catalysis. This deficiency is ascribable at least in part to the incompatibility of most typical nucleophilic carbon-centered functional groups with the strongly oxidizing conditions required for these reactions (see Chapter 8). Nonetheless, there are several successful reported examples of oxidative carbon–carbon bond-forming reactions promoted by hypervalent iodide reagents or intermediates. Fujita and coworkers achieved doubly intramolecular alkoxyarylation reactions promoted by

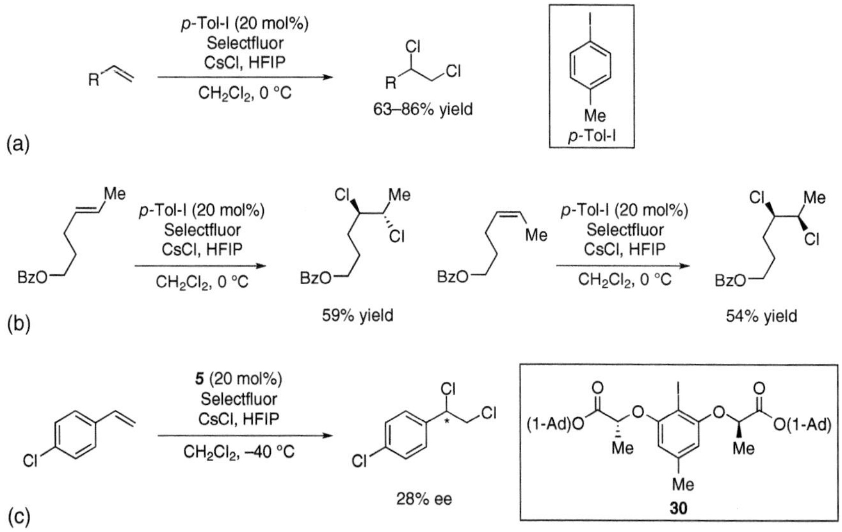

Figure 9.25 (a) Catalytic, vicinal dichlorination of terminal olefins reported by Gilmour. (b) Diastereospecific vicinal dichlorination of internal alkenes. (c) Initial catalytic, enantioselective dichlorination result.

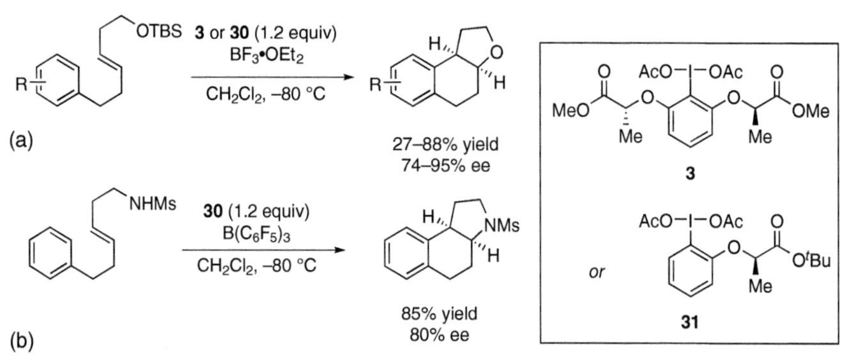

Figure 9.26 (a) Enantioselective alkoxyarylation of internal alkenes by stoichiometric iodine(III) reported by Fujita. (b) Enantioselective aminoarylation of internal alkenes promoted by a stoichiometric iodine(III) reagent.

stoichiometric iodine(III) reagents in the presence of boron trifluoride as the activator [85, 86]. Bicyclic tetrahydrofuran products were thus generated in up to 95% ee with syn diastereoselectivity (Figure 9.26a). An analogous mesylamide underwent aminoarylation to give the corresponding pyrrolidine in 80% ee (Figure 9.26b). The reactions were proposed to proceed through initial attack by the heteroatom nucleophile on the activated alkene•ArI(III) complex, with subsequent substitution of the resulting alkyl iodane by the aryl group to forge the C–C bond.

In 2020, Gilmour and coworkers extended the protocols developed in their laboratories for alkene difluorination (discussed in Section 9.2.1.2) to the catalytic, enantioselective fluoroarylation of allyl aryl ethers to furnish chromane

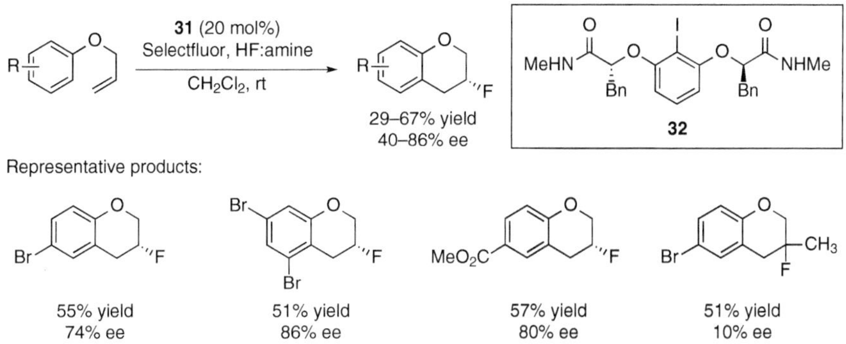

Figure 9.27 Catalytic, enantioselective fluoroarylation reported by Gilmour.

derivatives [87]. The chromane motif appears frequently in biologically active compounds, often with additional oxidation at the 3-position. The authors accessed stereodefined 3-fluorochromanes using HF buffered by Et_3N or pyridine as the fluoride source, Selectfluor as the stoichiometric oxidant, and the secondary amide Ishihara-type catalyst **31** (Figure 9.27). The HF-to-amine ratio was observed to be key for promoting the desired carbofluorination over vicinal difluorination. However, while a lower HF-to-amine ratio resulted in higher enantioselectivity and improved product distribution, reactivity was reduced. Therefore, two conditions were developed, one with a higher HF:amine ratio for increased reactivity and another with a lower HF:amine ratio for increased selectivity.

A variety of allyl aryl ethers underwent cyclization to afford the fluorinated chromane derivatives in moderate yield and good enantioselectivity. A methallyl aryl ether was shown to undergo the cyclization reaction to afford a product bearing a tertiary fluoride stereocenter in 51% yield, albeit in only 10% ee. The syn stereospecificity of the carbofluorination reaction was demonstrated using isotopically labelled substrates, with the (*E*)-deuteroolefin providing the cis product and the (*Z*)-olefin leading to the trans product (Figure 9.28). A mechanism involving a Claisen rearrangement followed by fluoroaryloxylation of the alkene was ruled out by the observation that no fluorochromane product was formed when a 2-allylphenol – the putative intermediate in such a sequence – was subjected to the reaction conditions. In light of the experimental evidence, the authors proposed a mechanism for carbofluorination involving arylation of the activated π-iodonium complex with subsequent rearomatization. The resulting alkyl iodane is then displaced by fluoride to afford the net *syn* addition product.

In 2020, Jacobsen reported enantioselective catalytic difluorination reactions that incorporated Wagner–Meerwein rearrangements (Figure 9.29) [31]. Under conditions similar to those developed for vicinal and geminal difluorinations of cinnamate derivatives (Section 9.2.1.2), styrene derivatives bearing β-cumyl groups underwent reaction with the migration of the allylic aryl group to furnish the *anti*-diastereomers selectively and in up to 94% ee. The incorporation of a para methoxycarbonyl group on the iodoresorcinol core proved critical for reactivity, as none of the desired difluorinated product was obtained using an analog of **9** lacking

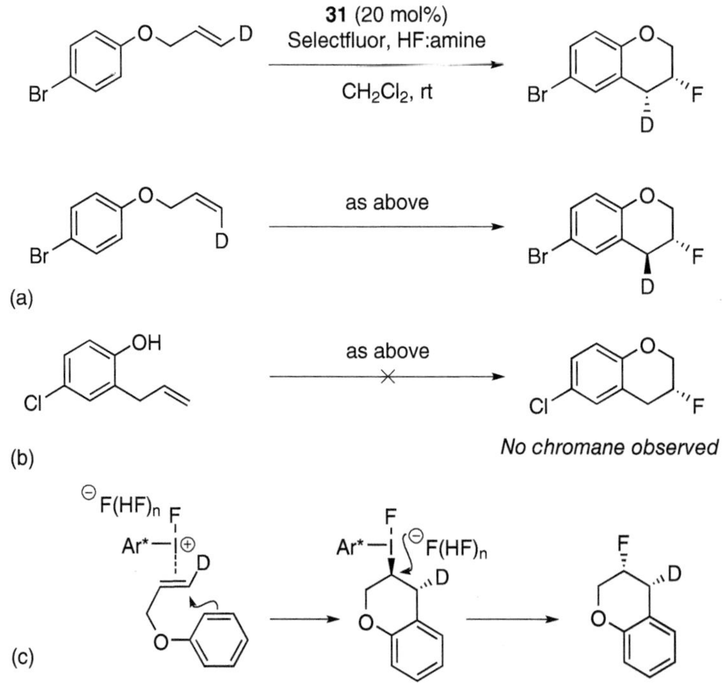

Figure 9.28 (a) Stereospecificity of the fluoroarylation of allyl aryl ethers established through deuterium-labeling experiments. (b) The absence of 3-fluorochromane product observed with 2-allylphenols rules out a mechanism involving a Claisen/fluoroaryloxylation sequence. (c) Proposed mechanism for the fluoroarylation reaction.

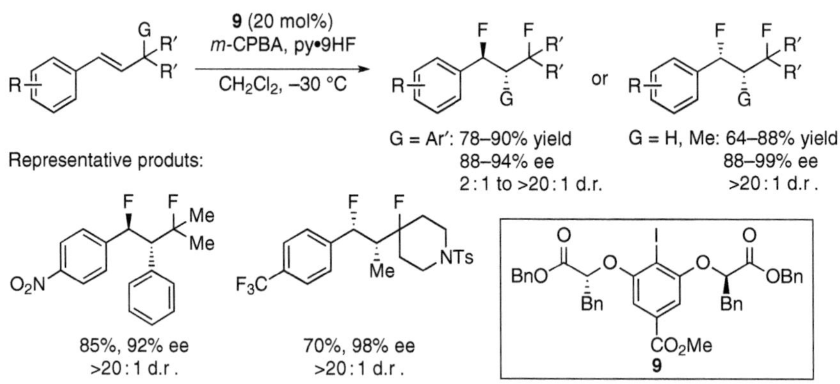

Figure 9.29 Catalytic, enantioselective Wagner–Meerwein rearrangements reported by Jacobsen.

the central ester. β-*tert*-alkylstyrenes also participated in the rearrangement reaction with methyl migration to afford 1,3-difluoride products in high enantio- and diastereoselectivities, although with the *syn* diastereomers generated selectively.

As noted in the General Mechanistic Considerations (Section 9.2), alkene difunctionalization reactions promoted by hypervalent iodine reagents are generally

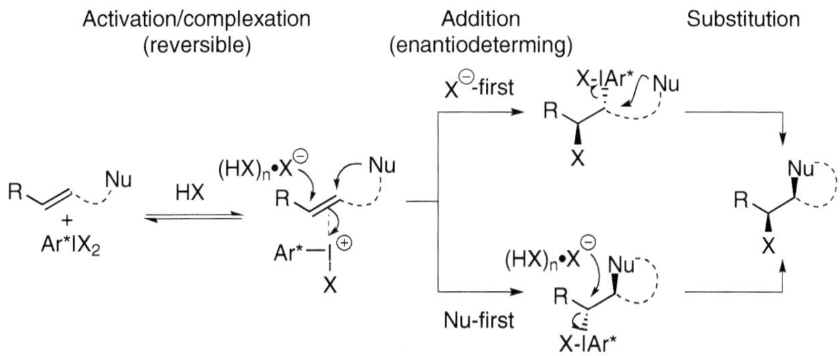

Figure 9.30 Alternative mechanisms for enantioselective *syn* difunctionalizations of alkenes by hypervalent iodoarene reagents. *Source*: Adapted with permission from Sharma et al. [31]. Copyright 2020 American Chemical Society.

understood to proceed via dissociative substitution to generate an electrophilic π-alkene complex, followed by a sequence of stereospecific nucleophilic addition and substitution steps (Figure 9.30). The identity of the components participating in the latter two steps (X^- and Nu) determine the nature of the difunctionalization products discussed throughout this chapter. Furthermore, the π-alkene complex may react with either the dissociated anionic nucleophile (X^--first) or with the other nucleophilic component (Nu-first) in the formation of the alkyl iodane intermediate, and the sequence can be reaction-dependent and, in many cases, difficult to determine. The question of which component is involved in the addition step is not only of fundamental interest, but also at the heart of the mechanism of enantioinduction since alkyl iodane formation is likely irreversible and therefore enantiodetermining.

The Wagner–Meerwein reaction described above provided a clear opportunity to distinguish the addition and substitution steps through systematic manipulation of the migrating groups. The aryl rearrangement reactions were analyzed through a Hammett study of substituent effects on the migrating aryl group. A strong electronic effect on the reaction rate was observed ($\rho = -3.7$), consistent with rate-determining aryl migration via phenonium ion formation (Figure 9.31). In this case, aryl migration precedes fluoride substitution (i.e. Nu-first). The observed *anti*-configuration of the fluoroarylated products can be rationalized by a second phenonium ion formation from the alkyl iodane intermediate followed by opening by fluoride.

The mechanism of the methyl rearrangement reaction was analyzed using heavy-atom kinetic isotope effect (KIE) experiments (Figure 9.32a). The intrinsic

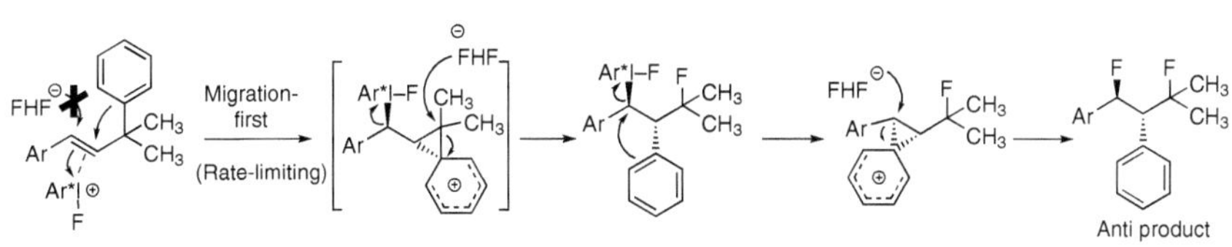

Figure 9.31 Proposed mechanism of the *anti*-selective aryl rearrangement reaction.

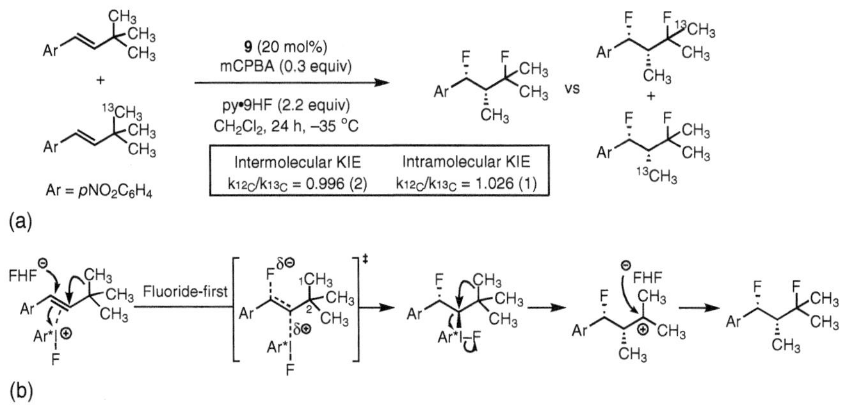

Figure 9.32 (a) Kinetic isotope effect (KIE) experiments used to establish the fluoride-first mechanism for methyl migration of β-*tert*-butylstyrenes. (b) Proposed mechanism of the *syn*-selective alkyl rearrangement reaction. *Source*: Adapted with permission from Sharma et al. [31]. Copyright 2020 American Chemical Society.

primary KIE for the methyl migration was shown in an intramolecular KIE experiment to be in the expected range (1.026). However, a negligible $^{12}C/^{13}C$ KIE on the migrating carbon was determined in an intermolecular competition experiment, ruling out methyl migration in the substrate-committing step. The possibility that alkene complexation might be substrate-committing was also ruled out with the observation of no significant KIE on the substrate alkene carbons. These KIE experiments support a mechanism involving fluoride-first addition, followed by methyl migration to afford syn-difunctionalized products (Figure 9.32b). Taken together, these studies demonstrate that the mechanism of catalysis and the identity of the enantiodetermining event in asymmetric aryl iodide alkene difunctionalization reactions can be dependent on relatively minor changes to the reacting partners and that high enantioselectivity is achievable by a given chiral catalyst despite the possible participation of competing mechanisms.

9.7 Conclusion and Outlook

Hypervalent iodine chemistry provides a powerful and versatile platform for the efficient introduction of molecular complexity from simple alkene-starting materials. With the discovery of simple protocols for achieving reoxidation of the aryl iodide by-products under relevant reaction conditions, the door has been opened to the discovery of a rich assortment of asymmetric, catalytic alkene difunctionalization reactions. While several different chiral aryl iodide frameworks have been developed for effecting enantioselective catalysis (see Chapter 7), the greatest success in alkene difunctionalization has been achieved thus far with the resorcinol-derived Ishihara-type catalysts. Considerable progress remains to be made to increase the practicality of the catalytic reactions, which, at this stage, typically require high (10–20 mol%) catalyst loadings and often caustic acidic conditions. Recent advances in the mechanistic elucidation of some of the reactions identified to date have shed light on the very subtle and interesting bases for enantiocontrol with these catalysts, and future work is certain both to improve our understanding of these catalysts and to lead to the discovery of new applications.

References

1 Margarita, C. and Andersson, P.G. (2017). Evolution and prospects of the asymmetric hydrogenation of unfunctionalized olefins. *J. Am. Chem. Soc.* 139: 1346–1356.

2 McDonald, R.I., Liu, G., and Stahl, S.S. (2011). Palladium(II)-catalyzed alkene functionalization via nucleopalladation: stereochemical pathways and enantioselective catalytic applications. *Chem. Rev.* 111: 2981–3019.

3 Coombs, J.R. and Morken, J.P. (2016). Catalytic enantioselective functionalization of unactivated terminal alkenes. *Angew. Chem. Int. Ed.* 55: 2636–2649.

4 Wu, S., Zhou, Y., and Li, Z. (2019). Biocatalytic selective functionalisation of alkenes via single-step and one-pot multi-step reactions. *Chem. Commun.* 55: 883–896.

5 Liu, R.Y. and Buchwald, S.L. (2020). CuH-catalyzed olefin functionalization: from hydroamination to carbonyl addition. *Acc. Chem. Res.* 53: 1229–1243.

6 Jacobsen, E.N., Pfaltz, A., and Yamamoto, H. (eds.) (1999). *Comprehensive Asymmetric Catalysis*, 1e. Berlin, Germany: Springer.

7 Zhdankin, V.V. and Stang, P.J. (2002). Recent developments in the chemistry of polyvalent iodine compounds. *Chem. Rev.* 102: 2523–2584.

8 Ghosh, S., Pradhan, S., and Chatterjee, I. (2018). A survey of chiral hypervalent iodine reagents in asymmetric synthesis. *Beilstein J. Org. Chem.* 14: 1244–1262.

9 Zhdankin, V.V. and Stang, P.J. (2008). Chemistry of polyvalent iodine. *Chem. Rev.* 108: 5299–5358.

10 Stang, P.J. and Zhdankin, V.V. (1996). Organic polyvalent iodine compounds. *Chem. Rev.* 96: 1123–1178.

11 Varvoglis, A. (2014). *Hypervalent Iodine in Organic Synthesis* (eds. O. Meth-Cohn, A.R. Katritzky and C.S. Rees). London: Academic Press.

12 Wirth, T. (ed.) (2016). *Hypervalent Iodine Chemistry*, 1e. Cham, Switzerland: Springer International Publishing.

13 Zhdankin, V.V. (2013). *Hypervalent Iodine Chemistry: Preparation, Structure, and Synthetic Applications of Polyvalent Iodine Compounds*. Nashville, TN: John Wiley & Sons.

14 Martínez, C. and Muñiz, K. (2018). Dioxygenation of alkenes. In: *Catalytic Oxidation in Organic Synthesis* (ed. K. Muñiz). Stuttgart: Georg Thieme Verlag.

15 Andries-Ulmer, A. and Gulder, T. (2018). Halogenation and halocyclization of alkenes. In: *Catalytic Oxidation in Organic Synthesis* (ed. K. Muñiz). Stuttgart: Georg Thieme Verlag.

16 Yoshimura, A. and Zhdankin, V.V. (2106). Advances in synthetic applications of hypervalent iodine compounds. *Chem. Rev.* 116: 3328–3435.

17 Lee, J.H., Choi, S., and Hong, K.B. (2019). Alkene difunctionalization using hypervalent iodine reagents: progress and developments in the past ten years. *Molecules* 24: 2634.

18 Parra, A. (2019). Chiral hypervalent iodines: active players in asymmetric synthesis. *Chem. Rev.* 119: 12033–12088.

19 Flores, A., Cots, E., Bergès, J., and Muñiz, K. (2019). Enantioselective iodine(I/III) catalysis in organic synthesis. *Adv. Synth. Catal.* 361: 2–25.

20 Berthiol, F. (2015). Reagent and catalyst design for asymmetric hypervalent iodine oxidations. *Synthesis* 47: 587–603.

21 Li, X., Chen, P., and Liu, G. (2018). Recent advances in hypervalent iodine(III)-catalyzed functionalization of alkenes. *Beilstein J. Org. Chem.* 14: 1813–1825.

22 Romero, R.M., Wöste, T.H., and Muñiz, K. (2014). Vicinal difunctionalization of alkenes with iodine(III) reagents and catalysts. *Chem. Asian J.* 9: 972–983.

23 Fujita, M. (2017). Mechanistic aspects of alkene oxidation using chiral hypervalent iodine reagents. *Tetrahedron Lett.* 58: 4409–4419.

24 Ochiai, M., Ito, T., Takaoka, Y., and Masaki, Y. (1991). Generation of allenyliodinanes and their reductive iodonio-Claisen rearrangement. *J. Am. Chem. Soc.* 113: 1319–1323.

25 Zefirov, N.S., Koz'min, A.S., Kasumov, T. et al. (1992). Interaction of an allene with polyvalent iodine derivatives. Preparation, x-ray molecular structure, and some reactions of phenyl(2,2-dimethyl-4-(diethylphosphono)-2,5-dihydro-3-furyl)iodonium salts. *J. Org. Chem.* 57: 2433–2437.

26 Izquierdo, S., Essafi, S., del Rosal, I. et al. (2016). Acid activation in phenyliodine dicarboxylates: direct observation, structures, and implications. *J. Am. Chem. Soc.* 138: 12747–12750.

27 Kang, Y.-B. and Gade, L.H. (2011). The nature of the catalytically active species in olefin dioxygenation with $PhI(OAc)_2$: metal or proton? *J. Am. Chem. Soc.* 133: 3658–3667.

28 Zhong, W., Yang, J., Meng, X., and Li, Z. (2011). $BF_3 \cdot OEt_2$-promoted diastereoselective diacetoxylation of alkenes by $PhI(OAc)_2$. *J. Org. Chem.* 76: 9997–10004.

29 Kita, Y., Gyoten, M., Ohtsubo, M. et al. (1996). Non-phenolic oxidative coupling of phenol ether derivatives using phenyliodine(III) bis(trifluoroacetate). *Chem. Commun.*: 1481–1482.

30 Sreenithya, A. and Sunoj, R.B. (2019). On the activation of hypercoordinate iodine(iii) compounds for reactions of current interest. *Dalton Tran.* 48: 4086–4093.

31 Sharma, H.A., Mennie, K.M., Kwan, E.E., and Jacobsen, E.N. (2020). Enantioselective aryl-iodide-catalyzed Wagner–Meerwein rearrangements. *J. Am. Chem. Soc.* 142: 16090–16096.

32 Zhou, B., Haj, M.K., Jacobsen, E.N. et al. (2018). Mechanism and origins of chemo- and stereoselectivities of aryl iodide-catalyzed asymmetric difluorinations of β-substituted styrenes. *J. Am. Chem. Soc.* 140: 15206–15218.

33 Ochiai, M. (2003). Reactivities, properties and structures. In: *Hypervalent Iodine Chemistry: Modern Developments in Organic Synthesis.* (ed. T. Wirth). Berlin, Germany: Springer Berlin.

34 Okuyama, T., Takino, T., Sueda, T., and Ochiai, M. (1995). Solvolysis of cyclohexenyliodonium salt, a new precursor for the vinyl cation: remarkable nucleofugality of the phenyliodonio group and evidence for internal return from an intimate ion-molecule pair. *J. Am. Chem. Soc.* 117: 3360–3367.

35 Wirth, T. and Hirt, U.H. (1997). Chiral hypervalent iodine compounds. *Tetrahedron Asymm.* 8: 23–26.

36 Hirt, U.H., Schuster, M.F.H., French, A.N. et al. (2001). Chiral hypervalent organo-iodine(III) compounds. *Eur. J. Org. Chem.*: 1569–1579.

37 Fujita, M., Wakita, M., and Sugimura, T. (2011). Enantioselective Prévost and Woodward reactions using chiral hypervalent iodine(III): switchover of stereochemical course of an optically active 1,3-dioxolan-2-yl cation. *Chem. Commun.* 47: 3983–3985.

38 Uyanik, M., Yasui, T., and Ishihara, K. (2010). Enantioselective Kita oxidative spirolactonization catalyzed by in situ generated chiral hypervalent iodine(III) species. *Angew. Chem. Int. Ed.* 49: 2175–2177.

39 Haubenreisser, S., Wöste, T.H., Martínez, C. et al. (2016). Structurally defined molecular hypervalent iodine catalysts for intermolecular enantioselective reactions. *Angew. Chem. Int. Ed.* 55: 413–417.

40 Muñiz, K. and Wöste, T. (2016). Enantioselective vicinal diacetoxylation of alkenes under chiral iodine(III) catalysis. *Synthesis* 48: 816–827.

41 Fujita, M., Miura, K., and Sugimura, T. (2018). Enantioselective dioxytosylation of styrenes using lactate-based chiral hypervalent iodine(III). *Beilstein J. Org. Chem.* 14: 659–663.

42 Fujita, M., Mori, K., Shimogaki, M., and Sugimura, T. (2013). Total synthesis of (12*R*)- and (12*S*)-12-hydroxymonocerins: stereoselective oxylactonization using a chiral hypervalent iodine(III) species. *RSC Adv.* 3: 17717–17725.

43 Takesue, T., Fujita, M., Sugimura, T., and Akutsu, H. (2014). A series of two oxidation reactions of ortho-alkenylbenzamide with hypervalent iodine(III): a concise entry into (3*R*,4*R*)-4-hydroxymellein and (3*R*,4*R*)-4-hydroxy-6-methoxymellein. *Org. Lett.* 16: 4634–4637.

44 Fujita, M., Okuno, S., Lee, H.J. et al. (2007). Enantiodifferentiating tetrahydrofuranylation of but-3-enyl carboxylates using optically active hypervalent iodine(III) reagents via a 1,3-dioxan-2-yl cation intermediate. *Tetrahedron Lett.* 48: 8691–8694.

45 Fujita, M., Yoshida, Y., Miyata, K. et al. (2010). Enantiodifferentiating *endo*-selective oxylactonization of *ortho*-alk-1-enylbenzoate with a lactate-derived aryl-λ3-iodane. *Angew. Chem. Int. Ed.* 49: 7068–7071.

46 Fujita, M., Mori, K., Shimogaki, M., and Sugimura, T. (2012). Asymmetric synthesis of 4,8-dihydroxyisochroman-1-one polyketide metabolites using chiral hypervalent iodine(III). *Org. Lett.* 14: 1294–1297.

47 Shimogaki, M., Fujita, M., and Sugimura, T. (2013). Enantioselective oxidation of alkenylbenzoates catalyzed by chiral hypervalent iodine(III) to yield

4-hydroxyisochroman-1-ones: enantioselective oxidation of alkenylbenzoates. *Eur. J. Org. Chem.*: 7128–7138.

48 Woerly, E.M., Banik, S.M., and Jacobsen, E.N. (2016). Enantioselective, catalytic fluorolactonization reactions with a nucleophilic fluoride source. *J. Am. Chem. Soc.* 138: 13858–13861.

49 Mennie, K.M., Banik, S.M., Reichert, E.C., and Jacobsen, E.N. (2018). Catalytic diastereo- and enantioselective fluoroamination of alkenes. *J. Am. Chem. Soc.* 140: 4797–4802.

50 Gelis, C., Dumoulin, A., Bekkaye, M. et al. (2017). Chiral hypervalent iodine(III) catalyst promotes highly enantioselective sulfonyl- and phosphoryl-oxylactonizations. *Org. Lett.* 19: 278–281.

51 Alhalib, A., Kamouka, S., and Moran, W.J. (2015). Iodoarene-catalyzed cyclizations of unsaturated amides. *Org. Lett.* 17: 1453–1456.

52 Boye, A.C., Meyer, D., Ingison, C.K. et al. (2003). Novel lactonization with phenonium ion participation induced by hypervalent iodine reagents. *Org. Lett.* 5: 2157–2159.

53 Singh, F.V., Rehbein, J., and Wirth, T. (2012). Facile oxidative rearrangements using hypervalent iodine reagents. *ChemistryOpen* 1: 245–250.

54 Qurban, J., Elsherbini, M., and Wirth, T. (2017). Electron-deficient chiral lactic acid-based hypervalent iodine reagents. *J. Org. Chem.* 82: 11872–11876.

55 Brown, M., Kumar, R., Rehbein, J., and Wirth, T. (2016). Enantioselective oxidative rearrangements with chiral hypervalent iodine reagents. *Chem. Eur. J.* 22: 4030–4035.

56 Farid, U., Malmedy, F., Claveau, R. et al. (2013). Stereoselective rearrangements with chiral hypervalent iodine reagents. *Angew. Chem. Int. Ed.* 52: 7018–7022.

57 Zhang, D.-Y., Zhang, Y., Wu, H., and Gong, L.-Z. (2019). Organoiodine-catalyzed enantioselective alkoxylation/oxidative rearrangement of allylic alcohols. *Angew. Chem. Int. Ed.* 58: 7450–7453.

58 Mizar, P., Laverny, A., El-Sherbini, M. et al. (2014). Enantioselective diamination with novel chiral hypervalent iodine catalysts. *Chem. Eur. J.* 20: 9910–9913.

59 Röben, C., Souto, J.A., González, Y. et al. (2011). Enantioselective metal-free diamination of styrenes. *Angew. Chem. Int. Ed.* 50: 9478–9482.

60 Muñiz, K., Barreiro, L., Romero, R.M., and Martínez, C. (2017). Catalytic asymmetric diamination of styrenes. *J. Am. Chem. Soc.* 139: 4354–4357.

61 Kong, W., Feige, P., de Haro, T., and Nevado, C. (2013). Regio- and enantioselective aminofluorination of alkenes. *Angew. Chem. Int. Ed.* 52: 2469–2473.

62 Suzuki, S., Kamo, T., Fukushi, K. et al. (2014). Iodoarene-catalyzed fluorination and aminofluorination by an Ar-I/HF·pyridine/mCPBA system. *Chem. Sci.* 5: 2754–2760.

63 Wang, Q., Lübcke, M., Biosca, M. et al. (2020). Enantioselective construction of tertiary fluoride stereocenters by organocatalytic fluorocyclization. *J. Am. Chem. Soc.* 142: 20048–20057.

64 Willgerodt, C. (1885). Ueber einige aromatische Jodidchloride. *J. Prakt. Chem.* 33: 154–160.

65 Garvey, B.S.J., Halley, L.F., and Allen, C.F.H. (1937). Aryliododihalides as halogenating agents. *J. Am. Chem. Soc.* 59: 1827–1829.

66 Cristol, S.J., Stermitz, F.R., and Ramey, P.S. (1956). Mechanisms of elimination reactions. XVII. The *cis*- and *trans*-1,2-dichloroacenaphthenes; Trans-1,2-dibromoacenaphthene. *J. Am. Chem. Soc.* 78: 4939–4941.

67 Hara, S., Nakahigashi, J., Ishi-I, K. et al. (1998). Difluorination of alkenes with iodotoluene difluoride. *Synlett*: 495–496.

68 Molnár, I.G. and Gilmour, R. (2016). Catalytic difluorination of olefins. *J. Am. Chem. Soc.* 138: 5004–5007.

69 Banik, S.M., Medley, J.W., and Jacobsen, E.N. (2016). Catalytic, diastereoselective 1,2-difluorination of alkenes. *J. Am. Chem. Soc.* 138: 5000–5003.

70 Uyanik, M., Yasui, T., and Ishihara, K. (2013). Hydrogen bonding and alcohol effects in asymmetric hypervalent iodine catalysis: enantioselective oxidative dearomatization of phenols. *Angew. Chem. Int. Ed.* 52: 9215–9218.

71 Uyanik, M., Sasakura, N., Mizuno, M., and Ishihara, K. (2017). Enantioselective synthesis of masked benzoquinones using designer chiral hypervalent organoiodine(III) catalysis. *ACS Catal.* 7: 872–876.

72 Uyanik, M., Yasui, T., and Ishihara, K. (2017). Chiral hypervalent organoiodine-catalyzed enantioselective oxidative spirolactonization of naphthol derivatives. *J. Org. Chem.* 82: 11946–11953.

73 Olah, G.A., Shih, J.G., and Prakash, G.K.S. (1986). Fluorine-containing reagents in organic synthesis. *J. Fluorine Chem.* 33: 377–396.

74 Scheidt, F., Schäfer, M., Sarie, J.C. et al. (2018). Enantioselective, catalytic vicinal difluorination of alkenes. *Angew. Chem. Int. Ed.* 57: 16431–16435.

75 Haj, M.K., Banik, S.M., and Jacobsen, E.N. (2019). Catalytic, enantioselective 1,2-difluorination of cinnamamides. *Org. Lett.* 21: 4919–4923.

76 Patrick, T.B., Scheibel, J.J., Hall, W.E., and Lee, Y.H. (1980). Substituent effects in the fluorination-rearrangement of 1,1-diarylethenes with aryliodine(III) difluorides. *J. Org. Chem.* 45: 4492–4494.

77 Gregorcic, A. and Zupan, M. (1977). Fluorination with substituted (Difluoroiodo)arenes. *Bull. Chem. Soc. J.* 50: 517–520.

78 Carpenter, W. (1966). Aryliodosodifluorides. *J. Org. Chem.* 31: 2688–2689.

79 Zupan, M. and Pollak, A. (1975). Preparation of a polymer-supported aryliodine(III) difluoride and its use to fluorinate olefins to difluorides. *J. Chem. Soc., Chem. Commun.*: 715–716.

80 Gregorčič, A. and Zupan, M. (1977). Fluorination of norbornadiene and 1,4-dihydro-1,4-methanonaphthalene with substituted (difluoroiodo)benzenes. *J. Chem. Soc., Perkin Trans. 1*: 1446–1449.

81 Šket, B., Zupan, M., and Zupet, P. (1984). Role of the polymer backbone on the reactivity of polymer-supported (dichloroiodo)benzene. *Tetrahedron* 40: 1603–1606.

82 Banik, S.M., Medley, J.W., and Jacobsen, E.N. (2016). Catalytic, asymmetric difluorination of alkenes to generate difluoromethylated stereocenters. *Science* 353: 51–54.

83 Levin, M.D., Ovian, J.M., Read, J.A. et al. (2020). Catalytic enantioselective synthesis of difluorinated alkyl bromides. *J. Am. Chem. Soc.* 142: 14831–14837.

84 Sarie, J.C., Neufeld, J., Daniliuc, C.G., and Gilmour, R. (2019). Catalytic vicinal dichlorination of unactivated alkenes. *ACS Catal.* 9: 7232–7237.

85 Shimogaki, M., Fujita, M., and Sugimura, T. (2016). Metal-free enantioselective oxidative arylation of alkenes: hypervalent-iodine-promoted oxidative C–C bond formation. *Angew. Chem. Int. Ed.* 55: 15797–15801.

86 Shimogaki, M., Fujita, M., and Sugimura, T. (2017). Enantioselective C–C bond formation during the oxidation of 5-Phenylpent-2-enyl carboxylates with hypervalent iodine(III). *J. Org. Chem.* 82: 11836–11840.

87 Sarie, J.C., Thiehoff, C., Neufeld, J. et al. (2020). Enantioselective synthesis of 3-fluorochromanes via iodine(I)/iodine(III) catalysis. *Angew. Chem. Int. Ed.* 59: 15069–15075.

10

Catalytic Oxidative α-Functionalization of Carbonyls

Muhammet Uyanik

Graduate School of Engineering, Nagoya University, Nagoya, Japan

10.1 Introduction

Carbonyl compounds play a central role in synthetic organic chemistry. The oxidative α-functionalization of carbonyl compounds enables access to a large number of synthetic building blocks and high-value molecules. In contrast to the traditional coupling process, oxidative coupling reactions proceed *via* the *in situ* activation of substrates to form reactive intermediates. To date, a number of oxidation systems using transition metal catalysis as well as iodine-based redox organocatalysis have been developed for this purpose [1–3].

Over the past three decades, hypervalent organoiodines(III and V) have attracted great attention due to their mild and chemoselective oxidizing properties and environmentally benign characteristics in contrast to heavy metal-based oxidants [4]. In particular, the reactivities of organoiodines(III) resemble those of Hg(II), Tl(III), and Pb(IV). However, the use of hypervalent iodines as stoichiometric oxidants has been limited because some are shock-sensitive or explosive, and/or show poor solubility in organic solvents. Therefore, the catalytic use of hypervalent iodine is strongly demanded from both economical and environmental perspectives. The first catalytic use of hypervalent organoiodine(III) compounds using chemical oxidants was reported in 2005 [5, 6]. In these reactions, iodine(III) active species can be generated *in situ* from a catalytic amount of iodoarene and *meta*-chloroperbenzoic acid (*m*-CPBA) as a stoichiometric oxidant in the presence of Lewis or Brønsted acids as a coactivator (Scheme 10.1). Since these breakthroughs, rapid progress has been made in the development of hypervalent iodine catalysis [4, 7–10].

However, relatively expensive and potentially explosive co-oxidants such as *m*-CPBA have often been used in the presence of acid activators for the organoiodine(III) catalysis. In addition, side-products derived such as *meta*-chlorobenzoic acid (*m*-CBA) from the oxidants are generated. To overcome these limitations, in 2010, Uyanik, Ishihara, and colleagues have developed ammonium hypoiodite catalysis for the oxidative coupling reactions [11]. The hypoiodite salts, catalytic active species, are generated *in situ* from the corresponding iodides in the presence

Iodine Catalysis in Organic Synthesis, First Edition. Edited by Kazuaki Ishihara and Kilian Muñiz.

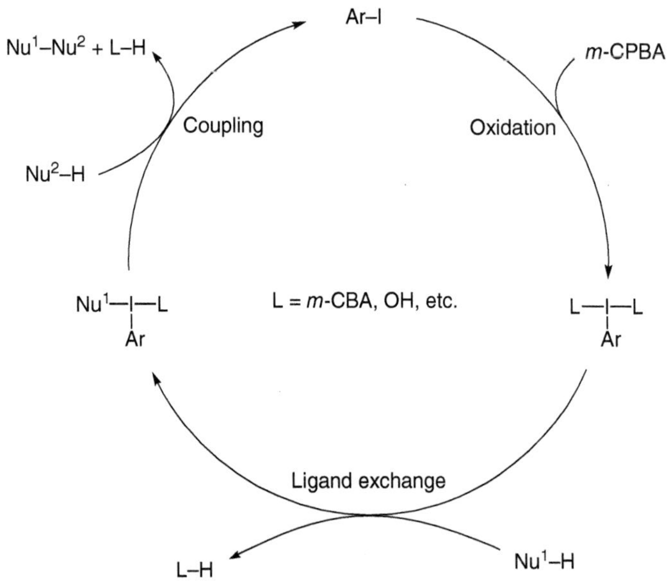

Scheme 10.1 General catalytic cycle of organoiodine(III) catalysis for oxidative coupling.

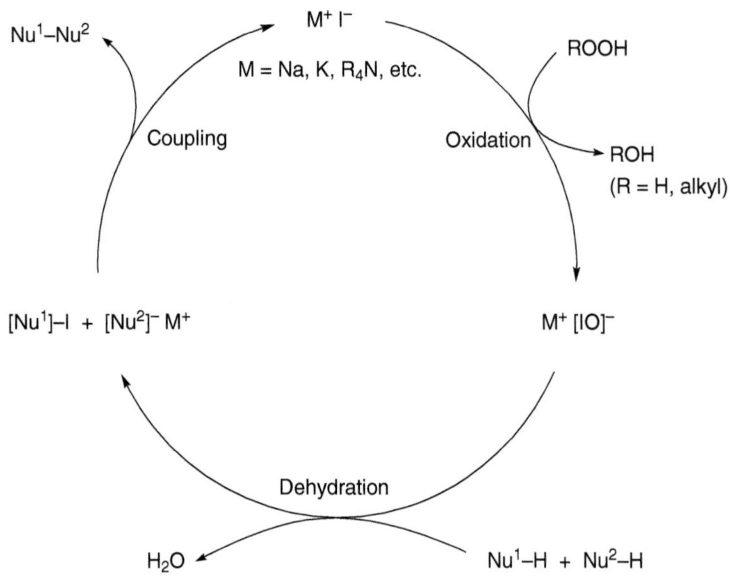

Scheme 10.2 General catalytic cycle of hypoiodite catalysis for oxidative coupling.

of relatively inexpensive and mild oxidants such as hydrogen peroxide or alkylhydroperoxides (Scheme 10.2). In contrast to organoiodine(III) catalysis, this catalytic oxidation system proceeds under milder conditions and water or alcohol is the only side-product derived from the oxidant used. After these findings, rapid progress has been made in the development of inorganic iodine-catalyzed oxidative transformations [9, 10, 12–14]. In this chapter, we discuss both iodine-based catalysis for the oxidative α-functionalization of carbonyl compounds. Specially, we focus on the enantioselective α-oxidative coupling reactions using chiral iodine-based catalysts. Please refer to Chapter 7 for details on the design concept of the chiral organoiodine(III) catalysts.

10.2 Organoiodine(III) Catalysis

Ochiai and colleagues have reported the first organoiodine(III)-catalyzed α-oxidative coupling of carbonyl compounds (Scheme 10.3) [5]. Iodobenzene is employed as a catalyst in the presence of *m*-CPBA and $BF_3{\cdot}Et_2O$ as an oxidant and activator, respectively. Acetic acid, a coupling partner, is used as a solvent quantity. $BF_3{\cdot}Et_2O$ accelerates both the enolization of ketones and the oxidation of iodobenzene to iodine(III) species by *m*-CPBA. The addition of water is also crucial to the success of α-oxyacylation, since most of the unreacted ketones are recovered in the absence of water. A variety of dialkyl and aryl alkyl ketones are smoothly α-oxidized to give the corresponding α-acetoxyketones in good-to-moderate yields. In the case of unsymmetrical ketones, oxidation of a methylene group of linear alkyl chains is favored over that of the methyl group. The reaction is proposed to proceed *via* α-*O*-bound iodine(III) intermediate, which on S_N2 displacement by acetic acid affords the desired α-acetoxyketone [5, 15].

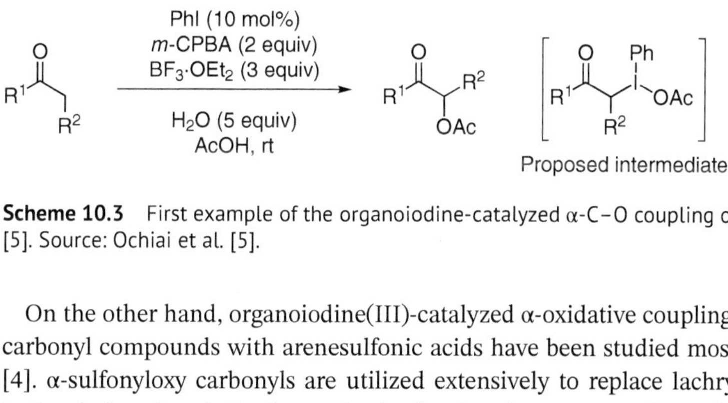

Scheme 10.3 First example of the organoiodine-catalyzed α-C–O coupling of carbonyls [5]. Source: Ochiai et al. [5].

On the other hand, organoiodine(III)-catalyzed α-oxidative coupling reactions of carbonyl compounds with arenesulfonic acids have been studied most extensively [4]. α-sulfonyloxy carbonyls are utilized extensively to replace lachrymatory and toxic α-halocarbonyls for the synthesis of various heteroaromatics such as thiazoles, oxazoles, imidazoles, pyrazoles, etc. [16, 17]. In 2006, Togo and colleagues have reported the first organoiodine(III)-catalyzed α-oxytosylation of ketones with TsOH by employing iodobenzene as a catalyst and *m*-CPBA as an oxidant [18]. After this report, several organoiodine(III) catalysts have been developed for α-oxysulfonylation of ketones [19–21].

The development of enantioselective α-oxytosylation of ketones using chiral organoiodine(III) reagents has been ongoing for almost a quarter of a century. In 1997, Wirth and colleagues have reported the first enantioselective α-oxytosylation of ketones using chiral hypervalent organoiodine(III) reagents [22]. In 2007, the first catalytic enantioselective α-oxytosylation of ketones has also been reported by Wirth's group [23]. By using chiral iodoarene **1** as a catalyst, α-tosyloxypropiophenone is obtained with 27% enantiomeric excess (ee) (Scheme 10.4). After this pioneering work, a number of structurally varied chiral iodoarene catalysts **2–6** have been developed; however, the enantioselectivities are moderate (<70% ee) (Scheme 10.4) [24–32].

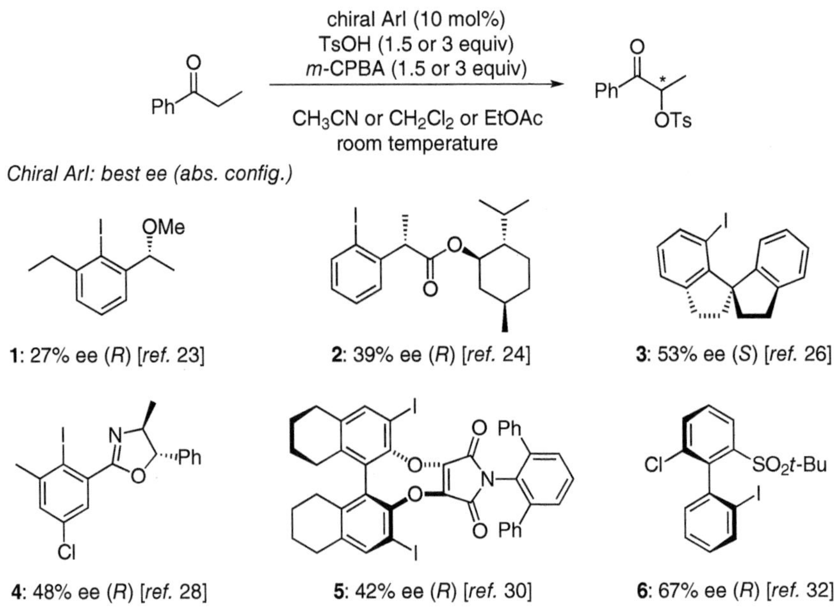

Scheme 10.4 Chiral organoiodine-catalyzed enantioselective α-oxytosylation of propiophenone.

The low-to-moderate enantioselectivity obtained for the α-oxytosylation of ketones may be attributed to the mechanistic pathway proceeding through an S_N2'-type reductive elimination involving an *O*-bonded iodane intermediate (**int-A**) generated from the iodine(III)-mediated enolization of the ketone (Scheme 10.5) [33]. In 2015, Legault and colleagues reported a breakthrough in the organoiodine-catalyzed enantioselective α-oxytosylation of enol ester derivatives [34]. The use of enol ester derivatives in place of ketones may prevent passage through the *O*-bonded intermediate pathway and instead would favor proceeding through an α-*C*-bonded iodane intermediate (**int-B**) (Scheme 10.5) [33, 35].

Indeed, enantioselective α-oxytosylation of enol esters affords the corresponding α-tosyloxy ketones with an unprecedented level of enantioselectivity up to 90% ee (Scheme 10.6a) [34]. The use of bis(lactate)-derived bisamide **7** [36, 37] as a chiral iodoarene gives the best results. The reaction can be performed under both catalytic and stoichiometric conditions. To simplify the investigation of the substrate scope, the authors focused on the development of stoichiometric reactions, since iodoarene **9** is stable under oxidative conditions and can be easily

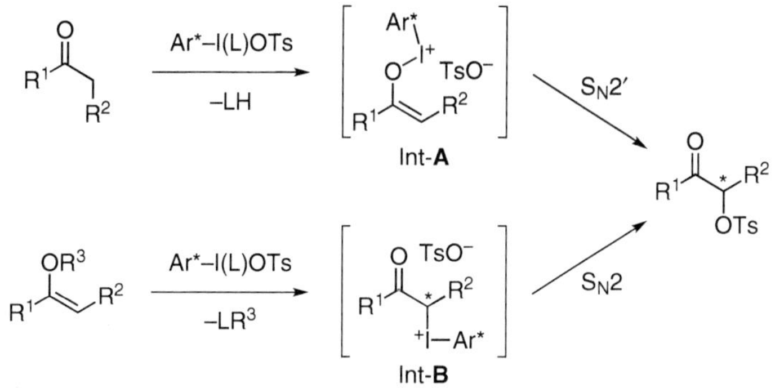

Scheme 10.5 Mechanistic consideration of the enantioselective α-oxytosylation of ketones and enol derivatives [33–35]. Source: Beaulieu and Legault [33], Basdevant and Legault [34], Basdevant and Legault [35].

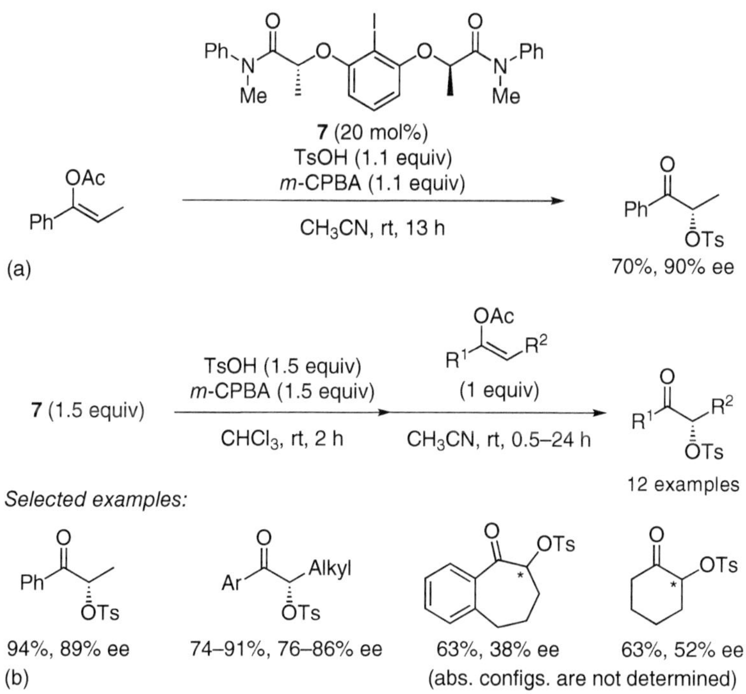

Scheme 10.6 Highly enantioselective α-oxytosylation of enol esters under catalytic and stoichiometric conditions [34]. (a) Catalytic conditions, (b) Stoichiometric conditions. Source: Basdevant and Legault [34].

recovered (>80%) (Scheme 10.6b). For stoichiometric reactions, the iodine(III) active species can be first prepared by the oxidation of **9** with *m*-CPBA in the presence of TsOH [38]. Several enol acetates can be converted to the corresponding α-tosyloxy ketones in high yields with high enantioselectivity. The stoichiometric conditions yielded the product with the same level of selectivity as the catalytic conditions. The stereochemistry of enol ester is highly important to induce high enantioselectivity. Cyclic substrates having (*E*)-*O*-enol stereochemistry afford only modest enantioselectivities.

Recently, Nachtsheim and colleagues have proved that high stereoinduction can be achieved for the enantioselective α-oxytosylation of ketones using designer organoiodine(III) catalysts [39]. The research group has developed *N*-heterocycle-stabilized organoiodines (NHIs) as stable and highly reactive hypervalent iodine catalysts [40, 41]. Specifically, a triazole-substituted aryl iodide **8** shows remarkable enantioselectivity up to 88% ee in the α-oxytosylation of propiophenone (Scheme 10.7) [39]. Notably, this catalyst can be defined as "omnipotent" since the observed enantioselectivity for mechanistically diverse enantioselective transformations such as α-oxygenation of carbonyls, oxidative dearomatization of arenols, and oxidative rearrangement of allyl alcohols is the highest ever reported. A significant role of the triazole unit of **8** has been rationalized by DFT calculations as a stabilizing donor both in a potential *N*-bound state (*N*–I interaction) or as a hydrogen-bond acceptor (*N*–HO interaction) in the iodine(III)-active species [39]. These secondary interactions may force the geometry of the iodine(III) center into a reactive bent state, "hypervalent twist" [27], with an unusual vertical alignment between the hypervalent bond and the arene.

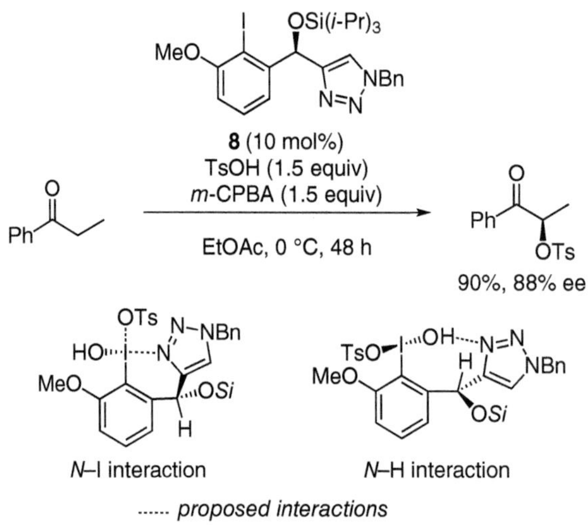

Scheme 10.7 Highly enantioselective α-oxytosylation using NHI **8** as an "omnipotent" catalyst [39]. Source: Abazid and Nachtsheim [39].

In contrast to α-oxysulfonylation reactions, the organoiodine(III)-catalyzed enantioselective intermolecular α-oxyacylation of carbonyls has not yet been achieved. Instead, a few examples of the catalytic enantioselective α-oxylactonization of ketocarboxylic acids as an intramolecular α-C–O coupling have been reported. For example, in 2012, Moran and colleagues reported the enantioselective oxylactonization of 5-oxo-5-phenylpentanoic acid using pseudoephedrine-derived iodoarene catalyst **9** in the presence of *m*-CPBA as an oxidant and TsOH as an activator to give γ-benzoyl-γ-butyrolactone with 51% ee (Scheme 10.8) [42]. Again, as in the α-oxytosylation reaction, the highest enantioselectivity up to 81% ee for α-oxylactonization reaction is achieved by Nachtsheim's catalyst **8** (Scheme 10.8) [39]

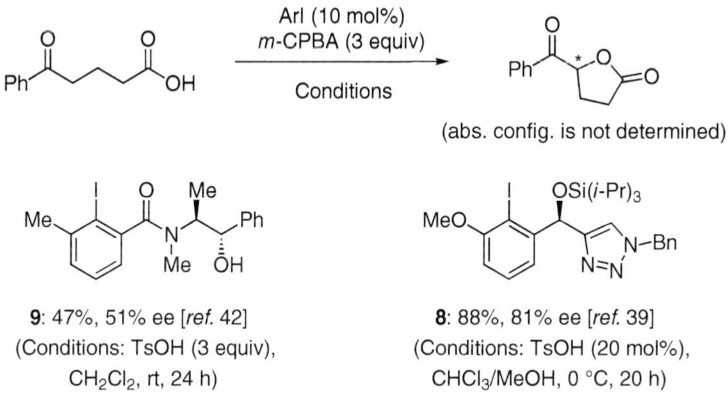

Scheme 10.8 Enantioselective oxylactonization of ketocarboxylic acid [39, 42]. Source: Abazid and Nachtsheim [39], Rodriguez and Moran [42].

On the other hand, α-halogenation, especially α-fluorination, of carbonyl compounds has also extensively investigated using hypervalent iodine(III) reagents such as (difluoroiodo)benzene ($PhIF_2$) [4]. In 2013, Kitamura and colleagues have reported the first organoiodine(III)-catalyzed α-fluorination of 1,3-dicarbonyls using aqueous HF as a fluorine source and *m*-CPBA as a terminal oxidant [43]. The key for this catalytic reaction is the *in situ* generation of an active species, $ArIF_2$, by the smooth oxidation of iodoarene in the presence of hydrogen fluoride.

The highly enantioselective electrophilic α-fluorination of carbonyl compounds is realized by transition metal catalysts or organocatalysts using Selectfluor or NFSI as an expensive electrophilic fluorinating reagent [44]. However, enantioselective nucleophilic α-fluorination of carbonyl compounds is less studied. In 2014, Shibata and colleagues reported the first chiral organoiodine(III)-catalyzed enantioselective α-fluorination of 1,3-dicarbonyl compounds using HF·pyridine as a nucleophilic fluorinating reagent in the presence of binaphthyl diiodide as a chiral catalyst (Scheme 10.9) [45]. The best selectivities are observed for the α-fluorination of indanone-derived β-keto esters having a sterically demanding adamantyl or menthyl ester, with 56% ee or 65% diastereomeric excess (de), respectively.

In 2018, Rueping and colleagues reported a highly enantioselective α-fluorination of indanone-derived β-ketoesters using C_2-symmetric iodoarene catalyst **10** having an electron-withdrawing substituent at *para*-position of the resorcinol unit (Scheme 10.10) [46]. The choice of hydrogen fluoride source is crucial for both chemical yields and enantioselectivity. Commercially available triethylamine pentahydrofluoride ($5HF{\cdot}Et_3N$) provides a compromise between good chemical

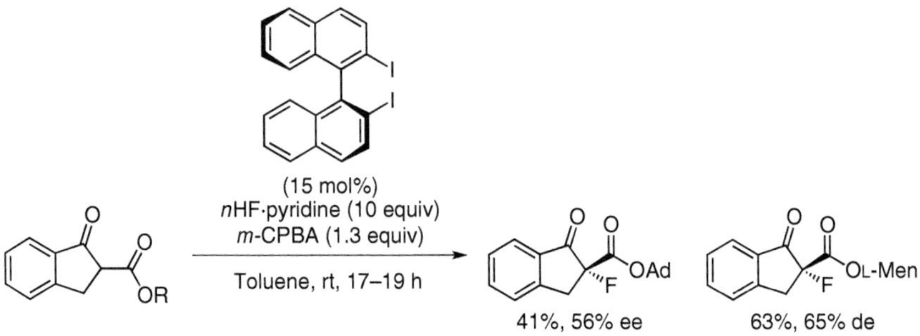

Scheme 10.9 The first chiral organoiodine(III)-catalyzed enantioselective α-fluorination [45]. Source: Suzuki et al. [45].

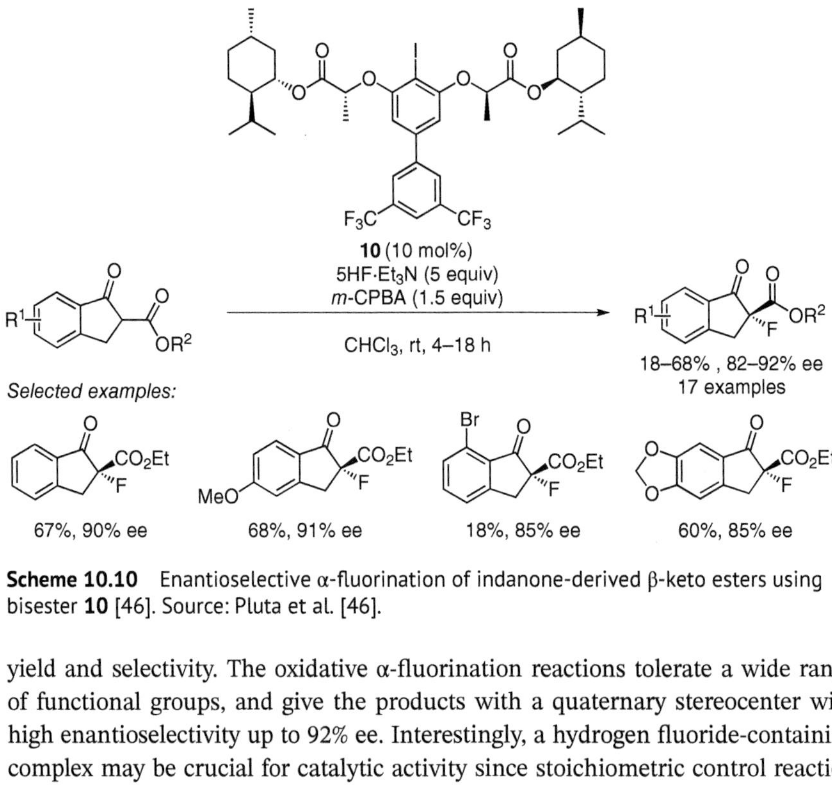

Scheme 10.10 Enantioselective α-fluorination of indanone-derived β-keto esters using bisester **10** [46]. Source: Pluta et al. [46].

yield and selectivity. The oxidative α-fluorination reactions tolerate a wide range of functional groups, and give the products with a quaternary stereocenter with high enantioselectivity up to 92% ee. Interestingly, a hydrogen fluoride-containing complex may be crucial for catalytic activity since stoichiometric control reaction with organoiodine(III) **10**·F_2 showed only traces of the product in the absence of hydrogen fluoride.

The authors performed DFT calculations to investigate the reaction mechanism in detail (Scheme 10.11) [46]. Experimental results reveal the *in situ* formation and activation of the iodine(III), $ArIF_2$ (**10**·F_2), by the HF molecule to give $[ArIF]^+[HF_2]^-$ as an active species. The first steps involve the reaction of β-keto ester with $[ArIF]^+/[HF_2]^-$ to give an *O*-bonded iodine(III) intermediate **11**. The reaction can proceed in a single step to the product through an S_N2'-type reductive elimination *via* a nucleophilic attack of a fluoride anion to the α-C atom of **11**.

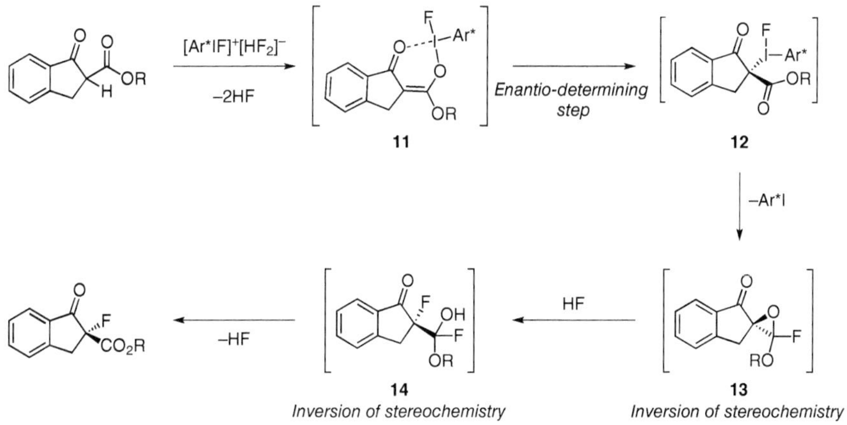

Scheme 10.11 Proposed mechanism of the enantioselective α-fluorination [46]. Source: Pluta et al. [46].

However, S_N2'-type mechanism would favor the formation of (*S*)-enantiomer of the product, which is in disagreement with the experimental selectivity of the observed (*R*)-enantiomer. The reaction can more easily proceed *via* transfer of the $[ArIF]^+$ fragment from the *O* atom to the α-C atom to give an α-C-bonded iodine(III) intermediates **12**. This is the enantioselectivity-determining step, and the formation of the (*R*)-enantiomer of intermediate **12** is favored. The intermediate **12** is very stable and the energy barrier (31.2 kcal/mol) for the reverse reaction from **12** to **11** is too high for the reaction conditions. From intermediate **12**, the reductive elimination of ArI and the formation of an epoxide ring occurs to give a hemiacetal intermediate **13** with an inversion of stereochemistry at α-C atom. Ring-opening of the epoxide **13** by a nucleophilic attack of fluoride anion affords intermediate **14** and restores the (*R*)-configuration at the α-C atom. Dissociation of the fluoride leads to (*R*)-enantiomer of the product.

In the same year, Zheng and colleagues developed a novel planar chiral iodoarene **15** based on [2,2]paracyclophane as a catalyst for the enantioselective oxidative α-fluorination of indanone-derived β-ketoesters with high enantioselectivity up to 89% ee (Scheme 10.12) [47]. Notably, α-fluorination of β-ketoamide also proceeds under similar conditions to afford the product with good enantioselectivity. Although this is the first highly enantioselective reaction using an organoiodine catalyst with planar chirality, catalyst **15** is inferior to resorcinol-derived catalyst **10** [46] with respect to enantioselectivity.

In 2014, Zhong and colleagues reported the first organoiodine(III)-catalyzed enantioselective oxidative α-C–C coupling of carbonyl compounds [48]. The sequential oxidative Friedel–Crafts-type spirocyclization of N^1,N^3-diarylmalonamides [49] using chiral organoiodine catalyst **16** gives the structurally diverse spirooxindoles in good-to-moderate yields with high enantioselectivity up to 90% ee (Scheme 10.13) [48]. The use of trifluoroacetic acid (TFA) as an activator is crucial to enhance the catalytic activity. Investigation of the sidearms of the resorcinol-derived C_2-symmetric chiral catalysts reveals the tertiary amide rather than secondary

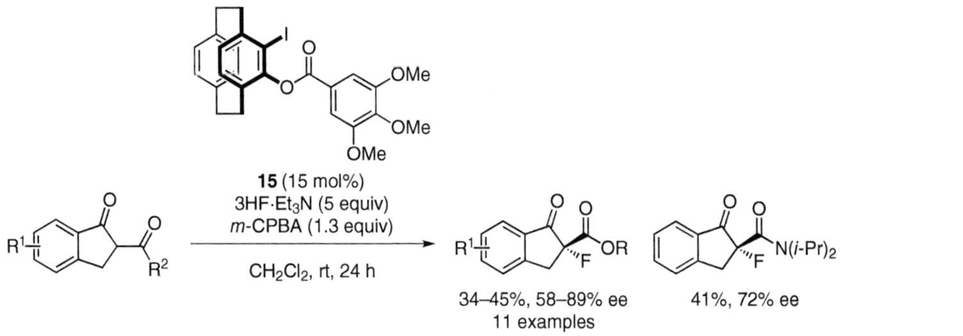

Scheme 10.12 Enantioselective α-fluorination using planar chiral catalyst **15** [47]. Source: Wang et al. [47].

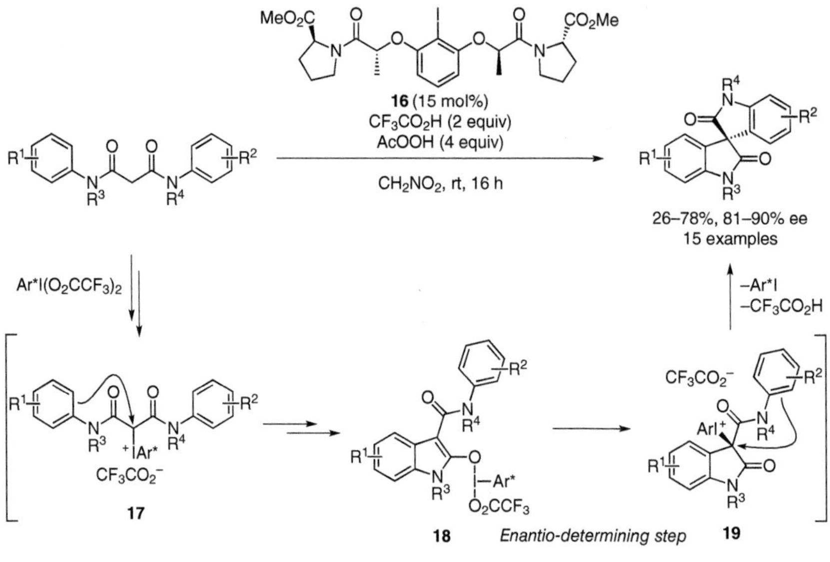

Scheme 10.13 Enantioselective oxidative Friedel–Crafts-type spirocyclizations [48]. Source: Wu et al. [48].

amide or carboxylic acid plays a key role to induce high enantioselectivity. Notably, the use of peracetic acid (AcOOH) as an oxidant instead of *m*-CPBA gives higher chemical yield and enantioselectivity. Sunoj and colleagues investigated the reaction mechanism of this cascade α-C–C coupling reactions (Scheme 10.13) [50, 51]. The multistep sequential mechanism of oxidative coupling of malonamide derivatives involves α-*C*-bonded iodonium species **17** and **19** as key intermediates with a critical trifluoroacetate counterion. DTF calculations suggest that trifluoroacetate anion offers the energetically most preferred transition state for the Friedel–Crafts-type cyclization. Additionally, 1,3-migration of the chiral iodine(III) in the *O*-iodonium enolate **18** to *C*-iodonium enolate **19** is identified as the enantio-determining step in the overall reaction. Several noncovalent secondary interactions (i.e. hydrogen-bonding interactions), particularly between amide sidearms and the iodine(III)-bound ligands (TFA), induce a helical fold [36, 52] around the iodine(III) center.

As a similar enantioselective cascade oxidative coupling reaction, in 2016, Du and colleagues reported a sequential oxidative spirocyclization of alkyl 3-oxopentanedioate monoamide derivatives catalyzed by a lactate-derived bis-*tert*-amide **20** to give biologically important spirofurooxindole with high enantioselectivity up to 91% ee (Scheme 10.14) [53]. The reaction conditions are similar to Zhang's procedure [48] using an excess amount of TFA as an activator. The reaction is proposed to proceed *via* oxidative C–O bond formation followed by oxidative C–C bond formation.

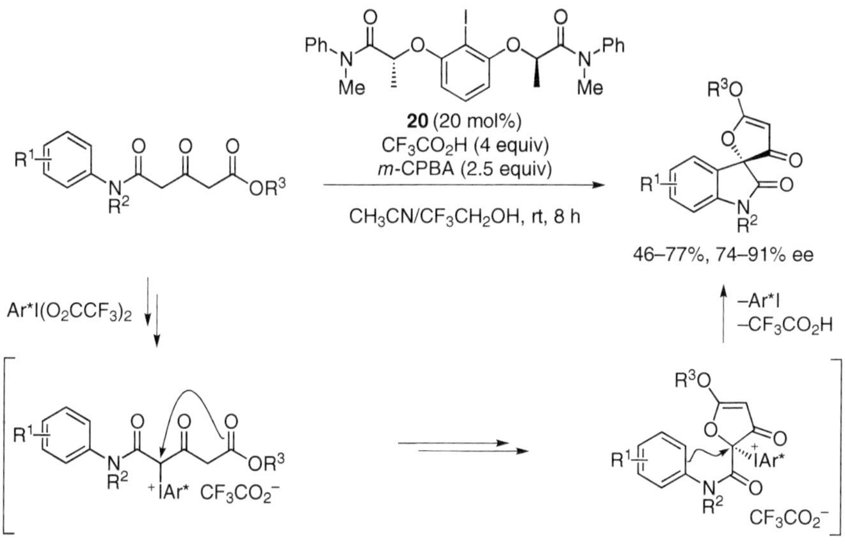

Scheme 10.14 Enantioselective cascade oxidative spirocyclization [53]. Source: Cao et al. [53].

10.3 Inorganic Iodine Catalysis

Uyanik, Ishihara, and colleagues have developed the organoiodine(III)-catalyzed oxylactonization of ketocarboxylic acids to ketolactones by using catalytic amounts of TsOH as an activator in the presence of *m*-CPBA as an oxidant (Scheme 10.15a) [54]. However, as often observed for organoiodine(III)/peracetic acid catalysis, low chemoselectivity is observed due to competition with Baeyer–Villiger oxidation with the terminal oxidant used.

To overcome the issues and limitations of the organoiodine(III)-catalyzed oxidative coupling reactions [4, 5, 54], the same research group has developed highly efficient and general intra- and intermolecular oxidative α-C–O coupling (α-oxyacylation) reactions catalyzed by *in situ*-generated ammonium hypoiodite species with hydrogen peroxide or *tert*-butyl hydroperoxide (TBHP) as

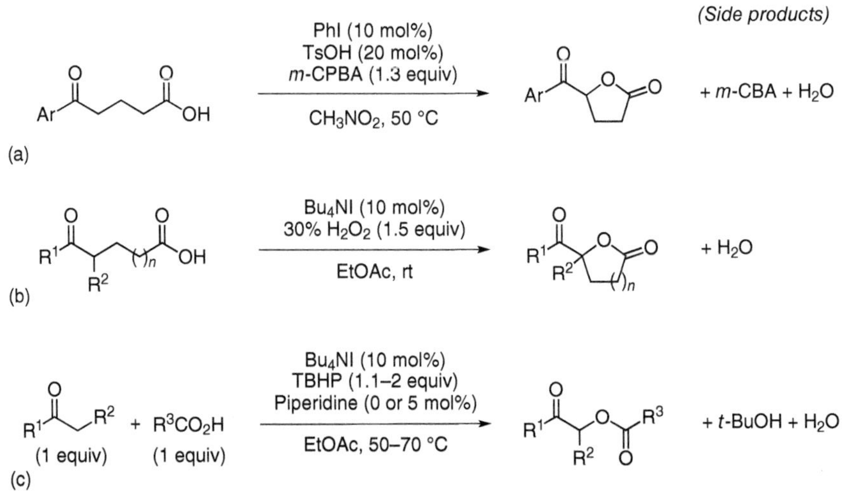

Scheme 10.15 Oxidative α-oxyacylation of carbonyls using iodine-based catalysis. (a) Organoiodine(III) catalysis [54], (b) Hypoiodite catalysis for intramolecular coupling [55, 56], (c) Hypoiodite catalysis for intramolecular coupling [55].

an environmentally benign oxidant [55, 56]. Tetrabutylammonium iodide (TBAI) as a catalyst and hydrogen peroxide or TBHP as an inexpensive and environmentally benign oxidant are used in place of iodobenzene and *m*-CPBA, respectively. In contrast to organoiodine-catalyzed reactions [54], the highly chemoselective oxylactonization (intramolecular coupling) of ketocarboxylic acids proceeds under milder conditions and water is the only by-product derived from the oxidant (Scheme 10.15b) [55, 56]. Importantly, no Baeyer–Villiger products are obtained. The use of anhydrous TBHP at elevated temperatures (70 °C) is much more effective than aqueous hydrogen peroxide as an oxidant for the intermolecular oxidative coupling of ketones with carboxylic acids (Scheme 10.15c) [54]. The oxidative coupling of several structurally diverse ketones, aldehydes, and 1,3-dicarbonyl compounds with carboxylic acids gives the corresponding α-acyloxycarbonyl compounds in good-to-excellent yields. Importantly, the addition of a catalytic amount of piperidine under milder conditions (50 °C) is found to be effective for enhancing chemoselectivity for the coupling of aldehydes. The reaction rate may be accelerated under milder conditions due to the enamine formation of aldehyde with piperidine. Several functional groups such as terminal or internal alkenyl, benzyloxy, silyloxy, acetal, halogen, and ester are tolerated under these conditions. Notably, polymerizable acrylic and methacrylic acids can be used as coupling partners under these conditions to give the corresponding α-acryloyloxy ketones, which are used industrially as starting materials for the production of important polymers.

Since 2010, rapid progress has been made in the development of inorganic iodine-based catalysis [9, 10, 12–14]. In Scheme 10.16, catalytic oxidative α-functionalizations of carbonyl compounds including α-C–O [55–70], α-C–N [71–86], α-C–C [87], α-C–S [88, 89], and α-C–Se [90] couplings are summarized.

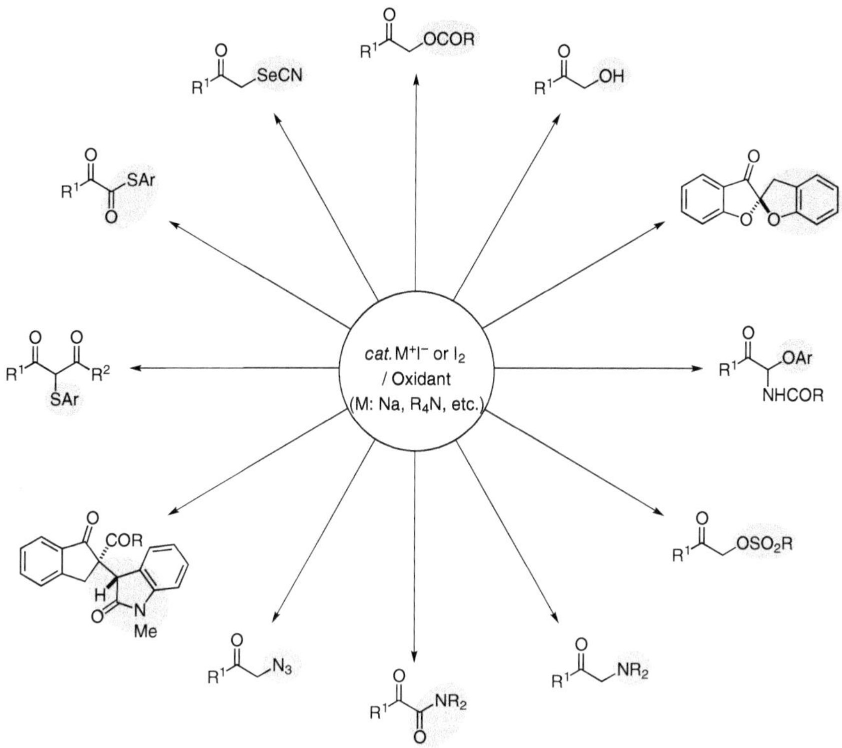

Scheme 10.16 Summarization of the oxidative α-coupling of carbonyls using inorganic iodine-based catalysis.

Tetraalkylammonium iodide or alkali metal iodides are often used as a catalyst in the presence of an appropriate oxidant such as hydrogen peroxide, alkyl hydroperoxides, sodium percarbonate, oxone, potassium persulfate, or *m*-CPBA. Depending on the coupling partners, and oxidant or reaction conditions, these reactions may proceed via ionic or radical mechanism [12–14]. Please refer to Chapter 4 for details on the inorganic iodine-based catalysis.

In 2010, Uyanik, Ishihara, and colleagues have developed a chiral ammonium hypoiodite-catalyzed oxidative cyclization of ketophenols with hydrogen peroxide, TBHP, or cumene hydroperoxide (CHP) as an oxidant (Scheme 10.17) [11]. The reaction proceeds with excellent chemoselectivity and no phenol oxidation products are detected. In sharp contrast, the oxidation of phenols with organoiodines(III) is well known [4]. The use of chiral binaphthyl-based quaternary ammonium [91] iodides **21** as catalysts and an *N*-phenylimidazol-2-yl (*Z*) group [92] as an auxiliary of ketones is effective for inducing high enantioselectivities. The enantioselective oxidative five-membered ring cyclization of β-(2-hydroxyphenyl)ketones **22** by using **21a** as a catalyst affords 2-acyl-2,3-dihydrobenzofurans **24** in excellent yields with high enantioselectivities [11]. Both aqueous hydrogen peroxide and anhydrous alkylhydroperoxides can be used as an oxidant for 5-membered cyclization. However, the enantioselective oxidative 6-membered ring cyclization

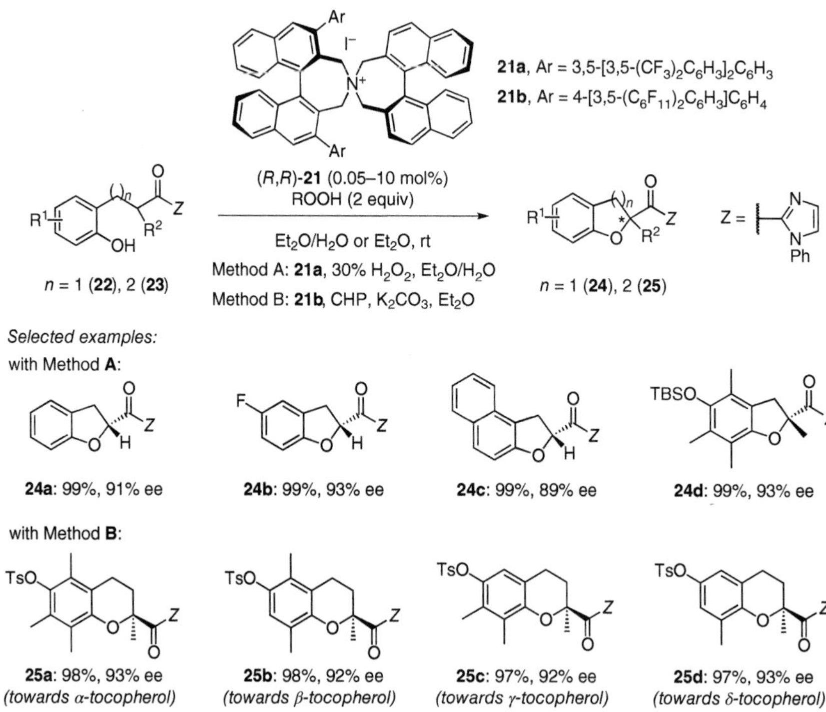

Scheme 10.17 Chiral ammonium hypoiodite-catalyzed oxidative cycloetherification of ketophenols [11, 93]. Source: Uyanik et al. [11], Uyanik et al. [93].

of hydroquinone-derived γ-(2-hydroxyphenyl)ketones **23** directed toward chromans **25** does not proceed under identical conditions that of 5-membered ring cyclization using hydrogen peroxide as an oxidant [93]. Since 6-membered oxidative cyclization is much slower than 5-membered cyclization, undesired side reactions such as the dearomatization of phenol moiety preferentially proceeded to give side-products. The tuning of acidity of the 2-hydroxyphenyl moieties of substrates with electron-withdrawing protective groups (i.e. tosyl group) is crucial for the chemoselective oxidative α-C–O coupling, since electron-rich 2-hydroxyphenyl moieties are easily dearomatized.

Chemo- and enantioselective oxidative 6-membered ring cyclization of tosyl-protected hydroquinone-derived γ-(2-hydroxyphenyl)ketones **23** gives the corresponding 2-acyl chromans **25** bearing a quaternary stereocenter as potent synthetic intermediates for all kinds of tocopherols (α, β, γ, δ) as well as all kinds of tocotrienols (α, β, γ, δ) quantitatively with excellent enantioselectivities (Scheme 10.17) [93]. Importantly, the use of perfluoroalkyl-substituted ammonium iodide **21b** as a catalyst and CHP as an oxidant is crucial to induce high enantioselectivities. Interestingly, the 3,3′-substituents (Ar) of the binaphthyl moiety of (*R*,*R*)-**21** has a dramatic effect on the absolute configuration of the products. The use of (*R*,*R*)-**21b** having biphenyl groups at the 3,3′ positions in place of (*R*,*R*)-**21a** gives (*S*)-**24** or (*S*)-**25** as an opposite enantiomer. This enantioselective hypoiodite catalysis can

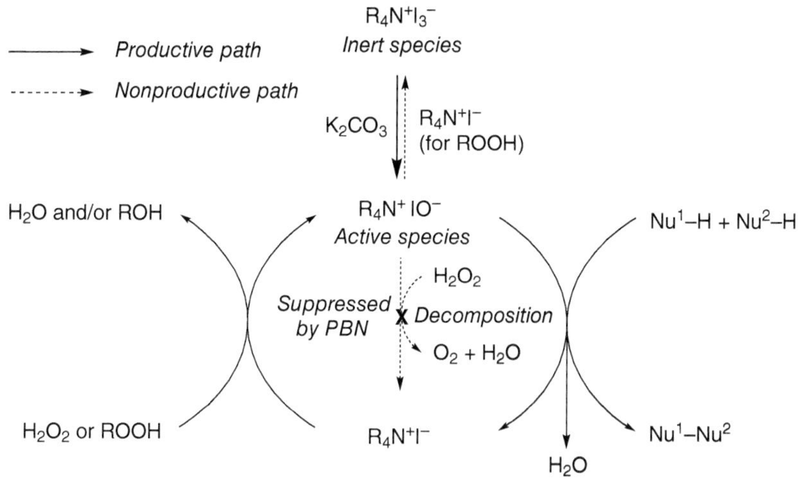

Scheme 10.18 Mechanistic consideration of hypoiodite/H_2O_2 or ROOH catalysis [11, 86, 93]. Source: Uyanik et al. [11], Uyanik et al. [93], Uyanik et al. [86].

be applied to oxidative cycloetherification of δ-hydroxyketone derivatives [94]. The corresponding 2-acyl tetrahydrofurans are obtained in high chemical yield with high enantioselectivity.

Hypoiodite, $[IO]^-$, would be readily generated *in situ* as an unstable catalytic active species from tetraalkylammonium iodide and hydrogen peroxide or alkylhydroperoxide (TBHP or CHP) as an oxidant (Scheme 10.18) [86, 93]. If the desired oxidative coupling reaction is slow, triiodide, $[I_3]^-$, as a stable inert species would be generated when an alkyl hydroperoxide is used as an oxidant [93]. On the other hand, when hydrogen peroxide is used as an oxidant for challenging reactions, the oxidative decomposition of hydrogen peroxide to water and oxygen gas catalyzed by hypoiodite/iodide couple would proceed preferentially [95]. The non-productive pathways of I^+/ROOH catalysis (i.e. generation of I_3^- inert species) can be suppressed under mild basic conditions by using inorganic base additives such as potassium carbonate, and high-performance I^+/ROOH catalysis has been established especially for enantioselective cycloetherification reactions (Method B in Scheme 10.17, turnover number of the catalyst = $\leq$2000) [93]. On the other hand, recently, Uyanik, Ishihara and colleagues developed high-performance I^+/H_2O_2 catalysis especially for the oxidative α-azidation of carbonyl compounds [86]. If the desired oxidative coupling reaction proceeds *via* ionic intermediates, the non-productive pathways (i.e. the oxidative decomposition of hydrogen peroxide *via* radical intermediates [96]) of I^+/H_2O_2 catalysis can be suppressed by using a catalytic amount of a radical trapping agent such as α-phenyl-*N*-*tert*-butylnitrone (PBN) [86].

The same research group has also achieved the first example of chiral quaternary ammonium hypoiodite-catalyzed highly enantioselective intermolecular coupling reaction, namely α-azidation of 1,3-dicarbonyl compounds (Scheme 10.19) [86].

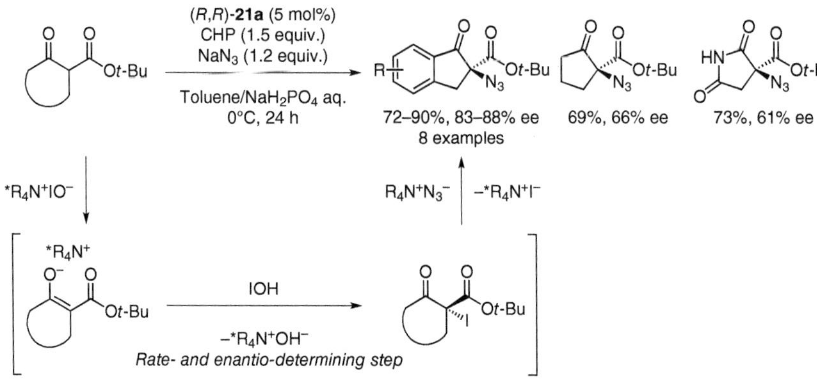

Scheme 10.19 Enantioselective oxidative α-azidation of β-ketoesters [86]. Source: Uyanik et al. [86].

Despite wide applications of α-azidocarbonyls in various fields [97], successful examples of enantioselective α-azidation are still rare [98–100]. A proposed mechanism of hypoiodite-catalyzed α-azidation of 1,3-dicarbonyl involves the α-iodination of 1,3-dicarbonyls followed by nucleophilic azidation with an azide anion *via* S_N2 manner. Mechanistic consideration reveals that, to render the reaction asymmetric, α-iodination step should proceed enantioselectively [86]. However, α-iodo intermediates are stereo-labile and readily epimerized as could not be isolated in enantioenriched form [86]. By taking advantage of the catalytic mechanism, in which nucleophilic azidation may proceed smoothly after the rate-determining α-iodination step, chiral hypoiodite catalysis can be applied to the enantioselective α-azidation. Enantioselective oxidative α-azidation of several indanone- as well as cyclopentanone- and succinimide-derived β-ketoesters with NaN_3 as an azide source by using chiral ammonium iodide **21a** proceeded smoothly to give the corresponding α-quaternary azido compounds with high enantioselectivity (up to 88% ee) (Scheme 10.19) [86]. Notably, the use of CHP is beneficial to slightly improve the enantioselectivity, especially at a lower temperature (0 °C). In addition, higher enantioselectivity can be achieved under slightly acidic conditions by using an aqueous solution of NaH_2PO_4 (pH 4.3).

Very recently, Waser and colleagues reported an unprecedented synergistic enantioselective hypoiodite/imine catalysis (Scheme 10.20) [101]. The use of chiral bifunctional ammonium iodide catalyst **26** in combination with a simple aldimine **27** allows the enantioselective α-hydroxylation reaction of β-ketoesters using aqueous hydrogen peroxide as an oxidant. Although aldimine **27** can be used in catalytic quantity, hydrolysis of **27** is also competed under the reaction conditions. The reaction of indanone- and tetralone-derived β-ketoesters are oxidized under the optimized conditions to afford the corresponding α-hydroxy adducts, generally, in good yield with high enantioselectivity up to 93% ee. The reaction is postulated to proceed *via* the *in situ* generation of an activated electrophilic oxygen transfer reagent from the oxidation of aldimine with (hypo)iodite species. Control experiments reveal that the bifunctional nature of the catalyst **26** would be important

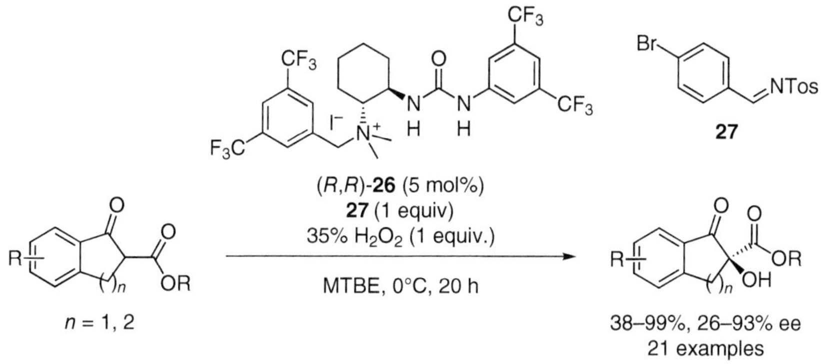

Scheme 10.20 Synergistic chiral ammonium hypoiodite/imine catalysis for the enantioselective α-hydroxylation of β-ketoesters [101]. Source: Mairhofer et al. [101].

to induce high reactivity. On the other hand, the structure of aldimines examined has a strong impact on the enantioselectivity, suggesting a fundamental role in the enantio-determining step. Notably, no *in situ* formation of the oxaziridines [102] from aldimines is observed.

10.4 Conclusion

The oxidative α-functionalization of carbonyl compounds has emerged as a promising tool for a large number of synthetic building blocks and high-value molecules. In this chapter, we reviewed the recent progress on the enantioselective iodine-based oxidation catalysis, i.e. organoiodine(III) and quaternary ammonium hypoiodite catalysis, for the oxidative α-functionalization of carbonyl compounds. The first organoiodine(III)-catalyzed enantioselective oxidative coupling reaction has been reported in 2007 [23]. After the breakthroughs of the development of the resorcinol-derived highly enantioselective organoiodine(III) catalyst in 2010 [36, 37], rapid progress has been made in the development of enantioselective organoiodine catalysis, especially, for oxidative α-coupling of carbonyls as well as oxidative dearomatization of arenols and oxidative difunctionalizaiton of alkenes. On the other hand, in 2010, the first enantioselective hypoiodite catalysis has been developed for the oxidative α-cycloetherification of ketones [11]. In contrast to organoiodine(III) catalysis, enantioselective hypoiodite catalysis has so far been explored modestly. Especially, the development of a general and highly enantioselective intermolecular coupling looks to be highly promising, and perhaps this is an area where major further developments might be expected.

References

1 Smith, A.M. and Hii, K.K. (2011). Transition metal catalyzed enantioselective α-heterofunctionalization of carbonyl compounds. *Chem. Rev.* 111: 1637–1656.

2 Guillena, G. and Ramón, D.J. (2006). Enantioselective α-heterofunctionalisation of carbonyl compounds: organocatalysis is the simplest approach. *Tetrahedron: Asymm.* 17: 1465–1492.
3 Vilaivan, T. and Bhanthumnavuin, W. (2010). Organocatalyzed asymmetric α-oxidation, α-aminoxylation and α-amination of carbonyl compounds. *Molecules* 15: 917–958.
4 Yoshimura, A. and Zhdankin, V.V. (2016). Advances in synthetic applications of hypervalent iodine compounds. *Chem. Rev.* 116: 3328–3435.
5 Ochiai, M., Takeuchi, Y., Katayama, T. et al. (2005). Iodobenzene-catalyzed α-acetoxylation of ketones. *In situ* generation of hypervalent (diacyloxyiodo)benzenes using *m*-chloroperbenzoic acid. *J. Am. Chem. Soc.* 127: 12244–12245.
6 Dohi, T., Maruyama, A., Yoshimura, M. et al. (2005). Versatile hypervalent-iodine(III)-catalyzed oxidations with *m*-chloroperbenzoic acid as a cooxidant. *Angew. Chem. Int. Ed.* 44: 6193–6196.
7 Ochiai, M. and Miyamoto, K. (2008). Catalytic version of and reuse in hypervalent organo-λ^3- and -λ^5-iodane oxidation. *Eur. J. Org. Chem.* 2008: 4229–4239.
8 Dohi, T. and Kita, Y. (2009). Hypervalent iodine reagents as a new entrance to organocatalysts. *Chem. Commun.* 45: 2073–2085.
9 Finkbeiner, P. and Nachtsheim, B.J. (2013). Iodine in modern oxidation catalysis. *Synthesis* 45: 979–999.
10 Yusubov, M.S. and Zhdankin, V.V. (2015). Iodine catalysis: a green alternative to transition metals in organic chemistry and technology. *Res.-Effic. Technol.* 1: 49–67.
11 Uyanik, M., Okamoto, H., Yasui, T., and Ishihara, K. (2010). Quaternary ammonium (hypo)iodite catalysis for enantioselective oxidative cycloetherification. *Science* 328: 1376–1379.
12 Uyanik, M. and Ishihara, K. (2012). Catalysis with in situ-generated (hypo)iodite ions for oxidative coupling reactions. *ChemCatChem* 4: 177–185.
13 Wu, X.-F., Gong, J.-L., and Qi, X. (2014). A powerful combination: recent achievements on using TBAI and TBHP as oxidation system. *Org. Biomol. Chem.* 12: 5807–5817.
14 Chen, R., Chen, J., Zhang, J., and Wan, X. (2018). Combination of tetrabutylammonium iodide (TBAI) with *tert*-butyl hydroperoxide (TBHP): an efficient transition-metal-free system to construct carious chemical bonds. *Chem. Rec.* 18: 1292–1305.
15 Wang, H.-Y., Zhou, J., and Guo, Y.-L. (2012). Study on the reactive transient α-λ^3-iodanyl-acetophenone complex in the iodine(III)/PhI(I) catalytic cycle of iodobenzene-catalyzed α-acetoxylation reaction of acetophenone by electrospray ionization tandem mass spectrometry. *Rapid Commun. Mass Spectrom.* 26: 616–620.
16 Hoffman, R.V. (1991). Synthetic transformations using arenesulfonyloxy groups, first as electrophiles, then as leaving groups. *Tetrahedron* 47: 1109–1135.
17 Prakash, O., Saini, N., and Shatma, P.K. (1994). Hypervalent iodine reagents in the synthesis of heterocyclic compounds. *Synlett* 1994: 221–227.

18 Yamamoto, Y. and Togo, H. (2006). PhI-catalyzed α-tosyloxylation of ketones with *m*-chloroperbenzoic acid and *p*-toluenesulfonic acid. *Synlett* 2006: 798–800.

19 Akiike, J., Yamamoto, Y., and Togo, H. (2007). Efficient conversion of ketones to α-tosyloxyketones with *m*-chloroperbenzoic acid and *p*-toluenesulfonic acid in the presence of catalytic amount of IL-supported PhI in [emim]OTs. *Synlett* 2007: 2168–2172.

20 Lex, T.R., Swasy, M.I., and Whitehead, D.C. (2015). Relative rate profiles of functionalized iodoarene catalysts for iodine(III) oxidations. *J. Org. Chem.* 80: 12234–12243.

21 Boelke, A. and Nachtsheim, B.J. (2020). Evolution of *N*-heterocycle-substituted iodoarenes (NHIAs) to efficient organocatalysts in iodine(I/III)-mediated oxidative transformations. *Adv. Synth. Catal.* 362: 184–191.

22 Wirth, T. and Hirt, U.H. (1997). Chiral hypervalent iodine compounds. *Tetrahedron: Asymm.* 8: 23–26.

23 Richardson, R.D., Page, T.K., Altermann, S.M. et al. (2007). Enantioselective α-oxytosylation of ketones catalysed by iodoarenes. *Synlett* 2007: 538–542.

24 Altermann, S.M., Richardson, R.D., Page, T.K. et al. (2008). Catalytic enantioselective α-oxysulfonylation of ketones mediated by iodoarenes. *Eur. J. Org. Chem.* 2008: 5315–5328.

25 Farooq, U., Schäfer, S., Shah, A.A. et al. (2010). Synthesis of new enantiomerically pure organoiodine catalysts and their application in the α-functionalization of ketones. *Synthesis* 47: 1023–1029.

26 Yu, J., Cui, J., Hou, X.-S. et al. (2011). Enantioselective α-tosyloxylation of ketones catalyzed by spirobiindane scaffold-based chiral iodoarenes. *Tetrahedron: Asymmetry* 22: 2039–2055.

27 Guilbault, A.-A. and Legault, C.Y. (2012). Drastic enhancement of activity in iodane-based α-tosyloxylation of ketones: iodine(III) does the hypervalent twist. *ACS Catal.* 2: 219–222.

28 Guilbault, A.-A., Basdevant, B., Wanie, V., and Legault, C.Y. (2012). Catalytic enantioselective α-tosyloxylation of ketones using iodoaryloxazoline catalysts: insights on the stereoinduction process. *J. Org. Chem.* 77: 11283–11295.

29 Thérien, M.-È., Guilbault, A.-A., and Legault, C.Y. (2013). New chiral iodooxazoline catalysts for the I(III)-mediated α-tosyloxylation of ketones: refining the stereoinduction model. *Tetrahedron: Asymmetry* 24: 1193–1197.

30 Brenet, S., Minozzi, C., Clarens, B. et al. (2015). 3,3′-Diiodobinaphthol and 3,3′-diiodobiphenol derivatives as hypervalent iodine organocatalysts for the α-oxytosylation of ketones. *Synthesis* 47: 3859–3873.

31 Feng, Y., Huang, R., Hu, L. et al. (2016). Chiral C_2-symmetric iodoarene-catalyzed asymmetric α-oxidation of β-keto esters. *Synthesis* 48: 2637–2644.

32 Levitre, G., Dumoulin, A., Retailleau, P. et al. (2017). Asymmetric α-sulfonyl- and α-phosphoryl-oxylation of ketones by a chiral hypervalent iodine(III). *J. Org. Chem.* 82: 11877–11883.

33 Beaulieu, S. and Legault, C.Y. (2015). Mechanistic insights on the iodine(III)-mediated α-oxidation of ketones. *Chem. Eur. J.* 21: 11206–11211.

34 Basdevant, B. and Legault, C.Y. (2015). Enantioselective iodine(III)-mediated synthesis of α-tosyloxy ketones: breaking the selectivity barrier. *Org. Lett.* 17: 4918–4921.

35 Basdevant, B. and Legault, C.Y. (2015). Study of the reactivity of [hydroxy(tosyloxy)iodo]benzene toward enol esters to access α-tosyloxy ketones. *J. Org. Chem.* 80: 6897–6902.

36 Uyanik, M., Yasui, T., and Ishihara, K. (2010). Enantioselective Kita oxidative spirolactonization catalyzed by *in situ* generated chiral hypervalent iodine(III) species. *Angew. Chem. Int. Ed.* 49: 2175–2177.

37 Fujita, M., Yoshida, Y., Miyata, K. et al. (2010). Enantiodifferentiating *endo*-selective oxylactonization of *ortho*-alk-1-enylbenzoate with a lactate-derived aryl-λ^3-iodane. *Angew. Chem. Int. Ed.* 49: 7068–7071.

38 Yamamoto, Y. and Togo, H. (2005). Facile one-pot preparation of [hydroxy(sulfonyloxy)iodo]arenes from iodoarenes with MCPBA in the presence of sulfonic acids. *Synlett* 2005: 2486–2488.

39 Abazid, A.H. and Nachtsheim, B.J. (2020). A triazole-substituted aryl iodide with omnipotent reactivity in enantioselective oxidations. *Angew. Chem. Int. Ed.* 59: 1479–1484.

40 Boelke, A., Lork, E., and Nachtsheim, B.J. (2018). *N*-Heterocycle-stabilized iodanes: from structure to reactivity. *Chem. Eur. J.* 24: 18653–18657.

41 Hempel, C., Maichle-Mössmer, C., Pericàs, M.A., and Nachtsheim, B.J. (2017). Modular synthesis of triazole-based chiral iodoarenes for enantioselective spirocyclizations. *Adv. Synth. Catal.* 359: 2931–2941.

42 Rodriguez, A. and Moran, W.J. (2012). Chiral aryl iodide-catalyzed enantioselective α-oxidation of ketones. *Synthesis* 2012: 1178–1182.

43 Kitamura, T., Muta, K., and Kuriki, S. (2013). Catalytic fluorination of 1,3-dicarbonyl compounds using iodoarene catalysts. *Tetrahedron Lett.* 54: 6118–6120.

44 Yang, X., Wu, T., Phippa, R.J., and Toste, F.D. (2015). Advances in catalytic enantioselective fluorination, mono-, di-, and trifluoromethylation, and trifluoromethylthiolation reactions. *Chem. Rev.* 115: 826–870.

45 Suzuki, S., Kamo, T., Fukushi, K. et al. (2014). Iodoarene-catalyzed fluorination and aminofluorination by an Ar-I/HF•pyridine/*m*CPBA system. *Chem. Sci.* 5: 2754–2760.

46 Pluta, R., Krach, P.E., Cavallo, L. et al. (2018). Metal-free catalytic asymmetric fluorination of keto esters using a combination of hydrogen fluoride (HF) and oxidant: experiment and computation. *ACS Catal.* 8: 2582–2588.

47 Wang, Y., Yuan, H., Lu, H., and Zheng, W.-H. (2018). Development of planar chiral iodoarenes based on [2.2]paracyclophane and their application in catalytic enantioselective fluorination of β-ketoesters. *Org. Lett.* 20: 2555–2558.

48 Wu, H., He, Y.-P., Xu, L. et al. (2014). Asymmetric organocatalytic direct $C(sp^2)$–H/$C(sp^3)$–H oxidative cross-coupling by chiral iodine reagents. *Angew. Chem. Int. Ed.* 53: 3466–3469.

49 Wang, J., Yuan, Y., Xiong, R. et al. (2012). Phenyliodine bis(trifluoroacetate)-mediated oxidative C–C bond formation: synthesis of 3-hydroxy-2-oxindoles and spirooxindoles from anilides. *Org. Lett.* 14: 2210–2213.

50 Sreenithya, A. and Sunoj, R.B. (2014). Mechanistic insights on iodine(III) promoted metal-free dual C–H activation involved in the formation of a spirocyclic bis-oxindole. *Org. Lett.* 16: 6224–6227.

51 Sreenithya, A., Patel, C., Hadad, C.M., and Sunoj, R.B. (2017). Hypercoordinate iodine catalysts in enantioselective transformation: the role of catalyst folding in stereoselectivity. *ACS Catal.* 7: 4189–4196.

52 Haubenreisser, S., Wöste, T.H., Martínez, C. et al. (2016). Structurally defined molecular hypervalent iodine catalysts for intermolecular enantioselective reactions. *Angew. Chem. Int. Ed.* 55: 413–417.

53 Cao, Y., Zhang, X., Lin, G. et al. (2016). Chiral aryliodine-mediated enantioselective organocatalytic spirocyclization: synthesis of spirofurooxindoles via cascade oxidative C–O and C–C bond formation. *Org. Lett.* 18: 5580–5583.

54 Uyanik, M., Yasui, T., and Ishihara, K. (2009). Hypervalent iodine-catalyzed oxylactonization of ketocarboxylic acids to ketolactones. *Bioorg. Med. Chem. Lett.* 48: 3848–3581.

55 Uyanik, M., Suzuki, D., Yasui, T., and Ishihara, K. (2011). *In situ* generated (hypo)iodite catalysts for the direct α-oxyacylation of carbonyl compounds with carboxylic acids. *Angew. Chem. Int. Ed.* 50: 5331–5334.

56 Uyanik, M., Suzuki, D., Watanabe, M. et al. (2015). High-performance hypoiodite/hydrogen peroxide catalytic system for the oxylactonization of aliphatic γ-oxocarboxylic acids. *Chem. Lett.* 44: 387–389.

57 Li, X., Zhou, C., and Xu, C. (2012). TBAI-catalyzed oxidative coupling of β-ketoesters with carboxylic acid: synthesis of α-carboxylic-β -ketoesters. *Arkivoc* 9: 150–158.

58 Tada, N., Ishigami, T., Cui, L. et al. (2013). Calcium iodide catalyzed photooxidative oxylactonization of oxocarboxylic acids using molecular oxygen as terminal oxidant. *Tetrahedron Lett.* 554: 256–258.

59 Guo, S., Yu, J.-T., Dai, Q. et al. (2014). The Bu_4NI-catalyzed α-acyloxylation of ketones with benzylic alcohols. *Chem. Commun.* 50: 6240–6242.

60 Zhou, Z., Cheng, J., and Yu, J.-T. (2015). Bu_4NI-catalyzed direct α-oxyacylation of diarylethanones with acyl peroxides. *Org. Biomol. Chem.* 13: 9751–9754.

61 Mondal, B., Sahoo, S.C., and Pan, S.C. (2015). $n$$Bu_4NI$-catalyzed α-benzoxylation of ketones with terminal aryl alkenes. *Eur. J. Org. Chem.* 2015: 3135–3140.

62 Kanai, N., Nakayama, H., Tada, N., and Itoh, A. (2010). Tandem oxidation/rearrangement of β-Ketoesters to tartronic esters with molecular oxygen catalyzed by calcium iodide under visible light irradiation with fluorescent lamp. *Org. Lett.* 12: 1948–1951.

63 Miao, C.-B., Wang, Y.-H., Xing, M.-L. et al. (2013). I2-catalyzed direct α-hydroxylation of β-dicarbonyl compounds with atmospheric oxygen under photoirradiation. *J. Org. Chem.* 78: 11584–11589.

64 Liang, Y.-F., Wu, K., Song, S. et al. (2015). I_2- or NBS-catalyzed highly efficient α-hydroxylation of ketones with dimethyl sulfoxide. *Org. Lett.* 17: 876–879.

65 Wei, W., Wang, Y., Yin, J. et al. (2012). Construction of aromatic [5,5] spiroketals via hypoiodite-catalyzed etherification combined in relay cascades. *Org. Lett.* 14: 1158–1161.

66 Wei, L., Xue, J., Liu, H. et al. (2012). Synthesis of (±)-γ-rubromycin via a new hypoiodite-catalytic oxidative cycloetherification. *Org. Lett.* 14: 5302–5305.

67 Wei, W., Li, L., Lin, X. et al. (2012). New fluoride-promoted hypoiodite-catalytic oxidative cycloetherification to aromatic spiroketals. *Org. Biomol. Chem.* 10: 3494–3499.

68 Xu, W. and Nachtsheim, B.J. (2015). TBAI-catalyzed oxidative cross-coupling of phenols and 2-aminoacetophenones. *Org. Lett.* 17: 1585–1588.

69 Guo, W., Vallcorba, O., Vallribera, A. et al. (2014). Oxidative breakdown of iodoalkanes to catalytically active iodine species: a case study in the a-tosyloxylation of ketones. *ChemCatChem* 6: 468–472.

70 Zhang, B., Han, L., Hu, J., and Yan, J. (2015). Novel α-tosyloxylation of ketones catalyzed by the *in situ* generated hypoiodous acid from alkyl iodide. *Tetrahedron Lett.* 55: 5851–5854.

71 Tian, J.-S., Ng, K.W.J., Wong, J.-R., and Loh, T.-P. (2012). α-Amination of aldehydes catalyzed by in situ generated hypoiodite. *Angew. Chem. Int. Ed.* 51: 9105–9109.

72 Zhang, Y.-X., Zhang, A.-Q., Tian, J.-S., and Loh, T.-P. (2013). Diastereoselective oxidative α-amination of aliphatic aldehydes catalyzed by iodine: synthesis of syn-γ-hydroxy-α-amino acetals. *Org. Biomol. Chem.* 11: 8387–8394.

73 Lamani, M. and Prabhu, K.R. (2012). NIS-catalyzed reactions: amidation of acetophenones and oxidative amination of propiophenones. *Chem. Eur. J.* 18: 14638–14642.

74 Zhang, J., Jiang, J., Li, Y. et al. (2013). A new strategy for the construction of α-amino acid esters via decarboxylation. *Org. Lett.* 15: 3222–3225.

75 Lv, Y., Li, Y., Xiong, T. et al. (2014). nBu_4NI-Catalyzed oxidative imidation of ketones with imides: synthesis of α-amino ketones. *Chem. Commun.* 50: 2367–2369.

76 Zhang, J., Shao, Y., Wang, Y. et al. (2015). Transition-metal-free decarboxylation of dimethyl malonate: an efficient construction of α-amino acid esters using TBAI/TBHP. *Org. Biomol. Chem.* 13: 3982–3987.

77 Jiang, Q., Xu, B., Zhao, A. et al. (2014). Transition-metal-free oxidative α-C−H amination of ketones via a radical mechanism: mild synthesis of α-amino ketones. *J. Org. Chem.* 79: 8750–8756.

78 Gao, Y.-T., Jin, X.-Y., Liu, Q. et al. (2018). Iodide/H_2O_2 catalyzed intramolecular oxidative amination for the synthesis of 3,2′-pyrrolidinyl spirooxindoles. *Molecules* 23: 2265.

79 Tan, C., Liu, Y., Liu, X. et al. (2020). Stereoselective synthesis of trans-aziridines via intramolecular oxidative $C(sp^3)$–H amination of β-amino ketones. *Org. Chem. Front.* 7: 780–786.

80 Nishiguchi, Y., Tomizuka, A., and Moriyama, K. (2020). Formal ring contraction of cyclic *N*-sulfonamides via C–N bond cleavage and α-amination by oxidation of halides. *Adv. Synth. Catal.* 362: 5518–5523.

81 Zhang, X. and Wang, L. (2012). TBHP/I_2-promoted oxidative coupling of acetophenones with amines at room temperature under metal-free and solvent-free conditions for the synthesis of α-ketoamides. *Green Chem.* 14: 2141–2145.

82 Wang, D., Zhang, K., Jia, L. et al. (2017). *n*Bu_4NI-Mediated oxidation of methyl ketones to α-ketoamides: using ammonium, primary and secondary amine-salt as an amine moiety. *Org. Biomol. Chem.* 15: 3427–3434.

83 Harschneck, T., Hummel, S., Kirsch, S.F., and Klahn, P. (2012). Practical azidation of 1,3-dicarbonyls. *Chem. Eur. J.* 18: 1187–1193.

84 Yasui, K., Kojima, K., Kato, T. et al. (2016). Guanidinium iodide/urea hydrogen peroxide-catalyzed azidation of β-dicarbonyl compounds with trimethylsilyl azide. *Tetrahedron* 72: 5350–5354.

85 Dhineshkumar, J. and Prabhu, K.R. (2016). An efficient tertiary azidation of 1,3-dicarbonyl compounds in water catalyzed by tetrabutylammonium iodide. *Eur. J. Org. Chem.* 2016: 447–452.

86 Uyanik, M., Sahara, N., Tsukahara, M. et al. (2020). Chemo- and enantioselective oxidative α-azidation of carbonyl compounds. *Angew. Chem. Int. Ed.* 59: 17110–17117.

87 Kato, T., Yasui, K., Odagi, M., and Nagasawa, K. (2017). Guanidinium hydroiodide/cumene hydroperoxide-mediated intermolecular oxidative coupling reaction of β-ketoamides with oxindoles. *Adv. Synth. Catal.* 359: 2881–2889.

88 Jiang, Y., Zou, J.-X., Huang, L.-T. et al. (2018). KI-catalyzed C–S bond formation via an oxidation relay strategy: efficient access to various α-thio-β-dicarbonyl compounds. *Org. Biomol. Chem.* 16: 1641–1645.

89 Hu, B., Zhou, P., Zhang, Q. et al. (2018). Metal-free oxidative thioesterification of methyl ketones with thiols/disulfides for the synthesis of α-ketothioesters. *J. Org. Chem.* 83: 14978–14986.

90 Sun, K., Lv, Y., Chen, Y. et al. (2017). A novel metal-free method for the selenocyanation of aromatic ketones to afford α-carbonyl selenocyanates. *Org. Biomol. Chem.* 15: 4464–4467.

91 Maruoka, K. (2019). Design of high-performance chiral phase-transfer catalysts with privileged structures. *Proc. Jpn. Acad., Ser. B* 95: 1–16.

92 Evans, D.A., Fandrick, K.R., Song, H.-J. et al. (2007). Enantioselective Friedel–Crafts alkylations catalyzed by bis(oxazolinyl)pyridine–scandium(III) triflate complexes. *J. Am. Chem. Soc.* 129: 10029–10041.

93 Uyanik, M., Hayashi, H., and Ishihara, K. (2014). High-turnover hypoiodite catalysis for asymmetric synthesis of tocopherols. *Science* 345: 291–294.

94 Uyanik, M., Hayashi, H., Iwata, H., and Ishihara, K. (2016). Chiral ammonium hypoiodite salt-catalyzed enantioselective oxidative cycloetherification to 2-acyl tetrahydrofurans. *Chem. Lett.* 45: 353–355.

95 Liebhafsky, H.A. (1932). The catalytic decomposition of hydrogen peroxide by iodine-iodide couple at 25 °C. *J. Am. Chem. Soc.* 54: 1792–1806.

96 Milenković, M. and Stanisavljev, D.R. (2012). Role of free radicals in modeling the iodide-peroxide reaction mechanism. *J. Phys. Chem. A* 116: 5541–5548.

97 Patonay, T., Kónya, K., and Juhász-Tóth, E. (2011). Syntheses and transformations of α-azido ketones and related derivatives. *Chem. Soc. Rev.* 40: 2797–2847.

98 Vita, M.V. and Waser, J. (2012). Azidation of β-keto esters and silyl enol ethers with a benziodoxole reagent. *Org. Lett.* 15: 3246–3249.

99 Deng, Q., Bleith, T., Wadepohl, H., and Gade, L.H. (2013). Enantioselective iron-catalyzed azidation of β-keto esters and oxindoles. *J. Am. Chem. Soc.* 135: 5356–5359.

100 Tiffner, M., Stockhammer, L., Schörgenhumer, J. et al. (2018). Towards an asymmetric organocatalytic α-azidation of β-ketoesters. *Molecules* 23: 1142.

101 Mairhofer, C., Novacek, J., and Waser, M. (2020). Synergistic ammonium (hypo)Iodite/imine catalysis for the asymmetric α-hydroxylation of β-ketoesters. *Org. Lett.* 22: 6138–6142.

102 Uraguchi, D., Tsutsumi, R., and Ooi, T. (2013). Catalytic asymmetric oxidation of *N*-sulfonyl imines with hydrogen peroxide–trichloroacetonitrile system. *J. Am. Chem. Soc.* 135: 8161–8164.

11

Oxidations with Iodine(V) Compounds – From Stoichiometric Compounds to Catalysts

Frederic Ballaschk and Stefan F. Kirsch

Organic Chemistry, Bergische Universität Wuppertal, Wuppertal, Germany

11.1 Introduction to Iodine(V) Compounds

Organic compounds containing iodine in the high oxidation state +V play a major role in today's chemical world. These pentavalent iodine compounds, or λ^5-iodanes, have a variety of applications ranging from simple oxidation of primary and secondary alcohols to more complex reactivities, e.g. rearrangements, dearomatizations, dehydrogenations, and many more. Due to their broad synthetic appeal, research in the field became a major objective over the last decades.

Hypervalent iodine(V) compounds can be roughly divided into the four heavily studied classes of oxidative agents, shown in Figure 11.1: (i) 2-Iodoxybenzoic acid (IBX) and derivatives thereof, (ii) Dess–Martin periodinane (DMP) and derivatives thereof, (iii) pseudocyclic systems, and (iv) iodylarenes.

These reagents are frequently used in all variants of synthetic organic chemistry, including the total synthesis of complex natural products. Based on their high reliability, easy handling, and broad applicability as well as their high tolerance to functional groups, these reagents have already become an integral part of the organic chemist's daily routine. However, these reagents also exhibit some good-to-know disadvantages: (i) Several hypervalent iodine(V) compounds have a hazardous character, which hampers their use in large-scale applications. (ii) Oxidative processes require stoichiometric amounts of the iodine(V) oxidants. As a result, a major research goal became to make the oxidative systems recyclable or catalytic, while keeping the oxidative power of the reagents. In the course of this chapter, we will discuss a personal selection of the key milestones of hypervalent iodine(V) chemistry, with a focus on alcohol oxidations while shortly touching other oxidative processes, until reaching the current state of catalytic applications.

Iodine Catalysis in Organic Synthesis, First Edition. Edited by Kazuaki Ishihara and Kilian Muñiz.

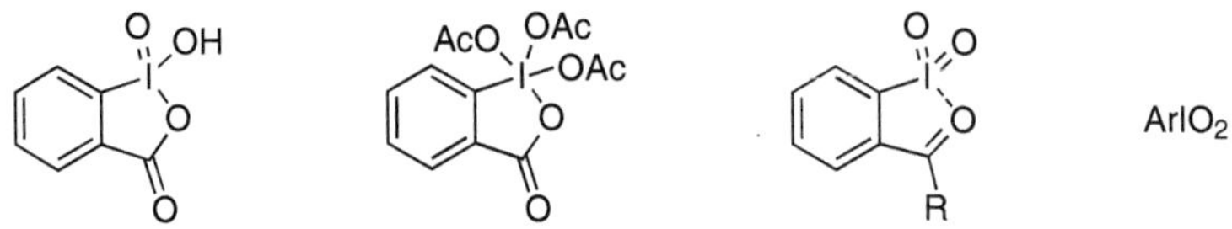

(1) (2) (3) (4)

Figure 11.1 Classes of iodine(V) compounds for oxidations.

11.2 Iodine(V) Compounds as Stoichiometric Reagents in Organic Syntheses

11.2.1 Synthesis of 2-Iodoxybenzoic Acid (IBX), Dess–Martin Periodinane (DMP), and Their Derivatives

The first mention of the synthesis and characterization of 2-iodoxybenzoic acid, IBX, originates from the works by Meyer and Hartmann in 1893, describing the alkaline oxidation of iodobenzoic acid with potassium permanganate [1]. IBX was then widely forgotten by synthetic chemists: "Its virtual insolubility in common organic solvents has discouraged the study of its chemical properties," as Dess and Martin mentioned in 1991 [2]. However, IBX became the main access point for the synthesis of the DMP, named after Dess and Martin, first reported in 1983. *o*-Iodobenzoic acid was used as the starting material, which was further converted in the presence of potassium bromate and sulfuric acid into IBX, which, in turn, could be then converted into DMP through heating in a mixture of acetic acid and acetic anhydride at 100 °C [3]. Alternatively, Liu and Ireland described the formation of DMP in 1993 by using catalytic amounts of *p*-toluenesulfonic acid instead of acetic acid, and the reaction proceeded in high yields up to 100 g scale to produce the DMP reagent [4] (Scheme 11.1).

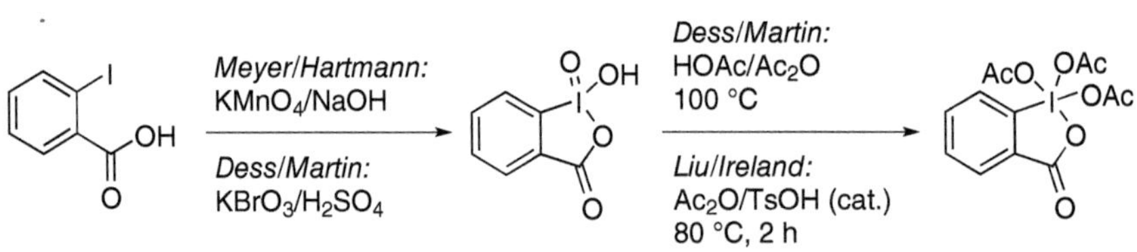

Scheme 11.1 Synthetic approach toward DMP using IBX intermediates.

The resulting DMP product was significantly more soluble in organic solvents than IBX, in particular, in the then common CH_2Cl_2 [3]. In addition, IBX was considered sensitive to heat and shock [5], which was probably due to contaminations with potassium bromate. However, DMP also had a decisive disadvantage since it was unstable to moisture, thus leading to rapid hydrolysis. Therefore, the use of DMP typically required an inert gas atmosphere or a slight excess (ca. 30%) of the reagent [3, 6].

In 1994, Santagostino and Frigerio revealed that IBX dissolves in dimethyl sulfoxide (DMSO) in quite high concentrations up to 1.5 mol/l. Furthermore, it was shown that IBX, contrary to DMP, is not sensitive to moisture, and the oxidation of alcohols can be achieved under open-flask conditions. The early studies also noted that the use of a cosolvent, e.g. tetrahydrofuran (THF), was possible for substances insoluble in DMSO or when employing reaction temperatures below 19 °C (that is the freezing point of DMSO). The oxidative properties of IBX were closely similar to those of DMP since it also oxidized alcohols to the corresponding aldehydes quickly and showed no signs of over-oxidation to possible carboxylic acids [7, 8]. Santagostino and coworkers also established an alternative route for the production of IBX using inexpensive Oxone® ($2KHSO_5$-$KHSO_4$-K_2SO_4): Under the conditions outlined in Scheme 11.2, the oxidation of the iodo compound took place within three hours in water at 70 °C. This method is the current standard method for the synthesis of IBX, combining good yields, inexpensive reagents, and easy scale-up with experimentally simple workup and purification protocols [9].

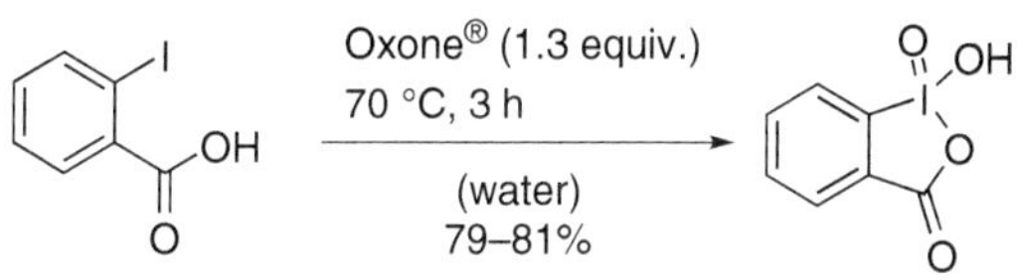

Scheme 11.2 Synthesis of IBX by Santagostino *et al.*

Over the last decades, a great realm of novel hypervalent iodine(V) compounds were reported, a small snippet of which is depicted in Figure 11.2. By varying the DMP and IBX parent compounds, the focus was put, for example, on gaining safer formulations [10, 11], stronger oxidative abilities [12–15], compounds with better solubility [16–18] and compounds for efficient recoveries [19]. Moreover, synthetic approaches for the production of hypervalent iodine(V) compounds were continuously optimized [20–22].

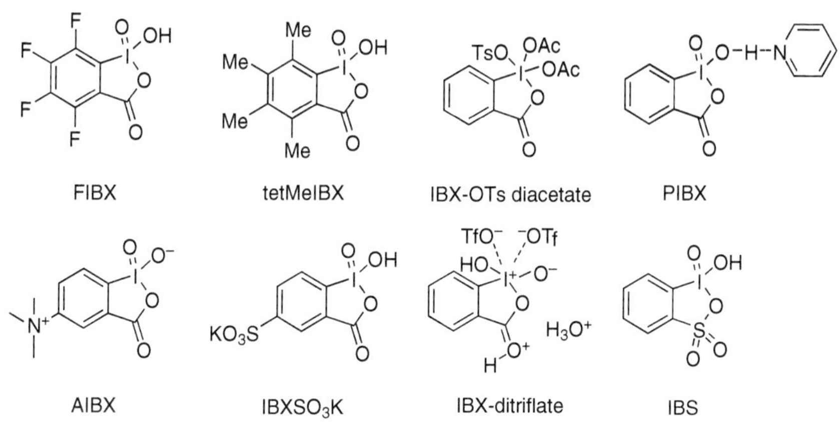

Figure 11.2 Selected examples of hypervalent iodine(V) compounds derived from IBX or DMP.

11.2.2 Oxidations of Alcohols with IBX, DMP, and Their Derivatives

IBX (in DMSO at room temperature) and DMP (in CH_2Cl_2 at room temperature) are primarily used for the oxidation of alcohols to give the corresponding aldehydes or ketones, a topic that was widely researched [23–59], and excellent summaries exist [60–67]. Herein, only a few representative examples are highlighted that diverge from the standard. As shown in Scheme 11.3a, Mal and coworkers achieved the IBX-oxidation of primary and secondary alcohols through ball-milling, with an experimental setup that excludes the use of solvents. Various substrates containing alkyl and aryl moieties were successfully converted into their carbonyls under solvent-free conditions providing yields between 77 and 99% [68]. It was also described that IBX-oxidations at elevated temperatures in solvents other than DMSO are sometimes advantageous, allowing for a simplified workup and purification [69]. The fact that solvent surrogates for DMSO were possible with IBX became important when developing Ugi-type multicomponent reactions where the aldehyde parts were generated through *in situ* oxidation with stoichiometric amounts of IBX. Zhu and coworkers used acetonitrile at 70 °C as a solvent for

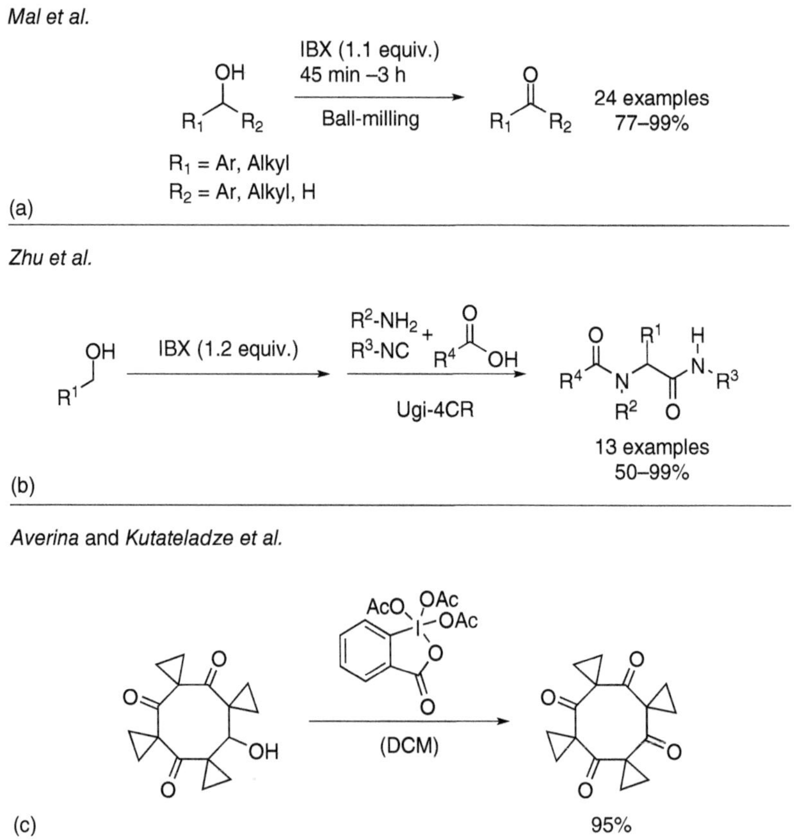

Scheme 11.3 Various applications for the IBX/DMP-mediated oxidation of alcohols. Source: Modified from Achar et al. [68], Zhu et al. [116], Sedenkova et al. [59].

this oxidation producing several α-acetamidoamides in yields between 50 and 99% (Scheme 11.3b) [53], or in a related reaction, THF at 60 °C was employed [70]. The mildness of the oxidation methods toward multiple functional groups is a major highlight of iodine(V) oxidants. In particular, DMP has advantages when tackling sensitive substrates, as exemplified in Scheme 11.3c by the high-yielding synthetic approach to a tetraketone, which was used as a precursor for the synthesis of [8]-rotane. It is worth mentioning that this oxidation failed when employing Swern conditions or pyridinium chlorochromate (PCC) instead of DMP [59].

Due to their outstanding reliability and mildness, IBX and DMP were widely used for the oxidation of alcohols in various total synthesis endeavors toward natural products, such as (±)-tetrodotoxin [71], (+)-lycoricidine [72], and many more [73–100]. Two examples of representative key steps are shown in Scheme 11.4: the total synthesis of (±)-tetrodotoxin using IBX (Scheme 11.4a) and the total synthesis of (+)-lycoricidine with DMP (Scheme 11.4b).

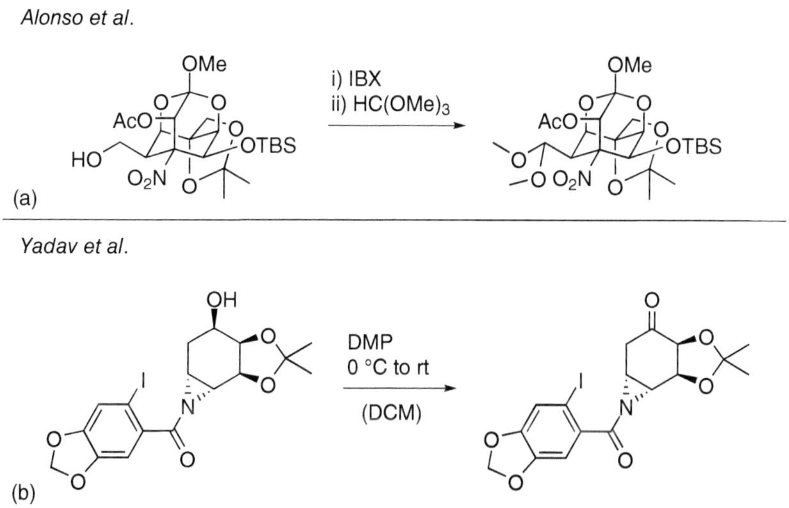

Scheme 11.4 Representative uses of IBX [71] and DMP [72] in total syntheses. Source: Lago-Santomé et al. [71], Yadav et al. [72].

11.2.3 Dehydrogenations with IBX, DMP, and Their Derivatives

Hypervalent iodine(V) reagents are versatile tools for the dehydrogenation of α,β-saturated ketones, following the breakthrough works by Nicolaou and coworkers [101, 102]. The formal abstraction of hydrogen (H_2) produces the α,β-unsaturated carbonyl compounds in a single step. When combined with other reactions including the carbonyl reduction, various 1,2- or 1,4-additions, epoxidations or cycloadditions, dehydrogenation becomes a great tool for complex transformations, as shown by various reports over the last decades [103–119]. The mildness of this transformation is exemplarily demonstrated in Scheme 11.5a. Therein, Compain and coworkers used the dehydrogenation with IBX to generate

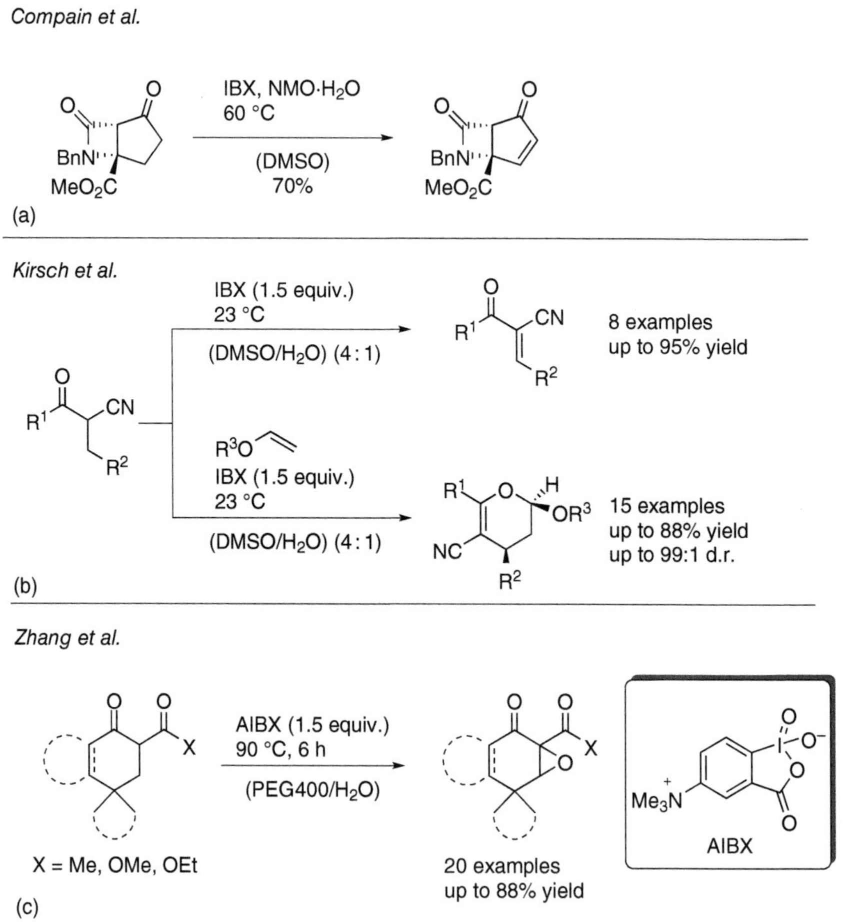

Scheme 11.5 Selected examples for carbonyl dehydrogenation with IBX. Source: Modified from Malinowski et al. [115], Modified from Klahn and Kirsch [109], Jiang et al. [112].

a cyclopentenone in a 70% yield. The bicyclic system was highly strained and the β-lactam was sensitive to hydrolysis. This smooth oxidation provides access to iminosugar mimics [115].

Kirsch and coworkers combined the room-temperature dehydrogenation of highly activated α-substituted β-oxonitrile compounds with Hetero–Diels–Alder reactions: Dihydropyrans in yields of up to 88% were formed in the presence of enol ethers, and excellent diastereoselectivities were obtained [109] (Scheme 11.5b). Zhang and coworkers connected the dehydrogenation of cyclic β-ketoesters with an epoxidation. In this rare example, trimethylammonio-1,3-dioxo-1,3-dihydro-1λ5-benzo[d][1,2]iodoxol-1-ol anion (AIBX) (the water-soluble variant of IBX) was employed for the initial dehydrogenation in a mixture of water and PEG400 (polyethylene glycol). Under the conditions, the epoxide products were directly formed through a mechanism, supported by DFT calculations, where upon dehydrogenation with AIBX, the resulting unsaturated system and the reduced form of AIBX (AIBA/iodine(III)) trigger the observed oxirane formation. Of note, *o*-iodobenzoic acid (iodine(I)) was found to be the final waste product of this reaction [112] (Scheme 11.5c).

The reliable dehydrogenation methodology was considered of use for various natural product syntheses. For example, the total syntheses of the (*S*)-pestalotiopsones D and E [120, 121] as well as the total synthesis of dihydroxerulin were based on dehydrogenation key steps [122] (Scheme 11.6). As shown in Scheme 11.6a, the pestalotiopsones were accessed from the chromanon system, which was easily dehydrogenated in 71% yield with IBX at 90 °C. In the case of dihydroxerulin (Scheme 11.6b), the unsaturated aldehyde was generated from the primary alcohol through an amine-catalyzed oxidation reaction with IBX as the oxidant. The powerful cascade reaction starts with the aldehyde formation followed by mild dehydrogenation at room temperature in the presence of the amine.

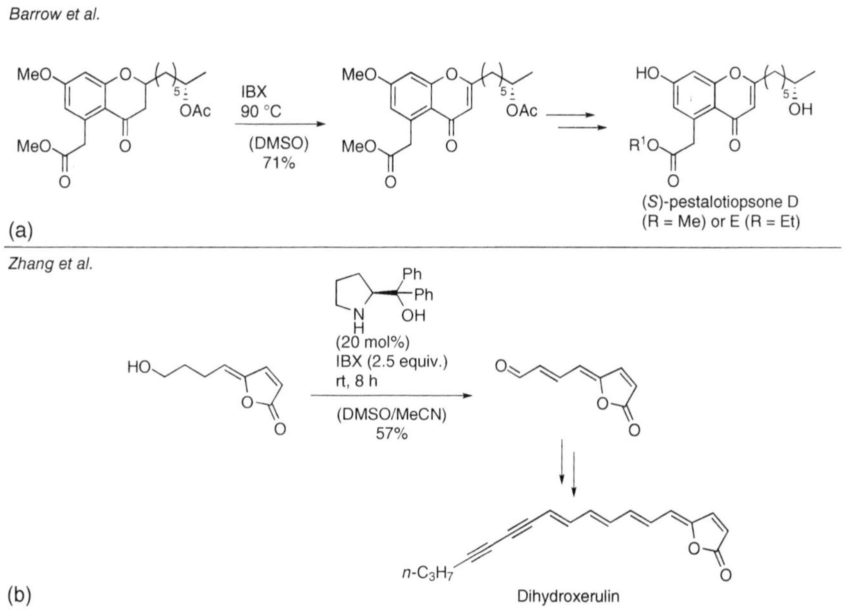

Scheme 11.6 Selected uses of the IBX-mediated dehydrogenation in total synthesis. Source: Beekman and Barrow [120], Modified from Xie et al. [122].

11.2.4 Oxidative Aromatizations and Dearomatizations with IBX, DMP, and Their Derivatives

Hypervalent iodine(V) compounds are capable of both oxidative aromatizations and oxidative dearomatizaions. In most cases, the main research focus is centered around the formation of nitrogen-containing aromatics through the direct oxidation of the nonaromatic cyclic precursors. Amongst others [123–129], this concept was used for the generation of pyrroles [130], pyridines [131, 132], dibenzooxazepines [133], and quinazolines [134]. In contrast, oxidative dearomatization relies mainly on phenol derivatives [135–146] that are converted into *o*- or *p*-quinones, as mechanistically studied *via* DFT calculations [147].

The oxidative dearomatization was recently incorporated into complex multistep cascades. For example, Moorthy and coworkers demonstrated a combination for the easy construction of highly complex aromatic heterocycles, which were then used for the detection of fluoride ions (Scheme 11.7a) [140]. In this transformation, naphthol underwent the aforementioned dearomatization with IBX. The resulting 1,2-naphthoquinone then reacted *in situ* with an indole-based diene in a Diels–Alder cycloaddition. Subsequent oxidation with atmospheric oxygen and reaction with *ortho*-phenylenediamine resulted in the final product with excellent chemoselectivities and moderate yields between 59 and 63%. In another example [144], IBX was again used for the *in situ* generation of 1,2-naphthoquinone, which reacted with an excess of the Wittig reagent in an olefination and cyclocondensation to provide the benzocoumarin product (Scheme 11.7b).

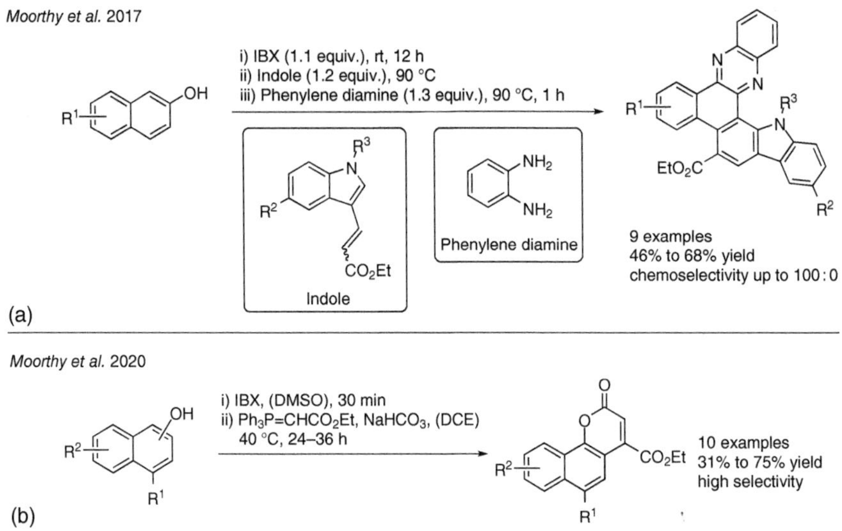

Scheme 11.7 Selected oxidative dearomatizations embedded in cascade reactions. Source: Mishra et al. [140], Chandra et al. [144].

Due to the experimental simplicity, IBX-initiated oxidative dearomatizations of phenols became a smart, albeit sparsely used tool for the total synthesis of natural products [148–150]. As shown in Scheme 11.8a, Franck and coworkers were able to obtain the core element of epicocconone from a relatively simple starting material [149]. After heavy optimization, they succeeded in obtaining the bicyclic structural motif from a phenolic precursor compound with a yield of 48%. Of note, excellent diastereoselectivity of 9:1 was achieved, using the quite far methyl group. Saikawa, Nakata, and coworkers demonstrated the potential of an oxidative dearomatization strategy in their total synthesis of kendomycin [150], employing an IBX oxidation to generate the desired quinone intermediate from the phenol (with concomitant decarboxylation). Direct conversion with concentrated hydrofluoric acid gave the natural product in a yield of 32% over two steps (Scheme 11.8b).

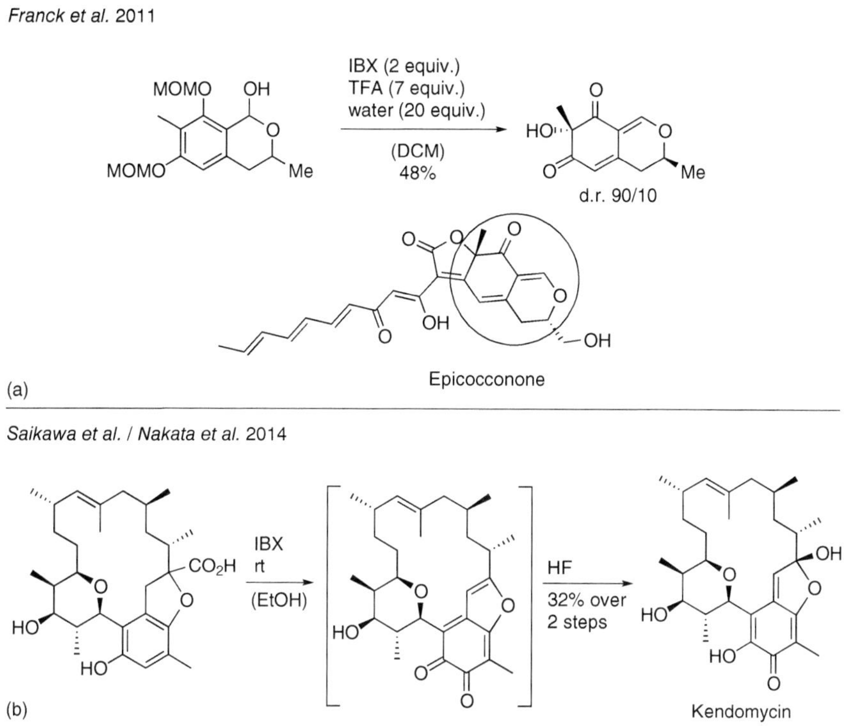

Scheme 11.8 Selected oxidative dearomatizations in natural product syntheses. Source: Modified from Boulangé et al. [149], Tanaka et al. [150].

With regard to oxidative aromatization processes (corresponding to dehydrogenations), the research focus is clearly on the synthesis of nitrogen-containing heterocycles. In 2018, for example, Kumar and coworkers were able to produce pyrazoles in a multicomponent reaction [130], starting with the condensation of aldehydes and 4-methoxyaniline $(PMP\text{-}NH)_2$. The resulting imine then reacted in a proline-catalyzed enamine reaction with the shown dialdehyde to form dihydropyrazole intermediate that gave the desired aromatic heterocycle upon treatment with IBX in yields up to 80% (Scheme 11.9a). Reddy and coworkers used *N*-acylhydrazones and aromatic aldehydes as starting materials for the synthesis of oxadiazoles [124]. The requisite aromatization was triggered by a mixture of IBX and potassium iodide. It is assumed that IBX produces the reactive iodonium ion (I^+), which plays an important role in the cyclization (Scheme 11.9b).

11.2.5 Fragmentations with IBX, DMP, and Their Derivatives

While oxidative fragmentations of organic molecules are commonly triggered with hypervalent iodine compounds, hypervalent iodine(III) compounds are the predominant devices in the field. For example, variants of the Hofmann rearrangement are possible with iodine(III) sources like phenyliodine(III) bis(trifluoroacetate) (PIFA), phenyliodine diacetate (PIDA) or others, converting primary amides

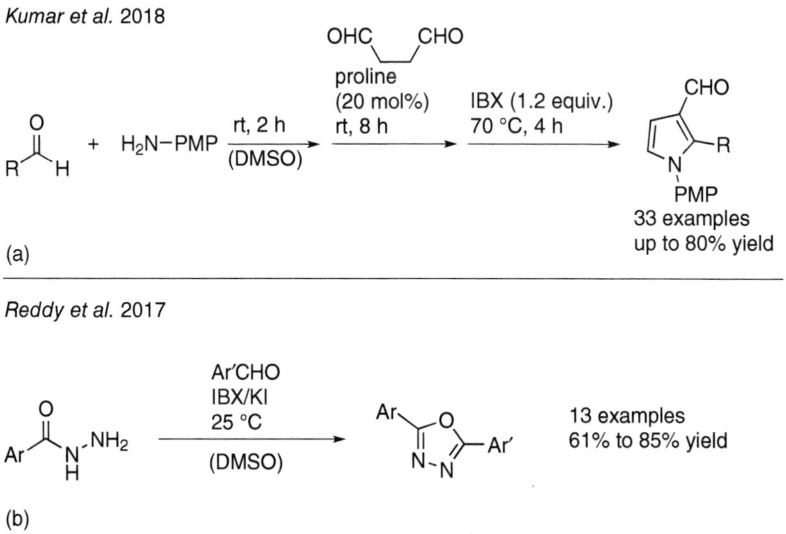

Scheme 11.9 IBX-mediated oxidative aromatizations. Source: Singh et al. [130], Khan et al. [124].

into the corresponding amines via intermediate formation of isocyanates and subsequent decarboxylation of the carbamic acid formed during hydrolysis. For iodine(V) compounds, the analogous reactivity is widely underdeveloped. However, Akamanchi and coworkers presented the direct transformation of primary amides into nitriles shortened by one carbon under oxidative conditions with IBX and tetraethylammonium bromide. Yields between 55 and 95% were achieved, accomplishing both aromatic and aliphatic nitriles (Scheme 11.10a/method A). Besides IBX, DMP and 2-iodosobenzoic acid (IBA) were also found to be capable of this transformation, whereas the inorganic iodine(V) compounds as HIO_3 and I_2O_5 showed no reactivity [151]. Of note, α,α-disubstituted primary amides undergo fragmentation via isocyanate intermediates to provide the corresponding ketones under the conditions (Scheme 11.10a/method B) [152]. A similar reaction was observed when submitting amino acids [153] or oxiranes [154] in aqueous ammonia to IBX or DMP. As summarized in Scheme 11.10b, oxidative fragmentation led to the formation of nitrile products.

Arimoto and coworkers showed that the combination of IBX and iodine under forcing conditions converts primary alcohols into dehomogolized carboxylic acids, rather than the expected aldehydes [155]. The fragmentation was only observed when using large excess of IBX (up to 8 equiv.) in the presence of overstoichiometric amounts of iodine and oxygen at 100 °C. It was concluded that iodine and IBX initially form a new iodine species, which plays a key role in this dehomologization reaction. However, the postulated active species was not evidenced by the crystal structure, and only hints on the structure with iodine(III) oxidation state using nuclear magnetic resonance (NMR), melting point, and decomposition experiments were obtained. Linear substrates gave the corresponding carboxylic acids in yields between 57 and 100% (Scheme 11.11a). When employing 2-branched alcohols,

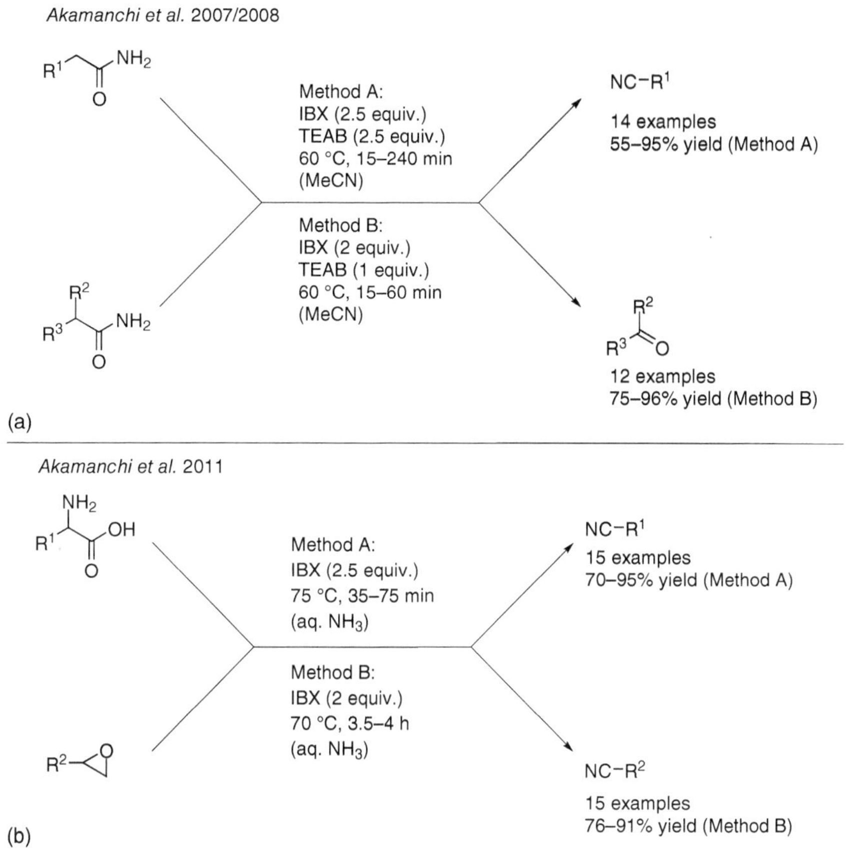

Scheme 11.10 Oxidative dehomologation of various substrates. Source: Bhalerao et al. [151], Bellale et al. [152], Deshmukh et al. [154].

the ketones were formed in reduced yields, alongside the non-dehomologized carboxylic acid (Scheme 11.11b).

11.2.6 Pseudocyclic Systems and Iodylarenes

Iodylarenes ($ArIO_2$) and their pseudocyclic congeners are iodine(V) species with a structural motif that is less popular for synthetic applications, in comparison to the ubiquitously used DMP and IBX reagents. Nevertheless, iodylarenes show a great potential for a variety of oxidative transformations, and the compound class is easily accessed using classical methods, starting from aryliodides through stepwise oxidation. Common oxidants for the formation of the iodine(V) target compounds are sodium hypochlorite, sodium periodate, dimethyldioxirane or Oxone® [156–159]. In addition, *Zhdankin* and coworkers developed a method based on the use of catalytic amounts of $RuCl_3$ and peracetic acid. This approach proved particularly strong for $ArIO_2$ compounds bearing strong electron-withdrawing substituents attached to the aromatic core (e.g. CF_3, Cl, Br, or F) [160]. With regard to the structure, X-ray crystallographic studies by Protasiewicz and coworkers showed in 2000, for example,

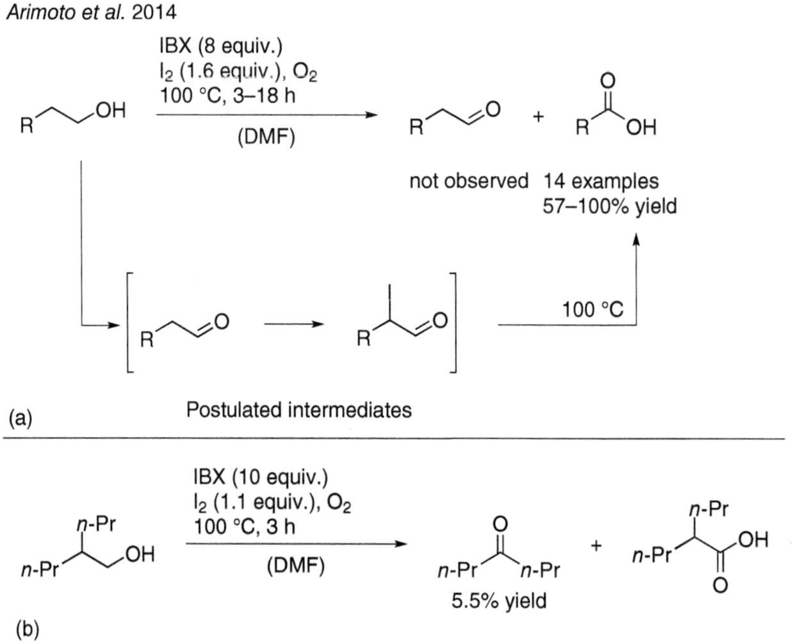

Scheme 11.11 Oxidative dehomologation of primary alcohols. Source: Xu et al. [94].

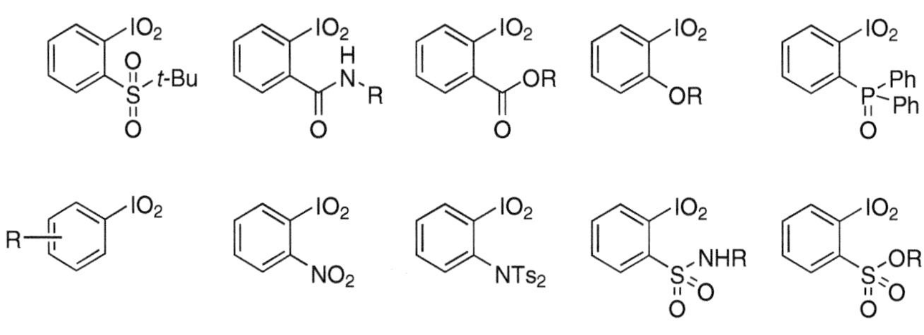

Figure 11.3 A selection of iodylarenes.

that in the case of a *tert*-butyl sulfonyl group at *ortho* position of the iodylarene moiety, iodine is located in pseudo-octahedral coordination: Intramolecular interactions between iodine and sulfonyl oxygen (2,693(3) Å) indicate a pseudocyclic motif that is accompanied by intermolecular interactions (2,777(3) Å) resulting in a chain-like polymer structure [158, 161]. Similar pseudocyclic structures are known for a range of iodylarenes bearing *ortho*-substituents with coordinating capabilities. Among them are amides [162, 163], esters [164, 165], ethers [166, 167], phosphine oxides [156], nitro compounds [168] as well as sulfonamides [169, 170] and sulfonate esters [171] (Figure 11.3).

11.2.7 Oxidative Transformations with Pseudocyclic Systems and Iodylarenes

The synthetic use of iodylarenes is still underdeveloped. As a general rule, iodylarenes and their pseudocyclic variants are less reactive, in comparison to the cyclic

DMP and IBX derivatives. For simple iodylarenes, the lowered reactivity results from their poor solubility in organic solvents; a quite rapid decomposition paired with a potential explosive power may also be problematic. This leads to a situation where oxidative reactions with noncyclic iodylarenes typically require elevated temperatures, or the addition of activating agents is mandatory. However, a range of pseudocyclic systems with appropriate *ortho*-substituents show an increased solubility in organic solvents [161]. Oxidations of alcohols and sulfides to the corresponding carbonyl compounds or sulfoxides are easily possible [162–164, 166, 167, 172–175].

Iodylarenes and some of the pseudocyclic hypervalent iodine(V) compounds were also found to be effective oxidants in a range of chemical transformations including oxidative dearomatizations [176–179], oxidative cyclizations [180] and oxidative cleavage of carbon–carbon double bonds [181]. For example, Chiba and coworkers successfully cyclized allylic enamines by using catalytic amounts of copper bromide and substoichiometric amounts of $PhIO_2$ (Scheme 11.12a). It was postulated that the initial activation of the enamine moiety by copper was followed by oxidative cyclization cyclopropane formation [180]. Çelik and coworkers reported the use of both PhIO and $PhIO_2$ for the oxidative cleavage of olefins (Scheme 11.12b). Several cyclic and acyclic alkenes with aryl substituents were successfully converted into the corresponding carbonyl compounds, as exemplified through the smooth reactions of indene, dihydronaphthalene, and *cis*- and *trans*-stilbene [181]. The asymmetric dearomatization of phenol derivatives under oxidative conditions became easily a key playground for iodine(V) compounds. Quideau and coworkers used a stoichiometric system based on a chiral binaphthyl corpse [178]. Yields of up to 77% and enantiomeric excess of up to 94% were achieved (Scheme 11.12c). The initially formed *o*-quinol typically reacts with a second equivalent of the *o*-quinol in a [4 + 2]-cycloaddition to form the rather complex product compounds. In a more recent variant, Yoshida and coworkers used another chiral hypervalent iodine(V) compound that resembles a crown ether structure, albeit delivering lowered yields and enantioselectivities (Scheme 11.12d) [176].

11.3 Iodine(V) Compounds as Recyclable Reagents in Organic Syntheses

A general disadvantage of employing hypervalent iodine(V) compounds in oxidative processes is the need for stoichiometric amounts of the oxidative agent. Besides the known risks attributed to the potential shock and heat sensitivity of IBX and DMP derivatives [5], the problem of "organic waste" production in the course of all types of oxidative processes with stoichiometric amounts of hypervalent iodine(V) compounds is a major drawback that makes the class of oxidants particularly less attractive in large-scale applications. In most cases, the oxidative methods relying on iodine(V) compounds provide the reduced iodine(III) or iodine(I) compounds, the reoxidation of which is often tedious and involves heavy synthetic efforts. In addition, the reisolation of the mix of possible iodine species for reoxidation is hampered by solubility discrepancies and thus mostly in the lower percentage range [69]. Those issues were widely addressed over the last decades, and one possible solution was found in the immobilization of the oxidant. In a simultaneous way, Giannis

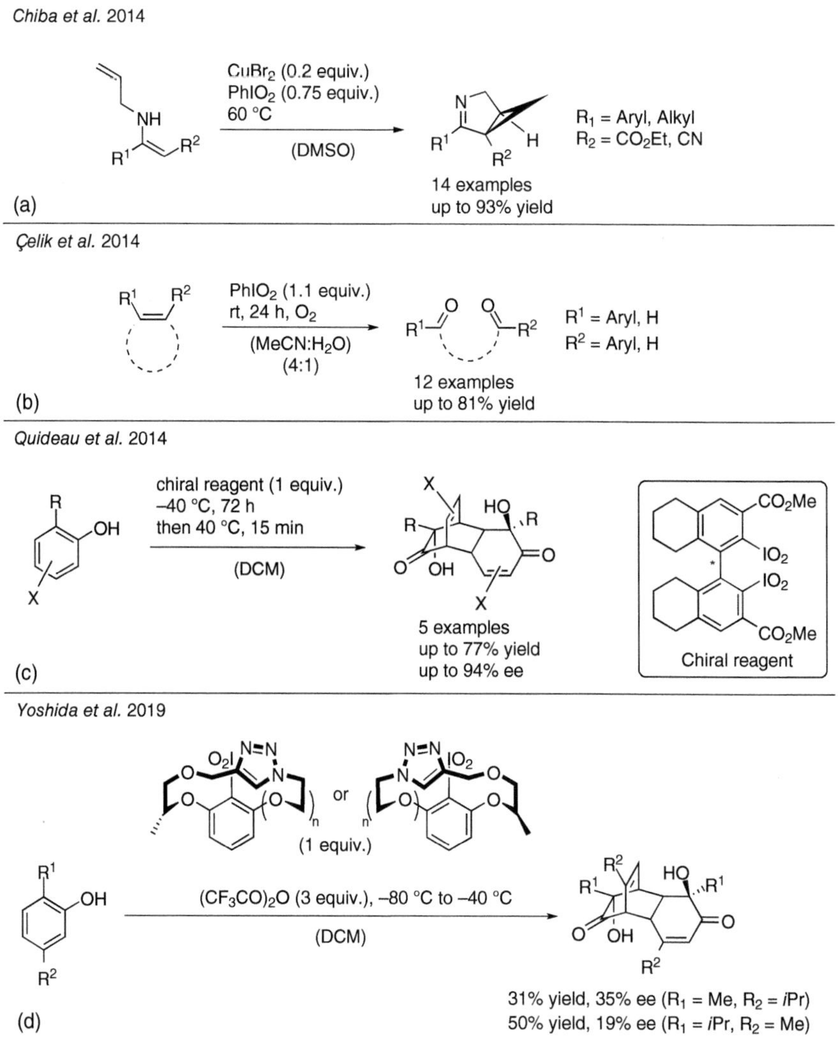

Scheme 11.12 Selected synthetic applications of $PhIO_2$ and its derivatives. Source: Modified from Toh et al. [180], Atmaca et al. [181], Modified from Bosset et al. [178], Yoshida et al. [176].

and Rademann and coworkers generated polymer-bound IBX derivatives for oxidation [182, 183]. While Giannis and coworkers utilized an aminopropyl silica gel as a solid phase, Rademann and coworkers relied on the (polystyrene-based) Merrifield resin. After completion of the oxidation reactions with the use of the highly active immobilized IBX variants, it was possible to remove the iodine species from the reaction mixture by simple filtration, and reoxidation allowed for easy reuse. It was demonstrated that multiple reuses are possible without loss of activity, which dramatically reduces the amount of waste produced. The overall approach is therefore in line with all the principles of green chemistry and holds potential for a plethora of

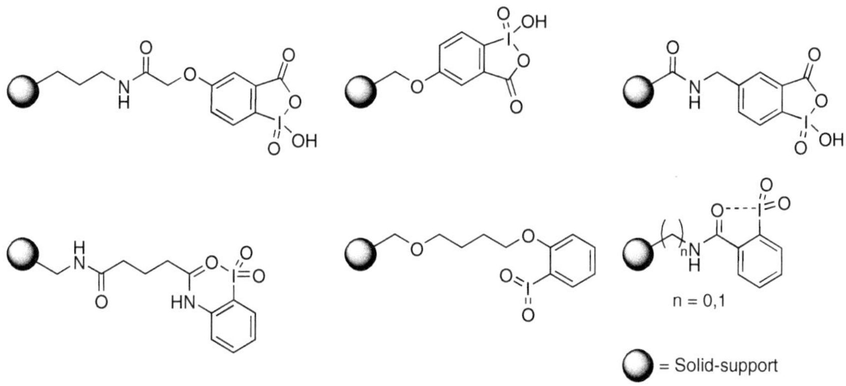

Figure 11.4 Selected examples for immobilized IBX derivatives.

applications. Following this pioneering lead, several analogous polymer-bound IBX derivatives were reported, all of which excel with easy recyclability (Figure 11.4). The connection to the polymer resin may be through attachment points at the aromatic core [184–193], or through IBX-esters [194], IBX-ethers [195] and IBX-amides [196–203].

The synthesis route toward the resin-bound hypervalent iodine(V) compounds depends strongly on the choice of the linkage. In most cases, etheric bridging is suitable; alternatively, amide bonds or other robust functional groups are introduced as stable linkers. As shown in Scheme 11.13, the aryl iodide derivatives are a standard starting point for synthesis. After attachment to the solid phase, the late-stage oxidation provides the reactive iodine(V) species. Of importance, the resin loading, i.e. the actual amount of active species attached to the resin, must be determined prior to use, for example, through elemental analysis [183, 198] or by performing test oxidation with an excess of alcohol [182, 197].

Besides resin-bound hypervalent iodine(V) compounds, several reagents were developed that appear to be highly advantageous during workup protocols, and thus also allow for the easy recovery and for the reduction of the "organic waste" (Figure 11.5). For example, Zhdankin and coworkers presented 3-alkoxy-2-iodylpyridines capable of sulfide and alcohol oxidation, the waste product of which was recovered by simple acid-base extraction [204]. A water-soluble iodylarene contained a sulfonate function in the *para* position of the aromatic core that allowed for the simple removal through aqueous extraction; reoxidation provided an entry into multiple uses [205]. Zhang and coworkers reported a variant of oxidative aromatization of cyclic β-ketoesters using AIBX, a water-soluble derivative of IBX bearing an ammonium group (Scheme 11.14b). AIBX was successfully recovered in 92% yield after the reaction was completed [16]. Bergbreiter and coworkers developed a recyclable polyisobutylene-bound iodine(V) reagent. The polyisobutylene (PIB)-IBX ester has a significant solubility in heptane and was shown to be an excellent reagent for the oxidation of secondary and primary alcohols. The recovery and reoxidation were possible due to the increased solubility

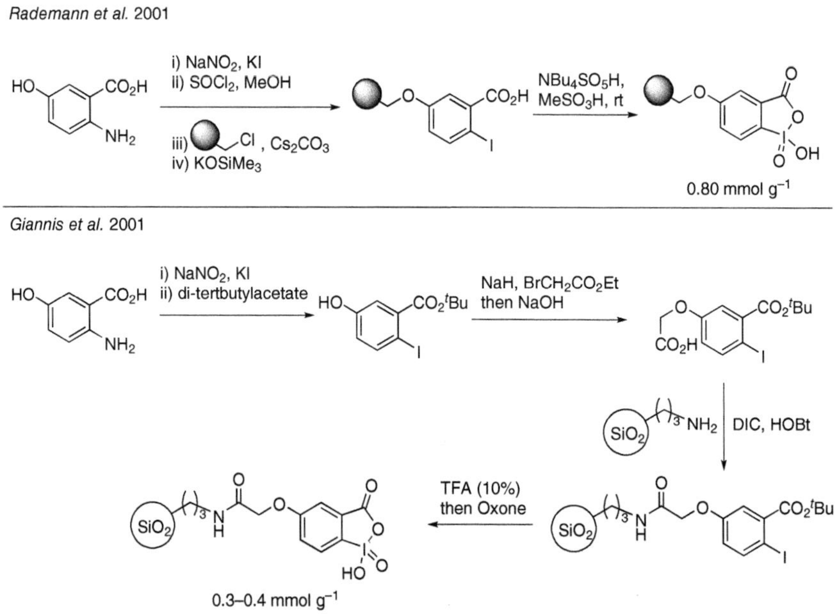

Scheme 11.13 Synthesis of solid-supported IBX derivatives by *Rademann* and *Giannis*. Source: Modified from Sorg et al. [183], Mülbaier and Giannis [182].

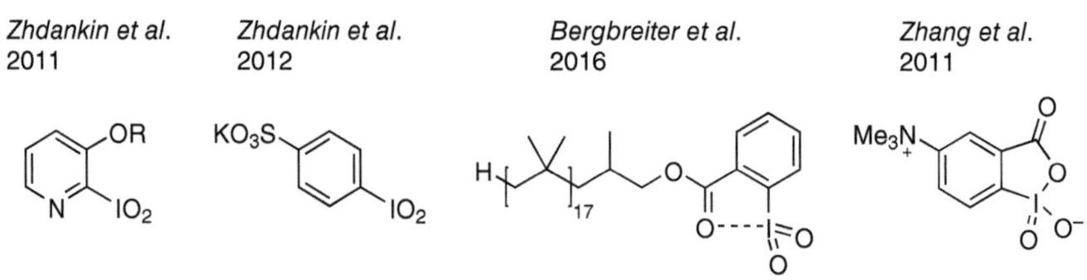

Figure 11.5 Recyclable hypervalente iodine(V) systems. Source: Yoshimura et al. [204], Yusubov et al. [205], Samunual and Bergbreiter [206], Cui et al. [227].

of the unpolar reagent in heptane; the oxidant was reused over at least 5 cycles (Scheme 11.14a) [206].

A noteworthy application of the polymer-bound IBX amide was demonstrated by Lee and coworkers. As shown in Scheme 11.14c, both fructose and hydroxymethylfurfural (HMF) were transformed into 2,5-diformylfuran (DFF) through the use of this oxidative agent. After complete conversion, regeneration of the reagent was easily achieved by filtration and subsequent reoxidation [200].

11.4 Iodine(V) Compounds as Catalytic Reagents in Organic Syntheses

The use of solid phase-supported hypervalent iodine compounds is a major step forward in reducing organic waste, and detailed isolation-reoxidation protocols allow

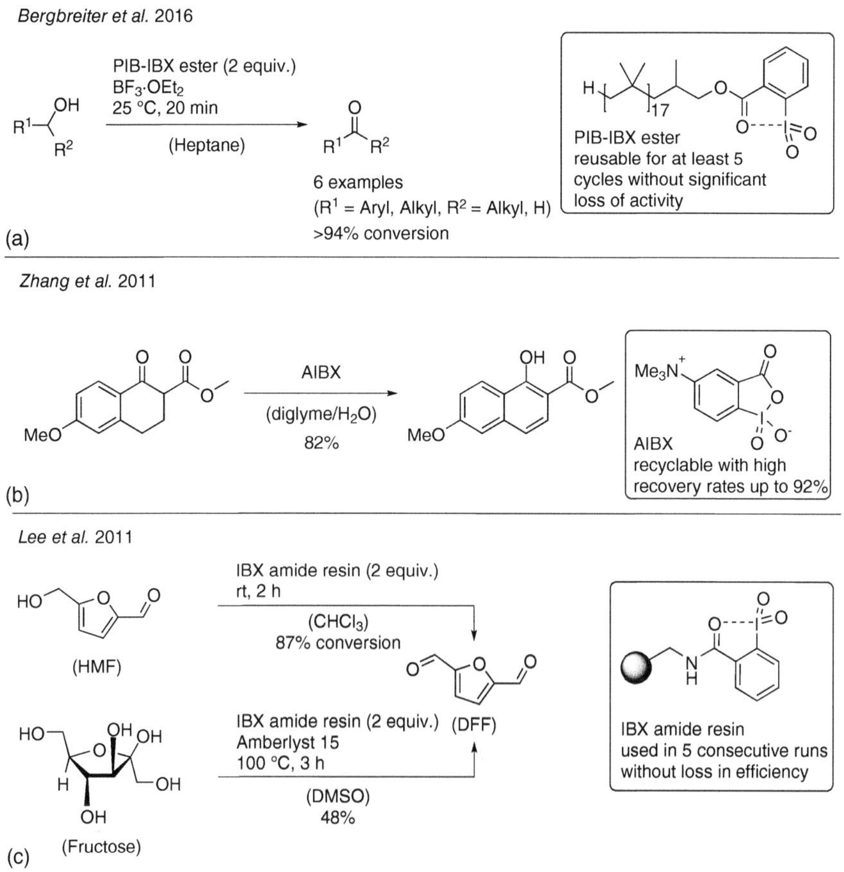

Scheme 11.14 Selected examples for the utilization of recyclable iodine(V) compounds. Source: Samunual and Bergbreiter [206], Cui et al. [227], Yoon et al. [200].

for the repetitive use over multiple cycles. However, a disadvantage remains since, after a completed run, the regeneration of the iodine(V) species is time-consuming and various reagents and solvents are required. A possible solution is due to the finding that the catalytic use of hypervalent iodine(V) reagents is also possible if *in situ* reoxidation can be triggered in the reaction mixture with external oxidants. In the perfect case, the external oxidant is cheap and easily separates from the iodine catalysts of choice.

11.4.1 Oxidative Transformations Based on IBX, DMP, and Iodylarenes

In this realm, Vinod and coworkers described the use of a catalytic iodine(V) system based on 2-iodobenzoic acid. Reoxidation was achieved with 1.3 equivalents of Oxone® in an aqueous acetonitrile system at 70 °C capable of oxidizing primary alcohols to the corresponding carboxylic acids, and secondary alcohols to the corresponding ketones [207]. Similarly, Giannis and coworkers reported the use of a catalytic IBX system for the oxidation of primary and secondary alcohols

[208]. Page and coworkers presented a system for the oxidation of alcohols using tetraphenylphosphonium monoperoxosulfate (TPPP) and 2-iodobenzoic acid. This reagent combination allowed the use of anhydrous reaction conditions. Of note, the oxidation of alcohols also took slowly a place under iodine-free conditions showing the competitive oxidative abilities of the TPPP reoxidant [209]. Several variants of this principle resulted in catalytic systems based on IBX [210–220], DMP [221], and iodylarenes [222–224].

A few representative examples are outlined in Scheme 11.15. Zhdankin and coworkers showed that the use of a cocatalyst system consisting of phenyliodide and $RuCl_3$ allows for the oxidation of alcohols, employing Oxone® as the stoichiometric oxidant. Interestingly, the direct C–H oxidation of simple hydrocarbons was also possible under the conditions, but benzylic positions were typically required (Scheme 11.15a) [223]. *Miura* and coworkers created an iodobenzoic acid derivative with a fluorinated side chain that was prone to catalytic use through reoxidation with Oxone®. In addition, the catalyst is easily recovered after each reaction, thus allowing for recyclability, and no loss of activity was observed for at least five cycles (Scheme 11.15b) [214]. A binuclear iodine system (DIDA) was introduced by Moorthy and coworkers for the oxidation of secondary and primary alcohol, which is oxidized *in situ* by Oxones to *bis*-IBX. By the appropriate choice of solvent, either the carboxylic acid (MeCN:water) or the aldehyde ($MeNO_2$) could be selectively produced, starting from primary alcohols (Scheme 11.15c) [211]. Rao and coworkers used catalytic amounts of IBA in the presence of Oxone® to produce β-keto esters from the respective Morita–Baylis–Hillman products (Scheme 11.15d) [212]. The catalytic use of DMP is also possible, as was shown by Donohoe and coworkers in 2016 [221]. The formal [2 + 2] cycloaddition of styrene derivatives shown in Scheme 11.15e requires only 10 mol% of DMP and proceeds most likely through radical intermediates, while the exact role of the hypervalent iodine(V) remains vague. An alternative concept for iodine(V) catalysis was revealed by Powers and coworkers in 2018. Therein various oxidative processes were described using aerobically generated pseudocyclic iodylarenes. While molecular oxygen was found to continuously form an iodine(III) species, the reactive iodine(V) species results from disproportionation of iodine (III) into iodine(I) and iodine(V), as supported by NMR spectroscopic experiments [224].

11.4.2 Oxidative Transformations Based on Iodoxybenzenesulfonic Acid (IBS)

While many of the iodine(V)-based catalysts are variants of the IBX core, Ishihara and coworkers reported a highly efficient catalyst in 2009 that relied on 2-iodoxybenzenesulfonic acid (IBS), rather than 2-iodoxybenzoic acid. The efficacy was outstanding for a range of oxidative transformations when employed in combination with tetrabutylammonium hydrogen sulfate and Oxone® as the external oxidant of choice. For example, a number of alcohols were readily oxidized, and IBX was readily outperformed allowing for catalyst loadings in the range of 1–2 mol% [225]. Under the reaction conditions, tertiary alcohols were found to undergo an appealing oxidative rearrangement that led to enone formation (Scheme 11.16a) [226].

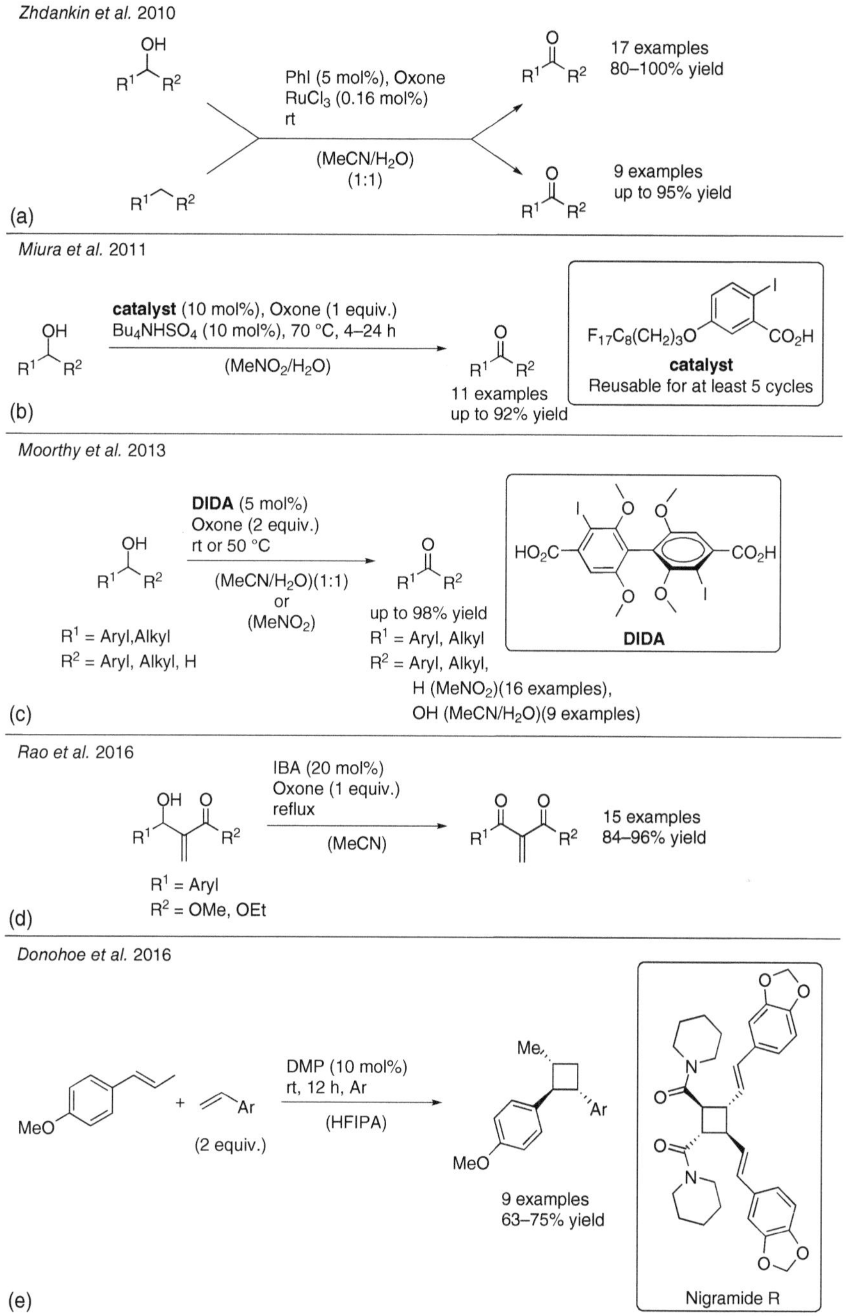

Scheme 11.15 Selected examples of catalytic processes with iodine(V) species. Source: Yusubov et al. [223], Miura et al. [214], Seth et al. [211], Bikshapathi et al. [212], Colomer et al. [221].

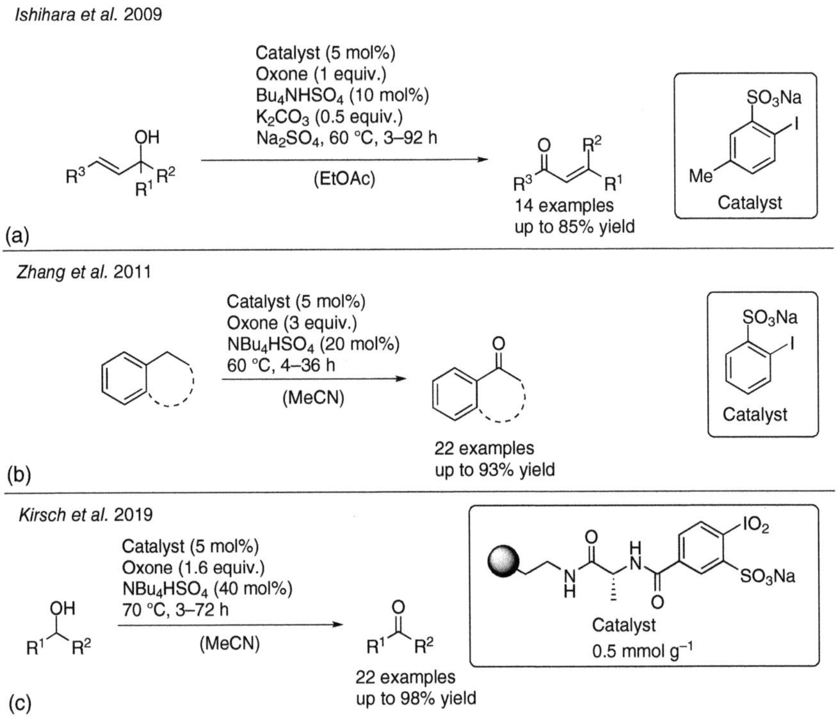

Scheme 11.16 Applications of catalytical systems based on IBS. Source: Uyanik and Ishihara [63], Zhang et al. [105], Ballaschk and Kirsch [230].

Following this lead, the Ishihara catalytic system was further advanced [227–231]. For example, Zhang and coworkers demonstrated the direct C-H oxidation of benzylic positions, a reaction that was previously challenging for IBX-based catalysts. When using 5 mol% of IBS precatalyst and 20 mol% of a suitable phase transfer catalyst, the benzylic oxidation provided several aryl ketones in good yields (Scheme 11.16b). Even turnover numbers (TON) of up to 285 were achieved, with a catalyst loading of 0.2 mol% [227]. *Ishihara* and coworkers used a methylated variant of IBS, 5-Me-IBS, for the smooth and catalytic dearomatization of phenol derivatives. This method gave 1,2-quinones on several phenols, naphthols, and phenanthroles [228]. Further improvements were possible with the 4,5-dimethyl-IBS precatalyst. Now, the catalytic site-selective hydroxylative dearomatization of phenol derivatives was exemplified [231].

Kirsch and coworkers studied the recyclability of the IBS catalysts by connecting it to a solid support consisting of an aminoethyl polystyrene resin with an additional amino acid as a placeholder. A range of secondary alcohols were successfully oxidized, and the corresponding ketones were obtained in excellent yields (Scheme 11.16c). Of note, catalyst loadings as low as 0.2 mol% were possible for complete conversions. In terms of recyclability, at least 5 cycles without reduced catalytic activity were tested, and simple filtration was sufficient for catalyst regeneration. In a 50 mmol batch experiment, 1 mol% of catalyst were effective to accomplish a yield of 80% and a recovery of the catalytic resin of 95%, making this system a prime example for the combination of catalysis and recyclability [230].

11.5 Conclusion

In summary, this chapter highlights the important role of iodine(V) compounds in contemporary synthetic chemistry. From stoichiometric methods to recyclability advantages because of solid support and true hypervalent iodine(V) catalysts, the trend goes to sustainable methods that are useful for industrial-scale oxidations. Until today, not all oxidative processes possible with hypervalent iodine(V) reagents were successfully transformed into catalytic methods. There is still room for development, aiming for the perfect oxidation catalyst with recyclability and the endless reactivity portfolio of IBX.

References

1 Hartmann, C. and Meyer, V. (1893). Ueber Jodobenzoësäure. *Ber. Dtsch. Chem. Ges.* 26: 1727–1732.

2 Dess, D.B. and Martin, J.C. (1991). A useful 12-I-5 triacetoxyperiodinane (the Dess–Martin periodinane) for the selective oxidation of primary or secondary alcohols and a variety of related 12-I-5 species. *J. Am. Chem. Soc.* 113: 7277–7287.

3 Dess, D.B. and Martin, J.C. (1983). Readily accessible 12-I-5 oxidant for the conversion of primary and secondary alcohols to aldehydes and ketones. *J. Org. Chem.* 48: 4155–4156.

4 Ireland, R.E. and Liu, L. (1993). An improved procedure for the preparation of the Dess–Martin periodinane. *J. Org. Chem.* 58: 2899.

5 Plumb, J.B. and Harper, D.J. (1990). 2-Iodoxybenzoic acid. *C&EN* 68: 3.

6 Meyer, S.D. and Schreiber, S.L. (1994). Acceleration of the Dess–Martin oxidation by water. *J. Org. Chem.* 59: 7549–7552.

7 Frigerio, M. and Santagostino, M. (1994). A mild oxidizing reagent for alcohols and 1,2-diols: o-iodoxybenzoic acid (IBX) in DMSO. *Tetrahedron Lett.* 35: 8019–8022.

8 Frigerio, M., Santagostino, M., Sputore, S., and Palmisano, G. (1995). Oxidation of alcohols with o-iodoxybenzoic acid in DMSO: a new insight into an old hypervalent iodine reagent. *J. Org. Chem.* 60: 7272–7276.

9 Frigerio, M., Santagostino, M., and Sputore, S. (1999). A user-friendly entry to 2-iodoxybenzoic acid (IBX). *J. Org. Chem.* 64: 4537–4538.

10 Kumanyaev, I.M., Lapitskaya, M.A., Vasiljeva, L.L., and Pivnitsky, K.K. (2012). Pyridinium o-iodoxybenzoate as a safe form of a famous oxidant. *Mendeleev Commun.* 22: 129–131.

11 Ozanne, A., Pouységu, L., Depernet, D. et al. (2003). A stabilized formulation of IBX (SIBX) for safe oxidation reactions including a new oxidative demethylation of phenolic methyl aryl ethers. *Org. Lett.* 5: 2903–2906.

12 Yusubov, M.S., Soldatova, N.S., Postnikov, P.S. et al. (2019). 2-Iodoxybenzoic acid ditriflate: the most powerful hypervalent iodine(v) oxidant. *Chem. Commun.* 55: 7760–7763.

13 Yusubov, M.S., Postnikov, P., Yoshimura, A., and Zhdankin, V.V. (2020). Benziodoxole-derived organosulfonates: the strongest hypervalent iodine electrophiles and oxidants. *Synlett* 31: 315–326.

14 Yusubov, M.S., Svitich, D.Y., Yoshimura, A. et al. (2013). 2-iodoxybenzoic acid organosulfonates: preparation, X-ray structure and reactivity of new, powerful hypervalent iodine(V) oxidants. *Chem. Commun.* 49: 11269–11271.

15 Mironova, I.A., Postnikov, P.S., Yusubova, R.Y. et al. (2018). Preparation and X-ray structure of 2-iodoxybenzenesulfonic acid (IBS) - a powerful hypervalent iodine(V) oxidant. *Beilstein J. Org. Chem.* 14: 1854–1858.

16 Cui, L.-Q., Dong, Z.-L., Liu, K., and Zhang, C. (2011). Design, synthesis, structure, and dehydrogenation reactivity of a water-soluble *o*-iodoxybenzoic acid derivative bearing a trimethylammonium group. *Org. Lett.* 13: 6488–6491.

17 Bredenkamp, A., Mohr, F., and Kirsch, S. (2015). Synthesis of isatins through direct oxidation of indoles with IBX-SO_3K/NaI. *Synthesis* 47: 1937–1943.

18 Moorthy, J.N., Senapati, K., Parida, K.N. et al. (2011). Twist does a twist to the reactivity: stoichiometric and catalytic oxidations with twisted tetramethyl-IBX. *J. Org. Chem.* 76: 9593–9601.

19 Richardson, R.D., Zayed, J.M., Altermann, S. et al. (2007). Tetrafluoro-IBA and-IBX: hypervalent iodine reagents. *Angew. Chem. Int. Ed.* 46: 6529–6532.

20 Bystron, T., Horbenko, A., Syslova, K. et al. (2018). 2-iodoxybenzoic acid synthesis by oxidation of 2-iodobenzoic acid at a Boron-Doped diamond anode. *ChemElectroChem* 5: 1002–1005.

21 Zhdankin, V.V., Nemykin, V.N., Karimov, R.R., and Kazhkenov, Z.-G. (2008). Preparation and X-ray crystal structure of 2-iodyl-N,N-dialkylaniline oxides: first entry into the heterocyclic system of benziodoxazole. *Chem. Commun.*: 6131–6133.

22 Koposov, A.Y., Litvinov, D.N., Zhdankin, V.V. et al. (2006). Preparation and reductive decomposition of 2-iodoxybenzenesulfonic acid. X-ray crystal structure of 1-hydroxy-1H-1,2,3-benziodoxathiole 3,3-dioxide. *Eur. J. Org. Chem.* 2006: 4791–4795.

23 Lin, C.-K. and Lu, T.-J. (2010). A simple method for the oxidation of primary alcohols with o-iodoxybenzoic acid (IBX) in the presence of acetic acid. *Tetrahedron* 66: 9688–9693.

24 Bartlett, S.L. and Beaudry, C.M. (2011). High-yielding oxidation of β-hydroxyketones to β-diketones using o-iodoxybenzoic acid. *J. Org. Chem.* 76: 9852–9855.

25 Moorthy, J.N. and Neogi, I. (2011). IBX-mediated one-pot synthesis of benzimidazoles from primary alcohols and arylmethyl bromides. *Tetrahedron Lett.* 52: 3868–3871.

26 Cheng, H., Pei, Y., Leng, F. et al. (2013). Highly efficient synthesis of aryl and heteroaryl trifluoromethyl ketones *via* o-iodobenzoic acid (IBX). *Tetrahedron Lett.* 54: 4483–4486.

27 Gade, N.R., Devendram, V., Pal, M., and Iqbal, J. (2013). IBX mediated reaction of β-enamino esters with allylic alcohols: a one pot metal free domino approach to functionalized pyridines. *Chem. Commun.* 49: 7926–7928.
28 Cong, H., Chen, Q., Geng, Q. et al. (2015). IBX oxidation of benzenedimethanols in the presence of cucurbit[8]uril. *Chin. J. Chem.* 33: 545–549.
29 Kotha, S.S. and Sekar, G. (2015). Metal free synthesis of α-keto amides from 2-hydroxy acetophenones through domino alcohol oxidation–oxidative amidation reaction. *Tetrahedron Lett.* 56: 6323–6326.
30 Mulani, S.K., Cheng, K.-C., and Mong, K.-K.T. (2015). General homologation strategy for synthesis of L-glycero- and D-glycero-heptopyranoses. *Org. Lett.* 17: 5536–5539.
31 Chen, X., Zhang, Y., Wan, H. et al. (2016). Stereoselective organocatalytic oxidation of alcohols to enals: a homologation method to prepare polyenes. *Chem. Commun.* 52: 3532–3535.
32 An, Q., Shen, J., Liu, D. et al. (2017). Construction of chiral-fused tricyclic γ-lactams via a *trans*-perhydroindolic acid-catalyzed asymmetric domino reaction. *Org. Lett.* 19: 2925–2928.
33 Yusubov, M.S., Postnikov, P.S., Yusubova, R.Y. et al. (2017). 2-iodoxybenzoic acid tosylates: the alternative to Dess–Martin periodinane oxidizing reagents. *Adv. Synth. Catal.* 359: 3207–3216.
34 Kumari, A., Gholap, S.P., and Fernandes, R.A. (2019). Tandem IBX-promoted primary alcohol oxidation/opening of intermediate β,γ-diolcarbonate aldehydes to (E)-γ-Hydroxy-α,β-enals. *Chem. Asian J.* 14: 2278–2290.
35 Liu, Y.-j., Huang, G.-l., Li, H.-l., and Chen, Y.-g. (2019). Synthesis of α,β-unsaturated ketones from natural triterpenoids and sterol by IBX mediated oxidation. *Chem. Nat. Compd.* 55: 370–372.
36 Mokenapelli, S., Gutam, M., Govu, R. et al. (2019). De novo consecutive chemo/regioselective IBX mediated oxidation of andrographolide and its derivatives. *Synth. Commun.* 49: 1241–1251.
37 Parasuraman, K., Chennaiah, A., Dubbu, S. et al. (2019). Stereoselective synthesis of substituted 1,2-annulated sugars by domino double-Michael addition reaction. *Carbohydr. Res.* 477: 26–31.
38 Shen, H.-J., Duan, Y.-N., Zheng, K., and Zhang, C. (2019). Redetermination of the structure of a water-soluble hypervalent iodine(V) reagent AIBX and its synthetic utility in the oxidation of alcohols and synthesis of isoxazoline *N*-oxides. *J. Org. Chem.* 84: 14381–14393.
39 Gao, P., Chen, H.-J., Bai, Z.-J. et al. (2020). IBX-promoted oxidative cyclization of N-hydroxyalkyl enamines: a metal-free approach toward 2,3-disubstituted pyrroles and pyridines. *J. Org. Chem.* 85: 7939–7951.
40 Kumar, K., Kumar, P., Joshi, P., and Rawat, D.S. (2020). IBX-TfOH mediated oxidation of alcohols to aldehydes and ketones under mild reaction conditions. *Tetrahedron Lett.* 61: 151749.
41 Liang, Y.-Y., Huang, H., Li, Y. et al. (2020). Efficient synthesis of cholic acid derivates through stereoselective C–H functionalization from hyodeoxycholic acid. *Steroids* 157: 108594.

42 Nakajima, M., Nagasawa, S., Matsumoto, K. et al. (2020). A direct $S_0 \rightarrow T_n$ transition in the photoreaction of heavy-atom-containing molecules. *Angew. Chem. Int. Ed.* 59: 6847–6852.

43 Gupta, M.K., Li, Z., and Snowden, T.S. (2012). One-pot synthesis of trichloromethyl carbinols from primary alcohols. *J. Org. Chem.* 77: 4854–4860.

44 Caturelli Kuran, J.A. and Moglioni, A.G. (2014). Differential oxidation conditions of substituted cyclobutanols derived from terpenes. *Synth. Commun.* 44: 2393–2400.

45 Gupta, M.K., Li, Z., and Snowden, T.S. (2014). Preparation of one-carbon homologated amides from aldehydes or primary alcohols. *Org. Lett.* 16: 1602–1605.

46 Hazlitt, R.A., Tran, Q.-L., Sowaileh, M.F., and Colby, D.A. (2017). Generation of magnesium pentafluoropropen-2-olate from hexafluoroisopropanol and synthesis of 2,2,4,4,4-Pentafluoro-3,3-dihydroxyketones. *J. Org. Chem.* 82: 2231–2236.

47 Tsizorik, N.M., Hrynyshyn, Y.V., Bol'but, A.V., and Vovk, M.V. (2018). Synthesis of 1-functionalized imidazo[1,5-a]pyrazolo[5,1-c]pyrazines. *Chem. Heterocycl. Compd.* 54: 1075–1078.

48 Lin, S., Wang, L., Aminoleslami, N. et al. (2019). A modular and concise approach to MIDA acylboronates via chemoselective oxidation of unsymmetrical geminal diborylalkanes: unlocking access to a novel class of acylborons. *Chem. Sci.* 10: 4684–4691.

49 Mangla, P., Maity, J., Rungta, P. et al. (2019). Synthesis of 6′-Methyl-2′-*O*,4′-*C*-methylene-α-L-ribofuranosyl-pyrimidine nucleosides. *ChemistrySelect* 4: 3241–3246.

50 Dikošová, L., Laceková, J., Záborský, O., and Fischer, R. (2020). Synthesis of 3-substituted isoxazolidin-4-ols using hydroboration-oxidation reactions of 4,5-unsubstituted 2,3-dihydroisoxazoles. *Beilstein J. Org. Chem.* 16: 1313–1319.

51 Ivon, Y.M., Kuchkovska, Y.O., Voitenko, Z.V., and Grygorenko, O.O. (2020). Aliphatic α-Boryl-α-bromoketones: synthesis and reactivity. *Eur. J. Org.Chem.* 2020: 3367–3377.

52 Sonousi, A., Vasella, A., and Crich, D. (2020). Synthesis of a pseudodisaccharide suitable for synthesis of ring I modified 4,5-2-deoxystreptamine type aminoglycoside antibiotics. *J. Org. Chem.* 85: 7583–7587.

53 Drouet, F., Masson, G., and Zhu, J. (2013). Ugi four-component reaction of alcohols: stoichiometric and catalytic oxidation/MCR sequences. *Org. Lett.* 15: 2854–2857.

54 Wang, Y.-H., Cong, H., Zhao, F.-F. et al. (2011). Selective catalysis for the oxidation of alcohols to aldehydes in the presence of cucurbit[8]uril. *Catal. Commun.* 12: 1127–1130.

55 Cong, H., Li, Z.-J., Wang, Y.-H. et al. (2013). Substituent effect of substrates on cucurbit[8]uril-catalytic oxidation of aryl alcohols. *J. Mol. Catal. A Chem.* 374–375: 32–38.

56 Cong, H., Yamato, T., and Tao, Z. (2013). Chemo-selective oxidation of hydroxybenzyl alcohols with IBX in the presence of hemicucurbit[6]uril. *New J. Chem.* 37: 3778.

57 Xie, A., Zhou, X., Feng, L. et al. (2014). The oxidation of alcohols with O-iodoxybenzoic acid (IBX) in aqueous nanomicelles at room temperature. *Tetrahedron* 70: 3514–3519.

58 Liu, Y., Xie, A., Cao, M. et al. (2015). Mild oxidation of alcohols with o-iodoxybenzoic acid in liquid tetrabutylammonium bromide. *Asian J. Chem.* 27: 587–590.

59 Sedenkova, K.N., Averina, E.B., Grishin, Y.K. et al. (2016). Beyond the dimer and trimer: tetraspiro[$2.1.2^5.1.2^9.1.2^{13}.1^3$]hexadecane-1,3,5,7-tetraone--the cyclic tetramer of carbonylcyclopropane. *Chem. Eur. J.* 22: 3996–3999.

60 Silva, L.F. and Olofsson, B. (2011). Hypervalent iodine reagents in the total synthesis of natural products. *Nat. Prod. Rep.* 28: 1722–1754.

61 Wirth, T. (2001). IBX—new reactions with an old reagent. *Angew. Chem. Int. Ed.* 40: 2812–2814.

62 Ladziata, U. and Zhdankin, V.V. (2005). Hypervalent iodine(V) reagents in organic synthesis. *Arkivoc* 2006: 26.

63 Uyanik, M. and Ishihara, K. (2009). Hypervalent iodine-mediated oxidation of alcohols. *Chem. Commun.*: 2086–2099.

64 Uyanik, M. and Ishihara, K. (2009). Oxidation of alcohols and amines. In: *PATAI'S Chemistry of Functional Groups* (ed. Z. Rappoport), 1–45. Chichester, UK: John Wiley & Sons, Ltd.

65 Zhdankin, V.V. (2011). Organoiodine(V) reagents in organic synthesis. *J. Org. Chem.* 76: 1185–1197.

66 Yoshimura, A. and Zhdankin, V.V. (2016). Advances in synthetic applications of hypervalent iodine compounds. *Chem. Rev.* 116: 3328–3435.

67 Duschek, A. and Kirsch, S.F. (2011). 2-Iodoxybenzoic acid--a simple oxidant with a dazzling array of potential applications. *Angew. Chem. Int. Ed.* 50: 1524–1552.

68 Achar, T.K., Maiti, S., and Mal, P. (2014). IBX works efficiently under solvent free conditions in ball milling. *RSC Adv.* 4: 12834–12839.

69 More, J.D. and Finney, N.S. (2002). A simple and advantageous protocol for the oxidation of alcohols with O-iodoxybenzoic acid (IBX). *Org. Lett.* 4: 3001–3003.

70 Ngouansavanh, T. and Zhu, J. (2007). IBX-mediated oxidative Ugi-type multicomponent reactions: application to the N and C1 functionalization of tetrahydroisoquinoline. *Angew. Chem. Int. Ed.* 46: 5775–5778.

71 Lago-Santomé, H., Meana-Pañeda, R., and Alonso, R. (2014). A convergent approach to the dioxaadamantane core of (±)-tetrodotoxin. *J. Org. Chem.* 79: 4300–4305.

72 Yadav, J.S., Satheesh, G., and Murthy, C.V.S.R. (2010). Synthesis of (+)-lycoricidine by the application of oxidative and regioselective ring-opening of aziridines. *Org. Lett.* 12: 2544–2547.

73 McLaughlin, N.P., Butler, E., Evans, P. et al. (2010). A short synthesis of (+) and (−)-falcarinol. *Tetrahedron* 66: 9681–9687.

74 Li, H., Wang, X., Hong, B., and Lei, X. (2013). Collective synthesis of lycopodium alkaloids and tautomer locking strategy for the total synthesis of (−)-lycojapodine A. *J. Org. Chem.* 78: 800–821.

75 Lei, H., Yan, J., Yu, J. et al. (2014). Total synthesis and stereochemical reassignment of mandelalide A. *Angew. Chem. Int. Ed.* 53: 6533–6537.

76 Bera, S., Das, S.K., Saha, T., and Panda, G. (2015). Total synthesis of 3-epi-(+)-lycoricidine from Garner aldehyde via intramolecular aldol cyclization. *Tetrahedron Lett.* 56: 146–149.

77 Grodner, J., Gołębiewski, W., Willis, M. et al. (2015). The first stereoselective synthesis of a dithiane derivative of the C18 β-diketodiene system proposed for an active compound isolated from cantharellus cibarius (chanterelle). *Synthesis* 47: 1181–1189.

78 Hartrampf, F.W.W. and Trauner, D. (2017). Total synthesis of lycopladine A and carinatine A via a base-mediated carbocyclization. *J. Org. Chem.* 82: 8206–8212.

79 Gaddam, V., Nadella, L., Sukumar, G. et al. (2018). Stereoselective synthesis of northern fragment of eribulin mesylate from *d*-mannose. *Synthesis* 50: 1901–1906.

80 Shi, Y., Xiao, Q., Lan, Q. et al. (2019). A synthesis of cephalostatin 1. *Tetrahedron* 75: 1722–1738.

81 Hampel, T., Neubauer, T., van Leeuwen, T., and Bach, T. (2012). Stereoselective preparation of (*E*)-configured 1,2-disubstituted propenes from two aldehydes by a two-carbon stitching strategy: convergent synthesis of 18,21-diisopropyl-geldanamycin hydroquinone and its C7 epimer. *Chem. Eur. J.* 18: 10382–10392.

82 Chen, J.-H., Levine, S.R., Buergler, J.F. et al. (2012). Total syntheses of (±)-securinine and (±)- allosecurinine. *Org. Lett.* 14: 4531–4533.

83 Adams, G.L., Carroll, P.J., and Smith, A.B. (2013). Access to the akuammiline family of alkaloids: total synthesis of (+)-scholarisine A. *J. Am.Chem. Soc.* 135: 519–528.

84 Itoh, N., Iwata, T., Sugihara, H. et al. (2013). Total syntheses of (±)-fawcettimine, (±)-fawcettidine, (±)-lycoflexine, and (±)-lycoposerramine-Q. *Chem. Eur. J.* 19: 8665–8672.

85 Jhillu Singh, Y., Rao Yarrapothu, G., and Vemula, R. (2013). Synthesis of the C1–C15 fragment of elaiolide. *RSC Adv.* 3: 55–58.

86 Kretschmer, M., Dieckmann, M., Li, P. et al. (2013). Modular total synthesis of rhizopodin: a highly potent G-actin dimerizing macrolide. *Chem. Eur. J.* 19: 15993–16018.

87 Li, L., Liu, Y., and Wang, Q. (2013). Regioselective oxidative dehydrogenation under nonenzymatic conditions: a synthetic route to gossypol. *Eur. J. Org. Chem.* 2013: 8014–8021.

88 Mohapatra, S., Nayak, S., Mishra, K., and Pattanaik, P. (2013). Synthesis of key fragments of 19-membered cytotoxic macrolide amphidinolide E. *Lett. Org. Chem.* 10: 65–69.

89 Raghavan, S. and Kumar, V.V. (2013). Stereoselective synthesis of the C1–C8 subunit of peloruside A. *Tetrahedron* 69: 4835–4844.

90 Tsuna, K., Noguchi, N., and Nakada, M. (2013). Enantioselective total synthesis of (+)-ophiobolin A. *Chem. Eur. J.* 19: 5476–5486.

91 Schleicher, K.D., Sasaki, Y., Tam, A. et al. (2013). Total synthesis and evaluation of vinblastine analogues containing systematic deep-seated modifications in the vindoline subunit ring system: core redesign. *J. Med. Chem.* 56: 483–495.

92 Zaimoku, H., Nishide, H., Nishibata, A. et al. (2013). Syntheses of (±)-serratine, (±)-lycoposerramine T, and (±)-lycopoclavamine B. *Org. Lett.* 15: 2140–2143.

93 Zhang, H., Ma, X., Kang, H. et al. (2013). The enantioselective formal synthesis of rhynchophylline and isorhynchophylline. *Chem. Asian J.* 8: 542–545.

94 Xu, S., Unabara, D., Uemura, D., and Arimoto, H. (2014). Enantioselective total synthesis of pinnaic acid and halichlorine. *Chem. Asian J.* 9: 367–375.

95 Huang, Q., Wang, P., Tian, Y. et al. (2015). Synthesis of (±)-pterocarpin and its Thia- and Aza-analogues in a modular manner. *Synlett* 26: 1385–1390.

96 Suzuki, K., Matsumoto, T., Kitamura, K. et al. (2015). Toward the pluramycins: route exploration from dihydroxyanthrone tricyclic platform to an aglycon, saptomycinone B. *Heterocycles* 90: 1240.

97 Reddy, C.R., Dilipkumar, U., Mallesh, K. et al. (2016). Total synthesis of (+)-cladospolide D. *Tetrahedron: Asymmetry* 27: 222–225.

98 Li, Z. and Tong, R. (2017). Asymmetric total syntheses of (−)-hedycoropyrans A and B. *J. Org. Chem.* 82: 1127–1135.

99 Kobayashi, T., Tanaka, K., Ishida, M. et al. (2018). Asymmetric total synthesis of pleurospiroketals A and B. *Chem. Commun.* 54: 10316–10319.

100 Yadav, J., Reddy, A., and Shankar, K. (2013). Concise total synthesis of helicascolides A, B, and C. *Synthesis* 45: 1034–1038.

101 Nicolaou, K.C., Zhong, Y.-L., and Baran, P.S. (2000). A new method for the one-step synthesis of α,β-unsaturated carbonyl systems from saturated alcohols and carbonyl compounds. *J. Am. Chem. Soc.* 122: 7596–7597.

102 Nicolaou, K.C., Montagnon, T., and Baran, P.S. (2002). Modulation of the reactivity profile of IBX by ligand complexation: ambient temperature dehydrogenation of aldehydes and ketones to α,β-unsaturated carbonyl compounds. *Angew. Chem. Int. Ed.* 41: 993–996.

103 Li, Z., Wang, Y., Tang, C. et al. (2010). Synthesis of 3-benzoyl acrylates/acrylamides *via* dehydrogenation of 3-benzoyl propionates/propionamides using IBX/*p*-TsOH. *Chin. J. Chem.* 28: 1301–1305.

104 Li, Q. and Tochtrop, G.P. (2011). A stereoselective synthesis of the allo-bile acids from the 5β-isomers. *Tetrahedron Lett.* 52: 4137–4139.

105 Zhang, S., Xie, H., Song, A. et al. (2011). Efficient preparation of trans-α,β-unsaturated aldehydes from saturated aldehydes by oxidative enamine catalysis. *Sci. China Chem.* 54: 1932–1936.

106 Deng, X., Kong, L.-M., Zhao, Y. et al. (2012). Exploring of drug leads from diversity-oriented Michael-acceptor library derived from natural products. *Nat. Prod. Bioprospect.* 2: 210–216.

107 Ogawa, S., Zhou, B., Kimoto, Y. et al. (2013). An efficient synthesis of 7α,12α-dihydroxy-4-cholesten-3-one and its biological precursor 7α-hydroxy-4-cholesten-3-one: key intermediates in bile acid biosynthesis. *Steroids* 78: 927–937.

108 Iida, T., Omura, K., Sakiyama, R., and Kodomari, M. (2014). Regioselective dehydrogenation of 3-keto-steroids to form conjugated enones using o-iodoxybenzoic acid and trifluoroacetic acid catalysis. *Chem. Phys. Lipids* 178: 45–51.

109 Klahn, P. and Kirsch, S.F. (2014). IBX-mediated dehydrogenation of substituted β-oxonitriles. *Eur. J. Org.Chem.* 2014: 3149–3155.

110 Ramaraju, P., Mir, N.A., Singh, D. et al. (2015). Enantioselective synthesis of N-PMP-1,2-dihydropyridines *via* formal [4 + 2] cycloaddition between aqueous glutaraldehyde and imines. *Org. Lett.* 17: 5582–5585.

111 Ren, J., Lu, L., Xu, J. et al. (2015). Selective oxidation of 1-tetralones to 1,2-naphthoquinones with IBX and to 1,4-naphthoquinones with oxone® and 2-iodobenzoic acid. *Synthesis* 47: 2270–2280.

112 Jiang, S., Yan, T.-S., Han, Y.-C. et al. (2017). Hypervalent-iodine-mediated formation of epoxides from carbon(sp^2)-carbon(sp^3) single bonds. *J. Org. Chem.* 82: 11691–11702.

113 Liu, Y., Zhang, Z., Wan, Y. et al. (2017). Oxidation of β-ketoamides: the synthesis of vicinal tricarbonyl amides. *J. Org. Chem.* 82: 3901–3907.

114 Choudhary, S., Pawar, A.P., Yadav, J. et al. (2018). One-pot synthesis of chiral tetracyclic dibenzo[*b,f*][1,4]oxazepine-fused 1,2-dihydropyridines (DHPs) under metal-free conditions. *J. Org. Chem.* 83: 9231–9239.

115 Malinowski, M., Hensienne, R., Kern, N. et al. (2018). Stereocontrolled synthesis of polyhydroxylated bicyclic azetidines as a new class of iminosugars. *Org. Biomol. Chem.* 16: 4688–4700.

116 Zhu, L., Zhang, L., and Luo, S. (2018). Catalytic asymmetric β-C-H functionalizations of ketones via enamine oxidation. *Org. Lett.* 20: 1672–1675.

117 Ramaraju, P., Pawar, A.P., Iype, E. et al. (2019). Enantio- and diastereoselective two-pot synthesis of isoquinuclidines from glutaraldehyde and N-aryl imines with DFT calculations. *J. Org. Chem.* 84: 12408–12419.

118 Nicolaou, K.C. and Mathison, C.J.N. (2005). Synthesis of imides, *N*-acyl vinylogous carbamates and ureas, and nitriles by oxidation of amides and amines with Dess–Martin periodinane. *Angew. Chem. Int. Ed.* 44: 5992–5997.

119 Teskey, C.J., Adler, P., Gonçalves, C.R., and Maulide, N. (2019). Chemoselective α,β-dehydrogenation of saturated amides. *Angew. Chem. Int. Ed.* 58: 447–451.

120 Beekman, A.M. and Barrow, R.A. (2013). Stereochemical assignment of the fungal metabolites pestalotiopsones D and E through enantiopure synthesis. *J. Nat. Prod.* 76: 2054–2059.

121 Beekman, A.M., Castillo Martinez, E., and Barrow, R.A. (2013). First syntheses of the biologically active fungal metabolites pestalotiopsones A, B, C and F. *Org. Biomol. Chem.* 11: 1109–1115.

122 Xie, H., Zhang, S., Li, H. et al. (2012). Total synthesis of polyene natural product dihydroxerulin by mild organocatalyzed dehydrogenation of alcohols. *Chem. Eur. J.* 18: 2230–2234.

123 Makra, Z., Puskás, L.G., and Kanizsai, I. (2019). A convenient approach for the preparation of imidazo1,2-a-fused bicyclic frameworks via IBX/NIS promoted oxidative annulation. *Org. Biomol. Chem.* 17: 9001–9007.

124 Khan, P.R., Durgaprasad, M., Reddy, S.G. et al. (2017). IBX/KI promoted synthesis of 2,5-disubstituted 1,3,4-oxadiazoles. *Lett. Org. Chem.* 15: 64–69.
125 Hati, S. and Sen, S. (2017). Cerium chloride catalyzed, 2-iodoxybenzoic acid mediated oxidative dehydrogenation of multiple heterocycles at room temperature. *Eur. J. Org. Chem.* 2017: 1277–1280.
126 Narayana Murthy, S. and Nageswar, Y.V.D. (2011). *o*-Iodoxybenzoic acid (IBX): a versatile reagent for the synthesis of N-substituted pyrroles mediated by β-cyclodextrin in water. *Tetrahedron Lett.* 52: 4481–4484.
127 Chen, F., Shen, C., and Yang, D. (2011). A simple protocol for the synthesis of 2-arylbenzoxazoles by oxidation with *o*-iodoxybenzoic acid (IBX) and its application in the synthesis of arylbenzoxazole-containing amino acids. *Tetrahedron Let.* 52: 2128–2131.
128 Panarese, J.D. and Waters, S.P. (2010). Room-temperature aromatization of tetrahydro-β-carbolines by 2-iodoxybenzoic acid: utility in a total synthesis of eudistomin U. *Org. Lett.* 12: 4086–4089.
129 Gamapwar, S.V., Tale, N.P., and Karade, N.N. (2012). Dess-Martin periodinane–mediated oxidative aromatization of 1,3,5-trisubstituted pyrazolines. *Synth. Commun.* 42: 2617–2623.
130 Singh, A., Mir, N.A., Choudhary, S. et al. (2018). One-pot sequential multicomponent reaction between in situ generated aldimines and succinaldehyde: facile synthesis of substituted pyrrole-3-carbaldehydes and applications towards medicinally important fused heterocycles. *RSC Adv.* 8: 15448–15458.
131 Chen, J.-M. and Zeng, X.-M. (2009). β-cyclodextrin-catalyzed mild aromatization of hantzsch 1,4-dihydropyridines with *o*-iodoxybenzoic acid in water/acetone. *Synth. Commun.* 39: 3521–3526.
132 Kupwade, R.V., Khot, S.S., Kulkarni, M.A. et al. (2017). Diethylamine Dess-Martin periodinane: an efficient catalyst–oxidant combination in a sequential, one-pot synthesis of difficult to access 2-amino-3,5-dicarbonitrile-6-sulfanyl-pyridines at ambient temperature. *RSC Adv.* 7: 38877–38883.
133 Choudhary, S., Singh, A., Yadav, J. et al. (2019). A simple route to tetracyclic oxazepine-fused pyrroles *via* metal-free [3+2] annulation between dibenzo[*b*, *f*][1,4]oxazepines and aqueous succinaldehyde. *New J. Chem.* 43: 953–962.
134 Hati, S. and Sen, S. (2016). Synthesis of quinazolines and dihydroquinazolines: *o*-iodoxybenzoic acid mediated tandem reaction of *o*-aminobenzylamine with aldehydes. *Synthesis* 48: 1389–1398.
135 Pouységu, L., Sylla, T., Garnier, T. et al. (2010). Hypervalent iodine-mediated oxygenative phenol dearomatization reactions. *Tetrahedron* 66: 5908–5917.
136 Wu, A., Duan, Y., Xu, D. et al. (2010). Regiospecific oxidation of polycyclic aromatic phenols to quinones by hypervalent iodine reagents. *Tetrahedron* 66: 2111–2118.
137 Bizzarri, B.M., Pieri, C., Botta, G. et al. (2015). Synthesis and antioxidant activity of DOPA peptidomimetics by a novel IBX mediated aromatic oxidative functionalization. *RSC Adv.* 5: 60354–60364.
138 Mishra, A.K. and Moorthy, J.N. (2016). *o*-Iodoxybenzoic acid-initiated one-pot synthesis of 4-arylthio-1,2-naphthoquinones, 4-arylthio-

1,2-diacetoxynaphthalenes, and 5-arylthio-/5-aminobenzoaphenazines. *J. Org. Chem.* 81: 6472–6480.

139 Yoshida, Y., Magara, A., Mino, T., and Sakamoto, M. (2016). Facile synthesis of amino acid-derived novel chiral hypervalent iodine(V) reagents and their applications. *Tetrahedron Lett.* 57: 5103–5107.

140 Mishra, A.K., Mukhopadhyay, A., and Moorthy, J.N. (2017). One-pot multistep synthesis of bipolar carbazolo-phenazines: hydrogen bond control of Diels-Alder cycloaddition and application for fluoride sensing. *Tetrahedron* 73: 2210–2216.

141 Stack, D.E. and Mahmud, B. (2018). Efficient access to bisphenol A metabolites: synthesis of monocatechol, mono-*o*-quinone, dicatechol, and di-*o*-quinone of bisphenol A. *Synthetic Communications* 48: 161–167.

142 Viault, G., Helesbeux, J.-J., Richomme, P., and Séraphin, D. (2018). Concise semisynthesis of novel phenazine-vitamin E hybrids *via* regioselective tocopheryl *ortho*-quinone formation. *Tetrahedron Lett.* 59: 2627–2630.

143 Caprioglio, D., Mattoteia, D., Pollastro, F. et al. (2020). The oxidation of phytocannabinoids to cannabinoquinoids. *J. Nat. Prod.* 83: 1711–1715.

144 Chandra, A., Jana, K., and Moorthy, J.N. (2020). One-pot synthesis of 4-carboalkoxy-substituted benzohcoumarins from α- and β-naphthols and their excited-state properties. *ACS Omega* 5: 207–218.

145 Dobrotă, C., Graeupner, J., Dumitru, I. et al. (2010). Expedient access to fused quinoxalines *via* Dess–Martin periodinane-mediated cyclization of unsymmetrical phenylenediamide derivatives. *Tetrahedron Lett.* 51: 1262–1264.

146 Ma, H., Wu, S., Sun, Q. et al. (2010). A new method for the synthesis of iminoquinones via DMP-mediated oxidative reaction. *Synthesis* 2010: 3295–3300.

147 Kaur, A. and Ariafard, A. (2020). Mechanistic investigation into phenol oxidation by IBX elucidated by DFT calculations. *Org. Biomol. Chem.* 18: 1117–1129.

148 Tada, M., Ohkanda, T., and Kurabe, J. (2010). Syntheses of carnosic acid and carnosol, anti-oxidants in Rosemary, from pisiferic acid, the major constituent of Sawara. *Chem. Pharm. Bull.* 58: 27–29.

149 Boulangé, A., Peixoto, P.A., and Franck, X. (2011). Diastereoselective IBX oxidative dearomatization of phenols by remote induction: towards the epicocconone core framework. *Chem. Eur. J.* 17: 10241–10245.

150 Tanaka, K., Matsuyama, H., Watanabe, M. et al. (2014). Synthesis and biological evaluation of kendomycin and its analogues. *J. Org. Chem.* 79: 9922–9947.

151 Bhalerao, D.S., Mahajan, U.S., Chaudhari, K.H., and Akamanchi, K.G. (2007). *o*-Iodoxybenzoic acid- and tetraethylammonium bromide-mediated oxidative transformation of primary carboxamides to one-carbon dehomologated nitriles. *J. Org. Chem.* 72: 662–665.

152 Bellale, E.V., Bhalerao, D.S., and Akamanchi, K.G. (2008). Oxidative conversion of alpha, alpha-disubstituted acetamides to corresponding one-carbon-shorter ketones using hypervalent iodine (lambda^5) reagents in combination with tetraethylammonium bromide. *J. Org. Chem.* 73: 9473–9475.

153 Bellale, E.V., Huddar, S.N., Mahajan, U.S., and Akamanchi, K.G. (2011). Oxidative decarboxylation of α-amino acids to nitriles using *o*-iodoxybenzoic acid in aqueous ammonia. *Pure Appl. Chem.* 83: 607–612.

154 Deshmukh, S.S., Huddar, S.N., Jadhav, R.R., and Akamanchi, K.G. (2011). Oxidative fragmentation of oxiranes to nitriles with hypervalent iodine(V) reagents in aqueous ammonia. *Tetrahedron Lett.* 52: 4533–4536.

155 Xu, S., Itto, K., Satoh, M., and Arimoto, H. (2014). Unexpected dehomologation of primary alcohols to one-carbon shorter carboxylic acids using o-iodoxybenzoic acid (IBX). *Chem. Commun.* 50: 2758–2761.

156 Meprathu, B.V., Justik, M.W., and Protasiewicz, J.D. (2005). *ortho*-Phosphoryl stabilized hypervalent iodosyl- and iodyl-benzene reagents. *Tetrahedron Lett.* 46: 5187–5190.

157 Kazmierczak, P., Skulski, L., and Kraszkiewicz, L. (2001). Syntheses of (diacetoxyiodo)arenes or iodylarenes from iodoarenes, with sodium periodate as the oxidant. *Molecules* 6: 881–891.

158 Macikenas, D., Skrzypczak-Jankun, E., and Protasiewicz, J.D. (2000). Redirecting secondary bonds to control molecular and crystal properties of an iodosyl- and an iodylbenzene. *Angew. Chem. Int. Ed.* 39: 2007–2010.

159 Soldatova, N., Postnikov, P., Troyan, A.A. et al. (2016). Mild and efficient synthesis of iodylarenes using oxone as oxidant. *Tetrahedron Lett.* 57: 4254–4256.

160 Koposov, A.Y., Karimov, R.R., Pronin, A.A. et al. (2006). RuCl3-catalyzed oxidation of iodoarenes with peracetic acid: new facile preparation of iodylarenes. *J. Org. Chem.* 71: 9912–9914.

161 Meprathu, B.V. and Protasiewicz, J.D. (2010). Enhancing the solubility for hypervalent *ortho*-sulfonyl iodine compounds. *Tetrahedron* 66: 5768–5774.

162 Zhdankin, V.V., Koposov, A.Y., Netzel, B.C. et al. (2003). IBX amides: a new family of hypervalent iodine reagents. *Angew. Chem. Int. Ed.* 42: 2194–2196.

163 Ladziata, U., Koposov, A.Y., Lo, K.Y. et al. (2005). Synthesis, structure, and chemoselective reactivity of *N*-(2-iodylphenyl)acylamides: hypervalent iodine reagents bearing a pseudo-six-membered ring scaffold. *Angew. Chem. Int. Ed.* 44: 7127–7131.

164 Zhdankin, V.V., Koposov, A.Y., Litvinov, D.N. et al. (2005). Esters of 2-iodoxybenzoic acid: hypervalent iodine oxidizing reagents with a pseudobenziodoxole structure. *J. Org. Chem.* 70: 6484–6491.

165 Zhdankin, V.V., Litvinov, D.N., Koposov, A.Y. et al. (2004). Preparation and structure of 2-iodoxybenzoate esters: soluble and stable periodinane oxidizing reagents. *Chem. Commun.*: 106–107.

166 Koposov, A.Y., Karimov, R.R., Geraskin, I.M. et al. (2006). 2-iodylphenol ethers: preparation, X-ray crystal structure, and reactivity of new hypervalent iodine(V) oxidizing reagents. *J. Org. Chem.* 71: 8452–8458.

167 Mailyan, A.K., Geraskin, I.M., Nemykin, V.N., and Zhdankin, V.V. (2009). Preparation, X-ray structure, and oxidative reactivity of *N*-(2-iodylphenyl) tosylamides and 2-iodylphenyl tosylate: iodylarenes stabilized by ortho-substitution with a sulfonyl group. *J. Org. Chem.* 74: 8444–8447.

168 Nikiforov, V.A., Karavan, V.S., Miltsov, S.A. et al. (2003). Hypervalent iodine compounds derived from *o*-nitroiodobenzene and related compounds: syntheses and structures. *Arkivoc* 2003: 191.

169 Koposov, A.Y., Nemykin, V.N., and Zhdankin, V.V. (2005). Intra- and intermolecular interactions in the solid state structure of 2-iodylbenzenesulfonamides: a heptacoordinated organic iodine(v) compound. *New J. Chem.* 29: 998.

170 Koposov, A.Y., Litvinov, D.N., and Zhdankin, V.V. (2004). 2-Iodoxybenzenesulfamides: new pseudobenziodoxole-based hypervalent iodine reagents. *Tetrahedron Lett.* 45: 2719–2721.

171 Zhdankin, V.V., Goncharenko, R.N., Litvinov, D.N., and Koposov, A.Y. (2004). Derivatives of 2-iodoxybenzenesulfonic acid: new pseudocyclic hypervalent iodine reagents. *Arkivoc* 2005: 8.

172 Koposov, A. and Zhdankin, V.V. (2005). Selective oxidation of sulfides to sulfoxides using IBX-esters. *Synthesis* 2005: 22–24.

173 Moorthy, J.N., Senapati, K., and Parida, K.N. (2010). 6-Membered pseudocyclic IBX acids: syntheses, X-ray structural characterizations, and oxidation reactivities in common organic solvents. *J. Org. Chem.* 75: 8416–8421.

174 Geraskin, I.M., Luedtke, M.W., Neu, H.M. et al. (2008). Organic iodine(V) compounds as terminal oxidants in iron(III) phthalocyanine catalyzed oxidation of alcohols. *Tetrahedron Lett.* 49: 7410–7412.

175 Geraskin, I.M., Pavlova, O., Neu, H.M. et al. (2009). Comparative reactivity of hypervalent iodine oxidants in metalloporphyrin-catalyzed oxygenation of hydrocarbons: iodosylbenzene sulfate and 2-iodylbenzoic acid ester as safe and convenient alternatives to iodosylbenzene. *Adv. Synth. Catal.* 351: 733–737.

176 Yoshida, Y., Kanashima, Y., Mino, T., and Sakamoto, M. (2019). Asymmetric syntheses and applications of planar chiral hypervalent iodine(V) reagents with crown ether backbones. *Tetrahedron* 75: 3840–3849.

177 Xiao, X., Greenwood, N.S., and Wengryniuk, S.E. (2019). Dearomatization of electron-deficient phenols to ortho-quinones: bidentate nitrogen-ligated iodine(V) reagents. *Angew. Chem. Int. Ed.* 58: 16181–16187.

178 Bosset, C., Coffinier, R., Peixoto, P.A. et al. (2014). Asymmetric hydroxylative phenol dearomatization promoted by chiral binaphthylic and biphenylic iodanes. *Angew. Chem. Int. Ed.* 53: 9860–9864.

179 Boppisetti, J.K. and Birman, V.B. (2009). Asymmetric oxidation of *o*-alkylphenols with chiral 2-(o-iodoxyphenyl)-oxazolines. *Org. Lett.* 11: 1221–1223.

180 Toh, K.K., Biswas, A., Wang, Y.-F. et al. (2014). Copper-mediated oxidative transformation of *N*-allyl enamine carboxylates toward synthesis of azaheterocycles. *J. Am. Chem. Soc.* 136: 6011–6020.

181 Atmaca, U., Usanmaz, H.K., and Çelik, M. (2014). Oxidations of alkenes with hypervalent iodine reagents: an alternative ozonolysis of phenyl substituted alkenes and allylic oxidation of unsubstituted cyclic alkenes. *Tetrahedron Lett.* 55: 2230–2232.

182 Mülbaier, M. and Giannis, A. (2001). The synthesis and oxidative properties of polymer-supported IBX. *Angew. Chem. Int. Ed.* 40: 4393.

183 Sorg, G., Mengel, A., Jung, G., and Rademann, J. (2001). Oxidizing polymers: a polymer-supported, recyclable hypervalent iodine(V) reagent for the efficient conversion of alcohols, carbonyl compounds, and unsaturated carbamates in solution J.R. gratefully acknowledges generous support from Prof. M. E. Maier, Tübingen, the Strukturfonds of the University of Tübingen, the Fonds der Chemischen Industrie, and the DFG. We thank Graeme Nicholson, Dietmar Schmid, and Daniel Bischoff for analytical support. *Angew. Chem. Int. Ed.* 40: 4395.

184 Reed, N.N., Delgado, M., Hereford, K. et al. (2002). Preparation of soluble and insoluble polymer supported IBX reagents. *Bioorg. Med.Chem. Lett.* 12: 2047–2049.

185 Lei, Z., Denecker, C., Jegasothy, S. et al. (2003). A facile route to a polymer-supported IBX reagent. *Tetrahedron Lett.* 44: 1635–1637.

186 Sorg, G., Thern, B., Mader, O. et al. (2005). Progress in the preparation of peptide aldehydes via polymer supported IBX oxidation and scavenging by threonyl resin. *J. Pept. Sci.* 11: 142–152.

187 Lei, Z.Q., Ma, H.C., Zhang, Z., and Yang, Y.X. (2006). Synthesis and oxidation reactions of a polymer-supported IBX reagent. *React. Funct. Polym.* 66: 840–844.

188 Bromberg, L., Zhang, H., and Hatton, T.A. (2008). Functional organic–inorganic colloids modified by iodoxybenzoic acid. *Chem. Mater.* 20: 2001–2008.

189 Bernini, R., Mincione, E., Crisante, F. et al. (2009). A novel use of the recyclable polymer-supported IBX: an efficient chemoselective and regioselective oxidation of phenolic compounds. The case of hydroxytyrosol derivatives. *Tetrahedron Lett.* 50: 1307–1310.

190 Chen, J.-M., Zeng, X.-M., and Zhdankin, V. (2010). Preparation and reactivity of polystyrene-supported iodosylbenzene sulfate: an efficient recyclable oxidizing system. *Synlett* 2010: 2771–2774.

191 Zhang, J. and Phillips, J.A. (2010). The synthesis and application of polymer-supported hypervalent iodine reagent in the organic chemistry laboratory. *J. Chem. Educ.* 87: 981–984.

192 Bernini, R., Crisante, F., Barontini, M. et al. (2012). Synthesis and structure/antioxidant activity relationship of novel catecholic antioxidant structural analogues to hydroxytyrosol and its lipophilic esters. *J. Agric. Food Chem.* 60: 7408–7416.

193 Koguchi, S., Mihoya, A., and Mimura, M. (2016). Alcohol oxidation via recyclable hydrophobic ionic liquid-supported IBX. *Tetrahedron* 72: 7633–7637.

194 Chung, W.-J., Kim, D.-K., and Lee, Y.-S. (2003). Simple preparation of polymer supported IBX esters and amides and their oxidative properties. *Tetrahedron Lett.* 44: 9251–9254.

195 Karimov, R.R., Kazhkenov, Z.-G.M., Modjewski, M.J. et al. (2007). Preparation and reactivity of polymer-supported 2-iodylphenol ethers, an efficient recyclable oxidizing system. *J. Org. Chem.* 72: 8149–8151.

196 Chung, W.-J., Kim, D.-K., and Lee, Y.-S. (2005). Polymer-supported IBX-amide reagents: significant role of spacer and additive in alcohol oxidation. *Synlett*: 2175–2178.

197 Lecarpentier, P., Crosignani, S., and Linclau, B. (2005). A practical synthesis of a high-loading solid-supported IBX amide for the oxidation of alcohols. *Mol. Divers.* 9: 341–351.

198 Ladziata, U., Willging, J., and Zhdankin, V.V. (2006). Facile preparation and reactivity of polymer-supported N-(2-Iodyl-phenyl)-acylamide, an efficient oxidizing system. *Org. Lett.* 8: 167–170.

199 Jang, H.-S., Chung, W.-J., and Lee, Y.-S. (2007). Macroporous polystyrene-supported IBX amide: the improved oxidative properties in various solvents. *Tetrahedron Lett.* 48: 3731–3734.

200 Yoon, H.-J., Choi, J.-W., Jang, H.-S. et al. (2011). Selective oxidation of 5-hydroxymethylfurfural to 2,5-diformylfuran by polymer-supported IBX amide. *Synlett* 2011: 165–168.

201 Jang, H.-S., Kim, Y.-H., Kim, Y.-O. et al. (2014). Polymer-supported IBX amide for mild and efficient oxidation reactions. *J. Ind. Eng. Chem.* 20: 29–36.

202 Kim, Y.-H., Jang, H.-S., Kim, Y.-O. et al. (2013). Heterogeneous transition-metal-free alcohol oxidation by graphene oxide supported iodoxybenzoic acid in water. *Synlett* 24: 2282–2286.

203 Bizzarri, B.M., Abdalghani, I., Botta, L. et al. (2018). Iodoxybenzoic acid supported on multi walled carbon nanotubes as biomimetic environmental friendly oxidative systems for the oxidation of alcohols to aldehydes. *Nanomaterials* 8: 516.

204 Yoshimura, A., Banek, C.T., Yusubov, M.S. et al. (2011). Preparation, X-ray structure, and reactivity of 2-iodylpyridines: recyclable hypervalent iodine(V) reagents. *J. Org. Chem.* 76: 3812–3819.

205 Yusubov, M.S., Yusubova, R.Y., Nemykin, V.N. et al. (2012). Potassium 4-iodylbenzenesulfonate: preparation, structure, and application as a reagent for oxidative iodination of arenes. *Eur. J. Org. Chem.* 2012: 5935–5942.

206 Samunual, P. and Bergbreiter, D.E. (2016). Recyclable soluble polyisobutylene-bound oxidizing agents. *Tetrahedron Lett.* 57: 3272–3276.

207 Thottumkara, A.P., Bowsher, M.S., and Vinod, T.K. (2005). In situ generation of o-iodoxybenzoic acid (IBX) and the catalytic use of it in oxidation reactions in the presence of Oxone as a co-oxidant. *Org. Lett.* 7: 2933–2936.

208 Schulze, A. and Giannis, A. (2006). Oxidation of alcohols with catalytic amounts of IBX. *Synthesis*: 257–260.

209 Page, P., Appleby, L., Buckley, B. et al. (2007). In Situ generation of 2-iodoxybenzoic acid (IBX) in the presence of tetraphenylphosphonium monoperoxysulfate (TPPP) for the conversion of primary alcohols into the corresponding aldehydes. *Synlett* 2007: 1565–1568.

210 Takale, S., Parab, S., Phatangare, K. et al. (2011). IBX in aqueous medium: a green protocol for the Biginelli reaction. *Catal. Sci. Technol.* 1: 1128.

211 Seth, S., Jhulki, S., and Moorthy, J.N. (2013). Catalytic and chemoselective oxidation of activated alcohols and direct conversion of diols to lactones with In Situ-generated Bis-IBX catalyst. *Eur. J. Org. Chem.* 2013: 2445–2452.

212 Bikshapathi, R., Prathima, P.S., and Rao, V.J. (2016). Hypervalent iodine catalysis for selective oxidation of Baylis–Hillman adducts *via in situ* generation of *o*-iodoxybenzoic acid (IBX) from 2-iodosobenzoic acid (IBA) in the presence of oxone. *New J. Chem.* 40: 10300–10304.

213 Zhang, Z., Li, X., Song, M. et al. (2019). Selective removal of aminoquinoline auxiliary by IBX oxidation. *J. Org. Chem.* 84: 12792–12799.

214 Miura, T., Nakashima, K., Tada, N., and Itoh, A. (2011). An effective and catalytic oxidation using recyclable fluorous IBX. *Chem. Commun.* 47: 1875–1877.

215 Yakura, T., Fujiwara, T., Yamada, A., and Nambu, H. (2018). 2-Iodo-*N*-isopropyl-5-methoxybenzamide as a highly reactive and environmentally benign catalyst for alcohol oxidation. *Beilstein J. Org. Chem.* 14: 971–978.

216 Jhulki, S., Seth, S., Mondal, M., and Moorthy, J.N. (2014). Facile organocatalytic domino oxidation of diols to lactones by in situ-generated TetMe-IBX. *Tetrahedron* 70: 2286–2293.

217 Moorthy, J.N. and Parida, K.N. (2014). Oxidative cleavage of olefins by in situ-generated catalytic 3,4,5,6-tetramethyl-2-iodoxybenzoic acid/oxone. *J. Org. Chem.* 79: 11431–11439.

218 Colomer, I., Barcelos, R.C., Christensen, K.E., and Donohoe, T.J. (2016). Orthogonally protected 1,2-diols from electron-rich alkenes using metal-free olefin syn-dihydroxylation. *Org. Lett.* 18: 5880–5883.

219 Bindu, V.H., Parvathaneni, S.P., and Rao, V.J. (2017). Iodosobenzoic acid (IBA) catalysed benzylic and aromatic C–H oxidations. *Catal. Lett.* 147: 1434–1440.

220 Mishra, A.K. and Moorthy, J.N. (2017). Mechanochemical catalytic oxidations in the solid state with in situ-generated modified IBX from 3,5-di-*tert*-butyl-2-iodobenzoic acid (DTB-IA)/Oxone. *Org. Chem. Front.* 4: 343–349.

221 Colomer, I., Coura Barcelos, R., and Donohoe, T.J. (2016). Catalytic hypervalent iodine promoters lead to styrene dimerization and the formation of tri- and tetrasubstituted cyclobutanes. *Angew. Chem. Int. Ed.* 55: 4748–4752.

222 Yusubov, M.S., Zagulyaeva, A.A., and Zhdankin, V.V. (2009). Iodine(V)/ruthenium(III)-cocatalyzed oxidations: a highly efficient tandem catalytic system for the oxidation of alcohols and hydrocarbons with oxone. *Chem. Eur. J.* 15: 11091–11094.

223 Yusubov, M.S., Nemykin, V.N., and Zhdankin, V.V. (2010). Transition metal-mediated oxidations utilizing monomeric iodosyl- and iodylarene species. *Tetrahedron* 66: 5745–5752.

224 Maity, A., Hyun, S.-M., Wortman, A.K., and Powers, D.C. (2018). Oxidation catalysis by an aerobically generated Dess-Martin periodinane analogue. *Angew. Chem. Int. Ed.* 57: 7205–7209.

225 Uyanik, M., Akakura, M., and Ishihara, K. (2009). 2-Iodoxybenzenesulfonic acid as an extremely active catalyst for the selective oxidation of alcohols to aldehydes, ketones, carboxylic acids, and enones with oxone. *J. Am. Chem. Soc.* 131: 251–262.

226 Uyanik, M., Fukatsu, R., and Ishihara, K. (2009). IBS-catalyzed oxidative rearrangement of tertiary allylic alcohols to enones with oxone. *Org. Lett.* 11: 3470–3473.

227 Cui, L.-Q., Liu, K., and Zhang, C. (2011). Effective oxidation of benzylic and alkane C–H bonds catalyzed by sodium *o*-iodobenzenesulfonate with oxone as a terminal oxidant under phase-transfer conditions. *Org. Biomol. Chem.* 9: 2258–2265.

228 Uyanik, M., Mutsuga, T., and Ishihara, K. (2012). IBS-catalyzed regioselective oxidation of phenols to 1,2-quinones with Oxone®. *Molecules* 17: 8604–8616.

229 Liu, Y. and Wang, B. (2014). 2-Iodoxybenzenesulfonic acid-catalysed oxidation of primary and secondary alcohols with oxone in cetyl trimethylammonium bromide micelles at room temperature. *J. Chem. Res.* 38: 427–431.

230 Ballaschk, F. and Kirsch, S.F. (2019). Oxidation of secondary alcohols using solid-supported hypervalent iodine catalysts. *Green Chem.* 21: 5896–5903.

231 Uyanik, M., Mutsuga, T., and Ishihara, K. (2017). 4,5-Dimethyl-2-iodoxybenzenesulfonic acid catalyzed site-selective oxidation of 2-substituted phenols to 1,2-quinols. *Angew. Chem. Int. Ed.* 56: 3956–3960.

12

Sustainable Methods in Hypervalent Iodine Chemistry

Brandon Frey†*, Asim Maity*†*, Hao Tan, Pritam Roychowdhury, and David C. Powers*

Department of Chemistry, Texas A&M University, College Station, TX, USA

12.1 Introduction

Oxidation reactions, in which electron equivalents are removed from a substrate, are often critical steps in the synthesis of functional organic molecules. By definition, oxidation reactions require the use of an electron acceptor (i.e. terminal oxidant). Because the reduced by-products generated from the terminal oxidant are often lost as chemical waste, the chemical structure of the terminal oxidant dictates the sustainability of the oxidation reaction. Biological oxidation chemistry often utilizes dioxygen (O_2) as the terminal oxidant and generates water as the ultimate by-product [1, 2]. As such, these enzymatic oxidation reactions are nearly ideal from the perspective of sustainability. Analogous utilization of O_2 as a terminal oxidant in synthetic chemistry remains a significant challenge because (i) the triplet ground state of O_2 imposes kinetic barriers to O_2 utilization [3], (ii) O_2 often engages in poorly selective radical chain reactions [4], (iii) the electron inventories of O_2 reduction (four-electrons) and substrate oxidation (two-electron) are mismatched [5, 6]; and, (iv) O_2 utilization can present safety concerns depending on the specific reaction conditions employed [7]. The challenges associated with direct O_2 utilization as a terminal oxidant in selective synthesis have stimulated the development of alternative chemical oxidants for application in specific synthetic contexts. Implicit in the use of designer chemical oxidants is an increase in the waste associated with oxidation chemistry and, in turn, reduced sustainability.

The principles of green chemistry codify the concept of atom economy and the impact of the nature of waste on the efficiency of a chemical process [8, 9] . Active oxygen content, which is defined as the percent weight ratio of the oxygen transferred to the product divided by the formula weight of the reagent, can be used as a guiding parameter for evaluating the sustainability of oxidants [10]. Table 12.1 lists the active oxygen content and associated waste products generated from a variety of commonly encountered terminal oxidants.

† These authors contributed equally to this work.

Iodine Catalysis in Organic Synthesis, First Edition. Edited by Kazuaki Ishihara and Kilian Muñiz.

Table 12.1 List of general oxidants in order of their respective active oxygen content [10].

Substrate + Oxidant ⟶ Product + Reduced Oxidant (Stoichiometric waste)

Oxidant	Active oxygen content (wt %)	Waste generated
O_2 or O_2 / reductant[a)]	100 (50)	None (H_2O)
H_2O_2	47	H_2O
$KMnO_4$	30.4	Mn(II) salt
NaOCl	21.6	NaCl
O_2 / CH_3CHO	21.1	CH_3CO_2H
CH_3CO_3H	21.1	CH_3CO_2H
*t*BuOOH	17.8	*t*BuOH
H_2O_2-Urea	17.0	H_2O-Urea
$KHSO_5$	10.5	$KHSO_4$
m-CPBA	9.3	*m*-CBA
$NaIO_4$	7.5	$NaIO_3$
PhIO	7.3	PhI
$NaBO_3 \cdot 4H_2O$	7.2	$NaBO_2$
Oxone®[b)]	2.6	$K_2S_2O_8 \cdot KHSO_4 \cdot K_2SO_4$

a) This analysis assumes the reduced by-product is H_2O. If chemical reductants are employed (i.e. H_2, NADH, CH_3CHO, etc.), the active oxygen content will be lower depending on the mass of the generated by-products.
b) Oxone® ($2KHSO_5 \cdot KHSO_4 \cdot K_2SO_4$).
Source: Adapted from Goti and Cardona [10].

While O_2 is an attractive oxidant based on its active oxygen content, chemical strategies that utilize both oxygen atoms for substrate functionalization (i.e. dioxygenase reactivity) are rare [11]. More typically, O_2 utilization is accomplished in the presence of an appropriate reductant, which addresses the incongruous electron inventory of O_2 reduction and substrate functionalization. Consequently, only 50% of the active oxygen content of O_2 is utilized and the other 50% is lost as waste (ideally H_2O). Hydrogen peroxide (H_2O_2) is also considered a green oxidant and has 47% active oxygen content when utilized in direct substrate oxidation chemistry with an attendant generation of H_2O as the by-product. Utilization of H_2O_2 as an oxidant in hypervalent iodine chemistry often requires acetic anhydride, to generate peracetic acid *in situ*, or the use of urea-hydrogen peroxide (UHP) adduct, both of which reduce the active oxygen content of the reagent. Other commonly encountered chemical oxidants, such as NaOCl, CH_3CO_3H, *meta*-chloroperoxybenzoic acid (*m*-CPBA), $NaIO_4$, $NaBO_3 \cdot 4H_2O$, and Oxone® display significantly lower active oxygen contents. The recent renaissance of organic electrosynthesis offers an alternative strategy to sustainable oxidation chemistry: Anodic substrate oxidation provides the opportunity to avoid stoichiometric redox reagents and the associated waste streams

[12]. The frequent requirement for high concentrations of the supporting electrolyte can reduce the chemical efficiency of electrosynthetic approaches.

Hypervalent iodine compounds are a class of metal-free oxidants based on three-centered-four-electron I–L bonds [13, 14] that have emerged as important reagents in the functionalization of unsaturated hydrocarbons, transition metal-catalyzed chemistry, alcohol and amine oxidation, and group transfer chemistry [15–21]. The reactivity of hypervalent iodine compounds bears some similarities to that of transition metal species – namely the accessibility of multiple stable redox states, the facility of ligand exchange steps, and the participation in bond-forming processes [15–18, 22]. Both I(III) and I(V) compounds are important oxidation states in hypervalent iodine chemistry due to their complementary reactivities. Although hypervalent iodine compounds are attractive oxidants due to the abundance of iodoarene-starting materials and lower toxicity compared to transition metal catalysts, when employed as terminal oxidants, hypervalent iodine reagents display poor atom economy. For example, when reduced, iodosylbenzene (PhIO) generates a stoichiometric quantity of iodobenzene and only features 7.3% active oxygen content. In addition, hypervalent iodine species are typically prepared with stoichiometric metal-based oxidants ($KMnO_4$, Oxone®, $NaIO_4$, etc.) or organic peracids (*m*-CPBA and CH_3CO_3H) which amplifies the overall chemical inefficiency of using these reagents in synthesis.

The development of either (i) methods to synthesize hypervalent iodine compounds using higher active-oxygen-content reagents, or (ii) methods to employ hypervalent iodine species as catalysts in tandem with oxygen-dense oxidants could provide a platform for sustainable oxidation methods for a wide variety of reactions. Here, we summarize progress toward sustainable synthesis and the use of hypervalent iodine compounds. We first discuss progress toward greener methods to prepare hypervalent iodine compounds and this discussion is organized by the active oxygen content of the terminal oxidant employed. Catalytic application of hypervalent iodine compounds as a way to reduce waste generation and increase energy efficiency is discussed. Electrochemistry as an alternate sustainable method for the synthesis of hypervalent iodine compounds is described in the context of both synthesis of hypervalent iodine compounds and their utilization in electrocatalysis. Polymer-supported hypervalent iodine reagents and fluorous hypervalent iodine compounds, which have been developed to support reagent recycling, are also discussed. Only organic hypervalent iodine chemistry is presented; utilization of tetrabutylammonium iodide and other inorganic iodide sources will not be described.

12.2 Chemical Synthesis of Hypervalent Iodine Reagents

12.2.1 Synthesis of Hypervalent Iodine Reagents Using O_2

In concept, if hypervalent iodine reagents were generated using O_2 as the terminal oxidant, the array of chemistry that is available to these reagents could be achieved aerobically. Iodoarenes do not react directly with O_2 to afford hypervalent iodine compounds, and thus the development of aerobic methods of hypervalent iodine chemistry requires the identification of an appropriate reductant to enable aerobic oxidation to be realized.

12.2.1.1 Synthesis of Hypervalent Iodine(III) Reagents with O_2

In 2018, Powers *et al.* reported the synthesis of hypervalent iodine reagents by intercepting reactive oxidants generated during acetaldehyde autoxidation [23]. Initial studies of the oxidation of iodobenzene (**1**) in the presence of acetaldehyde and O_2 afforded variable yields of (diacetoxyiodo)benzene (**2**, 42–91% yield), presumably due to inconsistent initiation of the aldehyde autoxidation chain reaction. Co(II) additives, which have been previously used as initiators of aldehyde autoxidation chemistry [24, 25], were found to be critical in achieving reproducible chemical yields. The developed aerobic oxidation protocol provided ready access to a wide variety of hypervalent iodine reagents (Figure 12.1): Oxidation of 4-substituted iodoarenes provided access to hypervalent iodine species bearing both electron-donating and -withdrawing groups (**3**–**8**). Oxidation of 2-substituted iodoarenes afforded the corresponding cyclic iodanes (**9**–**12**) in high yields, which are an important class of reagents for hypervalent iodine-mediated group transfer chemistry. The addition of external acids provided direct entry to acid-exchanged I(III) products (**13**–**15**) instead of the corresponding (diacetoxyiodo)benzene derivatives; for example, the addition of TsOH, benzoic acid (BzOH), and trifluoroacetic

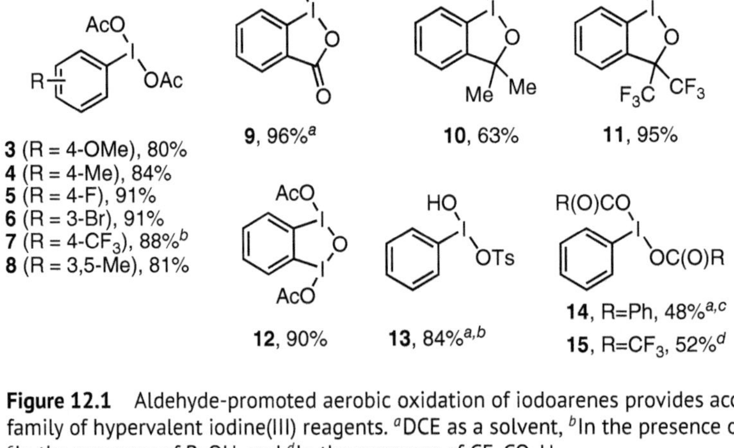

Figure 12.1 Aldehyde-promoted aerobic oxidation of iodoarenes provides access to a family of hypervalent iodine(III) reagents. [a]DCE as a solvent, [b]In the presence of $TsOH{\cdot}H_2O$, [c]In the presence of BzOH, and [d]In the presence of CF_3CO_2H.

acid (TFA) led to the formation of Koser's reagent (**13**), $PhI(OBz)_2$ (**14**), and $PhI(O_2CCF_3)_2$ (**15**) in 84, 48, and 52% yields, respectively.

A detailed investigation of the mechanism of aerobic oxidation revealed that iodoarene oxidation proceeds via a one-electron radical chain mechanism, which differs from the commonly observed two-electron oxidation/reduction reactions of these reagents [26]. The proposed mechanism was formulated based on (i) EPR spectroscopy of spin-trapped acetoxy radicals, (ii) comparative Hammett analyses of aerobic oxidation and peracetic acid oxidation, (iii) kinetic competition experiments, and (iv) computational analysis. These data indicated that aerobic oxidation of iodobenzene (**1**) in the presence of acetaldehyde and O_2 proceeds via the addition of aerobically generated acetoxy radicals (**16**) to iodoarenes to generate transient iodanyl radicals **17** (Figure 12.2). Subsequent chain propagation by reaction with peracetic acid (**18**, an intermediate from aldehyde autoxidation) affords I(III) product (**19**) and regenerates the acetoxy radical chain carrier (**16**). Chain termination via either combination of **17** with acetoxy radical or via iodanyl radical disproportionation leads to (diacetoxyiodo)benzene (**2**).

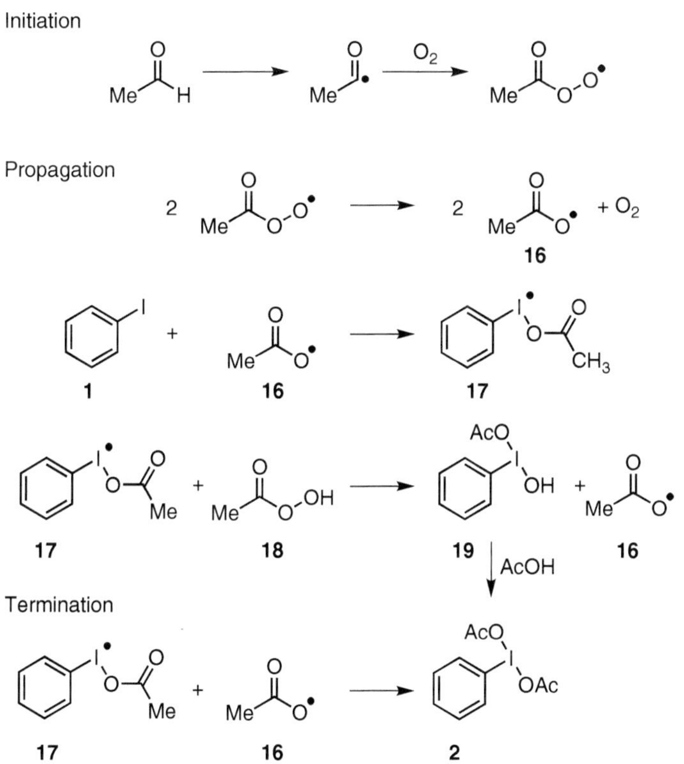

Figure 12.2 Aldehyde-promoted aerobic oxidation proceeds by the addition of aerobically generated acetoxy radical to iodobenzene to generate a transient iodanyl radical, which is a chain carrier in the synthesis of I(III) compounds.

12.2.1.2 Synthesis of Hypervalent Iodine(V) Reagents with O_2

The first aerobic synthesis of iodylbenzene (**20**) was reported by Jorissen *et al.* in 1938 [27], who demonstrated that an aerated acetaldehyde solution of iodobenzene (**1**) produced white crystals over the course of three weeks (Figure 12.3). These crystals were identified as iodylbenzene by iodometry.

In 2018, Powers and coworkers described a strategy for the aerobic synthesis of I(V) reagents by coupling aldehyde-promoted aerobic oxidation of iodoarenes with disproportionation of the initially generated I(III) species [28]. Aerobic oxidation of iodoarenes with weakly chelating *ortho*-substituents like sulfonyl, nitro, or pyridyl was found to furnish the corresponding I(V) compounds **21–23** in good yield (Figure 12.4) [22, 29].

12.2.2 Synthesis of Hypervalent Iodine Reagents with H_2O_2 or UHP

H_2O_2 is considered to be an environmentally benign oxidant due to its 47% active oxygen content and the potential generation of water as the by-product. H_2O_2 is produced via the Riedl–Pfleiderer process where 2-alkylanthraquinone is first reduced to 2-alkylhydroanthraquinone by hydrogenation and again oxidized aerobically to regenerate 2-alkylanthraquinone and H_2O_2 [30]. Decomposition of H_2O_2 to H_2O and O_2 is exothermic (−94.6 kJ/mol) [31], and the use of concentrated solutions presents significant safety hazards. Hence, H_2O_2 is typically employed as ≤30% aqueous solution. An alternative way to safely deliver H_2O_2 into the reaction is by using a urea adduct of hydrogen peroxide (UHP), which is an odorless, water-soluble, crystalline solid that has an active oxygen content of 17% [32].

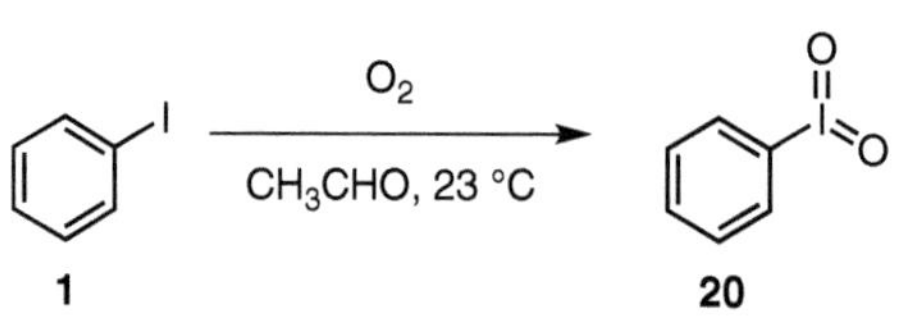

Figure 12.3 Upon standing for several weeks, an aerated solution of iodobenzene (**1**) in acetaldehyde resulted in the formation of crystalline iodylbenzene (**20**).

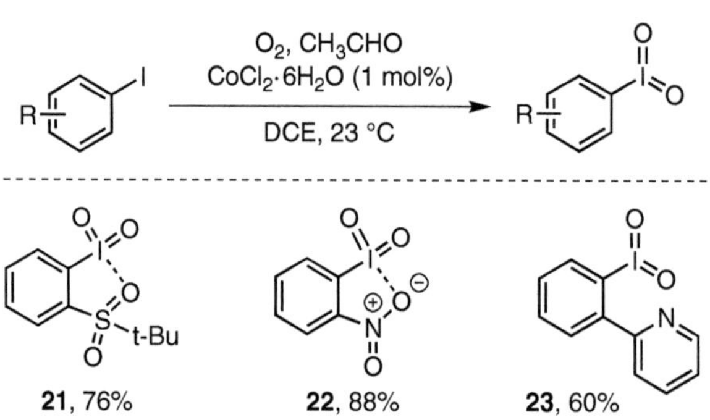

Figure 12.4 Aerobic synthesis of hypervalent iodine(V) reagents was achieved by coupling aldehyde-promoted aerobic oxidation iodoarenes with subsequent *in situ* disproportionation. *Source*: Zhdankin and Protasiewicz [22], Cardenal et al. [29].

12.2.2.1 Synthesis of Hypervalent Iodine(III) Reagents with H_2O_2 or UHP

There are no examples of direct oxidation of iodoarenes with H_2O_2 alone, which may be due to the spontaneous decomposition of H_2O_2 by I(III) compounds to form singlet oxygen ($^1O_2$) and reduced I(I) (Figure 12.5) [33]. H_2O_2 is often employed as an oxidant in combination with acid anhydrides and hydrohalic acids, which promote the *in situ* generation of peracids and hypohalous acids, respectively. Here, we organize the presentation of H_2O_2-based methods by the type of hypervalent iodine compounds generated.

Peracid-based oxidation of iodoarenes was first reported by Boeseken and Schneider in 1931 who synthesized (diacetoxyiodo)benzene by treatment of iodobenzene in chloroform with 55% peracetic acid [34]. In 1953, Pausacker reported a modified synthesis in which 30% H_2O_2 and acetic anhydride were stirred together at 40 °C to generate peracetic acid followed by the addition of iodobenzene (**1**) to the resulting solution (Figure 12.6) [35]. This method avoids the use of the highly concentrated peracetic acid solution and was the earliest example with H_2O_2 as the terminal oxidant for the synthesis of I(III) compounds. To date, this method is among the most widely used for the synthesis of $PhI(OAc)_2$ (**2**).

[Bis(trifluoroacetoxy)iodo]arenes are stronger oxidants than $PhI(OAc)_2$ due to the more electron-withdrawing trifluoroacetoxy groups on the iodine center [36]. Zhdankin *et al.* reported a direct synthetic route to [bis(trifluoroacetoxy)iodo]arenes using trifluoroperacetic acid, generated *in situ* by a combination of 80% H_2O_2 with trifluoroacetic anhydride [37]. Although the described procedure was more efficient than previous synthetic methods based on ligand exchange reactions of hypervalent iodine(III) compounds with trifluoroacetic acid [38, 39], silver trifluoroacetate [40], or trimethylsilyl trifluoroacetate [41], the use of such high concentration of H_2O_2 poses serious safety hazards. In 2006, Wirth *et al.* described a modification of the procedure using UHP as the terminal oxidant, which enabled *in situ* formation of trifluoroperacetic acid, leading to efficient formation of [bis(trifluoroacetoxy)iodo]benzene (**15**), [bis(trifluoroacetoxy)iodo]toluenes (**24, 25**), [bis(trifluoroacetoxy)iodo]arenes with electron-withdrawing substituents (**26–29**), and chiral I(III) product **30** (Figure 12.7) [42].

$$H_2O_2 \xrightarrow[-PhI]{PhI(OCOCF_3)_2} {}^1O_2 + 2\,CF_3COOH$$

Figure 12.5 Hydrogen peroxide reacts with (ditrifluoroacetoxy)iodobenzene to form singlet oxygen ($^1O_2$) and iodobenzene (**1**). *Source*: Catir et al. [33].

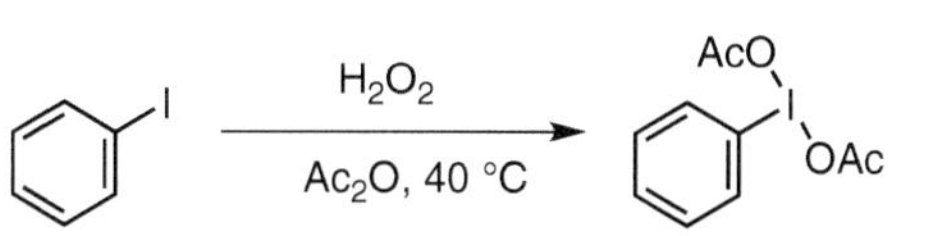

1 **2**

Figure 12.6 The combination of hydrogen peroxide and acetic anhydride generates peracetic acid *in situ*, which is responsible for the oxidation of iodobenzene (**1**) to (diacetoxyiodo)benzene (**2**). *Source*: Pausacker [35].

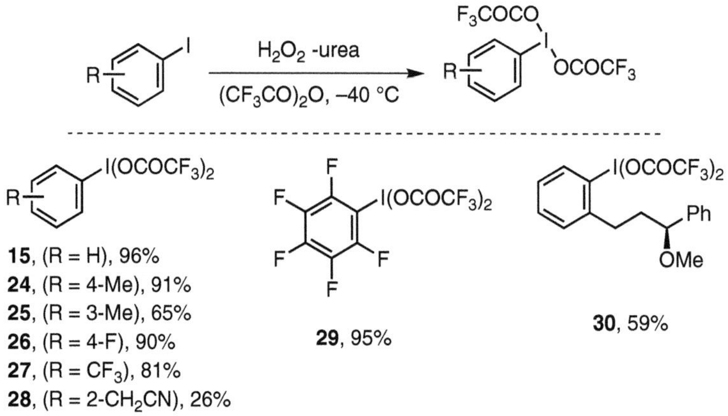

Figure 12.7 *In situ* generation of trifluoroperacetic acid from UHP and trifluoroacetic anhydride provides access to [bis(trifluoroacetoxy)iodo]arenes from the corresponding iodoarenes. *Source*: Page and Wirth [42].

(Dichloroiodo)arenes, $ArICl_2$, are widely used reagents for the chlorination of unsaturated hydrocarbons and also as terminal oxidants for transition metal catalysis [43]. Historically, (dichloroiodo)benzene (**31**) has been synthesized by bubbling chlorine gas into a solution of iodobenzene (**1**) at a low temperature. Use of toxic and corrosive chlorine gas can be avoided using hydrochloric acid in combination with oxidants like $KClO_3$ [44], $KMnO_4$ [45], concentrated HNO_3 [45], $Na_2S_2O_8$ [46], CrO_3 [47], $NaClO_2$, or NaClO [48]. These methods suffer from (super)stoichiometric use of terminal oxidants and poor substrate scope with respect to electron-withdrawing substituents on the iodoarenes. Jarnej and coworkers reported an alternative route for the synthesis of (dichloroiodo)arenes from a mixture of HCl/H_2O_2 (30% aqueous solution) in 1,1,1-trifluoroethanol (TFE) which acts both as a solvent and activator for H_2O_2 (Figure 12.8) [49]. In this protocol, hypochlorous acid (HOCl) is generated *in situ* and acts as the active oxidant and chlorinating agent. This method tolerates alkyl substituents (**32, 33**) as well as electron-withdrawing groups such as carboxyl-, nitro-, and chloro substituents (**34–37**). In the case of 4,4′-diiodo-1,1′-biphenyl, both the iodine centers are chlorinated to generate **38**. In the presence of electron-donating groups like dimethyl, trimethyl, or methoxy substituents, the corresponding (dichloroiodo)arenes (**39–41**) decompose to the chlorinated arene products.

Diaryliodonium salts are typically synthesized via a two-step process involving iodoarene oxidation followed by arylation of the I(III) with a suitable arene; thus, catalysis is often limited, and sustainable synthesis of these reagents is important to decrease the environmental impact. Building on the demonstration of UHP in the synthesis of iodine(III) reagents, Olofsson *et al.* developed a UHP-based synthesis of diaryliodonium triflates [50]. The authors utilized the *in situ* formation of triflic peroxide using UHP and triflic anhydride (Tf_2O) to oxidize the iodoarenes in the presence of suitable arenes to furnish diaryliodonium triflates in a single step. The developed methodology was effective in synthesizing both symmetric (**42**) and

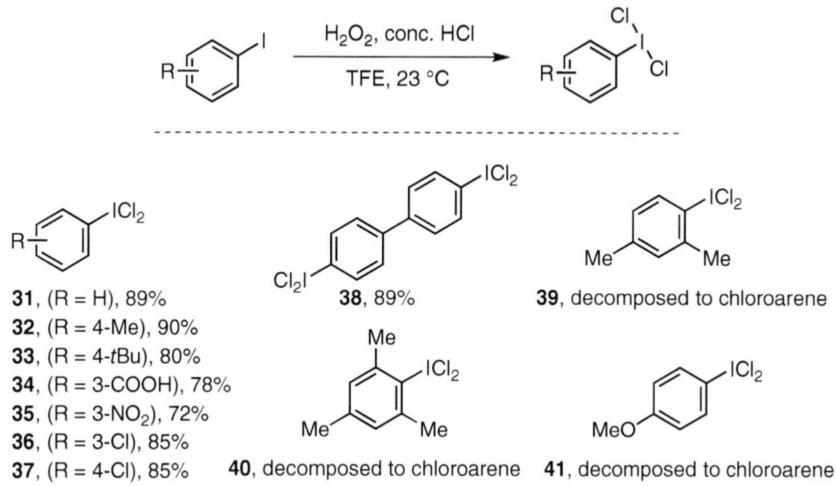

Figure 12.8 H_2O_2 was combined with concentrated HCl to form hypochlorous acid (HOCl) *in situ*, the active chlorinating reagent for the synthesis of (dicholoroiodo)arenes from iodoarenes.

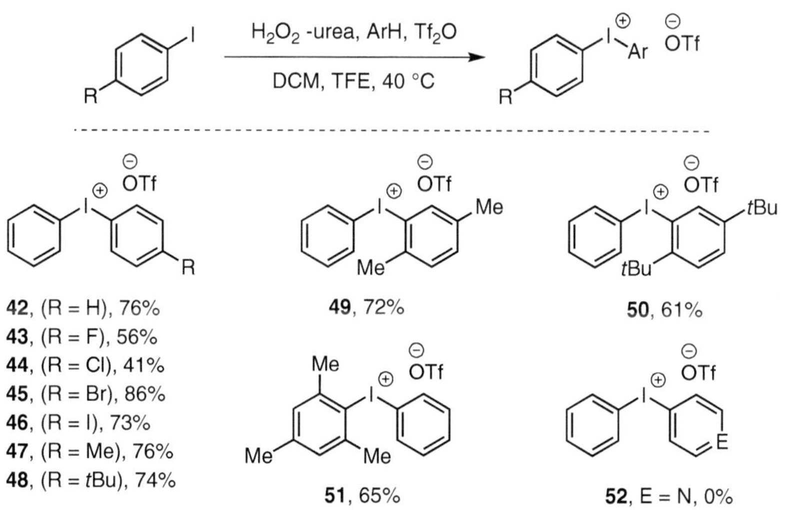

Figure 12.9 Diaryliodonium triflates were synthesized using UHP in a sustainable method leading to symmetric and unsymmetric iodonium salts.

unsymmetric iodonium (**43–51**) salts in good yield (42–86%) but, use of oxidatively labile substrate like anisole or pyridine led to undesired side products or pyridine *N*-oxides, respectively, without any desired product formation (**52, 53**) (Figure 12.9).

12.2.2.2 Synthesis of Hypervalent Iodine(V) Reagents from H_2O_2 or UHP

Zhdankin and coworkers reported a two-step, one-pot strategy to access iodylarenes using H_2O_2 as the terminal oxidant by coupling *in situ* generation

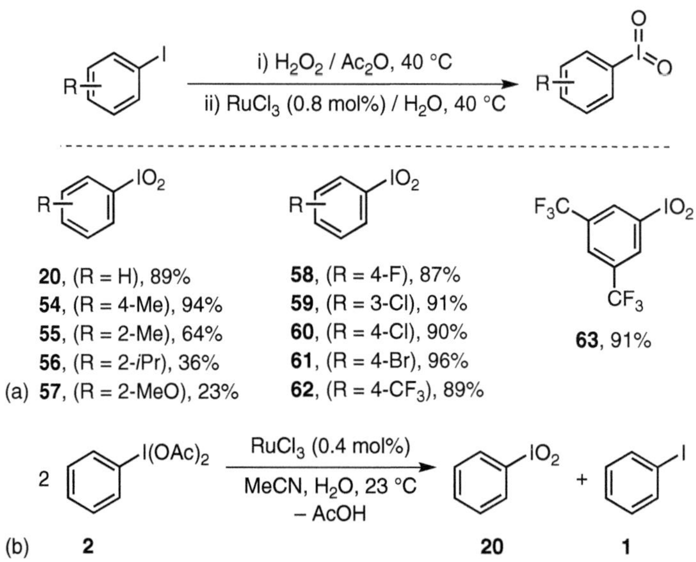

Figure 12.10 (a) $RuCl_3$-mediated disproportionation of initially formed hypervalent iodine(III) intermediates provided access to the corresponding iodylarenes. (b) Treatment of independently synthesized (diacetoxyiodo)benzene (**2**) with $RuCl_3$ resulted in disproportionation to an equimolar amount of iodobenzene (**1**) and iodylbenzene (**20**). *Source*: Yusubov et al. [52].

of peracid with subsequent disproportionation chemistry catalyzed by $RuCl_3$ (0.8 mol%) [51]. This method was effective in furnishing iodylarenes bearing both electron-donating (**54–57**) and electron-withdrawing groups (**58–63**), but the formation of 2-iodoxybenzoic acid (IBX) from 2-iodobenzoic acid was unsuccessful (Figure 12.10a). When independently synthesized $PhI(OAc)_2$ (**2**) was treated with $RuCl_3$ (0.4 mol%) in an aqueous acetonitrile solution at room temperature, immediate formation of an equimolar amount of iodylbenzene (**20**) and iodobenzene (**1**) were observed, which is consistent with disproportionation being operative in the synthesis of I(V) species (Figure 12.10b) [52].

12.3 Electrochemical Synthesis of Hypervalent Iodine Reagents

Anodic oxidation of iodoarenes represents a conceptually attractive approach to these species, in which stoichiometric oxidants are replaced by electrical potential. Electrosynthesis can broadly be classified as either *ex cell*, in which anodic oxidation of iodoarenes is accomplished during a first step and the resulting hypervalent iodine reagents are subsequently utilized for substrate functionalization in a separate step, or *in cell*, in which anodic oxidation of iodoarenes is carried out in the presence of substrate, such as is required to achieve electrocatalysis [53]. Both I(III) and I(V) reagents have been prepared electrochemically. Due to the substantial overpotential that is typically required for the anodic oxidation of iodoarenes, hypervalent iodine electrochemistry is often limited to *ex cell* applications (i.e. at the potential required to oxidize iodoarenes, many substrates participate in background oxidation).

12.3.1 Electrochemical Synthesis of Hypervalent Iodine(III) Reagents

The first example of the electrochemical synthesis of hypervalent I(III) reagents was reported in 1960 by Schmidt and Meinert, who described the anodic oxidation of iodobenzene in the presence of silver fluoride (AgF), which acts as both the fluorine source and supporting electrolyte, to generate (difluoroiodo)benzene ($PhIF_2$, **65**) [54]. Subsequent reports have described difficulty reproducing this synthesis [55]. In 1994, Fuchigami and coworkers discovered that direct electrolysis of 4-iodonitrobenzene (**64**) in the presence of $Et_3N{\cdot}3HF$ as the supporting electrolyte and fluorine source formed 4-(difluoroiodo)nitrobenzene (**66**) in 53% yield (Figure 12.11) [56]. But, electrolysis of iodobenzene and 4-iodotoluene did not lead to the desired I(III) compounds, but instead gave rise to benzylic fluorination and diaryliodonium products, respectively.

Hara *et al.* later discovered that using $Et_3N{\cdot}5HF$ as the supporting electrolyte and fluorine source resulted in the successful electrosynthesis of 4-(difluoroiodo)toluene (**68**) via potentiostatic electrolysis of 4-iodotoluene (**67**) at 1.5 V vs. Ag^+/Ag [57]. Electrochemically generated **68** has been applied to a variety of substrate functionalization reactions including fluorination of β-dicarbonyl substrates [57], fluorocyclization of *N*-allylbenzamides [58], and vicinal difluorination of olefins (Figure 12.12) [59].

Ex cell electrochemistry of hypervalent iodine reagents provides a means to generate unstable or dangerous reagents *in situ* without the need for isolation, which often leads to improved yields and safer reaction conditions [60, 61]. Nishiyama and coworkers developed anodic oxidation of iodoarenes in fluorinated solvents, such as TFE to afford [bis(trifluoroethoxy)iodo]benzene (**72**) [62–64]. Fluorinated alcohol solvents are known to help stabilize radical cation intermediates and consequently improve single-electron oxidation events of iodoarenes at anode surfaces [36, 65–67]. While many of the I(III) reagents supported by fluorinated alkoxide ligands are not stable to isolation (with the exception of the *p*-nitro-substituted derivative), these

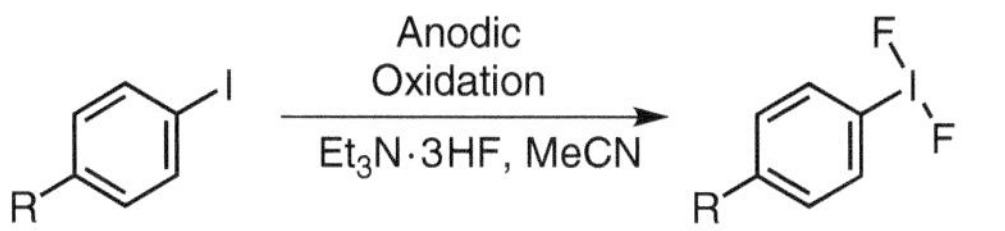

1, R = H
64, R = NO_2

65, R = H, 0%
66, R = NO_2, 53%

Figure 12.11 Electrochemical synthesis of 4-(difluoroiodo)nitrobenzene (**66**) was achieved using triethylamine adduct of hydrogen fluoride both as the source of fluoride and the supporting electrolyte under constant current electrolysis conditions.

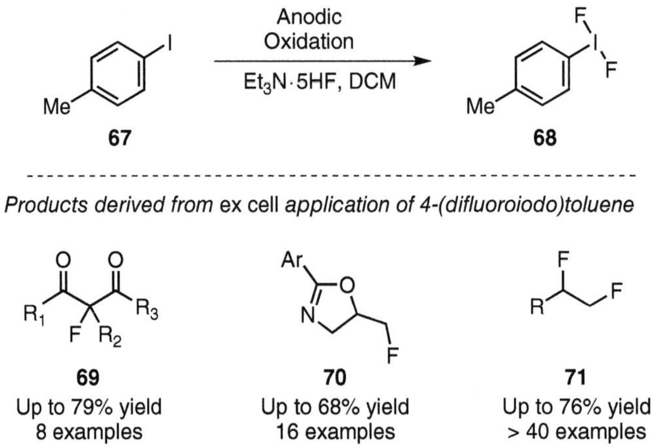

Figure 12.12 Electrochemical synthesis of 4-(difluoroiodo)toluene (**67**) was carried out under constant current electrolysis condition and then utilized in various *ex cell* fluorination reactions. *Source*: Doobary et al. [59].

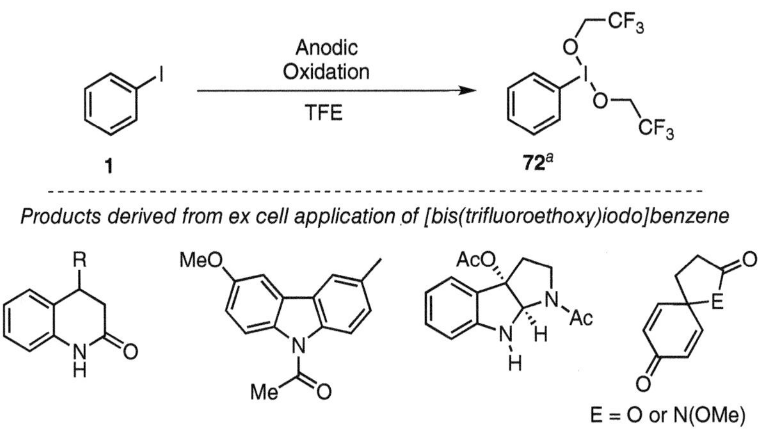

Figure 12.13 A fluorinated alcohol solvent such as trifluoroethanol (TFE) is used to stabilize anodically generated hypervalent iodine(III) intermediate and the resulting oxidizing solution is utilized for *ex cell* applications leading to various alkaloids, lactones, and lactams. [a]Only stable in solution, not isolable.

electrochemically derived reagents have been widely employed in subsequent *ex cell*-substrate functionalization (Figure 12.13) [62–64, 68–72].

The anodic oxidation of iodoarenes can also be used to electrochemically synthesize diaryliodonium salts. In 1967, Miller and Hoffmann disclosed that potentiostatic electrolysis of iodobenzene at 1.65 V vs. Ag^+/Ag, in acetonitrile, with lithium perchlorate as the supporting electrolyte afforded (4-iodophenyl)(phenyl)iodonium perchlorate in 45% yield [73]. Similarly, electrolysis of iodobenzene and 4-iodotoluene in the presence of benzene resulted in diphenyliodonium perchlorate (32% yield) and (4-methylphenyl)(phenyl)iodonium perchlorate (56%

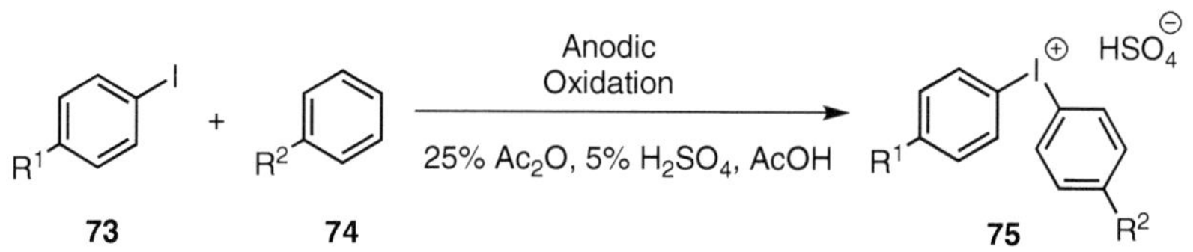

Figure 12.14 Electrochemical synthesis of both symmetric and unsymmetric diaryliodonium bisulfates under constant current conditions.

yield), respectively. Wendt and coworkers later investigated the mechanism of diaryliodonium formation. Because benzene does not oxidize below 1.9 V vs. Ag^+/Ag, they concluded that iodoarenes are oxidized at the anode surface [74].

In 2000, Peacock and Pletcher reported a method to expand the scope of diaryliodonium salts that could be prepared electrochemically [75, 76]. By using a mixture of acetic acid, acetic anhydride, and sulfuric acid as the reaction solvent, anodic synthesis of iodonium salts (**75**) was extended to electron-rich iodoarenes and arene-coupling partners (Figure 12.14). In addition, both symmetric and asymmetric iodonium salts were synthesized. Due to the substantial overpotential for anodic oxidation of iodoarenes, many functional groups were found to be incompatible with the developed conditions. Electron-withdrawing substituents on the iodoarene (**73**) increased the onset potential beyond the solvent window and thus prevented initial anodic oxidation. Although electron-donating substituents on the iodoarene, such as 4-methoxy, decrease the onset potential for iodoarene oxidation this ultimately gave rise to unproductive side reactions, such as anisole sulfonation by sulfuric acid. With respect to the arene-coupling partner (**74**), electron-withdrawing substituents decreased yields due to decreased nucleophilicity and electron-rich arene-coupling partners were limited due to direct anodic oxidation and subsequent radical polymerization.

One challenge in utilizing electrochemically derived hypervalent iodine reagents for substrate functionalization is the need to remove the iodoarene by-products and electrolytes from the reaction product. To address this, Francke and coworkers developed low-molecular-weight, charged iodoarenes as potential *ex cell* mediators in which the supporting electrolyte and the iodoarene mediator were combined [77–79]. The first examples developed were quaternary ammonium-substituted 4-iodoarenes. Anodic oxidation of iodoarene **76** in fluorinated solvents provided access to the corresponding hypervalent iodine compound **77**, which was subsequently utilized in *ex cell* C–N bond-forming reactions such as carbazole formation (**78**), *N*-arylation (**79**) [77], and oxazole syntheses (**80**) (Figure 12.15) [78].

Francke and coworkers advanced iodophenylsulfonates (**83**, **84**) and iodobenzoates (**85–87**) as redox-active supporting electrolytes [79]. The *ex cell* reactivity of the corresponding I(III) compounds $[ArI(OCH(CF_3)_2)_2]$ was evaluated by reaction with *N*-([1,1′-biphenyl]-2-yl)acetamide (**81**) to afford 1-(9*H*-carbazol-9-yl)ethan-1-one (**82**) (Figure 12.16). While both the iodophenylsulfonate isomers **83** and **84** could be used as *ex cell* mediators for the intramolecular C–H amination reaction,

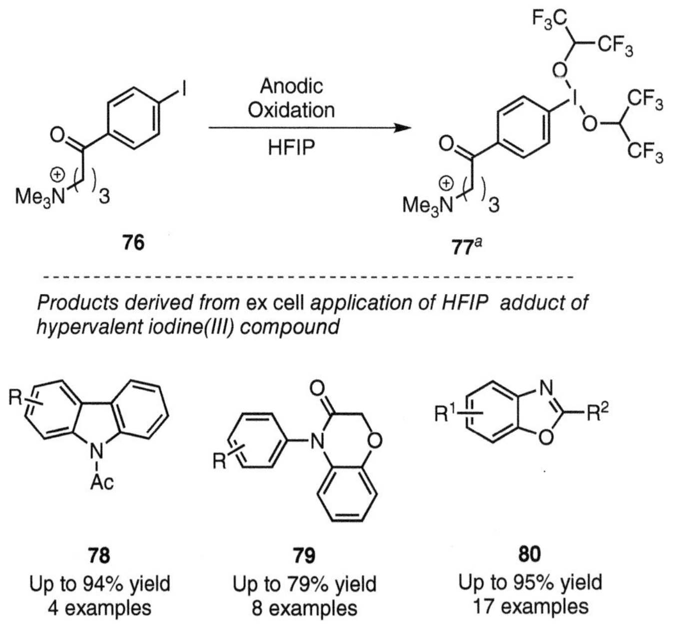

Figure 12.15 Iodoarene (**76**) with quaternary ammonium moiety obviates the need to use additional supporting electrolyte and acts as a mediator for C−N bond-forming reactions leading to carbazoles, *N*-aryl benzoxazinones and oxazoles. [a]Only stable in solution, not isolable. *Source*: Koleda et al. [78].

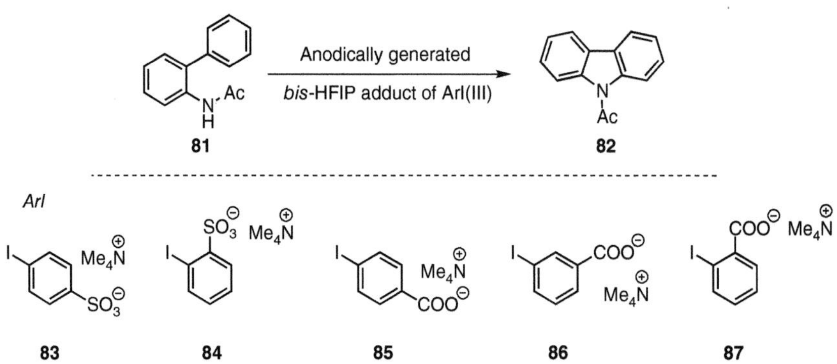

Figure 12.16 Tetramethylammonium salts of iodophenylsulfonates and iodobenzoates combine the roles of electrolyte and iodoarene mediator.

iodobenzoate compounds **85–87** did not promote substrate oxidation to form carbazole **82**.

Flow chemistry has also been advanced as an approach to achieve an anodic synthesis of hypervalent iodine reagents and utilize them in *ex cell*-substrate functionalization. Flow conditions offer advantages compared to mechanical stirring, such as better mixing, more efficient heat transfer, and easy scale-up [80]. Flow systems can be especially useful in industrial settings for energy conservation [81], multistep

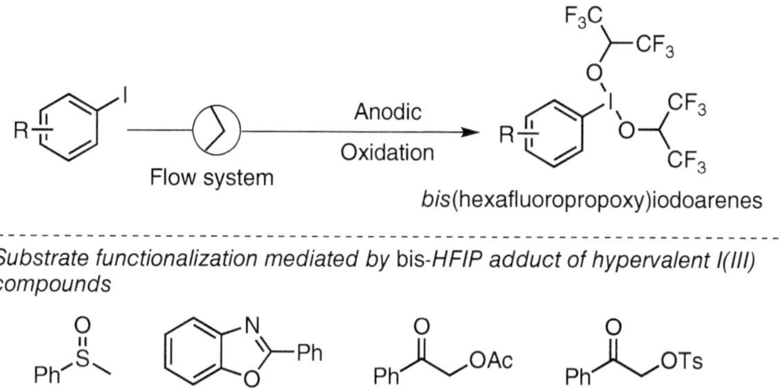

Figure 12.17 Anodically generated bis(hexafluoroisopropoxy) adduct under flow condition is used for various substrate functionalization reactions. *Source*: Elsherbini et al. [87].

synthesis [82] or end-to-end production [83]. Wirth and coworkers have studied the synthesis and utilization of hypervalent iodine species via flow technology [84–87]. Symmetrical and unsymmetrical iodonium salts have been synthesized by Wirth and coworkers using a microreactor with two platinum electrodes separated by a thin film (250 mm) [84]. This method was later elaborated to achieve *meta*-selective C–H arylation of anilines in an end-to-end production [85].

More recently, Wirth and coworkers extended the scope of electrochemically derived hypervalent iodine reagents generated in flow by preparing [bis(hexafluoroisopropoxy)iodo]arenes and utilizing them in a wide range of substrate oxidation reactions including sulfide oxidation, oxidative heterocyclization, and α-functionalization of carbonyl compounds (Figure 12.17) [87].

12.3.2 Electrochemical Synthesis of Hypervalent Iodine(V) Reagents

Anodic oxidation of 2-iodobenzoic acid (**88**) in 0.2 M H_2SO_4 aqueous solution with a boron-doped diamond electrode produced IBX (**89**) under potentiostatic electrolysis conditions [88, 89]. This system displayed potential-dependent product selectivity: Applied potentials between 1.6 and 1.8 V vs. SCE resulted in the formation of 1-hydroxy-λ^3-benzo[*d*][1,2]iodaoxol-3(1*H*)-one (**9**) while applied potentials above 1.8 V vs. SCE yielded IBX (**89**) (Figure 12.18). The boron-doped diamond electrode was found to be important due to both its stability and the high overpotential for O_2 evolution at this electrode.

12.4 Recyclable Hypervalent Iodine Reagents

In addition to developing more sustainable oxidation protocols for the synthesis of hypervalent iodine compounds, significant effort has been made to improve iodoarene recyclability. Separation of iodoarene by-products from the desired reaction products is often tedious, and thus iodoarenes are not typically reused.

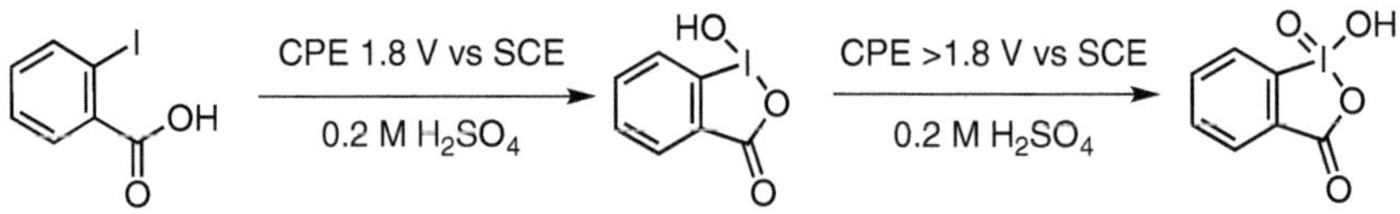

88 **9** **89**

Figure 12.18 Electrochemical oxidation of 2-iodobenzoic acid (**88**) leads to both 1-hydroxy-λ^3-benzo[*d*][1,2]iodaoxol-3(1*H*)-one (**9**) and 2-iodoxybenzoic acid (**89**) under constant potential electrolysis conditions.

Strategies based on polymer-supported hypervalent iodine compounds, which facilitate recovery via simple filtration, and based on fluorous alkyl iodoarenes, which facilitate recovery via biphasic separations with fluorous solvents, have been developed to enable iodoarene recycling. Here we highlight the synthesis and application of polymer-supported and fluorous I(III) and I(V) reagents.

12.4.1 Synthesis of Polymer-Supported Hypervalent Iodine(III) Reagents

Poly[4-(diacetoxyiodo)styrene] (**90**) was first prepared by Okawara via iodination of polystyrene followed by oxidation with peracetic acid [90, 91]. The loading of I(III) centers on the polystyrene backbone was determined to be 1.99 mmol/g by iodometry. In other examples, polystyrenes that display high molecular weight (poly-(diacetoxyiodo)styrene, ~45 000) and low molecular weight (poly-(diacetoxyiodo)(α-methylstyrene), ~6000) have been converted to the corresponding hydroxy(tosyloxy)iodine(III) polymers (**91** and **92**, respectively) by treatment with *p*-toluenesulfonic acid monohydrate [92, 93]. Alternately, iodinated polymers derived from coupling between poly[(aminomethyl)styrene] and either *p*-iodobenzoic acid or *p*-iodophenylacetic acid, were oxidized with *in situ*-generated peracetic acid to afford the polymer-supported (diacetoxyiodo)arenes (**93** and **94**, respectively) (Figure 12.19) [94]. In each of these cases, iodoarene recycling was achieved by simple filtration followed by a methanol wash of the polymer. In a similar context, highly insoluble small-molecule iodoarene mediators have also been demonstrated to be recyclable [95–98].

12.4.2 Synthesis of Polymer-Supported Hypervalent Iodine(V) Reagents

IBX is poorly soluble in many organic solvents, which often necessitates the use of dimethyl sulfoxide (DMSO) as a reaction solvent. In general, it is challenging to recover the corresponding I(III) by-product (i.e. **9**) from these reaction conditions [99, 100]. To alleviate the need for DMSO as a solvent, Mülbaier and Giannis developed silica-immobilized IBX reagent **95** (0.4 mmol/g) by tethering the appropriate iodoarene via a phenoxide linker (Figure 12.20) [101]. Oxidation to the corresponding immobilized I(V)-based material was accomplished with Oxone®. Oxidation of alcohols was achieved in good yield by **95** in tetrahydrofuran (THF) instead of

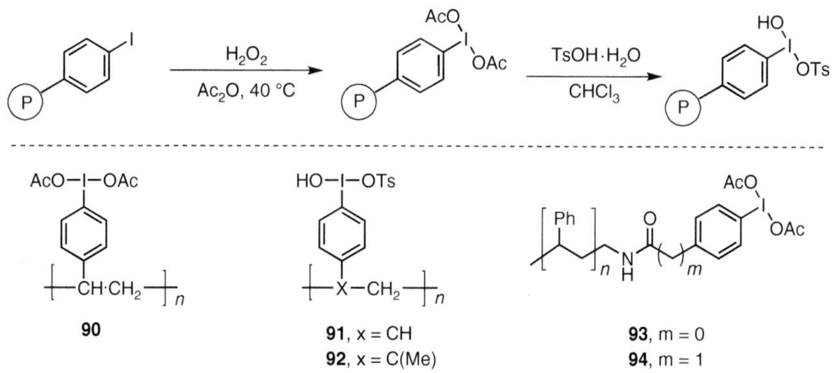

Figure 12.19 Synthesis of polymer-supported (diacetoxyiodo)arenes using H_2O_2 and acetic anhydride. These reagents participate in ligand exchange chemistry in the presence of *p*-toluenesulfonic acid to form tosylated I(III) reagents. These reagents can be recycled by simple filtration.

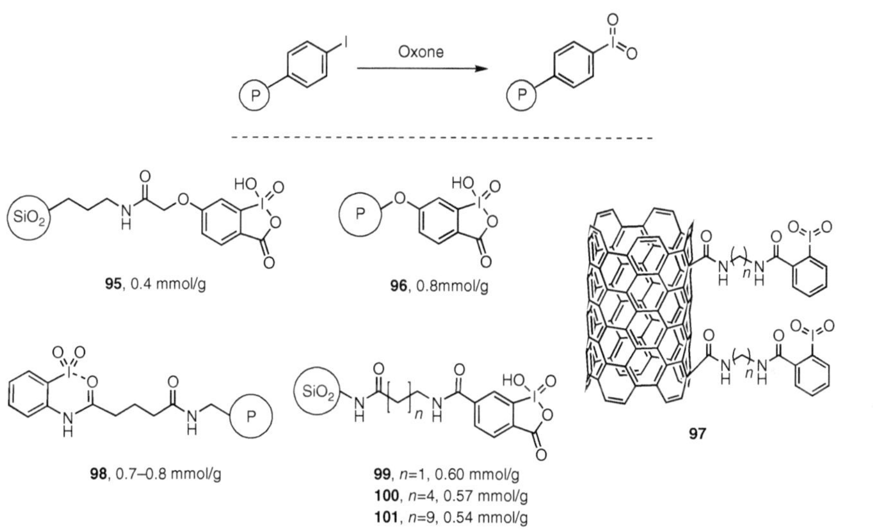

Figure 12.20 Examples of polymer-supported hypervalent iodine(V) reagents that facilitate recycling of the corresponding iodoarenes after substrate functionalization.

DMSO, the solvent required for IBX-mediated alcohol oxidation. Polychloromethyl styrene- [102] and multiwalled carbon nanotubes (MWCNT)-supported for I(V) reagents (**96** and **97**, respectively) have also been synthesized [103]. Similarly, amide linkages have been used to prepare polymer-supported I(V) reagents **98–101** [104, 105].

Yakura and coworkers reported magnetic nanoparticle-supported iodoarene **103** for the catalytic oxidation of phenols (**102**) to form *p*-quinones (**104**) and demonstrated recovery by application of an external magnetic field (Figure 12.21a) [106]. Kirsch and coworkers also reported solid-supported iodoarene catalysts **106** and **107**

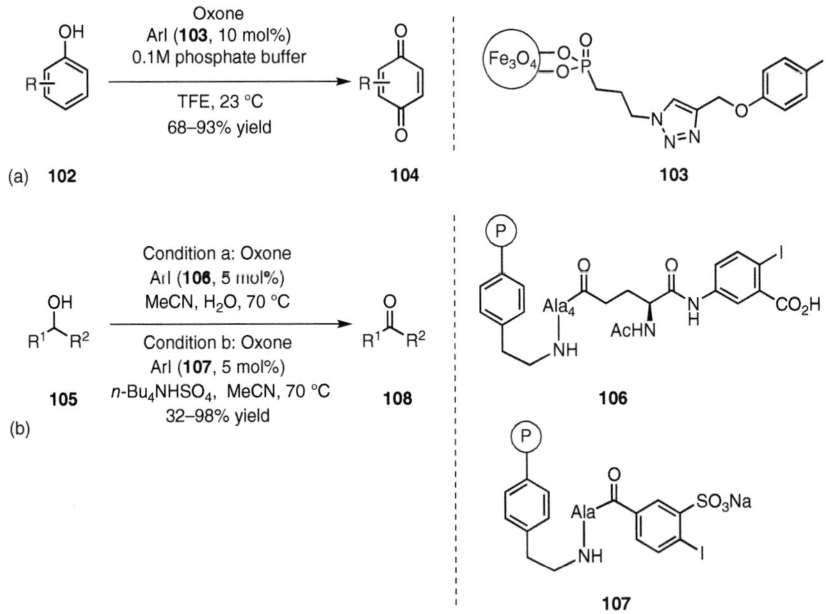

Figure 12.21 Oxone® is used as a terminal oxidant for (a) magnetic nanoparticle-supported iodoarene (**103**) catalyzed phenolic oxidation and *Source*: Nambu et al. [106]. (b) polymer-supported iodoarene (**106** and **107**) catalyzed oxidation of secondary alcohols to ketones. While **103** is recovered by the application of an external magnetic field, **106** and **107** were recycled by filtration after the reaction with ≥95% recovery. *Source*: Ballaschk and Kirsch [107].

and for catalytic oxidation of secondary alcohols (**105**) to ketones (**108**) in the presence of Oxone® and demonstrated that the recovered catalyst was catalytically active up to 5 cycles (Figure 12.21b) [107]. For additional discussion of iodoarene catalysis, see Section 12.5.

12.4.3 Fluorous Hypervalent Iodine(III) Reagents

Fluorous hypervalent iodine reagents have also been developed to enable extraction-based recovery of iodoarene by-products. Since the first synthesis of hypervalent iodine reagents with fluoroalkyl substituents in 1971 [108], these reagents have proved to be useful due to their high lipophobicity and hydrophobicity. In 2003, Gladysz and coworkers developed a series of hypervalent iodine reagents with fluoroalkyl substituents **110–113**, which were prepared by oxidation with $NaBO_3 \cdot H_2O$ [109]. This family of hypervalent iodine reagents participated in the oxidation of hydroquinones **109** to generate *p*-quinones **114** (Figure 12.22). The resulting iodoarene by-products displayed high fluorous-phase affinity, which facilitated separation and recycling.

In 2006, Gladysz and coworkers reported a family of hypervalent iodine reagents (**115**) derived from fluorous alkyliodides and applied them in alcohol oxidation chemistry (Figure 12.23a) [110]. Fluorous extraction enabled repeated use and

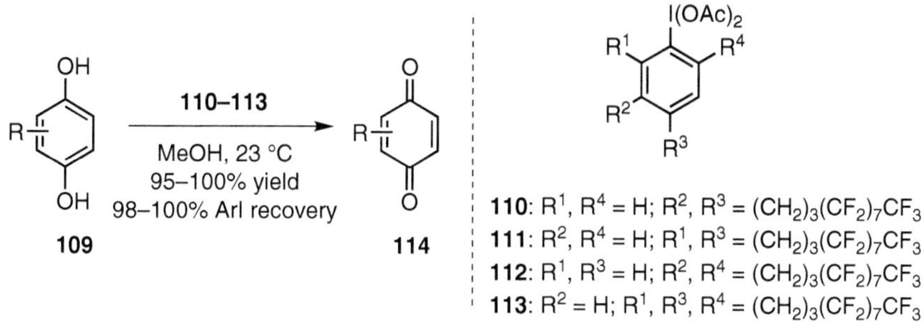

Figure 12.22 Fluorous hypervalent iodine-mediated oxidation of hydroquinone led to a high yield of benzoquinone. Iodoarene by-products were recovered by fluorous-phase extraction with $CF_3C_6F_{11}$.

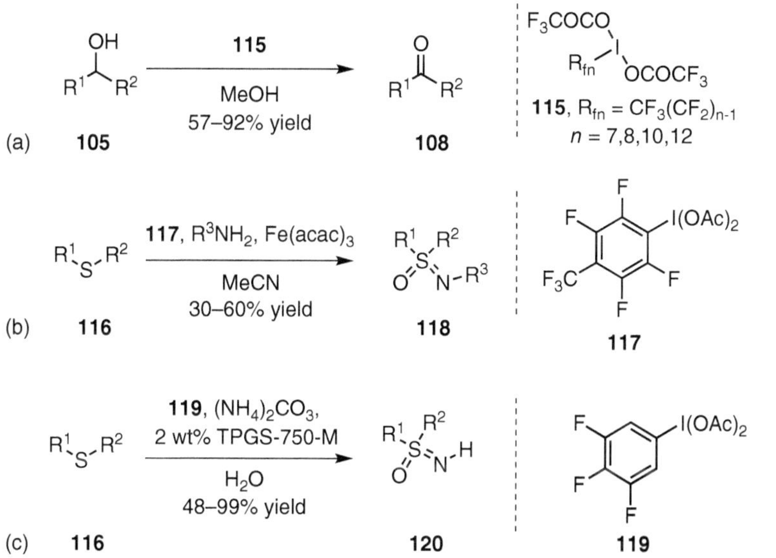

Figure 12.23 Fluorous hypervalent iodine compounds can be utilized for the oxidation of (a) alcohols to ketones, *Source*: Tesevic and Gladysz [110]. (b) sulfides to *N*-substituted sulfoximines, and *Source*: Schäfer and Wirth [111]. (c) sulfides to N–H sulfoximines. *Source*: Zhang et al. [112].

recovery; three cycles were achieved with high recovery of alkyliodides. More recently, polyfluorinated hypervalent iodine reagents have been utilized toward the synthesis of sulfoximines from sulfides under mild conditions [111, 112]. Wirth and coworkers developed oxidation of sulfides (**116**) with hypervalent iodine compound **117,** catalyzed by $Fe(acac)_3$ in acetonitrile, to afford *N*-substituted sulfoximines (**118**) (Figure 12.23b) [111]. Using the (diacetoxyiodo)arene **119** and tocopheryl polyethylene glycol succinate (TPGS)-750-M as a surfactant allowed for the formation of N–H sulfoximines (**120**) from **116** in aqueous media (Figure 12.23c) [112].

12.5 Catalytic Application of Hypervalent Iodine Compounds

Hypervalent iodine catalysis utilizes selective *in situ* oxidation of substoichiometric amounts of iodoarenes to accomplish hypervalent iodine-mediated substrate functionalization chemistry. Catalysis inherently reduces the iodoarene waste that is generated by hypervalent iodine reagents. The central challenge in achieving hypervalent iodine catalysis is the selective oxidation of the iodoarene catalyst in the presence of oxidatively labile substrates. Here we highlight recent progress in hypervalent iodine catalysis, and as above, organize the presentation by the nature of the terminal oxidant employed. The examples presented are not exhaustive but have been selected to highlight different approaches to achieving hypervalent iodine catalysis.

12.5.1 Catalytic Hypervalent Iodine Reactions Using O_2

12.5.1.1 Catalytic Hypervalent Iodine(III) Reactions Using O_2

Miyamoto and coworkers developed iodoarene-catalyzed aerobic glycol cleavage and Hofmann rearrangement chemistry under isobutyraldehyde autoxidation conditions [113]. Using pentamethyliodobenzene (**122**) as a catalyst, oxidative C–C cleavage of various diols (**121**) was achieved forming ketones **123** (Figure 12.24a). Notably, this aerobic oxidation condition was effective for glycol scission of cyclic *trans*-diols, which can be difficult to achieve using common oxidants like $NaIO_4$. The authors also carried out Hofmann rearrangement of carboxamides **124** to form carbamates **125** using the same catalysis conditions (Figure 12.24b). This was an improvement over existing protocols that use *m*-CPBA as the terminal oxidant because aromatic amides undergo background oxidation with *m*-CPBA to form *N*-oxides [114]. The reported hypervalent iodine catalysis can also be performed using air, in place of O_2, without a significant decrease in yield.

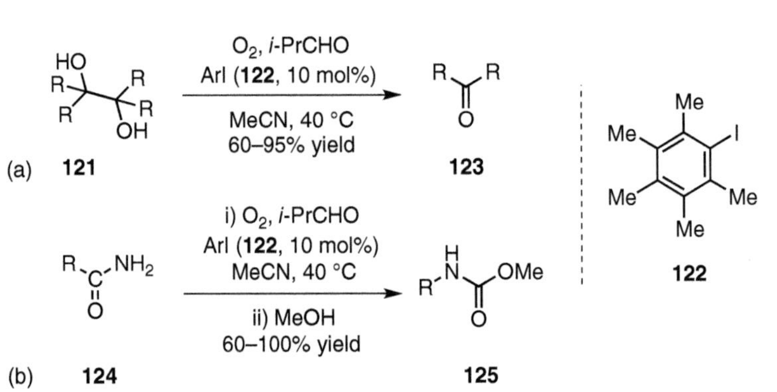

Figure 12.24 Aerobically generated hypervalent iodine(III) intermediates are applied to (a) oxidative C–C cleavage and (b) Hofmann rearrangement reactions using O_2 as the terminal oxidant.

Contemporaneously, Powers *et al.* utilized aerobically generated I(III) reagents (Figure 12.1) derived from acetaldehyde autoxidation in catalytic C–H amination, carbonyl α-functionalization, and spirocyclization reactions [23]. C–H amination was achieved using a catalytic amount of 1,2-diiodobenzene (**127**) [115] with O_2 as the terminal oxidant to furnish *N*-methoxy-4-methyl-*N*-phenylbenzenesulfonamide (**128**) from benzene and amine **126** (Figure 12.25a). Similar reaction conditions resulted in aerobic bromination of ethyl acetoacetate (**129**) and aerobic spirocyclization of *N*-methoxy-3-(4-methoxyphenyl)propenamide (**131**) to form **130** and **132**, respectively (Figure 12.25b–c). Control reactions in the absence of iodoarene did not furnish products in any of the reactions in Figure 12.25, which demonstrates that reactive intermediates from aldehyde autoxidation are not directly involved in substrate functionalization.

Similar aldehyde-promoted aerobic oxidation catalysis was reported by Sen and coworkers, who demonstrated the synthesis of substituted 1,3,4-oxadiazole **135** from *N*′-anilidene acetohydrazine **133** using 4-iodoanisole (**134**) as catalyst (Figure 12.26) [116].

In 2020, Cariou *et al.* developed photoinduced spirocyclization of *N*-oxy-amides **136** to *N*-fused spirolactams **138** using **137** as the catalyst, O_2 as the terminal oxidant, and mesityl-2,6-diphenylpyrylium tetrafluoroborate (MDPT) as the photocatalyst

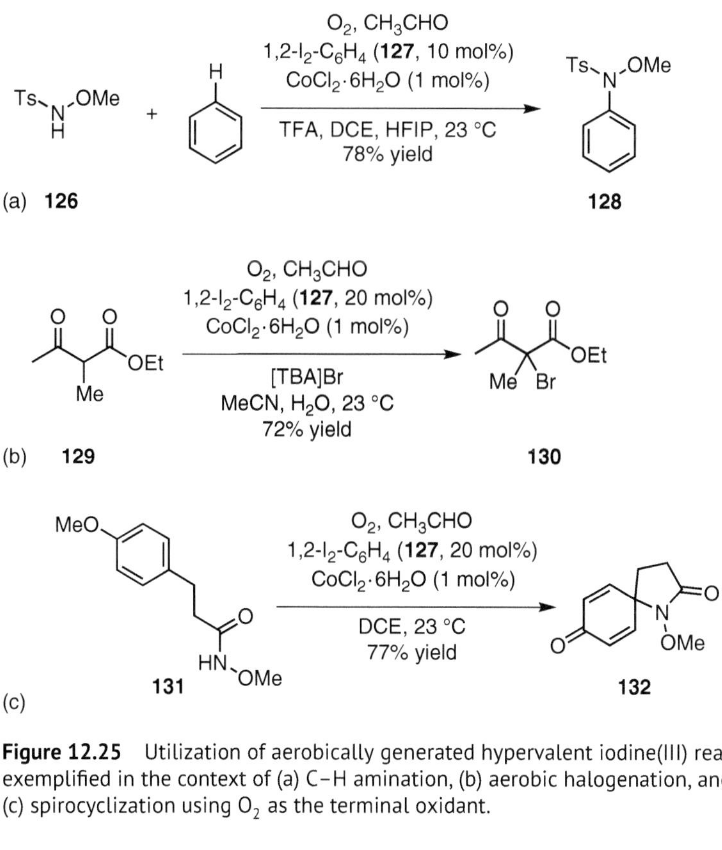

Figure 12.25 Utilization of aerobically generated hypervalent iodine(III) reagents are exemplified in the context of (a) C–H amination, (b) aerobic halogenation, and (c) spirocyclization using O_2 as the terminal oxidant.

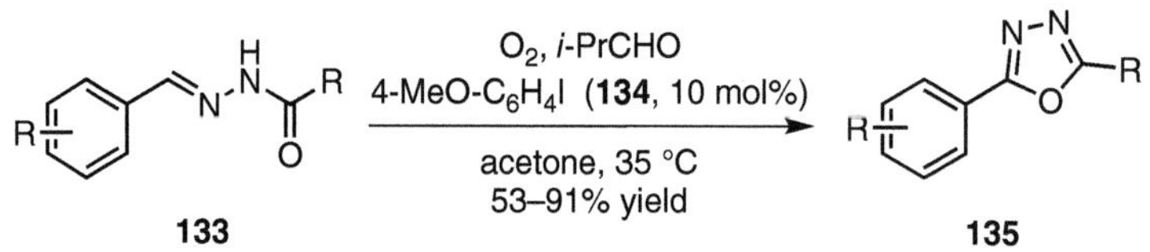

Figure 12.26 Synthesis of substituted oxadiazoles was carried out using O_2 as the terminal oxidant and 4-iodoanisole as the catalyst. *Source*: Chauhan et al. [116].

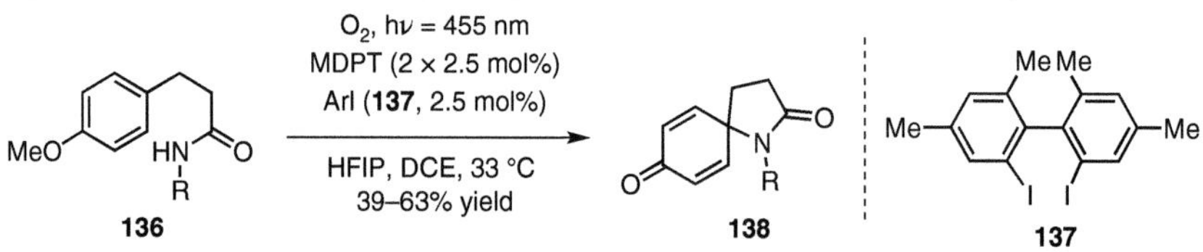

Figure 12.27 Spirocyclization of *N*-oxy-amides was achieved by photoinduced aerobic iodoarene catalysis. *Source*: Habert and Cariou [117].

under blue light irradiation (Figure 12.27) [117]. Control experiments in the absence of either catalysts, O_2, or light did not yield the desired spirocyclized product. Additionally, when radical trapping reagents like butylated hydroxytoluene (BHT) or (2,2,6,6-tetramethylpiperidin-1-yl)oxidanyl (TEMPO) were added to the reaction mixture, no desired product was observed.

12.5.1.2 Catalytic Hypervalent Iodine(V) Reactions Using O_2

In 2018, Powers and coworkers utilized *in situ* aerobic synthesis of (2-*tert*-butylsulfonyl)iodylbenzene (**21**, Figure 12.4) from (2-*tert*-butylsulfonyl)iodobenzene (**139**) to accomplish hypervalent-iodine-catalyzed alcohol oxidation [28]. Secondary alcohols were oxidized to ketones, whereas primary alcohols were oxidized to carboxylic acids because initially generated aldehydes were further oxidized under aldehyde autoxidation conditions (Figure 12.28a). Oxidative C–C cleavage was observed for 1,2-diols, which is characteristic of the reactivity of Dess–Martin Periodinane (DMP) (Figure 12.28b).

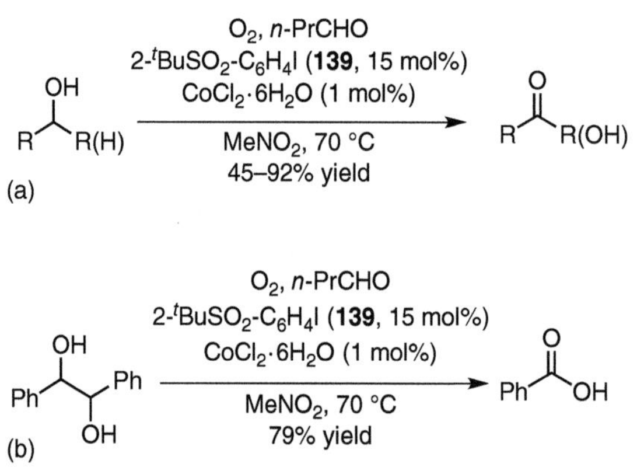

Figure 12.28 (a) Aerobically generated (2-*tert*-butylsulfonyl)iodylbenzene (**21**) was applied for oxidation of primary alcohols to carboxylic acids and secondary alcohols to ketones. (b) (2-*tert*-butylsulfonyl)iodobenzene (**139**)-catalyzed oxidative cleavage of hydroxybenzoin was realized under aldehyde autoxidation condition to form benzoic acid, analogous to DMP chemistry.

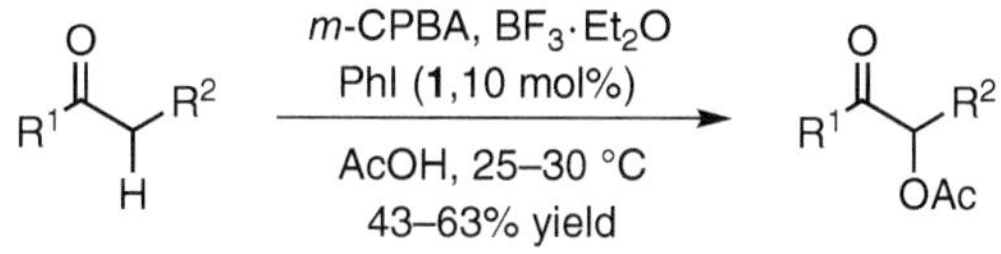

Figure 12.29 α-Functionalization of carbonyl compounds is exemplified by catalytic α-acetoxylation of ketones via hypervalent iodine intermediates generated using *m*-CPBA.

12.5.2 Catalytic Hypervalent Iodine Reactions Using Peracid Oxidants

Peracid oxidants are among the most utilized terminal oxidants in hypervalent iodine catalysis. The use of peracetic acid as a terminal oxidant generates acetic acid as the only by-product. While less oxygen-dense, *m*-CPBA has also become a popular terminal oxidant in hypervalent iodine catalysis. Because (i) similar reactions have been accomplished using peracetic acid and *m*-CPBA as terminal oxidants, and (ii) the chemical structures of these reagents are similar, hypervalent iodine catalysis using either peracid will be presented together here.

Carbonyl α-Functionalization. Some of the first examples of hypervalent iodine(III) catalysis utilized iodobenzene as a catalyst and *m*-CPBA as the terminal oxidant to achieve the α-oxidation of carbonyl compounds (i.e. conversion of **140–141**, Figure 12.29) [118]. Addition boron trifluoride diethyl etherate ($BF_3 \cdot Et_2O$), which was proposed to both induce the enolization of carbonyl-starting materials and promote ligand exchange at the *in situ* generated (diacetoxyiodo)benzenes, was found to be essential for the α-acetoxylation. Following the discovery of α-acetoxylation of ketones by Ochiai and coworkers, several research groups have reported additional examples of α-acetoxylation [119], as well as α-tosyloxylation [120–122], α-phosphoryloxylation [123], and α-fluorination reactions [124, 125].

Catalytic enantioselective α-oxidation reactions can be accomplished with chiral iodoarenes and *m*-CPBA as the terminal oxidant. In 2007, Wirth and coworkers reported the first catalytic, enantioselective α-tosyloxylation of **140** using chiral iodoarene **142** with *p*-toluenesulfonic acid as the tosylate source to form **144** (Figure 12.30, condition a) [126]. In contrast to the previously developed method, which used stoichiometric chiral I(III) to accomplish enantioselective α-tosyloxylation and required a reaction temperature of −30 °C for optimum enantioselectivity [127], the catalytic reaction was performed at room temperature to overcome slow reaction kinetics (lower enantioselectivity was observed). The triazole-substituted chiral iodoarene catalyst **143** was later identified to provide similar α-tosyloxylation with higher yields as well as increased enantioselectivity (Figure 12.30, condition b) [128–132]. Enantioselective α-fluorination of carbonyl compounds has also been accomplished by using chiral iodoarenes and replacing the tosylate nucleophile with triethylamine hydrogen fluoride as fluorine source [133, 134].

Spirocyclization. Hypervalent iodine catalysis has also been applied to oxidative spirocyclization chemistry (i.e. the conversion of phenol **145** to spirocycle **146**, Figure 12.31a) [135]. Trifluoroacetic acid was found to improve the yield of spirocyclized products, which was attributed to the *in situ* formation of the stronger oxidant

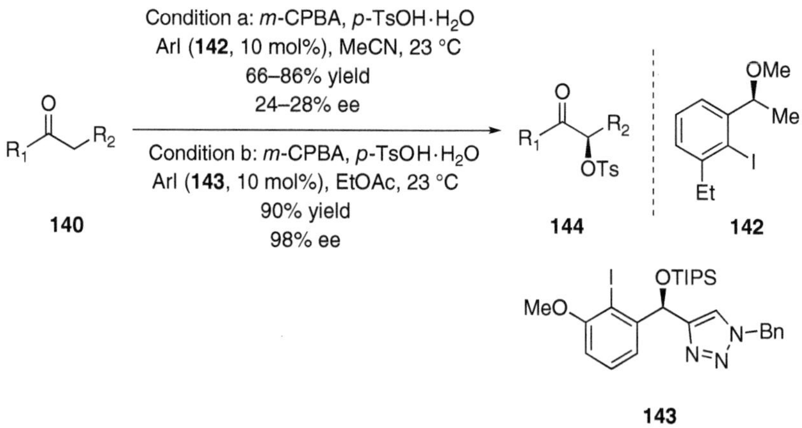

Figure 12.30 Enantioselective α-tosyloxylation of ketones was achieved using iodoarene catalysts bearing chiral *o*-substituents (**142** and **143**). Triazole-based catalyst **143** was observed to provide increased enantioselectivity.

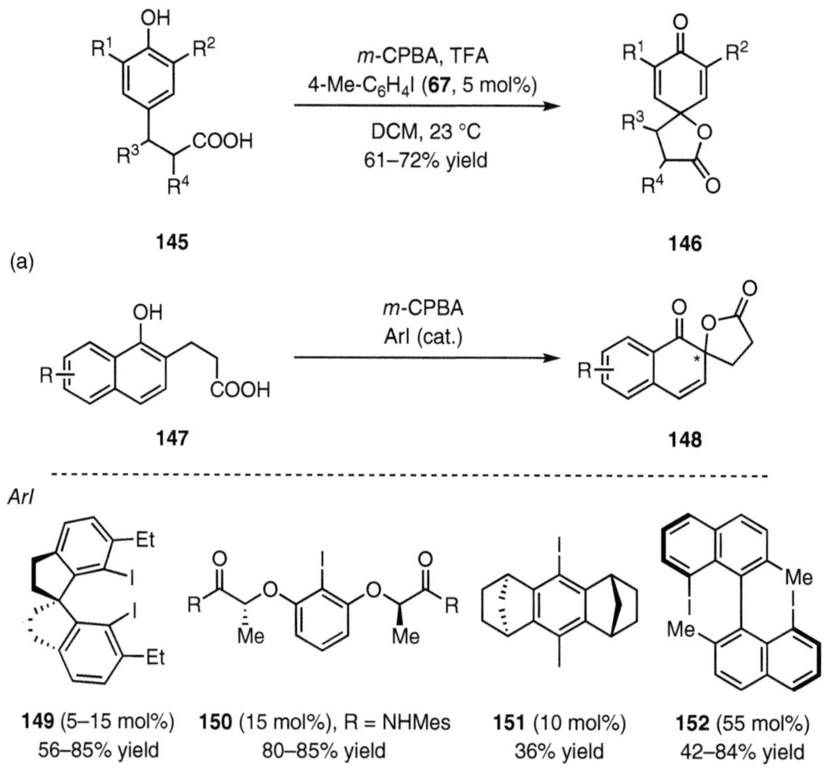

Figure 12.31 (a) Hypervalent iodine-catalyzed synthesis of lactones was achieved by dearomatization of 4-substituted phenolic derivatives. (b) Enantioselective spirocyclization of naphthyl propanoic acids can be achieved with various chiral iodoarene catalysts (**149–152**).

[bis(trifluoroacetoxy)iodo]benzene (**15**). Enantioselective examples of spirocyclization of phenolic substrates have also been studied [136–138]. Kita and coworkers reported chiral iodoarene **149** bearing an *ortho*-substituted spirobiindane backbone [136], Ishihara's group developed a conformationally flexible C_2-symmetric iodoarene catalyst **150** [137], and Ibrahim and coworkers prepared C_2-symmetric iodoarene catalyst **151** [138], all of which promote the spirocyclization of substrate **147** to form **148** in good enantioselectivity (Figure 12.31b). Other catalysts including atropisomeric binaphthyl-based chiral iodides **152** have also been developed for similar spirolactonization reactions [139–149].

Oxidative C–H Coupling Chemistry. In 2013, Kita and coworkers developed a hypervalent iodine-catalyzed oxidative cross-coupling reaction of aromatic sulfonanilides **153** with aromatic hydrocarbons to give the biaryl products **155** in high yields (Figure 12.32) [150]. While using iodobenzene as the catalyst resulted in low yields, 2,2′-diiodobiphenyl derivative **154** was found to be highly active toward the described C–C coupling chemistry.

Similarly, Gong and coworkers demonstrated intramolecular asymmetric oxidative C–C bond-forming chemistry, in the context of spirooxindole (**158**) synthesis from **156**, using chiral iodoarene **157** as the catalyst (Figure 12.33) [151]. In 2017, a computational study by Sunoj and coworkers reasoned that the observed enantioselectivity is due to the helical fold of the chiral amide arms on the iodoarene catalyst at *ortho*-positions [149].

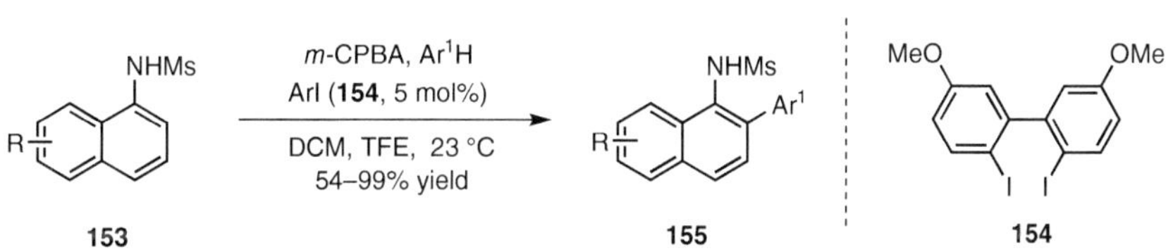

Figure 12.32 Hypervalent iodine-catalyzed oxidative C–C coupling with *m*-CPBA as the terminal oxidant. *Source*: Modified from Ito et al. [150].

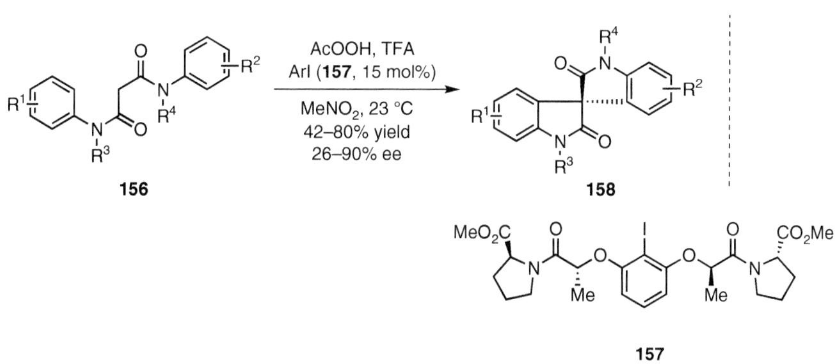

Figure 12.33 Asymmetric spirocyclization was achieved using chiral iodoarene catalyst **157** and peracetic acid as the terminal oxidant. *Source*: Wu et al. [151].

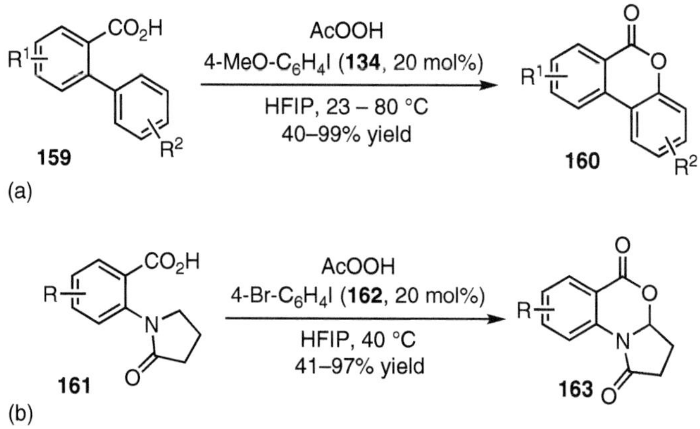

Figure 12.34 Iodoarene-catalyzed (a) C(sp^2)–H and (b) C(sp^3)–H functionalization of phenyl carboxylic acid derivatives resulted in the formation of corresponding C–O-coupled lactone products. *Source*: Wang et al. [152].

Martin and coworkers reported a C(sp^2)–O-coupling reaction of biphenyl-2-carboxylic acids **159** in the presence of catalytic amounts of iodoarene **134** with peracetic acid as the terminal oxidant to form lactones **160** (Figure 12.34a) [152]. Previously, similar chemistry had been accomplished under the action of either Pt or Pd catalysts in tandem with stoichiometric hypervalent iodine(III) or Ag(I) terminal oxidants [153–159]. The authors also described the formation of **163** via C(sp^3)–O coupling of pyrrolobenzoxazinones **161** using 4-bromoiodobenzene (**162**) as the catalyst (Figure 12.34b) [152].

A considerable amount of interest has focused on the development of hypervalent iodine catalysis for oxidative N–H/C–H coupling for both sp^2 and sp^3 C–H amination [115, 160–164]. The Muñiz group reported intermolecular C–H amination (i.e. the conversion of amine **164** to arylated product **165**) catalyzed by 1,2-diiodobenzene (**127**), which was identified as an exceptionally efficient catalyst that could be used in as low as 3 mol% loading (Figure 12.35a) [115]. The high reactivity of the corresponding μ-oxo iodine(III) species **12** was reasoned to be due to the instability of the 5-membered μ-oxo ring. Recently, Kita's group has also reported such reactivity using the 2,2′-diiodo-4,4′,6,6′-tetramethyl-1,1′-biphenyl (**137**) as catalyst (Figure 12.35b) [165]. Similar to **12**, the μ-oxo structure of the iodine(III) species generated from **137** results in high catalytic activity; catalyst loading as low as 0.5 mol% were found to be effective.

Difunctionalization of Unsaturated C–C Bonds. Difunctionalization of olefins is one of the main classes of transformations available to hypervalent iodine compounds [166]. Achieving catalysis using peracid-based oxidants requires that the rate of iodoarene oxidation competes successfully without direct olefin epoxidation chemistry, which can be facile with peracid-based oxidants.

Oxygenation Reactions. Iodomesitylene (**167**) catalyzed *syn*-diacetoxylation of alkenes was performed using *m*-CPBA as terminal oxidant, forming vicinal diacetates (Figure 12.36a) [167]. These methods complement previous methods

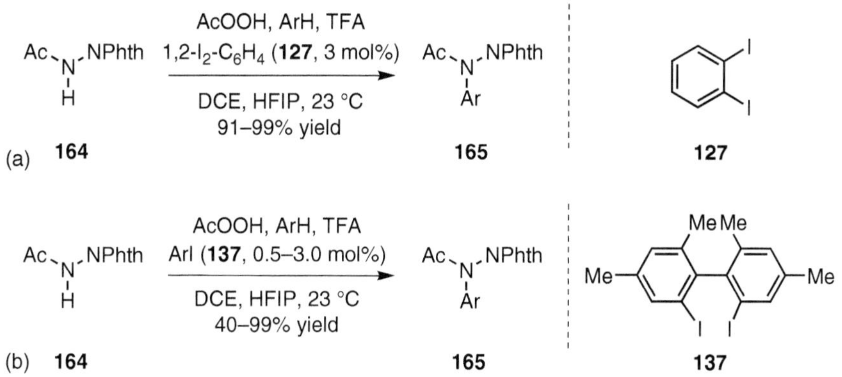

Figure 12.35 Hypervalent iodine-catalyzed C–H amination was achieved using (a) 1,2-diiodobenzene (**127**) and *Source*: Lucchetti et al. [115]. (b) 2,2′-diiodo-4,4′,6,6′-tetramethyl-1,1′-biphenyl (**137**) as catalysts and peracetic acid as terminal oxidant. These reactions are proposed to proceed through μ-oxo *bis*-iodine(III) intermediates (e.g. **12**). *Source*: Dohi et al. [165].

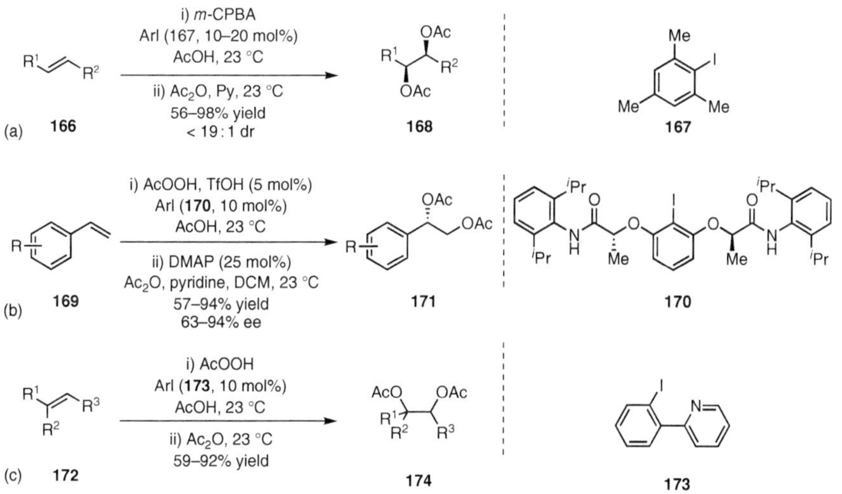

Figure 12.36 (a) Hypervalent iodine-catalyzed *syn*-diacetoxylation of alkenes was achieved using iodomesitylene (**167**) as a catalyst and *m*-CPBA as the terminal oxidant. *Source*: Zhong et al. [167]. (b) Enantioselective diacetoxylation of alkenes was achieved with C_2-symmetric iodoarene catalyst **170** and peracetic acid as the terminal oxidant. *Source*: Haubenreisser et al. [178]. (c) 2-Pyridyliodobenzene (**173**)-catalyzed diacetoxylation of alkenes. The proximal pyridyl group was found to accelerate catalysis as compared to iodobenzene. *Source*: Modified from Aertker et al. [179].

that have been developed with transition metal catalysts, including osmium [168], ruthenium [169, 170], manganese [171, 172], iron [173–175], and palladium [176, 177]. Muñiz and coworkers described an enantioselective diacetoxylation of styrenes using chiral iodoarene **170** and peracetic acid (Figure 12.36b) [178]. While the use of *m*-CPBA promoted significant epoxidation in this case, little background reaction was observed with peracetic acid as the terminal oxidant. The Muñiz group also demonstrated catalytic olefin diacetoxylation with iodoarene catalyst **173** bearing an *o*-pyridyl Lewis base, which proved to be kinetically superior to the previous acetoxylation methods (Figure 12.36c) [179]. In this case, other terminal oxidants including Selectfluor, perborate, and periodate were ineffective; only peracetic acid was found to be effective. Both activated and unactivated olefins were diacetoxylated in good yields with **173** as the catalyst, and the increased reactivity was attributed to the stabilizing effect of the pyridyl group on electrophilic iodine centers [180–182].

Fluorination Reactions. In 2016, Jacobsen and coworkers reported the enantioselective 1,2-*anti*-difluorination of alkenes with iodoarene catalyst **175**, *m*-CPBA as terminal oxidant, and an excess of Py·9HF as fluoride source (Figure 12.37a) [183]. The developed method is applicable to various substituted alkenes including terminal, internal, and acrylamide olefins. In the case of isolated alkenes, the addition of pyridine afforded improved yields, presumably due to reduced acidity and increased nucleophilicity of fluoride ions in the reaction medium. In the same year, the Jacobsen group also reported a method for the catalytic, asymmetric, migratory geminal difluorination of β-substituted styrenes to access products bearing difluoromethylated tertiary or quaternary stereocenters using a chiral iodoarene catalyst **178** (Figure 12.37b) [184].

Amination Reactions. Iodoarene derivatives with modified chiral lactic acid side chains have emerged as effective catalysts for intermolecular diamination reactions [185]. The first hypervalent iodine-catalyzed enantioselective intermolecular diamination (i.e. the conversion of **169–181**) was reported by the Muñiz group using iodoarene catalyst **180** (Figure 12.38). Background epoxidation/aminooxygenation reactions of alkenes by *m*-CPBA were reduced by changing the solvent from ethyl acetate to methyl *tert*-butyl ether (MTBE) or MTBE/HFIP mixtures, and by lowering

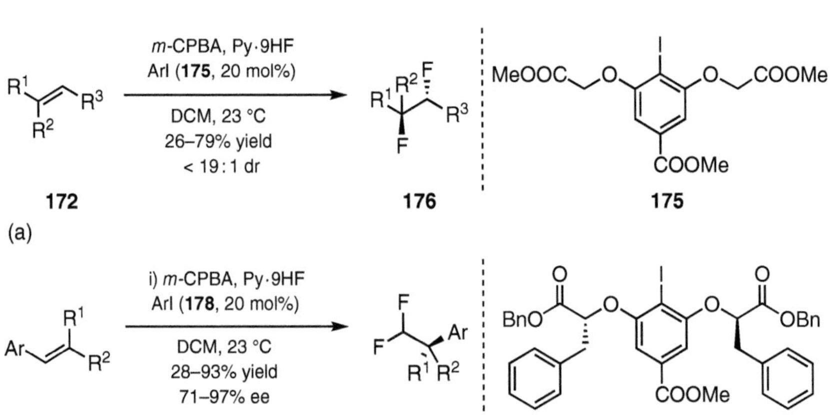

172 176 175

(a)

177 179 178

(b)

Figure 12.37 Asymmetric (a) vicinal difluorination and *Source*: Banik et al. [183]. (b) gem difluorination of alkenes via chiral hypervalent iodine catalysis were achieved using *m*-CPBA as the terminal oxidant. *Source*: Banik et al. [184].

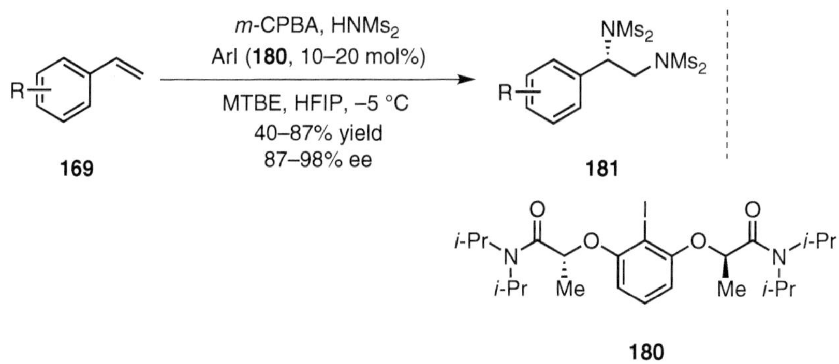

Figure 12.38 Hypervalent iodine-catalyzed asymmetric diamination of olefins was realized using C_2-symmetric catalyst (**180**) and *m*-CPBA as the terminal oxidant. *Source*: Barluenga et al. [180].

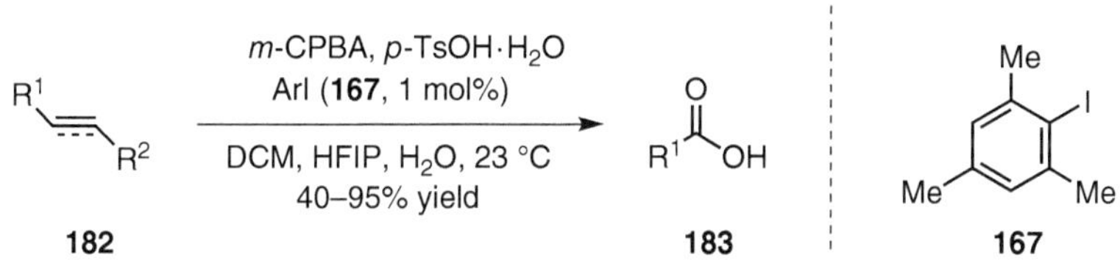

Figure 12.39 Iodomesitylene-catalyzed oxidative cleavage of C–C unsaturated bonds was developed as an alternative method to ozonolysis.

the reaction temperature to −5 °C. With a 10 mol % catalyst loading, the reaction provided 40–87% yield and up to 98% enantiomeric excess (ee).

Oxidative Cleavage Reactions. In 2009, Miyamoto and Ochiai developed an efficient method for iodoarene-catalyzed oxidative cleavage of carbon–carbon multiple bonds as an alternative to ozonolysis [186]. Under these conditions, various unsaturated compounds such as cyclic alkenes, acyclic alkenes, and aryl acetylenes are selectively cleaved to afford carboxylic acids (Figure 12.39).

12.5.3 Catalytic Hypervalent Iodine Reactions Using Oxone®

Although the active oxygen content of Oxone®, in which the active oxidant is the peroxymonosulfate anion (HSO_5^-), is lower than that of O_2 or H_2O_2, it offers several advantages, including stability, ease of transport, simple handling, controllable addition, and is nontoxic [187]. Oxone® is prepared from a mixture of H_2SO_4, H_2O_2, and potassium hydroxide (KOH) [188]. Both iodine(III) and iodine(V) species can be generated from iodoarene oxidation with Oxone®.

12.5.3.1 Catalytic Hypervalent Iodine(III) Reactions Using Oxone®

Hypervalent iodine-catalyzed oxidation of phenols using Oxone® as the terminal oxidant has been widely studied. Yakura and coworkers reported iodoarene (**185**)-catalyzed oxidation of phenolic derivatives to form *p*-quinones **186**

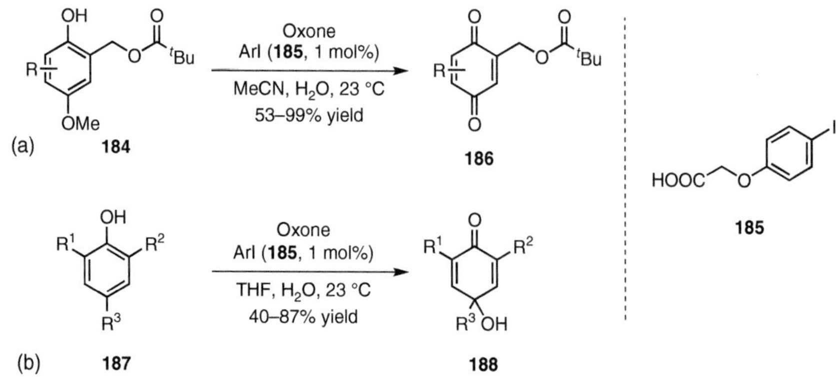

Figure 12.40 2-(4-Iodophenoxy)acetic acid (**185**)-catalyzed oxidation of phenolic derivatives provided access to (a) *p*-quinones and *Source*: Yakura and Konishi [189], Yakura et al. [190], Yakura et al. [191]. (b) *p*-quinols with Oxone® as the terminal oxidant. *Source*: Yakura et al. [192]. Yakura and Omoto [193].

(Figure 12.40a) [189–191], and *p*-quinols **188** (Figure 12.40b) [192, 193], depending on the substituents at *para*-positions of **184** and **187**, respectively. It was hypothesized that the *in situ*-generated I(III) intermediates react with phenolic compounds to form a cationic intermediate, which is then trapped by H_2O to form the desired quinols.

Zhdankin and coworkers demonstrated hypervalent iodine-catalyzed Hofmann rearrangements, which had been previously accomplished by Ochiai and coworkers using *m*-CPBA as the terminal oxidant [114], could also be accomplished using Oxone® as the terminal oxidant (Figure 12.41a) [194]. Using similar reaction conditions with 3,5-dimethyliodobenzene (**190**) as the catalyst, Zhdankin and coworkers reported cyclization of oxime **189** with alkenes or alkynes to form isoxazolines **192** or isoxazoles **193** (Figure 12.41b) [195]. In these reactions, *in situ*-generated iodine(III) reagents oxidize oximes into nitrile oxides **189**, which then undergo [3 + 2] cycloaddition with alkenes or alkynes to form the corresponding heterocycles. Synthesis of substituted oxazoles **195** via cycloisomerization-amination reactions of *N*-propargyl carboxamides **194** was also realized utilizing *in situ*-generated I(III) intermediate by Oxone® (Figure 12.41c) [196].

12.5.3.2 Catalytic Hypervalent Iodine(V) Reactions Using Oxone®

The first example of I(V) catalysis was reported by Vinod and coworkers in 2005, who reported oxidation of primary and secondary alcohols to carboxylic acids and ketones, respectively, by the *in situ* generation of IBX (**89**) from 2-iodobenzoic acid (**88**) using Oxone® as the terminal oxidant (Figure 12.42) [197]. 1,2-Diols were also cleanly oxidized to corresponding dicarbonyls without oxidative C–C cleavage. Control reactions in the absence of catalysts resulted in no product formation.

The Ishihara group reported oxidation of primary and secondary alcohols with iodophenylsulfonate **198** as the catalyst (Figure 12.43a) [198]. It is noteworthy that a significantly lower catalyst loading (as low as 1 mol%) was achievable with sodium 2-iodobenzenesulfonate (**198**) to generate 2-iodoxybenzenesulfonic

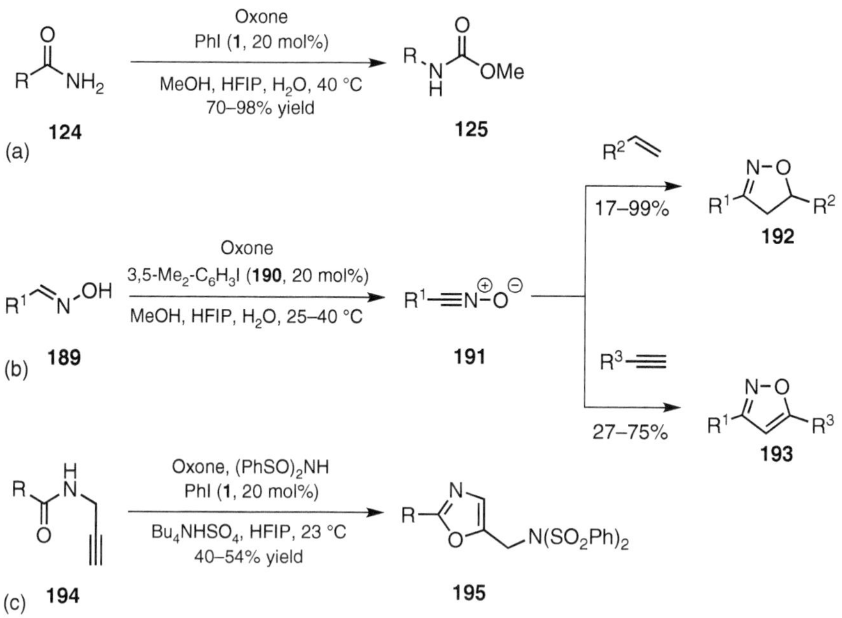

Figure 12.41 Oxone® is often used as the terminal oxidant for various hypervalent iodine-catalyzed reactions such as (a) Hofmann rearrangement of amides, *Source*: Yoshimura et al. [194]. (b) [3 + 2] cycloaddition of alkenes or alkynes with *in situ*-generated nitrile oxide, and *Source*: Yoshimura et al. [195]. (c) cycloisomerization-amination reactions of *N*-propargyl carboxamides. *Source*: Okamura et al. [196].

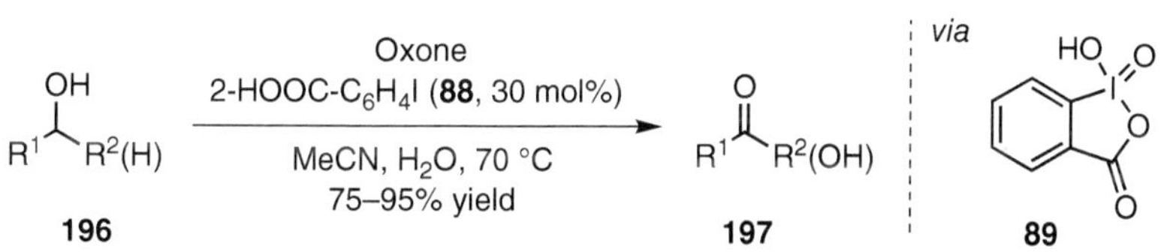

Figure 12.42 Catalytic oxidation of alcohols in the presence of 2-iodobenzoic acid and Oxone® leads to carboxylic acid from primary alcohols and ketones from secondary alcohols via the intermediacy of IBX (**89**). *Source*: Thottumkara et al. [197].

acid (IBS), which is an analogue of IBX. It was proposed that IBS might have had greater catalytic activity since the iodine(V) center has higher Lewis acidity than that of IBX due to a strong electron-withdrawing sulfonate group in the *ortho*-position. Also, the oxidative rearrangement of tertiary alcohols with sodium 5-methyl-2-iodobenzenesulfonate (**200**) as catalyst was accomplished to form enones with Oxone® as the terminal oxidant (Figure 12.43b) [199].

In contrast to the I(III)-mediated oxidation of phenolic derivatives in the presence of Oxone®, which gave rise to 1,4-quinones (section 12.5.3.1) [189–191], 5-methyl-2-iodobenzenesulfonate (**200**)-catalyzed oxidation of phenols leads to selective *ortho*-oxidation to form 1,2-quinones **203** via corresponding I(V) intermediate (Figure 12.44a) [200]. Although the reason for the selectivity remained

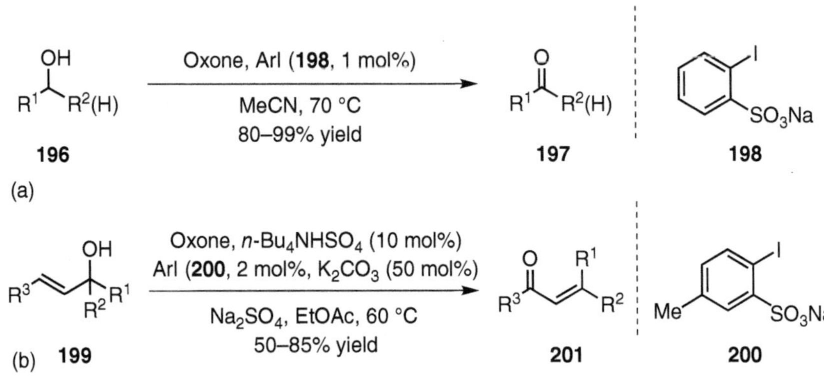

Figure 12.43 Iodoarene-catalyzed (a) oxidation of primary and secondary alcohols and *Source*: Uyanik et al. [198]. (b) rearrangement of tertiary alcohols were performed using Oxone® as the terminal oxidant. Selective oxidation of primary alcohols to aldehydes was observed. *Source*: Uyanik et al. [199].

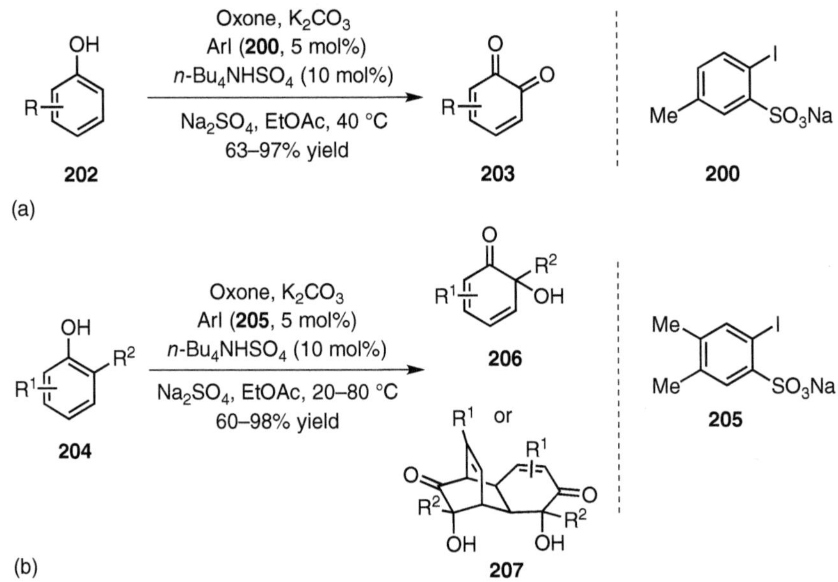

Figure 12.44 Catalytic oxidation of phenols by 2-iodobenzenesulfonic acid derivative and Oxone® leads to (a) *o*-quinone and *Source*: Uyanik et al. [200]. (b) *o*-quinol depending on the reaction conditions. *Source*: Uyanik et al. [201].

unclear, a mechanism was proposed where the phenol-iodine(V) adduct underwent [2, 3]-sigmatropic rearrangement to give the resulting 1,2-quinones. This reaction was further expanded to effect the formation of *o*-quinols (**206** and **207**) catalyzed by sodium 4,5-dimethyl-2-iodobenzenesulfonate (**205**) from *ortho*-substituted phenols **204** (Figure 12.44b) [201]. The introduction of a trialkylsilylmethyl substituent at the 2-position in this case further improved the reaction rate and selectivity for monomeric quinol product **206** over the dimerized product **207**.

12.5.4 Catalytic Hypervalent Iodine Reactions Using Electrophilic Halides

Electrophilic halogen-based reagents, such as Selectfluor and *N*-bromosuccinimide (NBS), can act as terminal oxidants in hypervalent iodine catalysis. While Selectfluor oxidation generates (difluoroiodo)arenes, which are stable to isolation [202], NBS oxidation affords (dibromoiodo)arenes, which are not generally stable to isolation (the presence of *ortho*-substituents that are Lewis basic can stabilize cyclic bromo-I(III) reagents) [203]. Among the oxidants discussed here, electrophilic halides are the least atom economical; the reduced products of Selectfluor and NBS are 1-chloromethyl-1,4-diazabicyclo[2.2.2]octane and succinimide, respectively.

12.5.4.1 Catalytic Hypervalent Iodine(III) Reactions Using Electrophilic Halides

(Difluoroiodo)arenes, generated *in situ* by the combined action of Selectfluor and hydrofluoric acid, have been applied in fluorination reactions. Moran and coworkers described the hypervalent iodine-catalyzed cyclization of *N*-allylamides **208** to the corresponding oxazoline compounds **210** with five- to seven-membered rings using *o*-iodoanisole (**209**) as a catalyst (Figure 12.45a) [204]. Saito and coworkers demonstrated metal-free and catalytic fluorinative cyclization of *N*-propargyl amides **211** to form substituted oxazoles **212** with a similar catalysis strategy (Figure 12.45b) [205].

Difunctionalization of olefins can also be achieved with Selectfluor as the terminal oxidant. In 2016, Muñiz and coworkers developed an enantioselective styrene diacetoxylation reaction using chiral iodoarene **213** as a catalyst (Figure 12.46a) [206]. The use of Selectfluor as the terminal oxidant avoided background reactions, such as epoxidation, that were observed with peracid-based oxidants. In addition, oxidation by Selecfluor generates arylfluoroiodonium species that have more electrophilic iodine(III) center and is highly reactive toward olefins compared to other

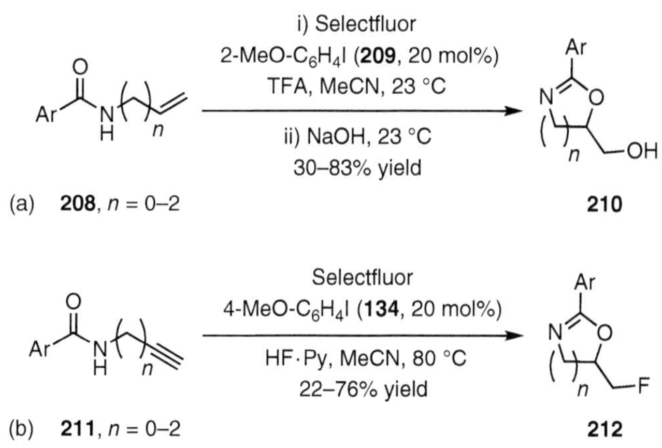

Figure 12.45 (a) Cyclization of *N*-allyamides in the presence of Selectflour and *o*-iodoanisole (**209**) furnished corresponding hydroxyl-substituted oxazoline compounds. *Source*: Alhalib et al. [204]. (b) Addition of fluoride source such as pyridine hydrogen fluoride to the previous reaction condition leads to fluorocyclized products **212**. *Source*: Asari et al. [205].

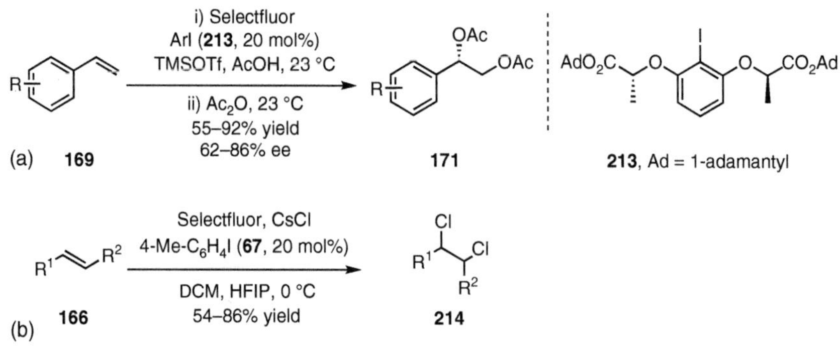

Figure 12.46 Selectflour is used as a terminal oxidant for iodoarene-catalyzed (a) diacetoxylation and *Source*: Wöste and Muñiz [206]. (b) vicinal dichlorination of alkenes. *Source*: Sarie et al. [207].

hypervalent iodine reagents like $PhI(OAc)_2$. Gilmour and coworkers elaborated on the developed method to achieve vicinal dichlorination of alkenes (i.e. the conversion of **166** to **214**) by *in situ* generation of 4-(dichloroiodo)toluene (**32**) from the I(III) fluoride species via ligand exchange reaction with cesium chloride (CsCl) (Figure 12.46b) [207].

NBS has been used for *in situ* generation of (dibromoiodo)arenes, which are efficient in bromine transfer to alkenes or arenes. Braddock and coworkers reported the *in situ* synthesis of cyclic brominated hypervalent iodine intermediates from a variety of *ortho*-substituted iodoarenes (e.g. **217–221**), and found that amidine **217** was superior in the synthesis of bromolactones **216** from alkenecarboxylic acids **215** (Figure 12.47) [203]. Although molecular bromine has a better atom economy, it failed to give rise to the formation of the expected bromoiodinane species. Instead, NBS was found to be a suitable reagent for such transformation in good yields.

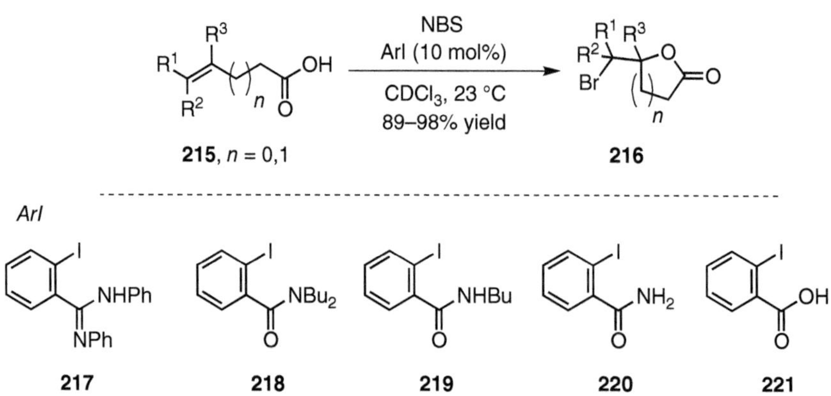

Figure 12.47 Bromocyclization of alkenecarboxylic acids to bromolactones was realized using (*E*)-2-iodo-*N*,*N*′-diphenylbenzimidamide (**217**) as a catalyst and Selectfluor as the terminal oxidant. *Source*: Braddock et al. [203].

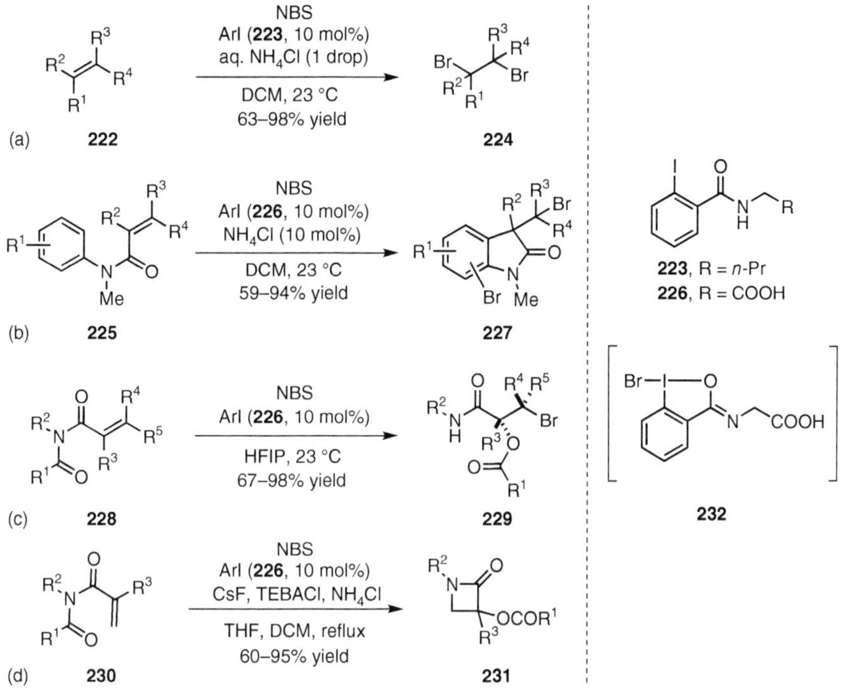

Figure 12.48 Hypervalent iodine-catalyzed substrate functionalizations with NBS as the terminal oxidant are exemplified in the context of (a) 1,2-dibromination of alkenes, *Source*: Stodulski et al. [208]. (b) bromocyclization of methacrylamide derivatives, *Source*: Fabry et al. [209]. (c) bromination/rearrangement, and *Source*: Ulmer et al. [210]. (d) cascade bromination/rearrangement/cyclization of *N*-alkyl amide derivatives. The formation of brominated hypervalent iodine intermediates is proposed in these transformations. *Source*: Patzelt et al. [211].

Increased reaction rate and yields were obtained by increasing the nucleophilicity of the *ortho*-substituents of the iodoarene.

Gulder and coworkers expanded the utility of olefin functionalization catalyzed by iodoarenes using NBS as the terminal oxidant to access bromination reactions with *o*-iodobenzamide derivatives **223** and **226** as the catalysts. Specifically, 1,2-dibromination of alkenes (i.e. the conversion of **222** to **224**, Figure 12.48a) [208] and bromocyclization of methacrylamide derivatives (i.e. the conversion of **225** to **227**, Figure 12.48b) [209] were accomplished by the iodoarene catalysts **223** and **226**, respectively. In a similar fashion, iodoarene-catalyzed bromination followed by rearrangement of *N*-alkyl derivatives **228** was performed to access α,α-dialkylated α-alkanoate carboxylamides **229** (Figure 12.48c) [210] as well as the cascade bromination/rearrangement/cyclization reaction to generate β-lactams **231** from **230** (Figure 12.48d) [211]. The generation of bromobenziodoxolone **232**, from *o*-iodobenzamide **226** and NBS, was observed by NMR spectroscopy and ESI-MS experiments, and was proposed to be the active brominating reagent in these transformations.

12.5.5 Hypervalent Iodine Electrocatalysis

As discussed in the context of electrochemical synthesis of hypervalent iodine reagents, *ex cell* methods require the stoichiometric generation of hypervalent iodine species, which obviates many of the potential advantages of electrochemistry vis-à-vis sustainability. *In cell* electrocatalytic methods, in which hypervalent iodine species are generated *in situ* during substrate functionalization, can be limited by the significant overpotential that is often required to anodically oxidize many iodoarenes; however, strategies to enable electrocatalysis have begun to emerge.

12.5.5.1 Catalytic Hypervalent Iodine(III) Electrocatalysis

In 1994, Fuchigami and coworkers reported *in cell* utilization of hypervalent iodine intermediates to accomplish 1,1-difluorination of dithioketals (i.e. the conversion of **233–234**, Figure 12.49) [56]. As discussed in section C.1, Fuchigami and coworkers were only able to electrosynthesize 4-(difluoroiodo)nitrobenzene upon anodic oxidation of 4-iodonitrobenzene in $Et_3N{\cdot}5HF$ (Figure 12.11). Since the oxidation peak potential of 4-iodonitrobenzene (2.3 V vs. SCE) was higher than the potential required for the oxidation of dithioketals (2.1–2.2 V vs. SCE), 4-iodoanisole was used as an electrocatalyst (1.9 V vs. SCE), and resulted in selective iodoarene oxidation to form the reactive 4-(difluoroiodo)anisole (**235**) *in situ*.

In 2019, Ackermann and coworkers reported an electrochemical co-catalytic system based on iodobenzene (**1**) and Ru-compound **237**, for the directed C–H hydroxylation of **236** to give **238** (Figure 12.50) [212]. The iodine(III)/ruthenium(II)-electrocatalyzed C–H functionalization was enabled by the electrochemical generation of hypervalent iodine(III) reagents. Complementary computational work indicated that the oxidation potential of iodobenzene is 200 mV lower than that of the ruthenium(II/IV) manifold, and most likely goes through anodically generated hypervalent iodine intermediates [213]. Replacement of applied potential with other oxidants, like *m*-CPBA or Oxone®, under these reaction conditions, resulted in considerably inferior yield (15 and 32%, respectively).

In 2020, Powers and coworkers developed hypervalent iodine electrocatalysis for intra- and intermolecular C–N bond-forming reactions [214]. In the intramolecular carbazole formation, either iodoarene catalyst **134** or **137** was used depending on the substrate (Figure 12.51a). 4-Iodoanisole was selected for its low onset potential and was successfully used as an electrocatalyst for electron-neutral or slightly electron-poor substrates. For more electron-poor substrates, bearing nitro- or

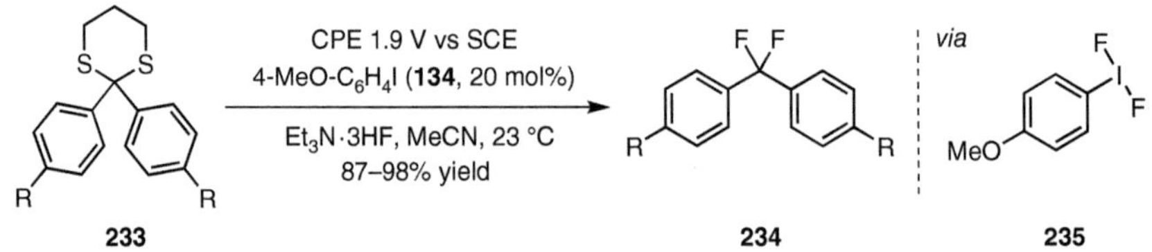

Figure 12.49 The first example of hypervalent iodine electrocatalysis was reported in the context of the difluorination of propane-1,3-dithiol-protected carbonyl compounds **233** under constant potential conditions. *Source*: Fuchigami and Fujita [56].

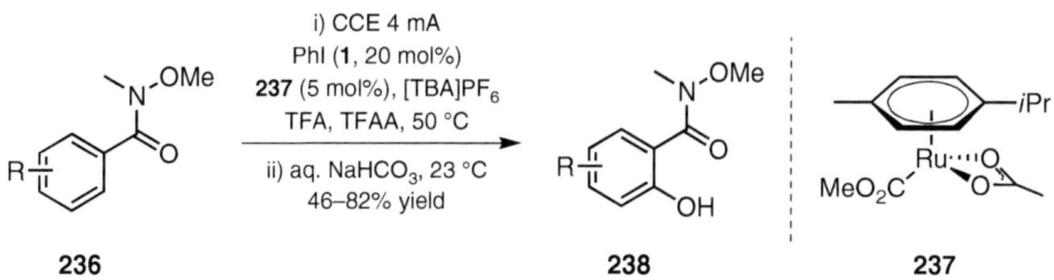

Figure 12.50 Iodoarene and Ru co-catalyzed C–H hydroxylation of benzamide derivatives. *Source*: Massignan et al. [212].

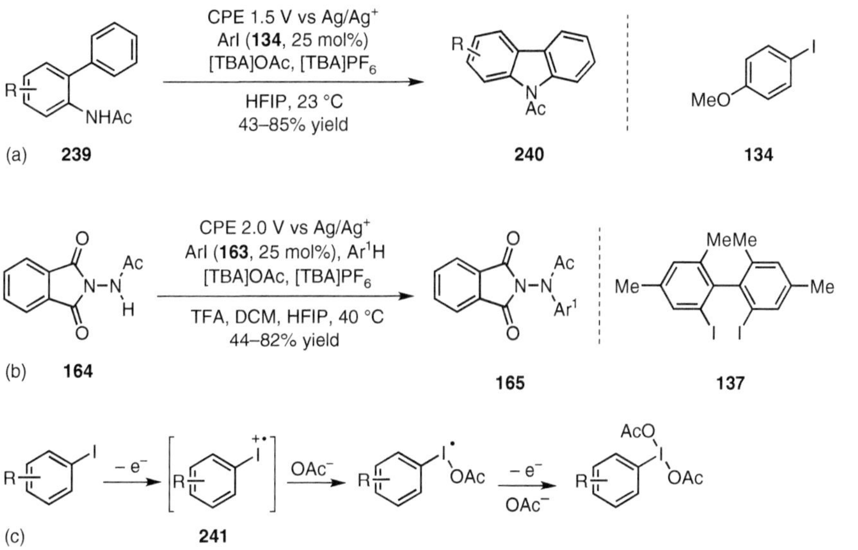

Figure 12.51 Anodically generated hypervalent iodine intermediates were used in (a) inter- and (b) intramolecular C–H amination reactions. (c) Stabilization of *in situ*-generated I(II) intermediate (**241**) by acetate anion was proposed to enable hypervalent iodine electrocatalysis.

carbonyl substituents, the more oxidizing iodoarene **137** was employed as a catalyst in potentiostatic electrolysis at 1.9 V vs. Ag^+/Ag. Electron-rich substrates with methyl or methoxy substituents underwent unselective anodic oxidative degradation due to lower onset potentials compared to iodoarene catalysts. Intermolecular C–H amination was also achieved between *N*-(1,3-dioxoisoindolin-2-yl)acetamide **164** and arenes using iodoarene **137** as catalyst (Figure 12.51b). The absence of an oxidation peak in the cyclic voltammogram (CV) for amine **164** at the high potentials required to oxidize iodoarene **137** allowed for its *in cell* utilization. Control experiments in the absence of iodoarene, acetate, or HFIP resulted in no desired products. Titration of increasing acetate concentration to iodoarene resulted in CVs with decreased reversibility and increased current response, which is indicative of trapping anodically generated I(II) intermediate by acetate ion and subsequent oxidation to I(III) intermediate (Figure 12.51c).

12.6 Conclusion

Hypervalent iodine compounds are an important class of oxidants in synthetic chemistry. The use of hypervalent iodine reagents is often not highly atom-economical due to the low active oxygen content of these species. Advances in the sustainable synthesis of these reagents promise to impact the sustainability of the substrate functionalization reactions that can be affected by these compounds. Here, we have discussed advances in the use of environmentally benign oxidants such as O_2 and H_2O_2 in the preparation of hypervalent iodine species. In addition, significant progress has been made toward the electrochemical oxidation of iodoarenes, which avoids the need for stoichiometric terminal chemical oxidants. Polymer-supported and fluorous hypervalent iodine compounds have emerged as recyclable platforms that could reduce the waste associated with the stoichiometric application of these reagents. More recently, the development of hypervalent iodine catalysis promises to significantly impact the sustainability of reaction chemistry effected at the hypervalent iodine center by enabling the use of substoichiometric iodoarene loading. To fully realize the potential of hypervalent iodine catalysis, new approaches to achieving selective oxidation of iodoarenes in the presence of oxidatively labile substrates is critical. Ongoing efforts to better define the mechanistic alternatives available for the synthesis of these species promises to impact the ongoing development of aerobic and electrochemical strategies to hypervalent iodine compounds and catalysis.

Acknowledgment

The authors thank the Welch Foundation (A-1907) and the National Science Foundation (CAREER 1848135) for financial support.

References

1 Fetzner, S. and Steiner, R.A. (2010). Cofactor-independent oxidases and oxygenases. *Appl. Microbiol. Biotechnol.* 86: 791–804.
2 Ho, R.Y., Liebman, J.F., and Valentine, J.S. (1995). Biological reactions of dioxygen: an introduction. In: *Active Oxygen in Biochemistry* (eds. J. Valentine, C.S. Foote, A. Greenberg and J.F. Liebman), 1–36. Springer.
3 Borden, W.T., Hoffmann, R., Stuyver, T., and Chen, B. (2017). Dioxygen: what makes this triplet diradical kinetically persistent? *J. Am. Chem. Soc.* 139: 9010–9018.
4 Filatov, M., Reckien, W., Peyerimhoff, S.D., and Shaik, S. (2000). What are the reasons for the knetic sability of a mixture of H_2 and O_2? *J. Phys. Chem. A* 104: 12014–12020.
5 Ho, R.Y.N., Liebman, J.F., and Valentine, J.S. (1995). Overview of the energetics and reactivity of oxygen. In: *Active Oxygen in Chemistry* (eds. C.S. Foote,

J.S. Valentine, A. Greenberg and J.F. Liebman), 1–23. Blackie Academic and Professional.

6 McCann, S.D. and Stahl, S.S. (2015). Copper-catalyzed aerobic oxidations of organic molecules: pathways for two-electron oxidation with a four-electron oxidant and a one-electron redox-active catalyst. *Acc. Chem. Res.* 48: 1756–1766.

7 Osterberg, P.M., Niemeier, J.K., Welch, C.J. et al. (2015). Experimental limiting oxygen concentrations for nine organic solvents at temperatures and pressures relevant to aerobic oxidations in the pharmaceutical industry. *Org. Pro. Res. Dev.* 19: 1537–1543.

8 Anastas, P. and Warner, J. (1998). *Green Chemistry: Theory and Practice.* Oxford University Press.

9 Anastas, P. and Eghbali, N. (2010). Green chemistry: principles and practice. *Chem. Soc. Rev.* 39: 301–312.

10 Goti, A. and Cardona, F. (2008). Hydrogen peroxide in green oxidation reactions: recent catalytic processes. In: *Green Chemical Reactions* (eds. P. Tundo and V. Esposito), 191–212. Springer.

11 Campbell, A.N. and Stahl, S.S. (2012). Overcoming the "oxidant problem": strategies to use O_2 as the oxidant in organometallic C–H oxidation reactions catalyzed by Pd (and Cu). *Acc. Chem. Res.* 45: 851–863.

12 Ibanez, J.G., Fitch, A., Frontana-Uribe, B.A., and Vasquez-Medrano, R. (2014). Green electrochemistry. In: *Encyclopedia of Applied Electrochemistry* (eds. G. Kreysa, K.-I. Ota and R.F. Savinell), 964–971. Springer.

13 Hach, R.J. and Rundle, R.E. (1951). The structure of tetramethylammonium pentaoxide. *J. Am. Chem. Soc.* 73: 4321–4324.

14 Pimentel, G.C. (1951). The bonding of trihalide abd bifluoride ions by the molecular orbital method. *J. Chem. Phys.* 19: 446–448.

15 Zhdankin, V.V. and Stang, P.J. (2002). Recent developments in the chemistry of polyvalent iodine compounds. *Chem. Rev.* 102: 2523–2584.

16 Stang, P.J. and Zhdankin, V.V. (1996). Organic polyvalent iodine compounds. *Chem. Rev.* 96: 1123–1178.

17 Zhdankin, V.V. and Stang, P.J. (2008). Chemistry of polyvalent iodine. *Chem. Rev.* 108: 5299–5358.

18 Yoshimura, A. and Zhdankin, V.V. (2016). Advances in synthetic applications of hypervalent iodine compounds. *Chem. Rev.* 116: 3328–3435.

19 Varvoglis, A. (1997). *Hypervalent Iodine in Organic Synthesis.* Academic Press.

20 Wirth, T. (2003). Oxidations and rearrangements. In: *Hypervalent Iodine Chemistry: Modern Developments in Organic Synthesis* (ed. T. Wirth), 185–208. Springer.

21 Singh, F.V. and Wirth, T. (2014). Hypervalent iodine-catalyzed oxidative functionalizations including stereoselective reactions. *Chem. Asian J.* 9: 950–971.

22 Zhdankin, V.V. and Protasiewicz, J.D. (2014). Development of new hypervalent iodine reagents with improved properties and reactivity by redirecting secondary bonds at iodine center. *Coord. Chem. Rev.* 275: 54–62.

23 Maity, A., Hyun, S.-M., and Powers, D.C. (2018). Oxidase catalysis via aerobically generated hypervalent iodine intermediates. *Nat. Chem.* 10: 200–204.

24 Larkin, D.R. (1990). The role of catalysts in the air oxidation of aliphatic aldehydes. *J. Org. Chem.* 55: 1563–1568.

25 Lehtinen, C. and Brunow, G. (2000). Factors affecting the selectivity of air oxidation of 2-ethyhexanal, an α-branched aliphatic aldehyde. *Org. Pro. Res. Dev.* 4: 544–549.

26 Hyun, S.-M., Yuan, M., Maity, A. et al. (2019). The role of iodanyl radicals as critical chain carriers in aerobic hypervalent iodine chemistry. *Chem* 5: 2388–2404.

27 Jorissen, W.P. and Dekking, A.C.B. (1938). On the induced oxidation of iodobenzene during the oxidation of acetaldehyde in an atmosphere of oxygen. *Recl. Trav. Chim. Pays-Bas* 57: 1125–1126.

28 Maity, A., Hyun, S.-M., Wortman, A.K., and Powers, D.C. (2018). Oxidation catalysis by an aerobically generated Dess-Martin periodinane analogue. *Angew. Chem. Int. Ed.* 57: 7205–7209.

29 Cardenal, A.D., Maity, A., Gao, W.-Y. et al. (2019). Iodosylbenzene coordination chemistry relevant to metal–organic framework catalysis. *Inorg. Chem.* 58: 10543–10553.

30 Hans-Joachim, R. and Georg, P. (1999). Production of hydrogen peroxide. Google Patents 1939.

31 Marzzacco, C.J. (1999). The enthalpy of decomposition of hydrogen peroxide: a general chemistry calorimetry experiment. *J. Chem. Educ.* 76: 1517–1518.

32 Lu, C.-S., Hughes, E., and Giguère, P.A. (1941). The crystal structure of the urea—hydrogen peroxide addition compound $CO(NH_2)_2{\cdot}H_2O_2$. *J. Am. Chem. Soc.* 63: 1507–1513.

33 Catir, M., Kilic, H., Nardello-Rataj, V. et al. (2009). Singlet oxygen generation from [Bis(trifluoroacetoxy)iodo]benzene and hydrogen peroxide. *J. Org. Chem.* 74: 4560–4564.

34 Boeseken, J. and Schneider, G.C.C. (1931). Über den Verlauf der oxydation der Doppelbindung mittels Peressigsäure und Perbenzoesäure. *J. Prakt. Chem.* 131: 285.

35 Pausacker, K. (1953). The oxidation of glycols by aryl iodosoacetates. A kinetic study. *J. Chem. Soc.*: 107–109.

36 Colomer, I., Batchelor-McAuley, C., Odell, B. et al. (2016). Hydrogen bonding to hexafluoroisopropanol controls the oxidative strength of hypervalent iodine reagents. *J. Am. Chem. Soc.* 138: 8855–8861.

37 Zhdankin, V.V., Scheuller, M.C., and Stang, P.J. (1993). A general approach to aryl(cyano)iodonium triflates-versatile iodonium transfer reagents. *Tetrahedron Lett.* 34: 6853–6856.

38 Spyroudis, S. and Varvoglis, A. (1975). Dehydrogenations with phenyliodine ditrifluoroacetate. *Synthesis*: 445–447.

39 White, J.D., Caravatti, G., Kline, T.B. et al. (1983). Biomimetic total synthesis of (−)-codeine. *Tetrahedron* 39: 2393–2397.

40 Alcock, N.W. and Waddington, T.C. (1963). 780. Chemistry of positive iodine. Part II. Reactions of iodobenzene dichloride with silver salts. *J. Chem. Soc.*: 4103–4109.

41 Zefirov, N., Safronov, S., Kaznacheev, A., and Zhdankin, V. (1989). General method for the synthesis of aryliodoso derivatives under aprotic conditions by the reaction of iodosobenzene with substituted trimethylsilanes. *Russ. J. Org. Chem.* 25: 1807–1808.
42 Page, T.K. and Wirth, T. (2006). Simple direct synthesis of [Bis (trifluoroacetoxy)iodo]arenes. *Synthesis*: 3153–3155.
43 Knight, D.W. and Russell, G.A. (2001). Phenyliodine(III) dichloride. *Encyclopedia of Reagents for Organic Synthesis*: 1–4.
44 Krassowska-Wiebocka, B., Prokopienko, G., and Skulski, L. (1999). Biphasic chlorination of iodoarenes to (Dichloroiodo)arenes. *Synlett*: 1409–1410.
45 Obeid, N. and Skulski, L. (2000). Novel oxidative, liquid-phase chlorination procedures for the preparation of (Dichloroiodo)arenes from iodoarenes. *Pol. J. Chem.* 74: 1609–1615.
46 Baranowski, A., Płachta, D., Skulski, L., and Klimaszewska, M. (2000). Liquid-phase and biphasic chlorination of some iodoarenes to form (Dichloroiodo)arenes with sodium peroxodisulfate as the oxidant. *J. Chem. Res.*: 435–437.
47 Kaźmierczak, P., Skulski, L., and Obeid, N. (1999). Oxidative chlorination of various iodoarenes to (Dichloroiodo)arenes with chromium (VI) oxide as the oxidant. *J. Chem. Res., Synop.* (1): 64–65.
48 Zhao, X.-F. and Zhang, C. (2007). Iodobenzene dichloride as a stoichiometric oxidant for the conversion of alcohols into carbonyl compounds; two facile methods for its preparation. *Synthesis*: 551–557.
49 Podgoršek, A. and Iskra, J. (2010). Conversion of aryl iodides into arylIodine(III) dichlorides by an oxidative halogenation strategy using 30% aqueous hydrogen peroxide in fluorinated alcohol. *Molecules* 15: 2857–2871.
50 Merritt, E.A., Malmgren, J., Klinke, F.J., and Olofsson, B. (2009). Synthesis of diaryliodonium triflates using environmentally benign oxidizing agents. *Synlett*: 2277–2280.
51 Koposov, A.Y., Karimov, R.R., Pronin, A.A. et al. (2006). $RuCl_3$-catalyzed oxidation of iodoarenes with peracetic acid: new facile preparation of iodylarenes. *The J. Org. Chem.* 71: 9912–9914.
52 Yusubov, M.S., Chi, K.-W., Park, J.Y. et al. (2006). Highly efficient RuCl3-catalyzed disproportionation of (diacetoxyiodo) benzene to iodylbenzene and iodobenzene; leading to the efficient oxidation of alcohols to carbonyl compounds. *Tetrahedron Lett.* 47: 6305–6308.
53 Elsherbini, M. and Wirth, T. (2018). Hypervalent iodine reagents by anodic oxidation powerful green synthesis. *Chem. Eur. J.* 24: 13399–13407.
54 Schmidt, H. and Meinert, H. (1960). Zum Mechanismus der Elektrochemischen Fluorierung und über die Bildung von Jod-monofluorid. *Angew. Chem.* 72: 109–110.
55 Rozhkov, I.N. (1976). Radical-cation mechanism of the anodic fluorination of organic compounds. *Russ. Chem. Rev.* 45: 615–619.
56 Fuchigami, T. and Fujita, T. (1994). Electrolytic partial fluorination of organic compounds. The first electrosynthesis of hypervalent iodobenzene difluoride

derivatives and its application to indirect anodic gem-difluorination. *J. Org. Chem.* 59: 7190–7192.

57 Hara, S., Hatakeyama, T., Chen, S.-Q. et al. (1998). Electrochemical fluorination of β-dicarbonyl compounds using *p*-iodotoluene difluoride as a mediator. *J. Fluor. Chem.* 87: 189–192.

58 Haupt, J.D., Berger, M., and Waldvogel, S.R. (2018). Electrochemical fluorocyclization of N-allylcarboxamides to 2-oxazolines by hypervalent iodine mediator. *Org. Lett.* 21: 242–245.

59 Doobary, S., Sedikides, A.T., Caldora, H.P. et al. (2020). Electrochemical vicinal difluorination of alkenes: scalable and amenable to electron-rich substrates. *Angew. Chem.* 132: 1171–1176.

60 Lund, H. and Hammerich, O. (2001). *Organic Electrochemistry*. Marcel Dekker Inc.

61 Simonet, J. and Pilard, J.-F. (2001). Electrogenerated reagents. In: *Organic Electrochemistry* (eds. H. Lund and O. Hammerich), 1163–1225. Marcel Dekker Inc.

62 Kajiyama, D., Saitoh, T., and Nishiyama, S. (2013). Application of electrochemically generated hypervalent iodine oxidant to natural products synthesis. *Electrochemistry* 81: 319–324.

63 Inoue, K., Ishikawa, Y., and Nishiyama, S. (2010). Synthesis of tetrahydropyrroloiminoquinone alkaloids based on electrochemically generated hypervalent iodine oxidative cyclization. *Org. Lett.* 12: 436–439.

64 Kajiyama, D., Inoue, K., Ishikawa, Y., and Nishiyama, S. (2010). A synthetic approach to carbazoles using electrochemically generated hypervalent iodine oxidant. *Tetrahedron* 66: 9779–9784.

65 Dohi, T., Ito, M., Yamaoka, N. et al. (2009). Hypervalent iodine(III): selective and efficient single-electron-transfer (SET) oxidizing agent. *Tetrahedron* 65: 10797–10815.

66 Dohi, T., Yamaoka, N., and Kita, Y. (2010). Fluoroalcohols: versatile solvents in hypervalent iodine chemistry and syntheses of diaryliodonium(III) salts. *Tetrahedron* 66: 5775–5785.

67 Eberson, L., Hartshorn, M.P., Persson, O., and Radner, F. (1996). Making radical cations live longer. *Chem. Comm.*: 2105–2112.

68 Amano, Y. and Nishiyama, S. (2006). Oxidative synthesis of azacyclic derivatives through the nitrenium ion application of a hypervalent iodine species electrochemically generated from iodobenzene. *Tetrahedron Lett.* 47: 6505–6507.

69 Faulkner, D.J. (2001). Marine natural products. *Nat. Prod. Rep.* 18: 1R–49R.

70 Izawa, T., Nishiyama, S., and Yamamura, S. (1994). Total syntheses of makaluvamines A, B, C, D and E, cytotoxic pyrroloiminoquinone alkaloids isolated from marine sponge bearing inhibitory activities against topoisomerase II. *Tetrahedron* 50: 13593–13600.

71 Amano, Y., Inoue, K., and Nishiyama, S. (2008). Oxidative access to quinolinone derivatives with simultaneous rearrangement of functional groups. *Synlett* 1: 134–136.

72 Amano, Y. and Nishiyama, S. (2008). Effects of aromatic substituents of electrochemically generated hypervalent iodine oxidant on oxidation reactions. *Heterocycles* 75: 1997–2003.

73 Miller, L.L. and Hoffmann, A.K. (1967). The electrochemical formation of carbonium and iodonium Ions from alkyl and aryl iodides. *J. Am. Chem. Soc.* 89: 593–597.

74 Hoffelner, H., Lorch, H., and Wendt, H. (1975). Anodic phenyl-onium cation formation: Part II. Reaction mechanism and optimization for the anodic formation of diphenyliodonium cations. *J. Electroanal. Chem. Interfacial Electrochem.* 66: 183–194.

75 Peacock, M.J. and Pletcher, D. (2001). The synthesis of diaryliodonium salts by the anodic oxidation of aryl iodide/arene mixtures. *J. Electrochem. Soc.* 148: D37–D42.

76 Peacock, M.J. and Pletcher, D. (2000). The electrosynthesis of diaryliodonium salts. *Tetrahedron Lett.* 41: 8995–8998.

77 Broese, T. and Francke, R. (2016). Electrosynthesis using a recyclable mediator-electrolyte system based on ionically tagged phenyl iodide and 1,1,1,3,3,3-hexafluoroisopropanol. *Org. Lett.* 18: 5896–5899.

78 Koleda, O., Broese, T., Noetzel, J. et al. (2017). Synthesis of benzoxazoles using electrochemically generated hypervalent iodine. *J. Org. Chem.* 82: 11669–11681.

79 Roesel, A.F., Broese, T., Májek, M., and Francke, R. (2019). Iodophenylsulfonates and iodobenzoates as redox-active supporting electrolytes for electrosynthesis. *ChemElectroChem* 6: 4229–4237.

80 Plutschack, M.B., Pieber, B.U., Gilmore, K., and Seeberger, P.H. (2017). The Hitchhiker's guide to flow chemistry ||. *Chem. Rev.* 117: 11796–11893.

81 Asprion, N., Mollner, S., Poth, N., and Rumpf, B. (2000). Energy management in chemical industry. In: *Ullmann's Encyclopedia of Industrial Chemistry*. Wiley-VCH.

82 Britton, J. and Raston, C.L. (2017). Multi-step continuous-flow synthesis. *Chem. Soc. Rev.* 46: 1250–1271.

83 Adamo, A., Beingessner, R.L., Behnam, M. et al. (2016). On-demand continuous-flow production of pharmaceuticals in a compact, reconfigurable system. *Science* 352: 61–67.

84 Watts, K., Gattrell, W., and Wirth, T. (2011). A practical microreactor for electrochemistry in flow. *Beilstein J. Org. Chem.* 7: 1108–1114.

85 Gemoets, H.P., Laudadio, G., Verstraete, K. et al. (2017). A modular flow design for the meta-selective C–H arylation of anilines. *Angew. Chem.Int. Ed.* 56: 7161–7165.

86 Gao, W.-C., Xiong, Z.-Y., Pirhaghani, S., and Wirth, T. (2019). Enantioselective electrochemical lactonization using chiral iodoarenes as mediators. *Synthesis* 51: 276–284.

87 Elsherbini, M., Winterson, B., Alharbi, H. et al. (2019). Continuous-flow electrochemical generator of hypervalent iodine reagents: synthetic applications. *Angew. Chem. Int. Ed.* 58: 9811–9815.

88 Bystron, T., Horbenko, A., Syslova, K. et al. (2018). 2-iodoxybenzoic acid synthesis by oxidation of 2-iodobenzoic acid at a Boron-Doped diamond anode. *ChemElectroChem* 5: 1002–1005.

89 Devadas, B., Svoboda, J., Krupička, M., and Bystron, T. (2020). Electrochemical and spectroscopic study of 2-iodobenzoic acid and 2-iodosobenzoic acid anodic oxidation in aqueous environment. *Electrochim. Acta* 342: 136080–136091.

90 Okawara, M. and Mizuta, K. (1961). Synthesis and some reactions of polystyrene iodosoacetate. *Kogyo Kagaku Zasshi* 64: 232–235.

91 Huang, X. and Zhu, Q. (2001). The synthesis of cross-linked poly[styrene(iodosodiacetate)] and application in preparation of pyrazoline. *Synth. Commun.* 31: 111–115.

92 Abe, S., Sakuratani, K., and Togo, H. (2001). Synthetic use of poly[4-hydroxy(tosyloxy)iodo]styrenes. *J. Org. Chem.* 66: 6174–6177.

93 Abe, S., Sakuratani, K., and Togo, H. (2001). Novel preparation and reactivity of poly[4-hydroxy(tosyloxy)iodo]styrenes. *Synlett* (12): 22–24.

94 Ficht, S., Mülbaier, M., and Giannis, A. (2001). Development of new and efficient polymer-supported hypervalent iodine reagents. *Tetrahedron* 57: 4863–4866.

95 Tohma, H., Maruyama, A., Maeda, A. et al. (2004). Preparation and reactivity of 1, 3, 5, 7-Tetrakis-[4-(diacetoxyiodo) phenyl]adamantane, a recyclable hypervalent iodine(III) reagent. *Angew. Chem. Int. Ed.* 43: 3595–3598.

96 Dohi, T., Maruyama, A., Yoshimura, M. et al. (2005). A unique site-selective reaction of ketones with new recyclable hypervalent iodine(III) reagents based on a tetraphenylmethane structure. *Chem. Commun.* (17): 2205–2207.

97 Dohi, T., Fukushima, K.I., Kamitanaka, T. et al. (2012). An excellent dual recycling strategy for the hypervalent iodine/nitroxyl radical mediated selective oxidation of alcohols to aldehydes and ketones. *Green Chem.* 14: 1493–1501.

98 Thorat, P.B., Bhong, B.Y., and Karade, N.N. (2013). 2, 4, 6-Tris-[(4-dichloroiodo)phenoxy]-1, 3, 5-triazine as a new recyclable hypervalent iodine(III) reagent for chlorination and oxidation reactions. *Synlett* 24: 2061–2066.

99 Liu, Z., Chen, Z.-C., and Zheng, Q.-G. (2003). Mild oxidation of alcohols with *o*-iodoxybenzoic acid (IBX) in ionic liquid 1-Butyl-3-methyl-imidazolium chloride and water. *Org. Lett.* 5: 3321–3323.

100 More, J.D. and Finney, N.S. (2002). A simple and advantageous protocol for the oxidation of alcohols with *o*-iodoxybenzoic acid (IBX). *Org. Lett.* 4: 3001–3003.

101 Mülbaier, M. and Giannis, A. (2001). The synthesis and oxidative properties of polymer-supported IBX. *Angew. Chem. Int. Ed.* 40: 4393–4394.

102 Sorg, G., Mengel, A., Jung, G., and Rademann, J. (2001). Oxidizing polymers: a polymer-supported, recyclable hypervalent iodine (V) reagent for the efficient conversion of alcohols, carbonyl compounds, and unsaturated carbamates in solution. *Angew. Chem. Int. Ed.* 40: 4395–4397.

103 Bizzarri, B.M., Abdalghani, I., Botta, L. et al. (2018). Iodoxybenzoic acid supported on multi walled carbon nanotubes as biomimetic environmental friendly oxidative systems for the oxidation of alcohols to aldehydes. *Nanomaterials* 8: 516–529.

104 Ladziata, U., Willging, J., and Zhdankin, V.V. (2006). Facile preparation and reactivity of polymer-supported *N*-(2-Iodyl-phenyl)-acylamide, an efficient oxidizing system. *Org. Lett.* 8: 167–170.

105 Chung, W.-J., Kim, D.-K., and Lee, Y.-S. (2005). Polymer-supported IBX-amide reagents: significant role of spacer and additive in alcohol oxidation. *Synlett*: 2175–2178.

106 Nambu, H., Shimokawa, I., Fujiwara, T., and Yakura, T. (2016). Recyclable magnetic nanoparticle-supported iodoarene catalysts for oxidation of 4-alkoxyphenols to quinones. *Asian J. Org. Chem.* 5: 486–489.

107 Ballaschk, F. and Kirsch, S.F. (2019). Oxidation of secondary alcohols using solid-upported hypervalent iodine catalysts. *Green Chem.* 21: 5896–5903.

108 Lyalin, V.V., Orda, V.V., Alekseeva, L.A., and Yagupolskii, L.M. (1971). Arylperfluoroalkyliodonium and arylpentafluorophenyliodonium compounds. *Zh. Org. Khim.* 7: 1473.

109 Rocaboy, C. and Gladysz, J.A. (2003). Convenient syntheses of fluorous aryl Iodides and hypervalent iodine compounds: $ArI(L)_n$ reagents that are recoverable by simple liquid/liquid biphase workups, and applications in oxidations of hydroquinones. *Chem. Eur. J.* 9: 88–95.

110 Tesevic, V. and Gladysz, J. (2006). Oxidations of secondary alcohols to ketones using easily recyclable Bis(trifluoroacetate) adducts of fluorous alkyl iodides, $CF_3(CF_2)_{n-1}I(OCOCF_3)_2$. *J. Org. Chem.* 71: 7433–7440.

111 Schäfer, S. and Wirth, T. (2010). A versatile and highly reactive polyfluorinated hypervalent iodine(III) compound. *Angew. Chem. Int. Ed.* 49: 2786–2789.

112 Zhang, G., Tan, H., Chen, W. et al. (2020). Synthesis of NH-sulfoximines using recyclable hypervalent iodine(III) reagents under aqueous micellar conditions. *ChemSusChem* 13: 922–928.

113 Miyamoto, K., Yamashita, J., Narita, S. et al. (2017). Iodoarene-catalyzed oxidative transformations using molecular oxygen. *Chem. Commun.* 53: 9781–9784.

114 Miyamoto, K., Sakai, Y., Goda, S., and Ochiai, M. (2012). A catalytic version of hypervalent Aryl-λ^3-Iodane-Induced hofmann rearrangement of primary carboxamides: iodobenzene as an organocatalyst and *m*-Chloroperbenzoic acid as a terminal oxidant. *Chem. Commun.* 48: 982–984.

115 Lucchetti, N., Scalone, M., Fantasia, S., and Muñiz, K. (2016). An improved catalyst for iodine(I/III)-catalysed intermolecular C–H amination. *Adv. Synth. Catal.* 358: 2093–2099.

116 Chauhan, J., Ravva, M.K., and Sen, S. (2019). Harnessing autoxidation of aldehydes: *In Situ* iodoarene catalyzed synthesis of substituted 1,3,4-oxadiazole, in the presence of molecular oxygen. *Org. Lett.* 21: 6562–6565.

117 Habert, L. and Cariou, K. (2020). Photoinduced aerobic iodoarene-catalyzed spirocyclization of *N*-Oxy-amides to N-Fused spirolactams. *Angew. Chem. Int. Ed.* https://doi.org/10.1002/anie.202009175.

118 Ochiai, M., Takeuchi, Y., Katayama, T. et al. (2005). Iodobenzene-catalyzed α-acetoxylation of ketones. *In Situ* generation of hypervalent (Diacyloxyiodo)benzenes using *m*-chloroperbenzoic acid. *J. Am. Chem. Soc.* 127: 12244–12245.

119 Sheng, J., Li, X., Tang, M. et al. (2007). An efficient method for the α-acetoxylation of ketones. *Synthesis*: 1165–1168.

120 Yamamoto, Y. and Togo, H. (2006). PhI-catalyzed α-tosyloxylation of ketones with *m*-chloroperbenzoic acid and *p*-toluenesulfonic acid. *Synlett*: 0798–0800.

121 Akiike, J., Yamamoto, Y., and Togo, H. (2007). Efficient conversion of ketones to α-tosyloxyketones with *m*-chloroperbenzoic acid and *p*-toluenesulfonic acid in the presence of catalytic amount of IL-supported PhI in [emim]OTs. *Synlett*: 2168–2172.

122 Yamamoto, Y., Kawano, Y., Toy, P.H., and Togo, H. (2007). PhI- and polymer-supported PhI-catalyzed oxidative conversion of ketones and alcohols to α-tosyloxyketones with *m*-chloroperbenzoic acid and *p*-toluenesulfonic acid. *Tetrahedron* 63: 4680–4687.

123 Pu, Y., Gao, L., Liu, H., and Yan, J. (2012). An effective catalytic α-phosphoryloxylation of ketones with iodobenzene. *Synthesis* 44: 99–103.

124 Kitamura, T., Muta, K., and Kuriki, S. (2013). Catalytic fluorination of 1,3-dicarbonyl compounds using iodoarene catalysts. *Tetrahedron Lett.* 54: 6118–6120.

125 Suzuki, S., Kamo, T., Fukushi, K. et al. (2014). Iodoarene-catalyzed fluorination and aminofluorination by an Ar-I/HF·pyridine/*m*CPBA system. *Chem. Sci.* 5: 2754–2760.

126 Richardson, R.D., Page, T.K., Altermann, S. et al. (2007). Enantioselective α-oxytosylation of ketones catalysed by iodoarenes. *Synlett*: 538–542.

127 Hirt, U.H., Schuster, M.F., French, A.N. et al. (2001). Chiral hypervalent organo-iodine(III) compounds. *Eur. J. Org. Chem.*: 1569–1579.

128 Abazid, A.H. and Nachtsheim, B.J. (2020). A triazole-substituted aryl iodide with omnipotent reactivity in enantioselective oxidations. *Angew. Chem. Int. Ed.* 59: 1479–1484.

129 Altermann, S.M., Richardson, R.D., Page, T.K. et al. (2008). Catalytic enantioselective α-oxysulfonylation of ketones mediated by iodoarenes. *Eur. J. Org. Chem.*: 5315–5328.

130 Guilbault, A.-A. and Legault, C.Y. (2012). Drastic enhancement of activity in iodane-based α-tosyloxylation of ketones: iodine(III) does the hypervalent twist. *ACS Catal.* 2: 219–222.

131 Guilbault, A.-A., Basdevant, B., Wanie, V., and Legault, C.Y. (2012). Catalytic enantioselective α-tosyloxylation of ketones using iodoaryloxazoline catalysts: insights on the stereoinduction process. *J. Org. Chem.* 77: 11283–11295.

132 Brenet, S., Berthiol, F., and Einhorn, J. (2013). 3,3′-Diiodo-BINOL-fused maleimides as chiral hypervalent iodine(III) organocatalysts. *Eur. J. Org. Chem.* (36): 8094–8096.

133 Pluta, R., Krach, P.E., Cavallo, L. et al. (2018). Metal-free catalytic asymmetric fluorination of keto esters using a combination of hydrogen fluoride (HF) and oxidant: experiment and computation. *ACS Catal.* 8: 2582–2588.

134 Wang, Y., Yuan, H., Lu, H., and Zheng, W.-H. (2018). Development of planar chiral iodoarenes based on [2.2]Paracyclophane and their application in catalytic enantioselective fluorination of β-ketoesters. *Org. Lett.* 20: 2555–2558.

135 Dohi, T., Maruyama, A., Yoshimura, M. et al. (2005). Versatile hypervalent-iodine(III)-catalyzed oxidations with *m*-chloroperbenzoic acid as a cooxidant. *Angew.Chem. Int. Ed.* 44: 6193–6196.

136 Dohi, T., Takenaga, N., Nakae, T. et al. (2013). Asymmetric dearomatizing spirolactonization of naphthols catalyzed by spirobiindane-based chiral hypervalent iodine species. *J. Am.Chem. Soc.* 135: 4558–4566.

137 Uyanik, M., Yasui, T., and Ishihara, K. (2010). Enantioselective kita oxidative spirolactonization catalyzed by *In Situ* generated chiral hypervalent iodine(III) species. *Angew. Chem.Int. Ed.* 49: 2175–2177.

138 Murray, S.J. and Ibrahim, H. (2015). Asymmetric kita spirolactonisation catalysed by anti-dimethanoanthracene-based iodoarenes. *Chem. Commun.* 51: 2376–2379.

139 Ogasawara, M., Sasa, H., Hu, H. et al. (2017). Atropisomeric chiral diiododienes (Z,Z)-2,3-Di(1-iodoalkylidene)tetralins: synthesis, enantiomeric resolution, and application in asymmetric catalysis. *Org. Lett.* 19: 4102–4105.

140 Dohi, T., Sasa, H., Miyazaki, K. et al. (2017). Chiral atropisomeric 8,8′-diiodobinaphthalene for asymmetric dearomatizing spirolactonizations in hypervalent iodine oxidations. *J. Org.Chem.* 82: 11954–11960.

141 Uyanik, M., Sasakura, N., Mizuno, M., and Ishihara, K. (2017). Enantioselective synthesis of masked benzoquinones using designer chiral hypervalent organoiodine(III) catalysis. *ACS Catal.* 7: 872–876.

142 Uyanik, M., Yasui, T., and Ishihara, K. (2017). Chiral hypervalent organoiodine-catalyzed enantioselective oxidative spirolactonization of naphthol derivatives. *J. Org. Chem.* 82: 1946–11953.

143 Hempel, C., Maichle-Mössmer, C., Pericàs, M.A., and Nachtsheim, B.J. (2017). Modular synthesis of triazole-based chiral iodoarenes for enantioselective spirocyclizations. *Adv. Synth. Catal.* 359: 2931–2941.

144 Zheng, H., Sang, Y., Houk, K.N. et al. (2019). Mechanism and origins of enantioselectivities in spirobiindane-based hypervalent iodine(III)-induced asymmetric dearomatizing spirolactonizations. *J. Am. Chem. Soc.* 141: 16046–16056.

145 Wang, Y., Zhao, C.-Y., Wang, Y.-P., and Zheng, W.-H. (2019). Enantioselective intramolecular dearomative lactonization of naphthols catalyzed by planar chiral iodoarene. *Synthesis* 51: 3675–3682.

146 Jain, N., Hein, J.E., and Ciufolini, M.A. (2019). Oxidative cyclization of naphtholic sulfonamides mediated by a chiral hypervalent iodine reagent: asymmetric synthesis versus resolution. *Synlett* 30: 1222–1227.

147 Jain, N., Xu, S., and Ciufolini, M.A. (2017). Asymmetric oxidative cycloetherification of naphtholic alcohols. *Chem. Eur. J.* 23: 4542–4546.

148 Cao, Y., Zhang, X., Lin, G. et al. (2016). Chiral aryliodine-mediated enantioselective organocatalytic spirocyclization: synthesis of spirofurooxindoles via cascade oxidative C–O and C–C bond formation. *Org. Lett.* 18: 5580–5583.

149 Sreenithya, A., Patel, C., Hadad, C.M., and Sunoj, R.B. (2017). Hypercoordinate iodine catalysts in enantioselective transformation: the role of catalyst folding in stereoselectivity. *ACS Catal.* 7: 4189–4196.

150 Ito, M., Kubo, H., Itani, I. et al. (2013). Organocatalytic C–H/C–H′ cross-biaryl coupling: C-selective arylation of sulfonanilides with aromatic hydrocarbons. *J. Am. Chem. Soc.* 135: 14078–14081.

151 Wu, H., He, Y.-P., Xu, L. et al. (2014). Asymmetric organocatalytic direct C(sp^2)–H/C(sp^3)–H oxidative cross-coupling by chiral iodine reagents. *Angew. Chem. Int. Ed.* 53: 3466–3469.

152 Wang, X., Gallardo-Donaire, J., and Martin, R. (2014). Mild arI-catalyzed C(sp^2)–H or C(sp^3)–H functionalization/C–O formation: an intriguing catalyst-controlled selectivity switch. *Angew. Chem. Int. Ed.* 53: 11084–11087.

153 Novák, P., Correa, A., Gallardo-Donaire, J., and Martin, R. (2011). Synergistic palladium-catalyzed C(sp^3)–H activation/C(sp^3)–O bond formation: a direct, step-economical route to benzolactones. *Angew. Chem. Int. Ed.* 50: 12236–12239.

154 Cheng, X.-F., Li, Y., Su, Y.-M. et al. (2013). Pd(II)-catalyzed enantioselective C–H activation/C–O bond formation: synthesis of chiral benzofuranones. *J. Am. Chem. Soc.* 135: 1236–1239.

155 Yang, M., Jiang, X., Shi, W.-J. et al. (2013). Direct lactonization of 2-arylacetic acids through Pd(II)-catalyzed C–H activation/C–O formation. *Org. Lett.* 15: 690–693.

156 Li, Y., Ding, Y.-J., Wang, J.-Y. et al. (2013). Pd-catalyzed C–H lactonization for expedient synthesis of biaryl lactones and total synthesis of cannabinol. *Org. Lett.* 15: 2574–2577.

157 Lee, J.M. and Chang, S. (2006). Pt-catalyzed sp^3 C–H bond activation of *o*-alkyl substituted aromatic carboxylic acid derivatives for the formation of aryl lactones. *Tetrahedron Lett.* 47: 1375–1379.

158 Gallardo-Donaire, J. and Martin, R. (2013). Cu-catalyzed mild C(sp^2)–H functionalization assisted by carboxylic acids en route to hydroxylated arenes. *J. Am. Chem. Soc.* 135: 9350–9353.

159 Wang, Y., Gulevich, A.V., and Gevorgyan, V. (2013). General and practical carboxyl-group-directed remote C–H oxygenation reactions of arenes. *Chem. Eur. J.* 19: 15836–15840.

160 Liang, D., Yu, W., Nguyen, N. et al. (2017). Iodobenzene-catalyzed synthesis of phenanthridinones via oxidative C–H amidation. *J. Org. Chem.* 82: 3589–3596.

161 Bal, A., Maiti, S., and Mal, P. (2018). Iodine(III)-enabled distal C–H functionalization of biarylsulfonanilides. *J. Org. Chem.* 83: 11278–11287.

162 Ding, Q., He, H., and Cai, Q. (2018). Chiral aryliodine-catalyzed asymmetric oxidative C–N bond formation via desymmetrization strategy. *Org. Lett.* 20: 4554–4557.

163 Zhu, C., Liang, Y., Hong, X. et al. (2015). Iodoarene-catalyzed stereospecific intramolecular sp^3 C–H amination: reaction development and mechanistic insights. *J. Am.Chem. Soc.* 137: 7564–7567.

164 Bose, A., Maiti, S., Sau, S., and Mal, P. (2019). An intramolecular C(sp^3)–H imination using PhI–*m*CPBA. *Chem. Commun.* 55: 2066–2069.

165 Dohi, T., Sasa, H., Dochi, M. et al. (2019). Oxidative coupling of *N*-methoxyamides and related compounds toward aromatic hydrocarbons by designer μ-oxo hypervalent iodine catalyst. *Synthesis* 51: 1185–1195.

166 Li, X., Chen, P., and Liu, G. (2018). Recent advances in hypervalent iodine(III)-catalyzed functionalization of alkenes. *Beilstein J. Org. Chem.* 14: 1813–1825.

167 Zhong, W., Liu, S., Yang, J. et al. (2012). Metal-free, organocatalytic syn-diacetoxylation of alkenes. *Org. Lett.* 14: 3336–3339.

168 Kolb, H.C., VanNieuwenhze, M.S., and Sharpless, K.B. (1994). Catalytic asymmetric dihydroxylation. *Chem. Rev.* 94: 2483–2547.

169 Ho, C.M., Yu, W.Y., and Che, C.M. (2004). Ruthenium nanoparticles supported on hydroxyapatite as an efficient and recyclable catalyst for cis-dihydroxylation and oxidative cleavage of alkenes. *Angew. Chem. Int. Ed.* 43: 3303–3307.

170 Yip, W.-P., Yu, W.-Y., Zhu, N., and Che, C.-M. (2005). Alkene cis-dihydroxylation by $[(Me_3tacn)(CF_3CO_2)\ Ru^{VI}O_2]ClO_4$ (Me_3tacn=1,4,7-trimethyl-1,4,7-triazacyclononane): structural characterization of [3 + 2] cycloadducts and kinetic studies. *J. Am. Chem. Soc.* 127: 14239–14249.

171 de Boer, J.W., Brinksma, J., Browne, W.R. et al. (2005). Cis-dihydroxylation and epoxidation of alkenes by $[Mn_2O(RCO_2)_2\ (tmtacn)_2]$: tailoring the selectivity of a highly H_2O_2-efficient catalyst. *J. Am. Chem. Soc.* 127: 7990–7991.

172 Chow, T.W.-S., Liu, Y., and Che, C.-M. (2011). Practical manganese-catalysed highly enantioselective cis-dihydroxylation of electron-deficient alkenes and detection of a cis-dioxomanganese(v) intermediate by high resolution ESI-MS analysis. *Chem. Commun.* 47: 11204–11206.

173 Oldenburg, P.D., Feng, Y., Pryjomska-Ray, I. et al. (2010). Olefin cis-dihydroxylation with bio-inspired iron catalysts. evidence for an Fe^{II}/Fe^{IV} catalytic cycle. *J. Am. Chem. Soc.* 132: 17713–17723.

174 Suzuki, K., Oldenburg, P.D., and Que, L. Jr., (2008). Iron-catalyzed asymmetric olefin cis-dihydroxylation with 97% enantiomeric excess. *Angew. Chem. Int. Ed.* 47: 1887–1889.

175 Bautz, J., Comba, P., Lopez de Laorden, C. et al. (2007). Biomimetic high-valent non-heme iron oxidants for the *cis*-dihydroxylation and epoxidation of olefins. *Angew. Chem. Int. Ed.* 46: 8067–8070.

176 Li, Y., Song, D., and Dong, V.M. (2008). Palladium-catalyzed olefin dioxygenation. *J. Am. Chem. Soc.* 130: 2962–2964.

177 Zhu, M.-K., Zhao, J.-F., and Loh, T.-P. (2010). Palladium-catalyzed oxime assisted intramolecular dioxygenation of alkenes with 1 atm of air as the sole oxidant. *J. Am. Chem. Soc.* 132: 6284–6285.

178 Haubenreisser, S., Wöste, T.H., Martínez, C. et al. (2016). Structurally defined molecular hypervalent iodine catalysts for intermolecular enantioselective reactions. *Angew. Chem. Int. Ed.* 55: 413–417.

179 Aertker, K., Rama, R.J., Opalach, J., and Muñiz, K. (2017). Vicinal difunctionalization of alkenes under iodine(III) catalysis involving lewis base adducts. *Adv. Synth. Catal.* 359: 1290–1294.

180 Barluenga, J., Trincado, M., Rubio, E., and Gonzalez, J.M. (2006). Direct intramolecular arylation of aldehydes promoted by reaction with IPy_2BF_4/HBF_4: synthesis of benzocyclic ketones. *Angew. Chem. Int. Ed.* 45: 3140–3143.

181 Barluenga, J., Vázquez-Villa, H., Ballesteros, A., and González, J.M. (2005). Synthesis of indoles upon sequential reaction of 3-Alkynylpyrrole-2-carboxaldehydes with iodonium ions and alkenes. Preparation of related benzofuran and benzothiophene derivatives. *Adv. Synth. Catal.* 347: 526–530.

182 Carlsson, A.-C.C., Mehmeti, K., Uhrbom, M. et al. (2016). Substituent effects on the $[N–I–N]^+$ halogen bond. *J. Am. Chem. Soc.* 138: 9853–9863.

183 Banik, S.M., Medley, J.W., and Jacobsen, E.N. (2016). Catalytic, diastereoselective 1,2-difluorination of alkenes. *J. Am. Chem. Soc.* 138: 5000–5003.

184 Banik, S.M., Medley, J.W., and Jacobsen, E.N. (2016). Catalytic, asymmetric difluorination of alkenes to generate difluoromethylated stereocenters. *Science* 353: 51–54.

185 Muñiz, K., Barreiro, L., Romero, R.M., and Martínez, C. (2017). Catalytic asymmetric diamination of styrenes. *J. Am. Chem. Soc.* 139: 4354–4357.

186 Miyamoto, K., Sei, Y., Yamaguchi, K., and Ochiai, M. (2009). Iodomesitylene-catalyzed oxidative cleavage of carbon–carbon double and triple bonds using *m*-chloroperbenzoic acid as a terminal oxidant. *J. Am. Chem. Soc.* 131: 1382–1383.

187 Travis, B.R., Sivakumar, M., Hollist, G.O., and Borhan, B. (2003). Facile oxidation of aldehydes to acids and esters with oxone. *Org. Lett.* 5: 1031–1034.

188 Hussain, H., Green, I.R., and Ahmed, I. (2013). Journey describing applications of oxone in synthetic chemistry. *Chem. Rev.* 113: 3329–3371.

189 Yakura, T. and Konishi, T. (2007). A novel catalytic hypervalent iodine oxidation of *p*-alkoxyphenols to *p*-quinones using 4-iodophenoxyacetic acid and oxone. *Synlett* (5): 765–768.

190 Yakura, T., Tian, Y., Yamauchi, Y. et al. (2009). Catalytic hypervalent iodine oxidation using 4-iodophenoxyacetic acid and oxone: oxidation of *p*-alkoxyphenols to *p*-benzoquinones. *Chem. Pharm. Bull.* 57: 252–256.

191 Yakura, T., Yamauchi, Y., Tian, Y., and Omoto, M. (2008). Catalytic hypervalent iodine oxidation of *p*-dialkoxybenzenes to *p*-quinones using 4-iodophenoxyacetic acid and oxone. *Chem. Pharm. Bull.* 56: 1632–1634.

192 Yakura, T., Omoto, M., Yamauchi, Y. et al. (2010). Hypervalent iodine oxidation of phenol derivatives using a catalytic amount of 4-iodophenoxyacetic acid and oxone as a Co-oxidant. *Tetrahedron* 66: 5833–5840.

193 Yakura, T. and Omoto, M. (2009). Efficient synthesis of *p*-quinols using catalytic hypervalent iodine oxidation of 4-arylphenols with 4-iodophenoxyacetic acid and oxone. *Chem. Pharm. Bull.* 57: 643–645.

194 Yoshimura, A., Middleton, K.R., Luedtke, M.W. et al. (2012). Hypervalent iodine catalyzed hofmann rearrangement of carboxamides using oxone as terminal oxidant. *J. Org. Chem.* 77: 11399–11404.

195 Yoshimura, A., Middleton, K.R., Todora, A.D. et al. (2013). Hypervalent iodine catalyzed generation of nitrile oxides from oximes and their cycloaddition with alkenes or alkynes. *Org. Lett.* 15: 4010–4013.

196 Okamura, Y., Sato, D., Yoshimura, A. et al. (2017). Iodine(III)-mediated/catalyzed cycloisomerization–amination sequence of *N*-propargyl carboxamides. *Adv. Synth. Catal.* 359: 3243–3247.

197 Thottumkara, A.P., Bowsher, M.S., and Vinod, T.K. (2005). *In Situ* generation of *o*-iodoxybenzoic acid (IBX) and the catalytic use of it in oxidation reactions in the presence of oxone as a Co-oxidant. *Org. Lett.* 7: 2933–2936.

198 Uyanik, M., Akakura, M., and Ishihara, K. (2009). 2-iodoxybenzenesulfonic acid as an extremely active catalyst for the selective oxidation of alcohols to aldehydes, ketones, carboxylic acids, and enones with oxone. *J. Am. Chem. Soc.* 131: 251–262.

199 Uyanik, M., Fukatsu, R., and Ishihara, K. (2009). IBS-catalyzed oxidative rearrangement of tertiary allylic alcohols to enones with oxone. *Org. Lett.* 11: 3470–3473.

200 Uyanik, M., Mutsuga, T., and Ishihara, K. (2012). IBS-catalyzed regioselective oxidation of phenols to 1, 2-quinones with oxone®. *Molecules* 17: 8604–8616.

201 Uyanik, M., Mutsuga, T., and Ishihara, K. (2017). 4,5-Dimethyl-2-Iodoxybenzenesulfonic acid catalyzed site-selective oxidation of 2-substituted phenols to 1,2-quinols. *Angew. Chem. Int. Ed.* 56: 3956–3960.

202 Ye, C., Twamley, B., and Shreeve, J.N.M. (2005). Straightforward syntheses of hypervalent iodine(III) reagents mediated by selectfluor. *Org. Lett.* 7: 3961–3964.

203 Braddock, D.C., Cansell, G., and Hermitage, S.A. (2006). Ortho-substituted iodobenzenes as novel organocatalysts for bromination of alkenes. *Chem. Commun.*: 2483–2485.

204 Alhalib, A., Kamouka, S., and Moran, W.J. (2015). Iodoarene-catalyzed cyclizations of unsaturated amides. *Org. Lett.* 17: 1453–1456.

205 Asari, N., Takemoto, Y., Shinomoto, Y. et al. (2016). Catalytic cycloisomerization–fluorination sequence of *N*-propargyl amides by iodoarene/HF·pyridine/selectfluor systems. *Asian J. Org. Chem.* 5: 1314–1317.

206 Wöste, T.H. and Muñiz, K. (2016). Enantioselective vicinal diacetoxylation of alkenes under chiral iodine(III) catalysis. *Synthesis* 48: 816–827.

207 Sarie, J.C., Neufeld, J., Daniliuc, C.G., and Gilmour, R. (2019). Catalytic vicinal dichlorination of unactivated alkenes. *ACS Catal.* 9: 7232–7237.

208 Stodulski, M., Goetzinger, A., Kohlhepp, S.V., and Gulder, T. (2014). Halocarbocyclization versus dihalogenation: substituent directed iodine(III) catalyzed halogenations. *Chem. Commun.* 50: 3435–3438.

209 Fabry, D.C., Stodulski, M., Hoerner, S., and Gulder, T. (2012). Metal-free synthesis of 3,3-disubstituted oxoindoles by iodine(III)-catalyzed bromocarbocyclizations. *Chem. Eur. J.* 18: 10834–10838.

210 Ulmer, A., Stodulski, M., Kohlhepp, S.V. et al. (2015). Iodine(III)-catalyzed rearrangements of imides: a versatile route to α,α-dialkylated α-hydroxy carboxylamides. *Chem. Eur. J.* 21: 1444–1448.

211 Patzelt, C., Pöthig, A., and Gulder, T. (2016). Iodine(III)-catalyzed cascade reactions enabling a direct access to β-lactams and α-hydroxy-β-amino acids. *Org. Lett.* 18: 3466–3469.

212 Massignan, L., Tan, X., Meyer, T.H. et al. (2020). C–H oxygenation reactions enabled by dual catalysis with electrogenerated hypervalent iodine species and ruthenium complexes. *Angew. Chem. Int. Ed.* 59: 3184–3189.

213 Roth, H.G., Romero, N.A., and Nicewicz, D.A. (2016). Experimental and calculated electrochemical potentials of common organic molecules for applications to single-electron redox chemistry. *Synlett* 27: 714–723.

214 Maity, A., Frey, B.L., Hoskinson, N.D., and Powers, D.C. (2020). Electrocatalytic C–N coupling via anodically generated hypervalent iodine intermediates. *J. Am. Chem. Soc.* 142: 4990–4995.

13

Industrial Application of Iodine Catalysis

Tatsuo Kaiho[1,2]

[1] *CIRIC, Chiba University, Chiba, Japan*
[2] *Godo Shigen Co. Ltd., Toshinshoji Bldg. 7F, 2-12-6, Kyobashi, Chuo-ku, Tokyo, Japan*

Iodine has a number of uses in our daily life. In the medical field, iodine is used as an X-ray contrast media, as a disinfectant in hospitals, and as an intermediate of pharmaceutical production. Iodine is also used as a stabilizer for the nylon fibers and in the production of polarizing film for the liquid crystal display (LCD) television and smartphone screens [1, 2] (Figure 13.1).

In addition to all of these uses, iodine and its derivatives are used as catalysts in a broad sense for industrial production. Some of the most common catalytic uses of iodine in the industry are detailed in this chapter [3].

13.1	Synthetic acetic acid	Hydrogen iodide
13.2	Production of tall oil	Metal iodide
13.3	Photo acid generator	Iodonium salt
13.4	Polymerization	Alkyl iodide
13.5	Dye-sensitized solar cell (DSSC)	Triiodide
13.6	Polyamide stabilizer	Copper iodide

13.1 Synthetic Acetic Acid

Acetic acid is an important chemical reagent and industrial material with an annual production of 12.1 million tons (2014) as shown in Figure 13.2 [4]. In 2018, the global production capacity of acetic acid reached almost 18 million metric tons [5].

Acetic acid is used in the production of chemical compounds such as vinyl acetate monomer (VAM), purified terephthalic acid (PTA), acetic anhydride, and acetate esters. Among all these applications, VAM and PTA are the most prominent applications accounting for more than 61% of total acetic acid consumption. VAM is used in the paints, adhesives and sealants industry. PTA is primarily consumed in the manufacturing of polyester fiber and polyethylene terephthalate (PET) that is used in textile, plastic bottles, and packaging materials. Acetic acid is produced industrially both synthetically (90%) and by bacterial fermentation (10%). Feedstock choices

Iodine Catalysis in Organic Synthesis, First Edition. Edited by Kazuaki Ishihara and Kilian Muñiz.

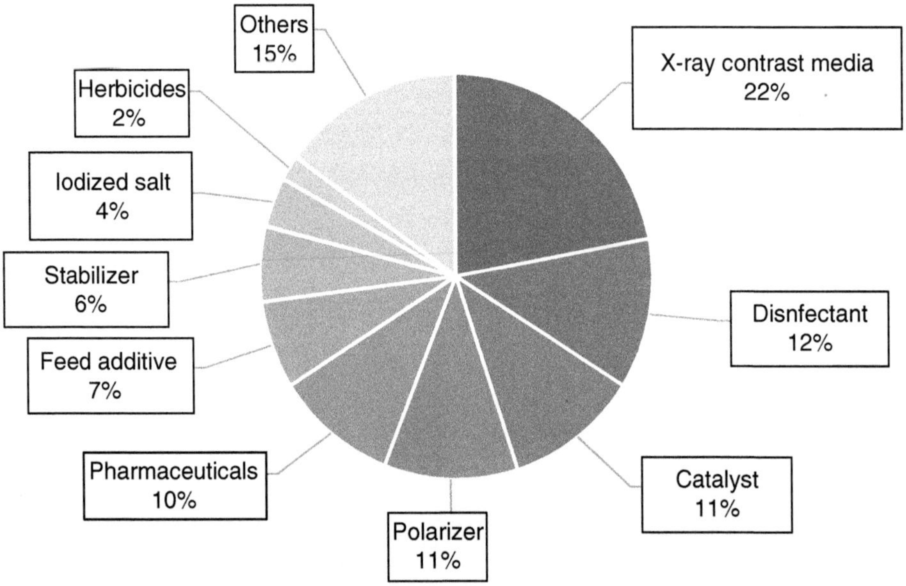

Figure 13.1 Applications of iodine.

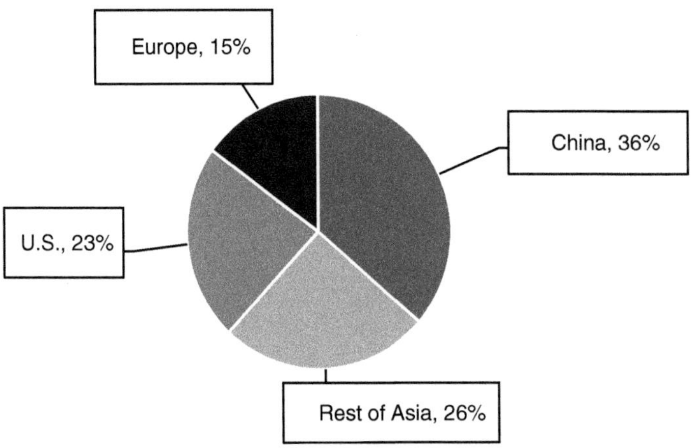

Figure 13.2 Annual production of acetic acid. Source: Data from The Essential Chemical Industry [4].

and process routes for acetic acid are summarized in Figure 13.3. About 75% of acetic acid for use in the chemical industry is made by the carbonylation of methanol using hydrogen iodide (Figure 13.4) [6].

13.1.1 BASF Process

The first methanol carbonylation process, commercialized in the 1960s by BASF, involved iodide-promoted active cobalt catalyst [$HCo(CO)_4$] but required very high

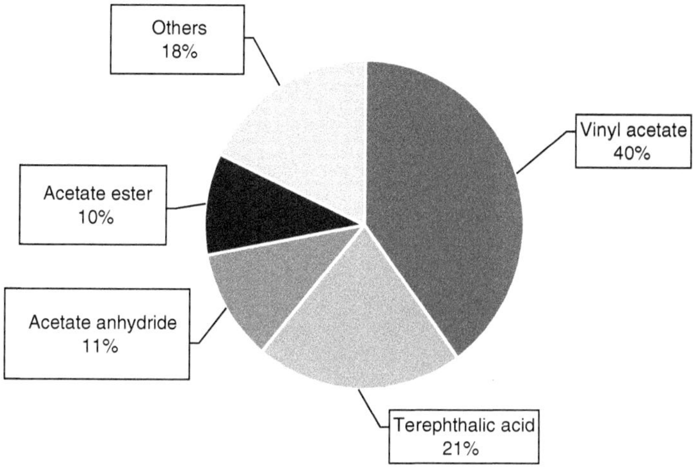

Figure 13.3 The uses of acetic acid.

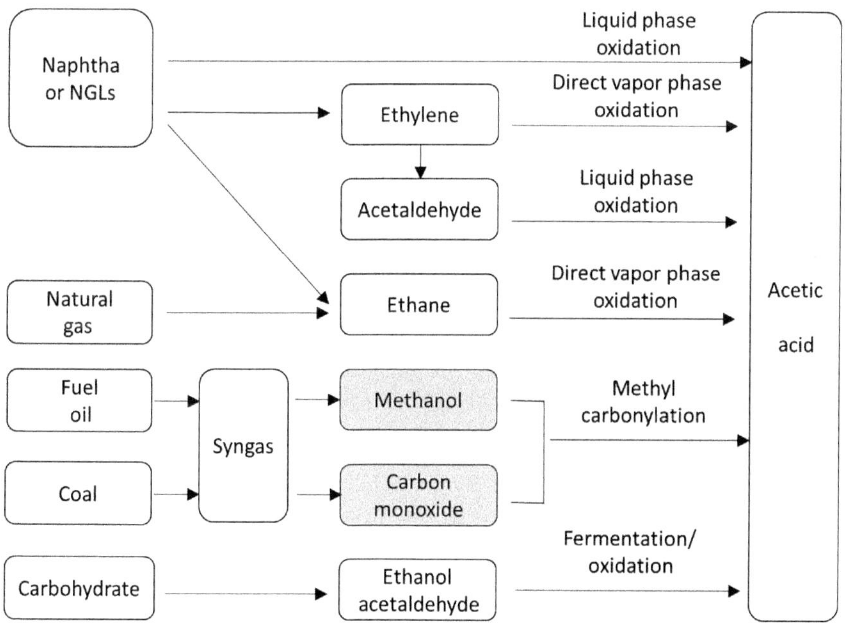

Figure 13.4 Feedstock choices and process routes for acetic acid [7]. Source: Data from Acetic Acid Production and Manufacturing Process [7].

pressure (~600 atm) as well as high temperatures (230 °C), and gave only ca. 90% selectivity based on methanol. In the Co-based catalytic carbonylation, the presence of iodide is necessary in order to convert methanol into methyl iodide prior to carbonylation. Therefore, the actual substrate of carbonylation is methyl iodide. The reaction rate depends on the carbon monoxide pressure which is responsible for the

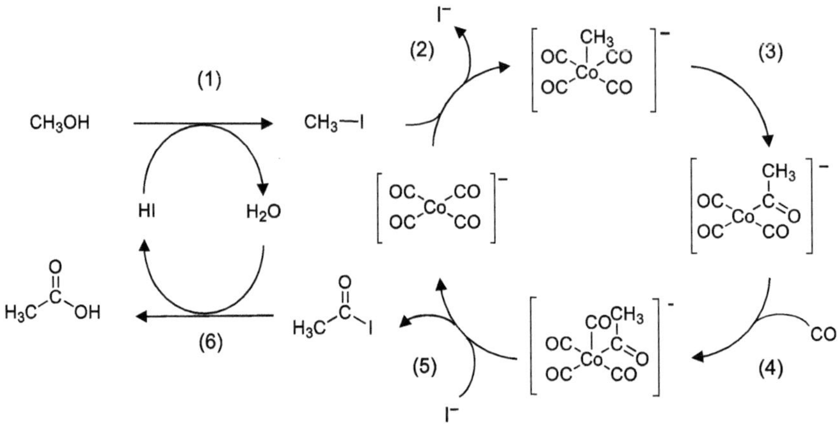

Figure 13.5 BASF acetic acid process. Source: Data from Acetic Acid Production and Manufacturing Process [7].

conversion. The mechanism of the cobalt-catalyzed process is deduced as shown in Figure 13.5 [7].

13.1.2 Monsanto Process

In order to reduce the severe process parameters with BASF process, Monsanto commercialized a new process in 1970, using rhodium-based homogeneous liquid-phase catalyst $[RhI_2(CO)_2]$, which is operated at a pressure of 30~60 atm and at temperatures of 150~200 °C with the high selectivity over 99% based on methanol. Methyl iodide is generated by the reaction of methanol with hydrogen iodide. The catalytically active species is the anion cis-$[Rh(CO)_2I_2]^-$ (Figure 13.6). The first organometallic step is the oxidative addition of methyl iodide to cis-$[Rh(CO)_2I_2]^-$ to form the hexacoordinate species $[(CH_3)Rh(CO)_2I_3]^-$. This anion rapidly transforms, via the migration of a methyl group to an adjacent carbonyl ligand, affording the pentacoordinate acetyl complex $[(CH_3CO)Rh(CO)I_3]^-$. This five-coordinate

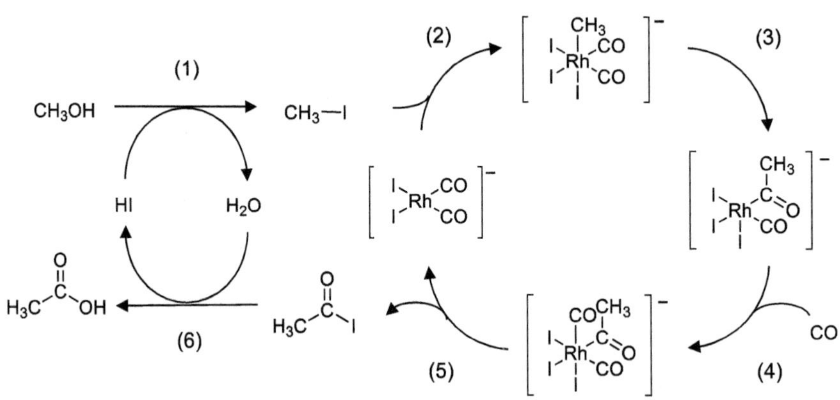

Figure 13.6 Monsanto acetic acid process.

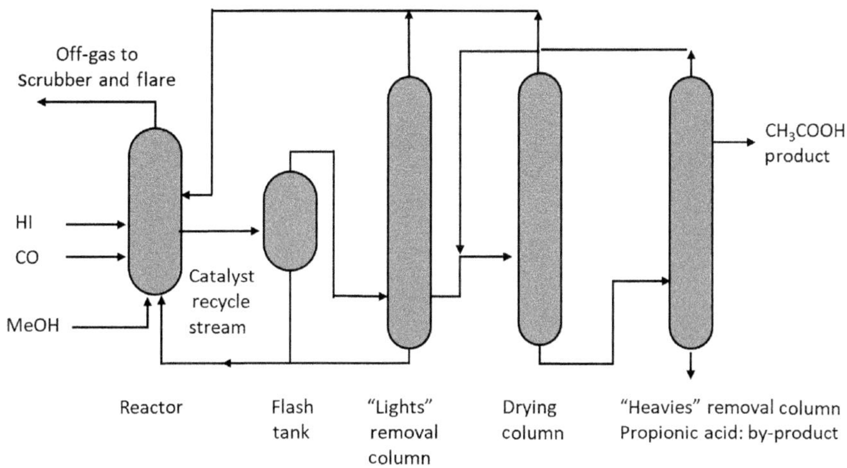

Figure 13.7 Flow diagram of the Monsanto process. Source: Based on Dutta [8].

complex then reacts with carbon monoxide to form the six-coordinate dicarbonyl complex, which undergoes reductive elimination to release acetyl iodide (CH_3COI). The catalytic cycle involves two non-organometallic steps: conversion of methanol to methyl iodide [step (1) in Figure 13.6] and the hydrolysis of the acetyl iodide to acetic acid and hydrogen iodide [step (6) in Figure 13.6]. The reaction has been shown to be first-order with respect to methyl iodide and $[Rh(CO)_2I_2]^-$ [step (2) in Figure 13.6]. Hence the oxidative addition of methyl iodide is proposed as the rate-determining step.

The flow diagram of the Monsanto process is shown in Figure 13.7 [8]. This process consists of three distillation columns to sequentially remove "Lights" (methyl iodide and methyl acetate), water, and "Hevies" (propionic acid) removal columns and deliver high-purity acetic acid product (Figure 13.8).

13.1.3 Cativa Process

The Cativa process, which is based on an iridium-iodine catalyst, was developed by BP Chemicals in 1996. The Cativa and Monsanto processes are sufficiently similar that they can use the same chemical plant. The switch from rhodium to iridium allows the use of less water in the reaction mixture (5 vs. 15%). This change decreases the formation of by-products, such as propionic acid, and reduces the number of drying columns necessary (Figure 13.9). The other process parameters are also improved and the new process is operated at a temperature of ~180 °C and a pressure of 30 atm.

The catalytic cycle for the Cativa process is essentially the same as with the Monsanto process as shown in Figure 13.9. The reaction of methyl iodide with the iridium catalyst is about 150 times faster than the analogous reaction with rhodium. This represents a dramatic improvement in reaction rates, as this step is now no longer rate-determining (as in the case of rhodium). Presently, most industrial acetic acid is produced using this process [9].

Figure 13.8 Acetic acid production facility (courtesy of Daicel Co.). Source: courtesy of Daicel Co.

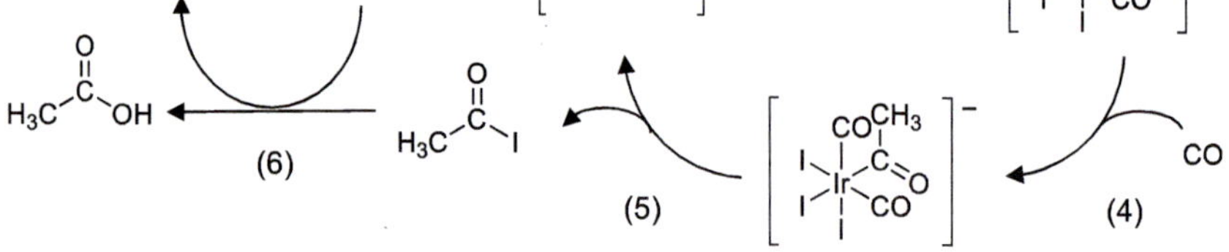

Figure 13.9 Cativa acetic acid process.

The rhodium and iridium-catalyzed carbonylation reactions depend on iodide, which is a good nucleophile, a very weak base, a good leaving group, and an effective ligand for rhodium or iridium. No alternative to iodide has been found yet. However, hydrogen iodide is corrosive. It is a drawback and a major engineering problem.

13.2 Tall Oil

Tall oil is a viscous yellow-black odorous liquid obtained as a by-product of the Kraft process of wood pulp manufacture. Tall oil has unique properties that differ from those of common edible vegetable oils. The name "tall oil" originated from the Swedish word for pine oil "Tallolja." Pine wood is separated into five major components, namely cellulose, hemicellulose, lignin, tall oil, and turpentine (Figure 13.10).

In the Kraft process (Figure 13.11), after the wood (pine) chips are digested and filtered, the filtrate, called "black liquor," is fed to an evaporator and skim tank for soap making. The soap is then acidulated to make crude tall oil (CTO). The CTO is then further fractionated at the refinery to get tall oil heads, tall oil fatty acids (TOFA), distilled tall oil (DTO) fatty acids, tall oil rosin (TOR), and tall oil pitch (TOP) (Figure 13.12).

The most important components in CTO are TOFA and TOR. A unique property of TOFA is that it contains various amounts of rosin (represented here are abietic and pimaric acids). The presence of rosin creates physical properties that could not be obtained from vegetable and animal fats.

Current applications of TOFA include adhesives, inks, surfactants, painting and coatings, mining, and metal-working. TOR is used in many applications, including adhesives, inks, tires, chewing gums, varnishes, electronics, papermaking, coating, and roadmaking. One of the common reactions in tall oil derivatization is "disproportionation" – an isomerization reaction of TOFA or rosin under heat, usually with an iodine catalyst.

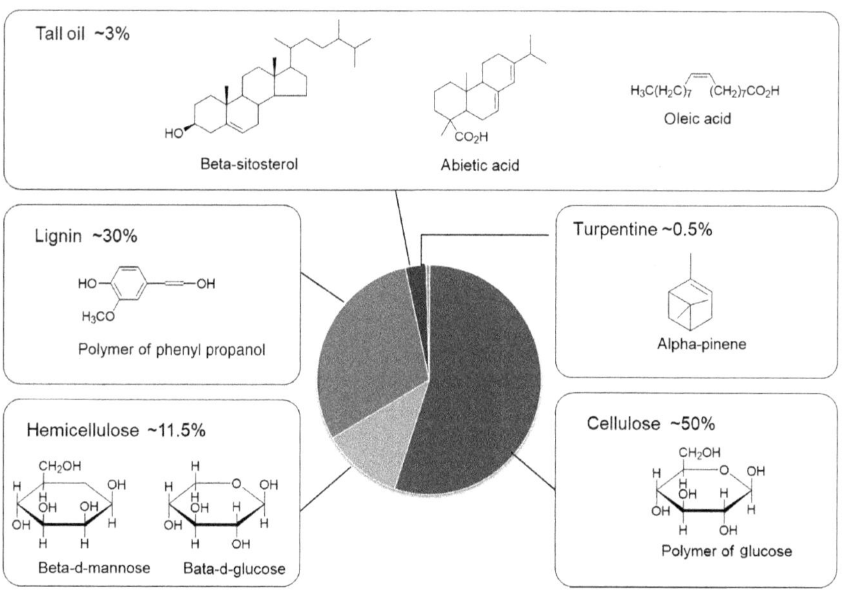

Figure 13.10 Composition of a pine tree.

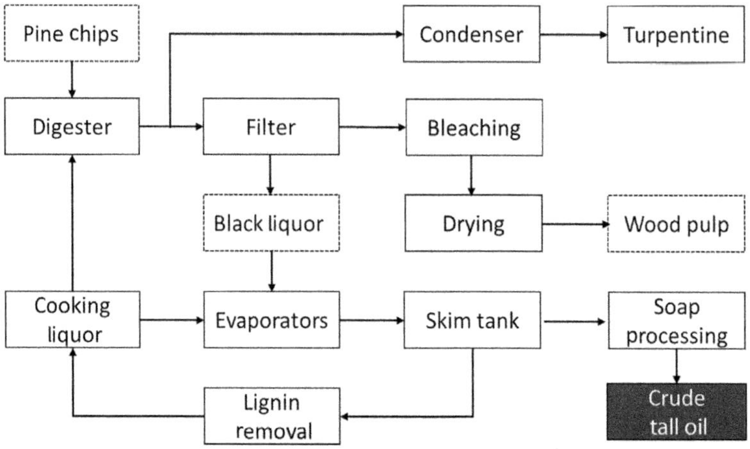

Figure 13.11 Flow diagram of a typical Kraft pulping process.

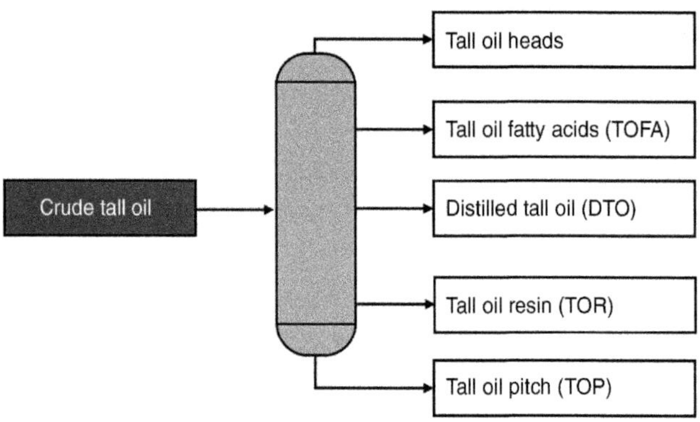

Figure 13.12 Components in tall oil fraction.

Iodine has been used as the catalyst for the disproportionation of TOFA; thus non-conjugated linoleic acid is converted to conjugated linoleic acid, and eventually to oleic acid (Figure 13.13).

Similarly, TOR can be disproportionated from its conjugated diene components, such as abietic acid and pimaric acid, to dehydroabietic acid, which is widely used in the rubber industry (Figure 13.14) [10, 11]. Iodine and iodide were excellent catalysts, and the optimum result was reported when tall oil was heated with 1% mixed iodide for 3–4 hours at 220–240 °C. In this reaction, the disproportionation was observed, and the product mainly contained dehydroabietic acid and octadecenoic acid (Table 13.1) [12].

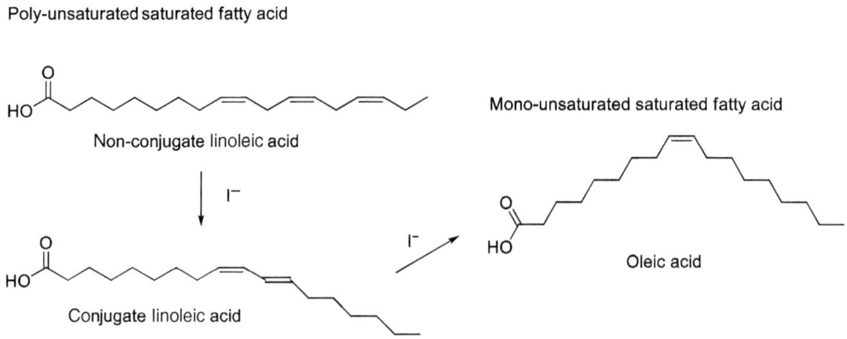

Figure 13.13 Disproportionation of fatty acids in tall oil with iodine catalyst.

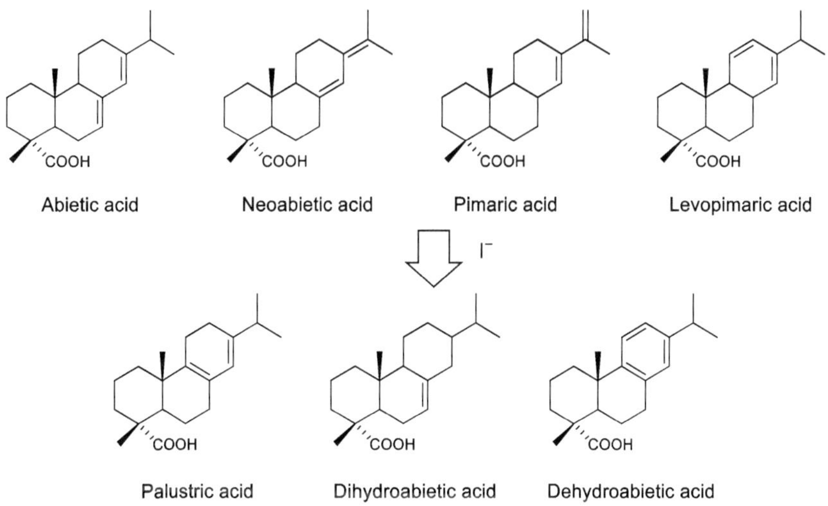

Figure 13.14 Disproportionation of resin acids.

Table 13.1 Disproportionation of Rosin[a].

Item	Unit	Untreated rosin	Disproportionated rosin
Acid number		178.1	171.5
Rosin acids	wt %	91.5	87.2
Fatty acids	wt %	4.2	4.8
Color	Gardner	5	6
Melting point	°C	72.2	65.2
Abietic acid	wt %	42.8	0.1
Dehydroabietic acid	wt %	16.6	49.1

a) 1000 g Rosin was treated with iodine catalyst (2 g–55% aq. LiI & 8 g–40% aq. FeI_2) at 220 °C
Source: Jadhav [12].

13.3 Photoinitiator

Diaryliodonium salts are widely used as photoinitiators of cationic photopolymerizations. Crivello and coworkers have first discovered that iodonium salts having BF_4^-, PF_6^-, AsF_6^-, or SbF_6^- counter-ions are efficient photoinitiators for the polymerization of a variety of cationically polymerizable monomers. A typical synthetic method is shown in Figure 13.15 [13].

The study of the photolysis mechanism of iodonium salts confirmed a pathway, involving radical cations and aryl radicals as key intermediates. The major pathway involves the facile decomposition of the excited iodonium compound to aryliodo radical cation, aryl radical, and anion (Figure 13.16).

This process should be highly efficient due to the very low bond energy of the C–I bond (26–27 kcal/mol). Interaction of the aryliodo radical cation with the solvent (R′–H) generates a protonated iodoaromatic compound, which rapidly deprotonates, and a radical R• derived from the solvent. During the photolysis, the BF_4^-, PF_6^-, AsF_6^-, or SbF_6^- counterions associated with the diaryliodonium salts remain unchanged and appear in the products as the corresponding Brønsted acids. These acids (HX) work as the true initiators of cationic polymerization of various monomers. According to this mechanism of action, diaryliodonium salts belong to an important class of photoacid generators (PAGs), which find broad applications in the manufacturing of photoresist, protective coatings, adhesives, microelectronics, graphic arts, dentistry, and 3D-printing, etc. [Figure 13.17].

From the industrial point of view, the solubility of photoinitiators in nonpolar monomers is a very important aspect. It has been found that diaryliodonium salts, asymmetrically substituted in the aromatic rings with long alkyl or alkoxy chains, show much better solubility than their symmetrical equivalents. Additionally, asymmetry of the iodonium system is also associated with the lower toxicity of these compounds compared to their symmetrically substituted analogs.

Furthermore, the solubility also depends on the type of anion used in these salts. In general, the bulkier the anion, the more soluble the corresponding salt becomes

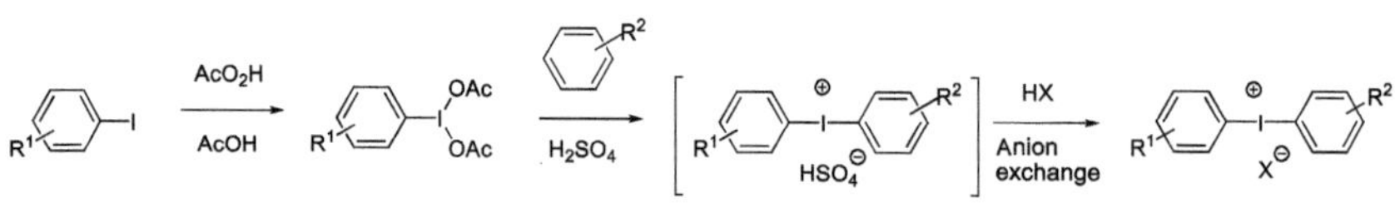

Figure 13.15 Synthesis of photoinitiators. Source: Crivello and Lam [13].

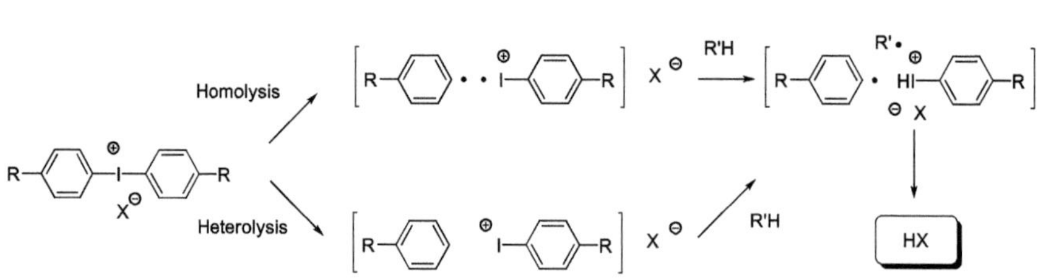

Figure 13.16 Decomposition of photoinitiator.

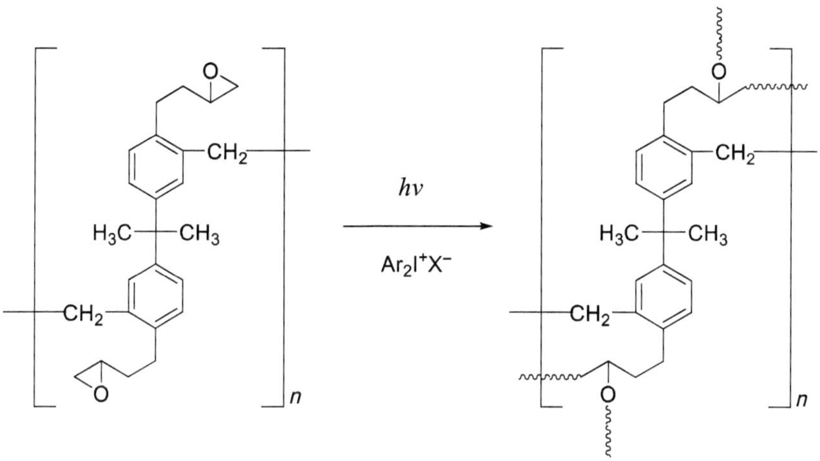

Positive photoresist

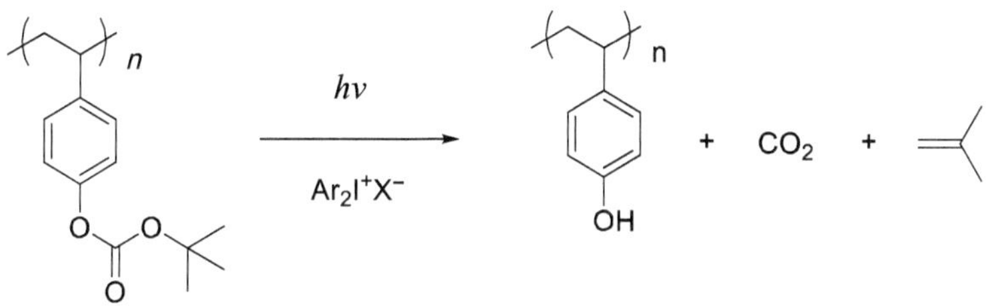

Figure 13.17 High-performance photoresists.

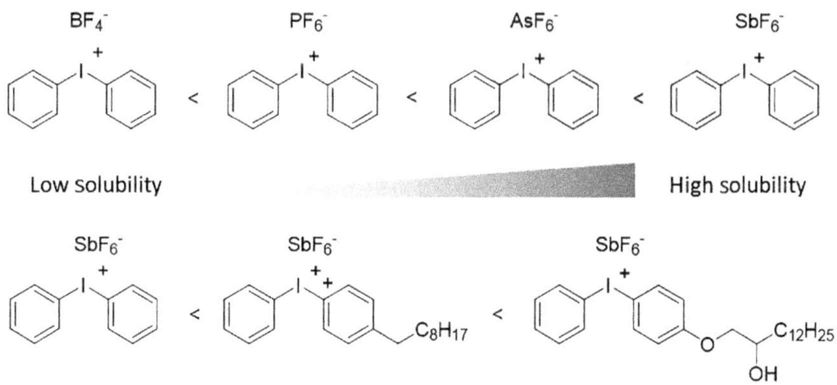

Figure 13.18 Solubility of photoinitiators.

because of a greater dissipation of the negative charge over the larger surface area of the anion, which lowers the anion hydrophilicity. Therefore, it is deduced that the solubility of the diaryliodonium salts increases in the following order: $BF_4^- < PF_6^- < AsF_6^- < SbF_6^-$. Hexafluoroantimonates and hexafluorophosphates are used most often in the industry as photoinitiators (Figure 13.18).

For the application of iodonium salt as a cationic polymerization photoinitiator, the anionic part of the diaryliodonium salt has to be non-nucleophilic. Otherwise, the formation of σ-bonds between anions and growing macro cations will terminate the polymerization process. This is the reason why perfluorinated anions are used most often. The degree of separation in the propagation ion pair is dependent on both the size and the electron density of the anion. Therefore, the larger anion has the lower nucleophilicity and the higher propagation rate of the photopolymerization. The reactivity order is analogous to the solubility order: $BF_4^- < PF_6^- < AsF_6^- < SbF_6^-$.

Crivello and Lee described the synthesis and characterization of a series of (4-alkoxyphenyl)phenyliodonium salts [14], which are excellent photo- and thermal initiators for the cationic polymerization of vinyl and heterocyclic monomers. Iodonium salts are conveniently prepared by the reaction of alkoxyphenols with [hydroxy(tosyloxy)iodo]benzene followed by the anion exchange with sodium hexafluoroantimonate (Figure 13.19). Products have very good solubility and photoresponse characteristics, which make them especially attractive for use in UV curing applications.

Compounds with alkoxy chains of eight carbons and longer are essentially nontoxic, compared to diphenyl iodonium hexafluoroantimonate, which has an oral LD_{50} of 40 mg/kg (rats) [15]. The commercially available photoinitiators are shown in Figure 13.20.

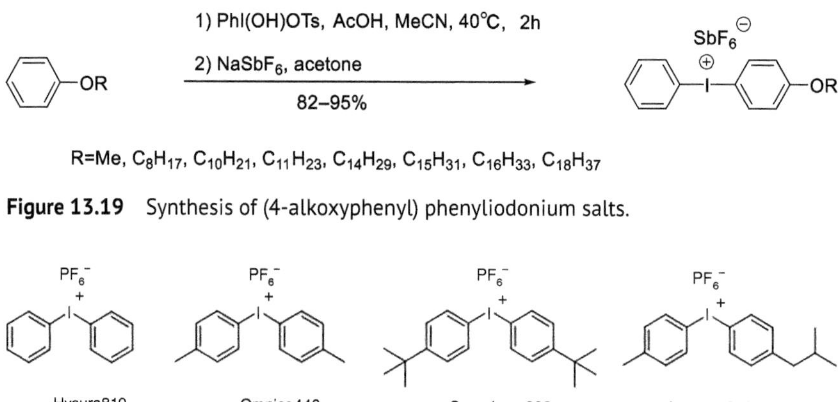

Figure 13.19 Synthesis of (4-alkoxyphenyl) phenyliodonium salts.

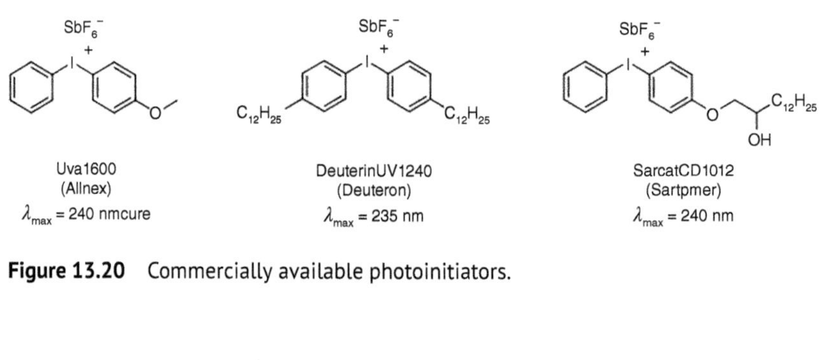

Figure 13.20 Commercially available photoinitiators.

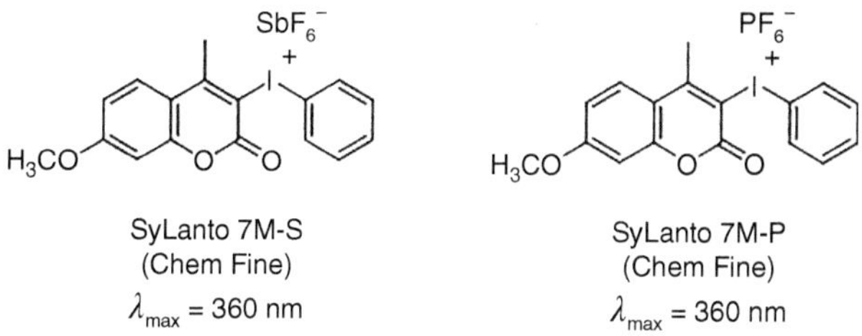

Figure 13.21 Photoinitiators based with a coumarin moiety. Source: Zhdankin [16], Kaiho et al. [17].

There is another type of photoinitiator based with a coumarin, which has an excellent absorption spectrum (λmax = 360 nm), under the name Sylanto. In this case, any anion can be selected as a counteranion. With these photoinitiators, there is no need to use co-initiators, like thioxanthone (Figure 13.21) [16, 17].

13.4 Polymerization (Chain Transfer Agent, Initiator, and Catalyst)

(Perfluoroalkyl)iodocompounds are highly attractive as potential intermediates for the synthesis of polymers because of the ability of the iodine atom to be a good leaving atom. In addition, they are efficient transfer agents in radical polymerization. Boutevin, B. et al. described iodine transfer polymerization (ITP) precisely in their review "Use of Iodocompounds in Radical Polymerization" in 2006 [18].

Perfluoroalkyl iodides are valuable intermediates of fluorinated repellents. Fluorinated compounds are used as oil and water repellents for clothes, shoes, and umbrellas; oil- and grease-resistant agents for fast food packaging and paper trays; and antifouling agents for carpeting. Chemicals containing the perfluoroalkyl group are also used as fire-extinguishing agents, leveling agents for floor polishing, and emulsifiers (Figure 13.22).

A typical production process of fluorinated repellents is known as telomerization (Figure 13.23). (i) Iodine pentafluoride (IF_5) is reacted with tetrafluoroethylene and iodine (I_2) to produce pentafluoroethyl iodide (C_2F_5I), known as telogen (chain transfer agent). (ii) Pentafluoroethyl iodide reacts with the tetrafluoroethylene and yields perfluoroalkyl iodide, the so-called telomer. (iii) The perfluoroalkyl iodides do not react with nucleophiles such as OH^- or NH_3 and cannot be converted directly to common intermediates. The perfluoroalkylethyl iodides are reacted with ethylene and converted to the corresponding alcohol. (iv) Ethyl acrylate is prepared from perfluoro-alkylethyl alcohol and acrylic acid, and by reacting this with other comonomers (acrylate, methacrylate, vinyl chloride, vinylidene chloride), water, and oil-repellent fluoropolymers are obtained. In addition to C_2F_5I, heptafluoropropyl iodide ($(CF_3)_2CFI$) is also used as a telogen. In the sequence of reactions, iodine does not remain in the polymers and can thus be recycled [2, 19].

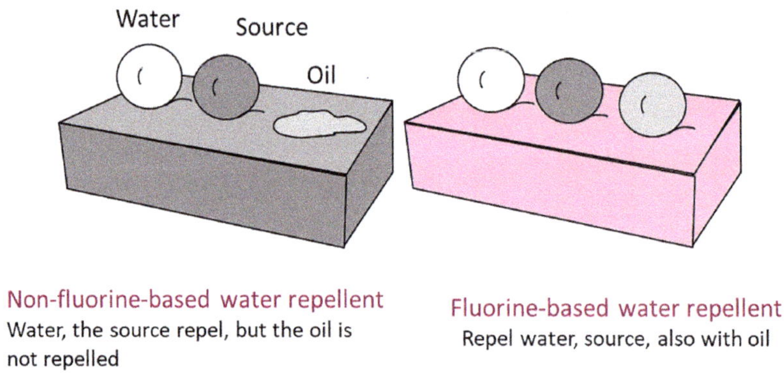

Figure 13.22 Fluorine-based repellent.

Synthesis of telogen

$$2IF_5 + 10C_2F_4 + I_2 \longrightarrow 10\boxed{C_2F_5I} \quad \text{Telogen}$$

Telomerization reaction

$$C_2F_5I + nCF_2{=}CF_2 \longrightarrow \boxed{C_2F_5(C_2F_2)_nI} \quad \text{Telomer}$$

Transformation of telomer and polymerizaton

$$R_fI \longrightarrow R_fCH_2CH_2I \longrightarrow R_fCH_2CH_2OH$$

$$CH_2{=}CH_2CO_2C_2H_4R_f \longrightarrow \text{Fluorine resin}$$

Figure 13.23 Typical production process of fluorinated repellents.

In the late 1970s, Tatemoto, M. and coworkers invented the ITP process at Daikin Industries. The first ITP-derived fluoroelastomers were commercialized by Daikin under the trade name Dai-El. Daikin's Dai-El is a family of high-performance fluoroelastomers. ITP is particularly suited for producing fluorinated thermoplastic elastomers (TPEs) starting from α, ω-diiodo-compounds ($I\text{-}(CF_2)_n\text{-}I = I\text{-}R_f\text{-}I$) which serve as chain transfer agents in aqueous heterogeneous processes. ABA triblock copolymers are produced according to a two-stage process, with a soft segment (T_g values slightly below 0 °C) based on monomers selected among vinylidene fluoride (VDF), chlorotrifluoroethylene (CTFE), hexafluoropropylene (HFP), tetrafluoroethylene (TFE), and polymethyl vinyl ether (PMVE), and hard segment (VDF or a mixture of ethylene and TFE, T_m values 160–267 °C) (Figure 13.24).

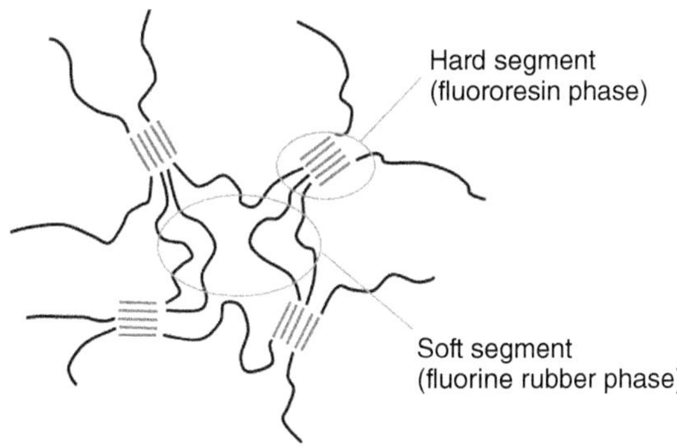

Figure 13.24 Schematic structural morphology of TPEs.

VDF: Vinylidene fluoride	$CF_2{=}CH_2$
CTFE: Chlorotrifluoroethylene	$ClFC{=}CF_2$
HFP: Hexafluoropropylene	$CF_2{=}CFCF_3$
TFE: Tetrafluoroethylene	$CF_2{=}CF_2$
PMVE: Polymethyl vinyl ether	$(CH_3OCH{=}CH_2)_n$

The mechanism of ITP involves thermal decomposition of the radical initiator, generating the initiating radical In•. This radical adds to the monomer M to form the species P_1•, which can propagate to P_m•. By exchange of iodine from the transfer agent R-I to the propagating radical P_m•, a new radical R• is formed and P_m• becomes dormant. This species can propagate with monomer M to P_n•. During polymerization, exchange between the different polymer chains and the transfer agent occurs, which is typical for a degenerative transfer process (Figure 13.25).

The combined TPE properties and fluorinated nature of these triblock copolymers give high thermal stability properties, chemical resistance, enhanced mechanical properties, transparency, and cleanliness. These characteristics are advantageous for the application in high-tech areas such as automotive (seals, O-rings, hoses, and gaskets), aeronautics, and aerospace because they provide excellent resistance against aggressive heat, chemicals, fuels, oils, and ozone. Some examples of fluoroelastomers produced by ITP at Daikin are shown in Figure 13.26. Many companies such as Dupon P.E., Solvay Solexis, Dyneon, etc., have shown interest in Daikin's work and developed various fluoroelastomers. Main commercially available fluoroelastomers today are summarized in Table 13.2 [18].

Living radical polymerization using organic catalysts (organo-catalyzed controlled polymerization (OCCP)) (Figure 13.27) has been explored as a low-cost manufacturing method for various functional polymers. The block polymers made by OCCP are innovative polymer dispersants that can finely disperse extremely aggregative nanoparticle pigments in liquids and resins. Such dispersants and dispersions find a wide variety of industrial applications such as inkjet pigment inks and adhesions.

A. Goto and Dainichiseika Color & Chemicals (Dainichiseika) developed a novel method to synthesize uniformly dispersed pigments in water and oil by OCCP where the carbon-iodide chain end is activated with organic iodine catalysts (Figure 13.28) [20].

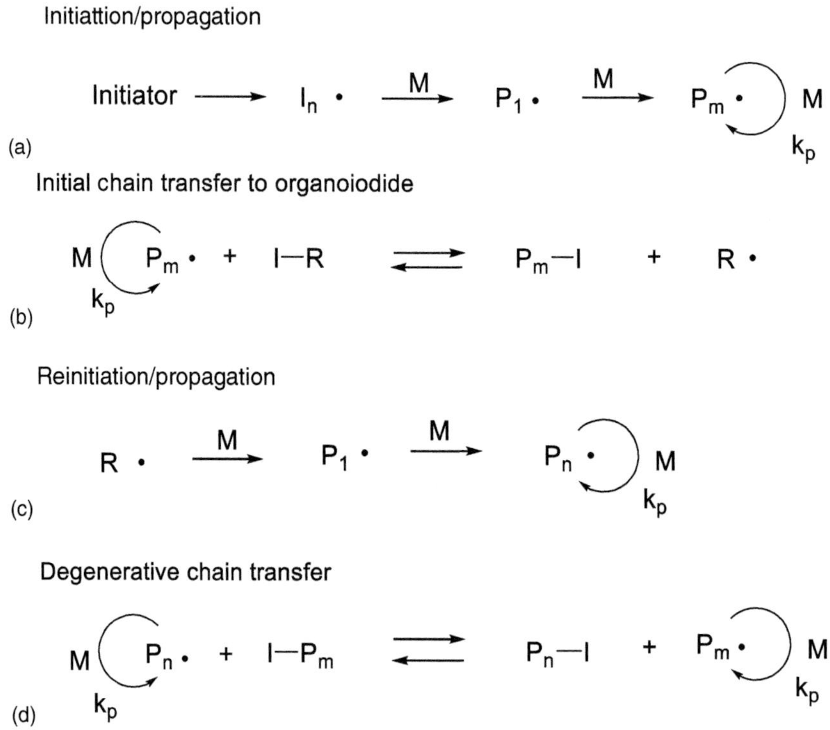

Figure 13.25 Iodine transfer polymerization mechanism.

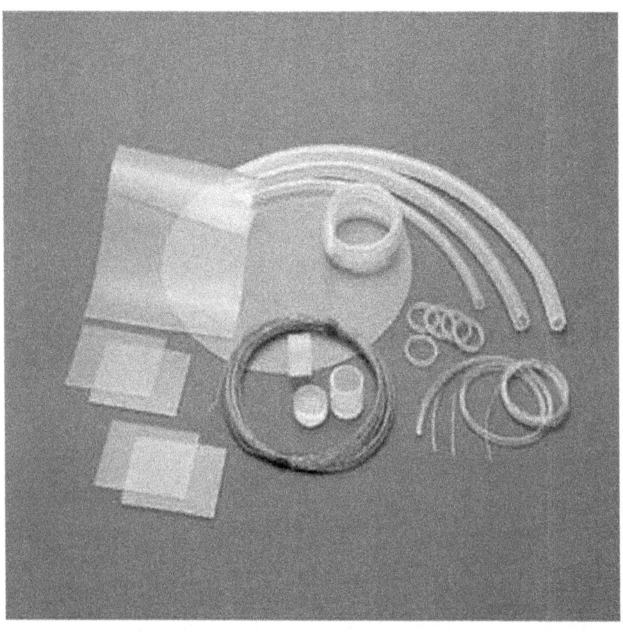

Figure 13.26 Several examples of items obtained by ITP (courtesy of Daikin). Source: courtesy of Daikin.

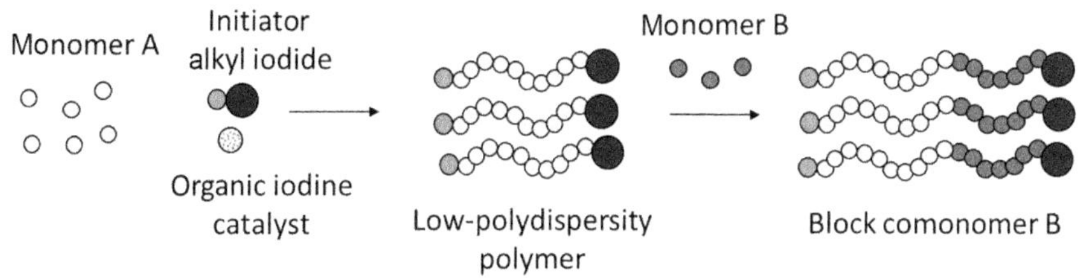

Figure 13.27 Organo-catalyzed controlled polymerization (OCCP).

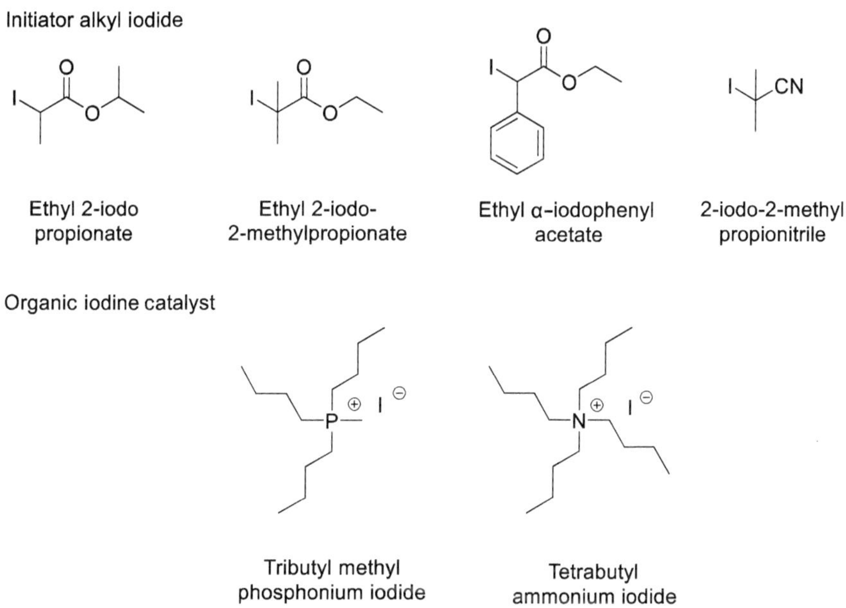

Figure 13.28 Initiators (up) and organic iodine catalysts (down). Source: Goto et al. [20].

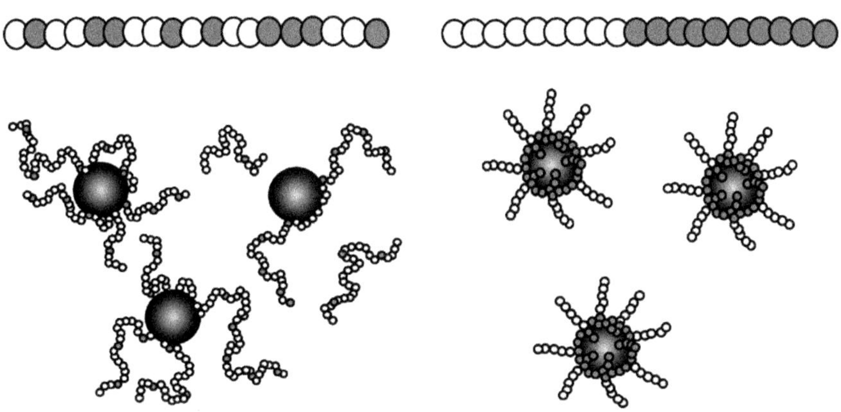

Figure 13.29 An image of dispersions of dispersant with random copolymer [left] and dispersant with block copolymer [right].

Table 13.2 Main commercially available fluoroelastomers.

Monomer	HFP	PMVE	CTFE	P (Propylene)
VDF	Daiel 801 (Daikin) Fluorel (3M/Dyneon) Kynar (Arkema) Tecnoflon (Solvay Solexis) SKF-26 (Rassian) Viton A (DuPont P. E.)		Kel F (Dyneon) SKF-32 (Rassian) Voltalef (Arkema)	
TFE		Kalrez (DuPont P.E.)		Aflas (Asahi Glass) Viton Extreme (DuPont P. E.)
VDF + TFE	Daiel 901 (Daikin) Fluorel (Dyneon) Tecnoflon (Solvay Sol.) Viton B (DuPont P.E.)	Viton GLT (DuPont P.E.)		
VDF + TFE + Ethylene	Tecnoflon (Solvay Sol.)			

Source: David et al. [18].

Figure 13.29 shows the structures of pigment dispersants. A polymer dispersant consists of two different units. One unit (dark grey dots) can be adsorbed onto pigments (black dots) and the other unit (light grey dots) can be dissolved in dispersive media such as water and oil. Conventional radical polymerization can provide only random copolymer dispersants (Figure 13.29, left) in which the two units (with different roles) are randomly distributed. In addition, the molecular weight is not uniform. On the other hand, the living radical polymerization method can provide block copolymer dispersants (Figure 13.29, right) in which the two units are separately distributed in two distinct segments. Random copolymer dispersants can attach onto pigments in a "dot-like" manner because the pigment-attaching

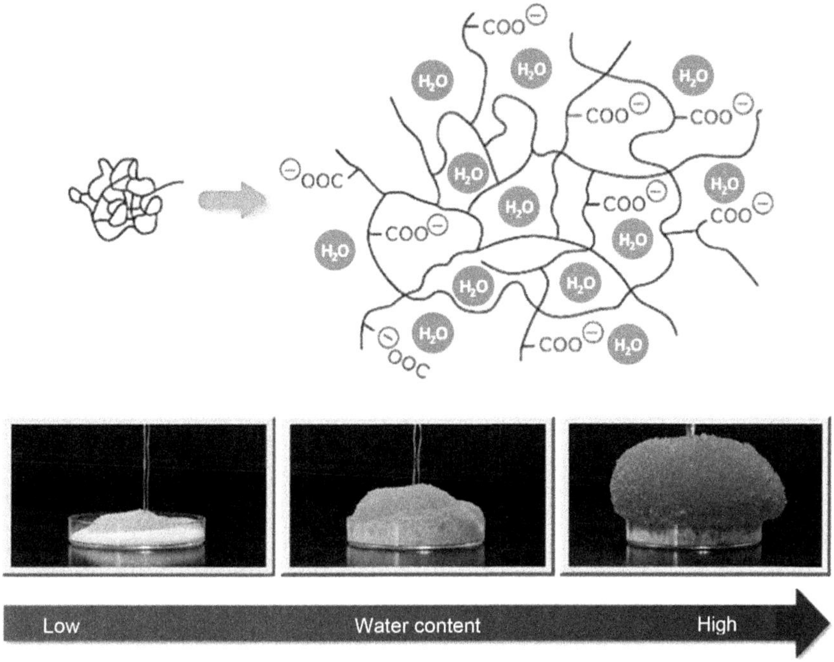

Figure 13.30 Superabsorbent polymer (SAP) (courtesy of Sanyo). Source: courtesy of Sanyo.

units (dark grey dots) are randomly distributed in a chain. On the other hand, block copolymer dispersants can attach onto pigments in a "line-like" manner because the pigment-attaching units form a segment. Therefore, block copolymer dispersants have strong adhesion of pigment. Through these researches and tests, Dainichiseika succeeded in the commercial production of dispersion materials in which nanoparticles are finely dispersed in stable forms [21, 22].

Superabsorbent polymer (SAP) (Figure 13.30) is a material capable of absorbing several hundred to 1000 times its own weight in water. Even when subject to pressure, this polymer securely retains the absorbed liquids [23]. SAP is used in a wide range of applications, from hygiene materials used mainly in disposable diapers and sanitary products to non-hygienic applications such as agriculture and horticulture, pet, construction, and civil engineering, etc. SAPs are crosslinked networks of flexible polymer chains. The polymer chains are shrunk before absorbing water. When water comes into contact with polymer chains, these chains are expanded. Water is trapped tightly within networks (Figure 13.30) [24]. In 2019, the annual global demand for SAPs is estimated to have reached ~3 million tons.

Sanyo Chemical Industries (Sanyo) and Godo Shigen Co. (Godo) applied ITP to develop a novel class of acrylic acid (AA)-based SAP with remarkably improved absorption performance.

Previously, ITP has been considered applicable only to specific monomers such as fluorinated monomers [25, 26] and considered to be unsuitable for monosubstituted hydrocarbon monomers such as AA. In addition, AA is known to generate easily ultrahigh molecular-weight polymers and to undergo readily self-cross-linking during polymerization due to its extremely fast polymerization rate.

Typically, alkyl iodides (R–I) are used as initial chain transfer agents and involved degenerative chain transfer as a key process. Since the protecting group in ITP is the iodine atom alone, the rate of degenerative chain transfer does not depend on the structure of the initial chain transfer agent but only on the structure of the polymer chain ends.

Considering the nature of the aqueous solution polymerization of AA, they thought that molecular weight control in the early stage of polymerization is very important, where a great deal of entanglements would be formed. Conversely, strict control over the entire period from the early to late stages of polymerization is not necessary. For these reasons, they employed a strategy to accelerate an initial chain transfer by organoiodine compounds.

The basic design concept of the new organoiodine chain transfer agent is as follows. First, to enhance chain transfer activity, radical-stabilizing groups, such as highly substituted alkyl and electron-withdrawing groups, were introduced. Second, the compounds are required to be water-soluble and hydrolysis-resistant to fully demonstrate their ability under actual use conditions. The structures of the newly designed and synthesized organoiodo compounds as chain transfer agents for SAPs are shown in Figure 13.31. They have selected the specific compound that satisfies all the above requirements at an acceptable level in these compounds.

They have succeeded in the precise control of the network structure of SAP through ITP. The polymer prepared via ITP has a narrower molecular weight distribution and a better structural homogeneity than that of comparative polymers via conventional free-radical polymerization (FRP) (Figure 13.32). New SAPs exhibit a good absorption capacity (Figure 13.33).

The developed ITP technology for SAP production does not involve productivity declines, such as prolongation of polymerization time and an increase in the residual monomer amount. They have confirmed through a series of careful tests that the addition of the developed compound does not adversely affect the coloration and safeness of the SAP and that no special postprocessing is required. The iodine atom derived from the developed compound is considered to be incorporated at the chain

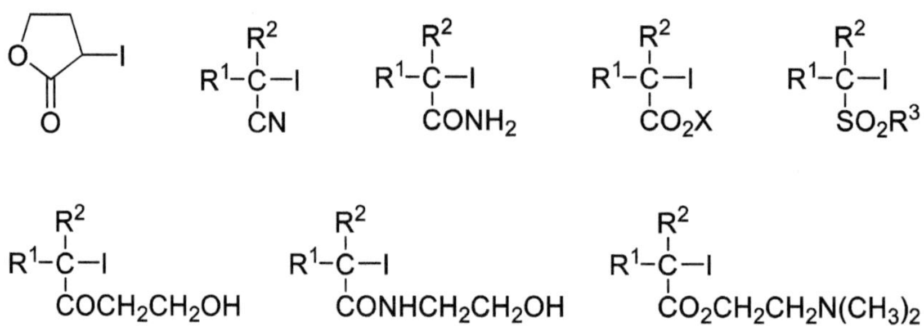

Figure 13.31 Organoiodine chain transfer agents.

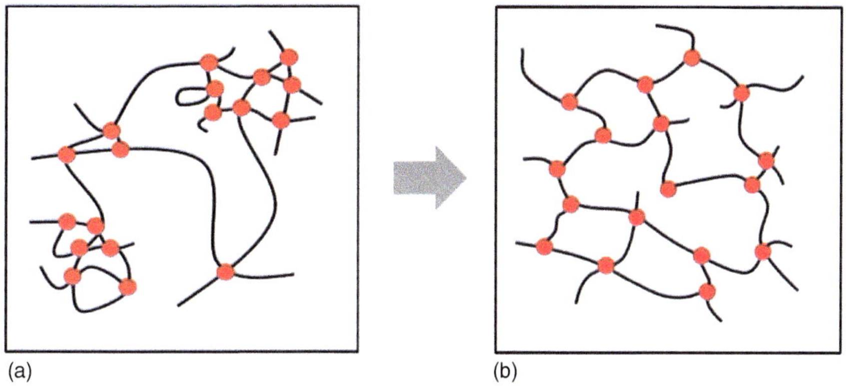

Figure 13.32 Network structure of SAP. (a) Conventional FRP, (b) ITP.

Figure 13.33 Retention capacity of the base SAP.

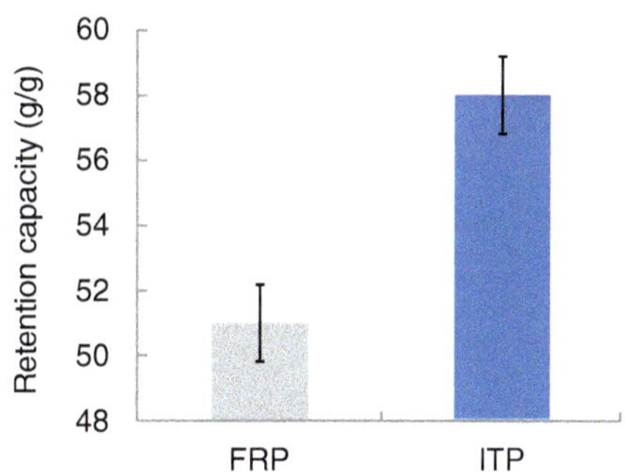

ends immediately after the polymerization but is assumed to be converted into an inactive form in the subsequent steps after the polymerization. A series of SAP products using this ITP technology has already been on the market. The newly developed organoiodine chain transfer agents were commercialized by Godo [27].

13.5 Dye-Sensitized Solar Cell (DSSC)

A dye-sensitized solar cell (DSSC) is based on a semiconductor formed between a photosensitized anode and an electrolyte, a photoelectrochemical system. DSSCs were invented in 1991 by Professor Michael Graetzel and Dr Brian O'Regan at École Polytechnique Fédérale de Lausanne (EPFL), Switzerland. The DSSC can be manufactured using a low-energy consumption, high-efficiency, roll-to-roll manufacturing technique. DSSC uses inexpensive and eco-friendly nano-materials without concern about the shortage of resources. Iodide/triiodide (I^-/I_3^-) has been used as an electrolyte from the very beginning of DSC research, and has proven to be one of the most versatile redox couples so far combining high overall conversion efficiency of up to 11–12% [2].

The following steps convert DSSC photon (light) to current: The incident photon is absorbed by Ru complex photosensitizers (Figure 13.34) adsorbed on the TiO_2

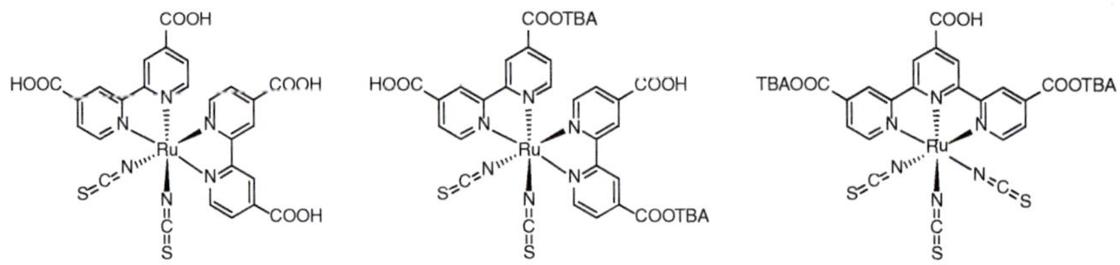

Figure 13.34 Dye for DSSC.

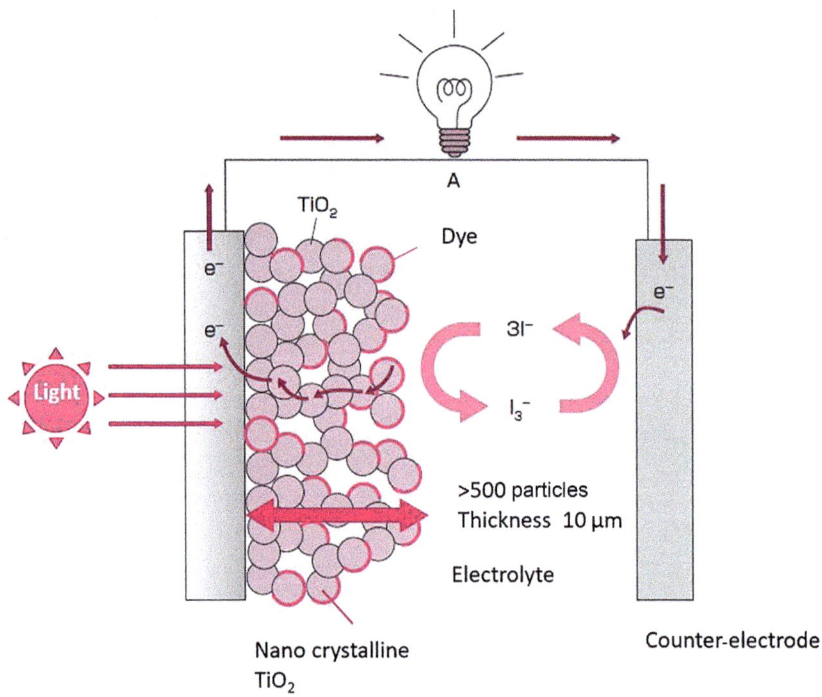

Figure 13.35 Dye-sensitized solar cells.

surface. The photosensitizers are excited from the ground state (S) to the excited state (S*), see Eq. 13.1. The excited electrons are injected into the conduction band of the TiO_2 electrode. This results in the oxidation of the photosensitizer (S^+), see Eq. 13.2. The oxidized dye molecule (S^+) is regenerated by electron donation from the iodide in the electrolyte, see Eq. 13.3. In return, iodide is regenerated by the reduction of triiodide on the cathode (see Eq. 13.4) to complete the cycle (Figure 13.35).

$$S + h\nu \rightarrow S^* \tag{13.1}$$

$$S^* \rightarrow S^+ + e^- \tag{13.2}$$

$$2S^+ + 3I^- \rightarrow 2S + I_3^- \tag{13.3}$$

$$I_3^- + 2e^- \rightarrow 3I^- \tag{13.4}$$

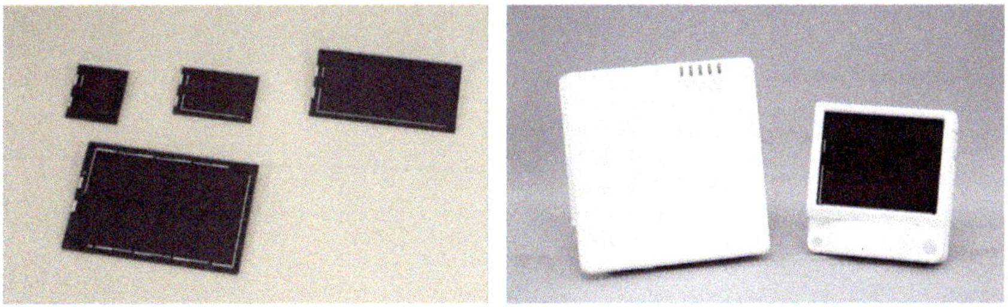

Figure 13.36 DSSC (left) and IoT device powered by DSSC (right) (courtesy of Fujikura Ltd.) Source: courtesy of Fujikura Ltd.

DSSCs are expected to be next-generation solar cells because of their environmentally friendly features compared to ISCs (inorganic solar cells). However, extensive investigation is necessary to resolve some of their critical issues such as environmental stability, lifetime, and large-area fabrication.

The Internet of Things (IoT) has been rapidly growing in the past few years. The IoT connects numerous devices, such as wireless sensors, actuators, and wearable devices, to optimize and monitor daily activities. Most of these devices require power in the microwatt range and operate indoors [28]. A DSSC (Figure 13.36) is capable of generating electric power highly efficiently in low light ranging from 10 to 5000 lux while other types of solar cells cannot generate enough electricity in the same conditions. This cell is most suitable for supplying power to sensor nodes or small electronic devices. Fujikura Ltd. has recently commercialized DSSC featuring high power-generation efficiency under lower-lighting conditions such as indoors or shady places as energy harvesters for the DSC power module [29].

13.6 Polyamide Stabilizer

Polyamides (nylons), such as polyamide 6 and polyamide 66, are the first engineering plastics [30, 31] and still represent the biggest and most important class of these types of material. Major properties such as resistance to chemicals, toughness, thermal stability, good appearance, and good processability are key considerations that make polyamides suitable for engineering plastics. Polyamide 6 is synthesized by ring-opening polymerization of caprolactam. On the other hand, polyamide 66 is synthesized by polycondensation of hexamethylenediamine and adipic acid. Polyamide 6 and polyamide 66 (Figure 13.37) continue to be the most popular type among commercial products, still accounting for 90% of nylon used in the global market [32]. According to industrial statistics, the global consumption of polyamide 6 and polyamide 66 in the world estimated around 6.8 million tons in 2011 [33].

One of the major applications of polyamide fibers is the manufacturing of tires, where heat and photoresistance of the polymer is essential (Figure 13.38) [34]. For example, the tires of a light vehicle while driving could reach a temperature near 120 °C. Potassium and cuprous iodides are used as heat-stabilizing agents in the manufacturing of polyamide 6 and polyamide 66.

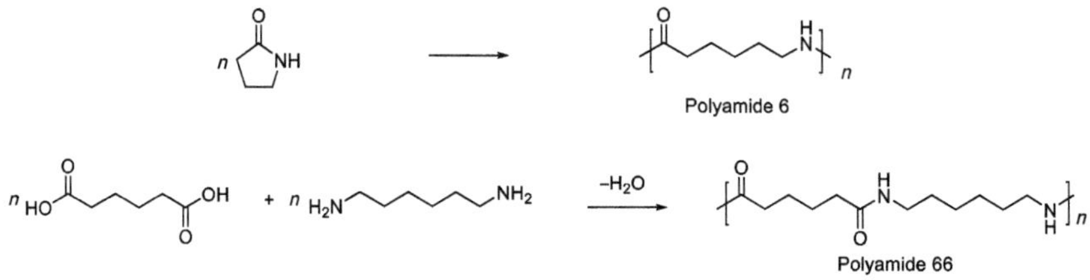

Figure 13.37 Synthesis of polyamide 66 and polyamide 6.

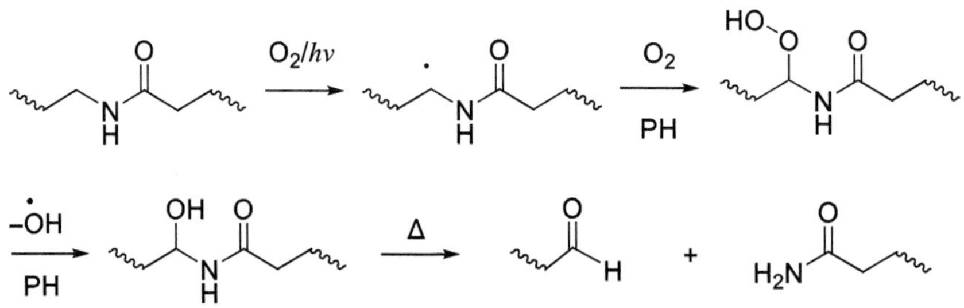

Figure 13.38 Degradation of polyamide by O_2 and light. Source: Lauterbach et al. [34a], El-Mazry et al. [34b].

The iodide and cuprous ions inserted within the nylon polymer net prevent the cracking of the chemical chain, keeping the properties of the material on heavy-duty conditions. On average, the use of potassium and cuprous iodides allows stabilization up to 170 °C, while with other combinations, it could be 25 °C less [35]. Aromatic amines are classical stabilizers for long-term thermal stability (LTTS) applications but lead to discoloration in polyamides. Phenolic antioxidants also help stabilize aliphatic polyamides by improving initial color after polycondensation. These antioxidants may be added to the condensation mass preferably prior to the termination of the condensation reaction (Table 13.3). At elevated aging temperatures, e.g. above 150 °C, the copper/iodide stabilizer systems show the best performance, but at low aging temperatures, the phenolic antioxidant is more effective (Table 13.4). The mechanism of stabilization with copper/halogen compounds is still subject to investigation. In the case of the stabilization of polyamides with Cu ions, it has been suggested by G. Scott that the antioxidant effect in catalytic Cu^{2+} oxidizes an alkyl radical to a carbonium ion, and the Cu^{+} formed in this reaction reduces alkylperoxy radicals [36].

However, according to the experiments using model compounds (see Figure 13.39) by K. Jannssen et al. [37], the mechanism postulated above (radical scavenging) cannot in itself be responsible for the stabilization of aliphatic polyamides. The mechanism of stabilization is probably based on the decomposition of hydroperoxides. KI is responsible for the decomposition of hydroperoxides into non-radical products, whereas the metal salt will catalyze this decomposition reaction by the complexation of metal salt onto amide and/or hydroperoxide group. It is confirmed that the combination of metal salts with KI results in a synergistic mixture. Consequently, the most effective stabilizers against the long-term thermo-oxidative degradation for

Table 13.3 Comparison of different polymer stabilizer systems.

AO system	Advantages	Disadvantages
Cu/I	Very effective at low concentrations. Good contribution to LTTS at aging temperature above 150 °C	Dispersability in a substrate is critical. Leaching in contact with water/solvent. May cause discoloration. Discoloration
Aromatic amines	Good contribution to LTTS	Need high concentration. Discoloration
Phenols	Good contribution to LTTS at aging temperature below 150°C. Good color performance. Can be added during condensation. No negative interaction with other polymers in lends	

Table 13.4 Effects of stabilizer systems on LTTS of polyamide 6[a] at different temperatures.

Stabilizer	Oven aging [b], Time to 50 retained elongation, Hours at			
	100 °C	120 °C	150 °C	165 °C
None	480	24	6	3
0.003% Cu/0.14% I	5520	624	324	44
0.5% AO-7[c]	6840	600	36	15

a) Polyamide 6; 1 mm injected molded dumbbells
b) Circulating air oven
c) AO-7: high molecular phenolic antioxidant

Figure 13.39 Stabilization mechanism of polyamide.

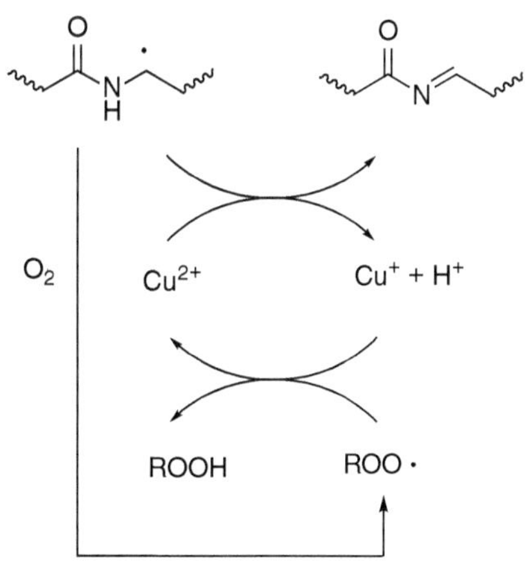

$$Cu^{2+} + 4I^- \longrightarrow 2CuI_2 \longrightarrow 2CuI + I_2$$

aliphatic polyamides are synergetic combinations of metal (mainly Cu and Mn) and halogen (mainly Br and I) salts [37].

References

1 Kaiho, T. (ed.) (2014). *Iodine Chemistry and Applications*. Hoboken, NJ: John Wiley & Sons, Inc.

2 Kaiho, T. (2017). *Iodine Made Simple*. Abingdon, Oxon, UK: CRC Press.

3 How iodine functions as a catalyst. [updated: 18 January 2016] https://www.ajay-sqm.com/how-iodine-functions-as-a-catalyst/ (accessed 24 August 2021).

4 The essential chemical industry – online. Ethanoic acid (acetic acid). [updated: 26 October 2016] https://www.essentialchemicalindustry.org/chemicals/ethanoic-acid.html (accessed 24 August 2021).

5 Garside, M. (2019). Global production capacity of acetic acid 2018 & 2023. [updated 6 November 2019] https://www.statista.com/statistics/1063215/acetic-acid-production-capacity-globally/(accessed 24 August 2021).

6 Spoilt for choice. [updated: 6 April 2003] https://www.icis.com/explore/resources/news/2003/04/04/193199/spoilt-for-choice/ (accessed 24 August 2021).

7 Acetic acid production and manufacturing process. [updated 23 April 2010] https://www.icis.com/explore/resources/news/2007/10/31/9074780/acetic-acid-production-and-manufacturing-process/ (accessed 24 August 2021).

8 Dutta, D.K. (2018). Carbonylation of methanol to produce acetic acid: a versatile reaction to study. *J. Indian Chem. Soc.* 95: 879–892.

9 Jones, J.H. (2000). The Cativa™ Process for the manufacture plant of acetic acid. *Platinum Metal Rev.* 44 (3): 94–105.

10 Wang, B. Tall oil, its chemistries and applications. [updated January 2018] http://www.informmagazine-digital.org/informmagazine/january_2018/MobilePagedArticle.action?articleId=1270074#articleId1270074 (accessed 24 August 2021).

11 Abdel-Raouf, M.E. and Abdul-Raheim, A.M. (2018). Rosin: chemistry, derivatives, and applications. *Review BAOJ Chem.* 4 (1): 39.

12 Jadhav, Y. J. (2000). Process to produce disproportionated rosin based emulsifier for emulsion polymerization. US Patent 6,087,318, July 11 to Georgia-pacific Resins Inc.

13 Crivello, J.V. and Lam, J.H.W. (1977). Diaryliodonium salts. A new class of photoinitiators for cationic polymerization. *Macromolecules* 10: 1307–1315.

14 Crivello, J.V. and Lee, J.L. (1989). Alkoxy-substituted diaryliodonium salt cationic photoinitiators. *J. Polym. Sci., Part A Polym. Chem.* 27: 3951.

15 Nowakc, D., Ortyl, J., Kamińska-Boreka, I. et al. (2017). Photopolymerization of hybrid monomers: Part I: comparison of the performance of selected photoinitiators in cationic and free-radical polymerization of hybrid monomers. *Polym. Test. J.* 64: 313–320.

16 Zhdankin, V.V. (2013). Hypervalent iodine chemistry: preparation, structure and synthetic applications of polyvalent iodine compounds. In: *Practical Applications*

of Polyvalent Iodine Compounds (ed. V.V. Zhdankin), 425–447. Chichester, U.K: John Wiley & Sons, Ltd.

17 Kaiho, T., Viktor, V., and Zhdankin, V.V. (2018). Industrial applications. In: *PATAI'S Chemistry of Functional Groups, Hypervalent Halogen Compounds* (eds. B. Olofsson, I. Marek and Z. Rappoport), 1–13. Chichester, U.K: John Wiley & Sons, Ltd.

18 David, G., Boyer, C., Tonnar, J. et al. (2006). Use of iodocompounds in radical polymerization. *Chem. Rev.* 106: 3936–3962.

19 Hintzer, K., Zipplies, T., Carlson, D.P., and Schmiegel, W. (2014). *Ullmann's Encyclopedia of Industrial Chemistry Fluoropolymers, Organic*, 1–55. Weinheim: Wiley-VCH.

20 Goto, A., Ohtsuki, A., Ohfuji, H. et al. (2013). Reversible generation of a carbon-centered radical from alkyl iodide using organic salts and their application as organic catalysts in living radical polymerization. *J. Am. Chem. Soc.* 135 (30): 11131–11139.

21 Goto, A., Kaji, H. (2015). Catalyst for living radical polymerization and polymerization method. US Patent 9,018,325 B2, April 28 to Kyoto University.

22 Goto, A. Versatile and economic polymerization method: Practical application of innovative, advanced color materials. https://www.jst.go.jp/EN/achievements/research/pdf/dainichiseika_color2016.pdf (accessed 24 August 2021).

23 Horie, K., Báron, M., Fox, R.B. et al. (2004). Definitions of terms relating to reactions of polymers and to functional polymeric materials (IUPAC Recommendations 2003). *Pure Appl. Chem.* 76 (4): 889–906.

24 Superabsorbent polymer. [updated July 2020] https://www.sanyo-chemical.co.jp/eng/products_info/superabsorbent-polymer (accessed 24 August 2021).

25 Oka, M. and Tatemoto, M. (1984). Vinylidene fluoride-hexafluoropropylene copolymer having terminal iodines. *Contemp. Top Polym. Sci.* 4: 763–777.

26 Tatemoto, M. (1992). Development of "iodine transfer polymerization" and its applications to telechelically reactive polymers. *Kobunshi Ronbunshu.* 49: 765–783.

27 Miyajima, T., Matsubara, Y., Komatsu, H. et al. (2020). Development of a superabsorbent polymer using iodine transfer polymerization. *Polym. J.* 52: 365–373.

28 Biswas, S. and Kim, H. (2020). Solar cells for indoor applications: progress and development. *Polymers* 12: 1338–1360.

29 Fujikura releases thin dye-sensitized solar cell module panels. [updated 27 July 2020] https://www.fujikura.co.jp/eng/newsrelease/products/2062445_11777.html (accessed 24 August 2021).

30 Carrothers, W.H.(1938). Linear polyamides and their production. US. Patent 2,130,523, September 20 to E.I. DuPont de Nemours and Co.

31 Schlack, P. (1941). Preparation of polyamide.US 2,241,321., May 6 to I. G. Farbenindustrie AG.

32 Process evaluation/research planning report Nylon 6 and Nylon 6,6 PERP07/08S6, White Plains, NY: Nexant Inc. January 2009.

33 MacDonald, B. (2011). *World PA6 & PA66 Supply/Demand Report 2011 by PCI Nylon*. Bad Homburg: PCI Nylon GmbH.

34 a) Lauterbach, A., Ober, G., Rios, S., Basinger, W., Shipp. A. (2001). Iodine, SQM company brochure. b) El-Mazry, C., Ben Hassine, M., Correc, O., and Colin, X. (2013). Thermal oxidation kinetics of additive free polyamide 6-6, polym. *Degrad. Stab.* 98 (1): 22–26.
35 Zweifel, H., Ralph, M.D., and Schiller, M. (2009). *Plastic Additive Handbook*, 6e, 80. Hanser Fachbuchverlag.
36 Scott, G. (1993). *Atomspheric Oxidation & Antioxidants*, vol. 2 (ed. G. Scott), 141–218. Amsterdam: Elsevier Science Publishers.
37 Janssen, K., Gijsman, P., and Tummers, D. (1995). Mechanistic aspects of the stabilization of polyamides by combinations of metal and halogen salts. *Polym. Degrad. Stab.* 49 (1): 127–133.

Index

Iodine Catalysis in Organic Synthesis, First Edition. Edited by Kazuaki Ishihara and Kilian Muñiz.

b

d

i

o

p

q

r